RING THEORY

RING THEORY

edited by

ROBERT GORDON

DEPARTMENT OF MATHEMATICS
THE UNIVERSITY OF UTAH
SALT LAKE CITY, UTAH

Proceedings of a Conference on Ring Theory held in Park City, Utah
March 2–6, 1971

Academic Press New York and London 1972

ACADEMIC PRESS, INC.
111 Fifth Avenue, New York, New York 10003

United Kingdom Edition published by
ACADEMIC PRESS, INC. (LONDON) LTD.
24/28 Oval Road, London NW1 7DD

LIBRARY OF CONGRESS CATALOG CARD NUMBER: 71-159614

PRINTED IN THE UNITED STATES OF AMERICA

CONTENTS

CONTENTS

CONTRIBUTORS

Armendariz, Efraim P., Department of Mathematics, University of Southwestern Louisiana, Lafayette, Louisiana 70501

Azumaya, Goro, Department of Mathematics, Indiana University, Bloomington, Indiana 47401

Bergman, George M., Department of Mathematics, University of California, Berkeley, Berkeley, California 94720

Buchsbaum, David A., Department of Mathematics, Brandeis University, Waltham, Massachusetts 02154

Chase, Stephen. U., Department of Mathematics, Cornell University, Ithaca, New York 14850

Cohn, P. M., Department of Mathematics, Bedford College, Regents Park, London N. W. 1, England

Eisenbud, David, Department of Mathematics, Brandeis University, Waltham, Massachusetts 02154

Fossum, R., Department of Mathematics, University of Illinois, Urbana, Illinois 61801

Griffith, P., Department of Mathematics, University of Illinois, Urbana, Illinois 61801

Gersten, S. M., Department of Mathematics, Rice University, Houston, Texas 77001

Goldie, A. W., Department of Mathematics, University of Leeds, Leeds, England

Hummel, K. E., Department of Mathematics, Trinity University, San Antonio, Texas 78212

Jategaonkar, Arun Vinayak, Department of Mathematics, Cornell University, Ithaca, New York 14850

CONTRIBUTORS

Lambek, J., Department of Mathematics, McGill University, Montreal, Quebec, Canada

Lanski, Charles, Department of Mathematics, University of Southern California, Los Angeles, California 90007

Mewborn, A. C., Department of Mathematics, University of North Carolina, Chapel Hill, North Carolina

Michler, Gerhard O., Department of Mathematics, University of Tubingen, Mathematics Institute, Tubingen, Federal Republic of Germany

Morita, Kiiti, Department of Mathematics, Faculty of Science, The University of Tokyo, Ohtsuka, Bunkyo-ku, Tokyo, 112, Japan

Procesi, C., Istituto Di Matematica, Facolta Di Scienze, Universita Degli Studi di Lecce, 73100 Lecce, Italy

Ramras, Mark, Department of Mathematics, Boston College, Chestnut Hill, Massachusetts 02167

Reiten, I., Department of Mathematics, University of Illinois, Urbana, Illinois 61801

Richman, Fred, Department of Mathematics, New Mexico State University, Las Cruces, New Mexico 88001

Robson, J. C., Department of Mathematics, University of Leeds, Leeds, England

Rutter, Edgar A., Jr., Department of Mathematics, University of Kansas, Lawrence, Kansas 66044

Sandomierski, F. L., Department of Mathematics, Kent State University, Kent, Ohio 44240

Small, Lance W., Department of Mathematics, University of California, San Diego, La Jolla, California 92037

Sweedler, Moss E., Department of Mathematics, Cornell University, Ithaca, New York 14850

Walker, Elbert A., Department of Mathematics, New Mexico State University, Las Cruces, New Mexico 88001

PREFACE

This volume contains a selection of articles centered around ring theory. Most were presented at the Ring Theory Symposium in Park City, Utah, March 2-6, 1971. Time limitations prevented all the papers from being presented.

Thanks are due to Academic Press for publishing these articles and for supporting their preparation. The conference itself was sponsored by the University of Utah with partial support from the National Science Foundation under Grant No. GP–25913.

The other members of the Organizing Committee, Professors John S. Alin and Timothy V. Fossum, and I would like to thank Professors Israel N. Herstein, Richard E. Johnson, Lawrence S. Levy, Alex Rosenberg, Lance W. Small, and Elbert A. Walker who served as Consultants to the Conference. I would like to express my particular thanks to Professors Rosenberg, Small, and Walker for their excellent advice and efforts on behalf of the symposium.

I am indebted to the authors of the papers herein both for their manuscripts and for their prompt preparation. Since some of the contributors did not have the opportunity to proofread their papers, I would like to take the responsibility for any misprints the reader finds.

The preparation of these Proceedings would have been impossible without the remarkable typing of Shawn Warriner ably assisted by Mrs. Ruth B. Anderson. Miss Warriner also took shorthand notes during the Problem Session held at Park City. This session was kindly tape recorded by Professor Don H. Tucker.

Finally, I must thank Ann Reed and my wife, Muriel Gordon, for their assistance in making the conference run smoothly.

RING THEORY

RESTRICTED SEMIPRIMARY RINGS

Efraim P. Armendariz

University of Texas

and

Kenneth E. Hummel

Trinity University

1. Introduction

In what follows A will denote a ring with 1 and modules will be unital left A–modules. A ring A is *semiprimary* if the Jacobson radical J(A) is nilpotent and A/J(A) is Artinian. If A has the property that A/I is semiprimary for each ideal $I \neq 0$ of A, we call A a *restricted semiprimary ring*, or RSP–ring for brevity. The concept of a commutative ring all of whose factors are Artinian (RM–rings) was introduced by I. S. Cohen in [4], and later noncommutative RM–rings were considered in [10] by A. J. Ornstein. Thus, just as the class of semiprimary rings extends the class of Artinian rings, RSP–rings encompass RM–rings. The main results obtained occur in Section 2 where we characterize RSP–rings and show that non–semiprimary RSP–rings are prime. Examples are provided of RSP–prime rings which are not RM; moreover, for commutative RSP–domains, integral extensions are RSP–domains and torsion modules have the primary decomposition property (in the sense of S. E. Dickson [5,6]). Section 3 contains two results related to the RM and RSP conditions.

2. Main Results

We begin with a characterization of RSP–rings.

Theorem 2.1. Let A be a ring. Then A is an RSP–ring if and only if

(a) A/I is Artinian for every nonzero maximal ideal I of A;

(b) Each non–zero ideal of A contains a product of maximal ideals of A.

Proof. If A is an RSP–ring and $I \neq 0$ is a maximal ideal of A, then A/I is a simple semiprimary ring and thus Artinian and so (a) holds. For any ideal $I \neq 0$ of A, since A/I is semiprimary, $J(A/I)$ is a finite intersection of maximal ideals of A and is nilpotent. Hence there exists maximal ideals $M_1,\cdots,M_k$ of A containing I and such that $(M_1 \cap \cdots \cap M_k)^t \subseteq I$ for some integer $t \geq 1$. Then $(M_1 \cdots M_k)^t \subseteq (M_1 \cap \cdots \cap M_k)^t \subseteq I$ and so (b) holds.

Conversely, assume (a) and (b) hold; let I be a nonzero ideal of A, and let $M_1,\cdots,M_k$ be maximal ideals of A such that $M_1 \cdots M_k \subseteq I$. By (a), $\oplus \Sigma_{j=1}^k A/M_j$ is a semisimple Artinian ring. We may assume that $I \subseteq M_j$ for each $1 \geq j \geq k$. Otherwise, if $I \not\subseteq M_j$ for some j, then $A = I + M_j$, hence $I \supseteq M_1 \cdots M_k + M_1 \cdots M_{j-1}IM_{j+1} \cdots M_k = M_1 \cdots M_{j-1}(M_j + I)M_{j+1} \cdots M_k = M_1 \cdots M_{j-1}AM_{j+1} \cdots M_k = M_1 \cdots M_{j-1}M_{j+1} \cdots M_k$, and M_j can be omitted. Now $J(A/I) \subseteq (M_1/I) \cap \cdots \cap (M_k/I)$ and so $J(A/I)^k \subseteq (M_1/I) \cdots (M_k/I) = 0$; hence $J(A/I)$ is nilpotent. Since $(M_1/I) \cap \cdots \cap (M_k/I)$ is nilpotent, $J(A/I) = (M_1/I) \cap \cdots \cap (M_k/I)$. Also $A/M_j \cong (A/I)/(M_j/I)$ and hence $(A/I)/J(A/I)$, being a finite subdirect sum of the simple Artinian rings A/M_j, is semisimple Artinian. This completes the proof.

An important special case of the above result is the following:

Theorem 2.2. Let A be a commutative ring. Then A is an RSP–ring if and only if each nonzero ideal of A contains a product of maximal ideals of A.

Corollary 2.3. Let A be a commutative RSP–ring. Then every nonzero prime ideal of A is maximal and each nonzero primary ideal of A contains a power of a maximal ideal of A.

We will show later that non–semiprimary RSP–rings are prime rings. However, before proceeding we give an example of an RSP–domain which is not Noetherian and hence not an RM–domain; the example is due to Krull and appears in [8, p. 780].

Let K be any field, X, Y indeterminates and D the set of all rational functions f/g in X and Y for which $X \nmid g(X, Y)$ and $f(0, Y)/g(0, Y) \in K$. Then D is a local domain with unique maximal ideal $M = \{ f/g : X \mid f(X, Y) \}$. For any $0 \neq f/g \in D$, if n is the highest power of X dividing f then $M^n \subseteq (f/g)D$. Hence every nonzero ideal of D contains a power of M and so D is RSP. However D is not Noetherian, since if $a_i = X(Y + 1)/Y^i$ for $i = 1,2,\cdots$ and $B_i = (a_1,a_2,\cdots,a_i)$, then we have the strictly increasing chain of ideals $B_1 \subset B_2 \subset B_3 \subset \cdots$.

Examples of non–commutative RSP–rings which are prime and not RM–rings can now be obtained by forming matrix rings. Specifically, we have the following:

Proposition 2.4. If A is an RSP–ring and $e \neq 0$ is an idempotent, then eAe is an RSP–ring.

Proof. It is readily verified that $0 \neq T$ is a maximal ideal of eAe if and only if $T = eMe$ for some maximal ideal M of A. Thus if $I \neq 0$ is an ideal of eAe, then $AIA \supseteq M_1 \cdots M_k$ for maximal ideals $M_1, \cdots, M_k$ of A and so $(eM_1e) \ldots (eM_ke) \subseteq I$; thus (b) of Theorem 2.1 holds. To show that (a) also holds, let $0 \neq T$ be maximal in eAe, $T = eMe$. Then there is a non–zero epimorphism $\theta : eAe \rightarrow \bar{e}\bar{A}\bar{e}$, where $\bar{e} = e + A$, $\bar{A} = A/M$, with $\operatorname{Ker} \theta$ containing T, and so $eAe/eMe \cong \bar{e}\bar{A}\bar{e}$ is Artinian.

Corollary 2.5. For any ring A the following are equivalent:

(i) A is an RSP–ring;

(ii) The matrix ring A_n is RSP for all $n \geqslant 1$;

(iii) The matrix ring A_n is RSP for some $n \geqslant 1$.

Corollary 2.6. If A is any RSP–ring and P is finitely generated projective (left or right) A–module, then $B = \operatorname{End}_A(P)$ is an RSP–ring.

We next show that non–prime RSP–rings are semiprimary. Recall that a ring A is (right) *perfect* [2] in case A/J(A) is semisimple Artinian and each nonzero left A–module contains a nonzero simple submodule. We let socle (M) = sum of all simple submodules of the left A–module M.

Theorem 2.7.

(a) If A is a (left or right) perfect RSP–ring, then A is semiprimary.

(b) If A is an RSP–ring which is not semiprimary, then A is a prime ring.

Proof. (a) Let $N = J(A)$. Since A/N is semisimple Artinian we can assume $N \neq 0$. Let $K = \text{left socle}(A) \neq 0$. Then A/K is semiprimary and $(N+K)/K \subseteq J(A/K)$; hence $(N+K)/K$ is nilpotent, say $N^t \subseteq K$. But then $N^{t+1} \subseteq NK = 0$ and so N is nilpotent and A is semiprimary.

(b) Since A is not semiprimary, A is not perfect by (a). If $N \neq 0$, then since A/N is semisimple Artinian there is a nonzero left A–module M with socle $(M) = 0$. Suppose $0 \neq B$ is a submodule of M with $(0 : B) = \{x \in A \mid xB = 0\} \neq 0$. Then B is a left $A/(0 : B)$–module; hence, since $A/(0 : B)$ is semiprimary, socle $(B) \neq 0$ as an $A/(0 : B)$–module. But then socle $(B) \neq 0$ is an A–module and so socle (M) $\neq 0$, a contradiction. It follows that $(0 : B) = 0$ for each nonzero submodule B of M. Now if I, J are ideals of A with $IJ = 0$ and $J \neq 0$, then $JM \neq 0$ and thus $I \subseteq (0 : JM) = 0$. Hence A is a prime ring in this case.

Now suppose $N = 0$. If A is not prime there exist nonzero ideals I, J with $IJ = 0$. By Theorem 2.1 we have nonzero maximal ideals $M_1,\cdots,M_k,M_{k+1},\cdots,M_t$ such that $M_1,\cdots,M_k \subseteq I$ and $M_{k+1}\cdots M_t \subseteq J$, hence $M_1\cdots M_t \subseteq IJ = 0$. Since $(M_1 \cap \cdots \cap M_t)^t \subseteq M_1\cdots M_t$ and $N = 0$, $M_1 \cap \cdots \cap M_t = 0$, and so A is then a direct sum of some or all the simple Artinian rings $A/M_1,\cdots,A/M_t$ and hence Artinian. This completes the proof.

Corollary 2.8. If A is a commutative RSP–ring which is not semiprimary, then A is an integral domain.

Because of Corollary 2.8, we will refer to commutative non–semiprimary RSP–rings as RSP–domains.

The example given previously of a non–Noetherian RSP–domain turns out to be a local ring. The next result provides a means of obtaining other RSP–domains which are not RM–domains. Recall that a ring A is a *valuation* ring if given $a,b \in A$ either $Aa \subseteq Ab$ or $Ab \subseteq Aa$; in particular every valuation ring is a local ring.

Proposition 2.9. Let A be a valuation domain with unique maximal ideal M. Then A is an RSP–domain if and only if $\cap_{n=1}^{\infty} M^n = M_\omega = 0$.

Proof. Suppose $M_\omega = 0$. If $I \neq 0$ is an ideal of A then $I \nsubseteq M^k$ for some $k \geqslant 1$. Since A is a valuation domain, $M^k \subseteq I$ and so by Theorem 2.2, A is an RSP–ring. Conversely, assume A is an RSP–domain. If $M_\omega \neq 0$, then M_ω contains a product of maximal ideals; i.e., $M^k \subseteq M_\omega$ for some $k \geqslant 1$ and so $M^k = M^{k+j}$ for all $j \geqslant 1$. Now if I is a nonzero ideal of A, then $M^t \subseteq I$ for some $t \geqslant 1$ and hence $M_\omega \subseteq I$. Thus M_ω is a minimal idempotent ideal of A and hence M_ω is a field. But then $M = J(A)$ contains a nonzero idempotent; it follows that $M_\omega = 0$.

Corollary 2.10. If A is a non–Noetherian valuation domain with unique maximal ideal M and $\cap_{n=1}^{\infty} M^n = 0$, then A is an RSP–domain which is not an RM–domain.

If K denotes the quotient field of an RSP–domain A, we do not know if every intermediate ring between A and K must be an RSP–domain, though we suspect this is not the case. In this connection, however, we have the following:

Proposition 2.11. Let A be an integral domain with quotient field K and let D be an overring of A in K which is integral over A. Then D is an RSP–domain if and only if A is an RSP–domain.

Proof. By [13, p. 259] if P is a prime ideal of A, then there is a prime ideal Q of D such that $Q \cap A = P$; moreover Q is maximal if and only if P is maximal. A straightforward application of Theorem 2.2 completes the proof.

A similar proof establishes the following "descent" property for RSP–domains.

Proposition 2.12. If D is an RSP–domain and A is a subdomain of D such that D is a finitely generated A–module, then A is an RSP–domain.

We make one more observation without proof concerning RSP–domains.

Proposition 2.13. Let A be an integral domain. Then $A[X]$ is an RSP–domain if and only if A is a field (and $A[X]$ is an RM–domain.)

The next result of this section gives necessary and sufficient conditions for a prime ring with nonzero socle to be an RSP–ring. For this and later use we say that a left A–module M has a *socle sequence* if there is a (possibly transfinite) sequence of submodules $0 = M_0 \subset M_1 \subset \cdots \subset M_\delta = M$ such that:

(i) $M_\alpha/M_{\alpha-1}$ is a completely reducible A–module whenever $\alpha-1$ exists

(ii) $M_\alpha = \cup_{\beta<\alpha} M_\alpha$ for limit ordinals α.

Note that M has a socle sequence if and only if every nonzero factor module of M contains a nonzero simple submodule.

Theorem 2.14. For a prime ring A the following statements are equivalent:

(a) A is an RSP–ring with nonzero socle.

(b) A has nonzero socle A_1 and A/A_1 is a semiprimary ring.

(c) A has a finite socle sequence $0 = A_0 \subset A_1 \subset \cdots \subset A_{k-1} \subset A_k = A$ such that $A_{k-1}/A_1 = J(A/A_1)$.

Proof. Clearly (a) $\Leftrightarrow$ (b). Assume (b) and let $N/A_1 = J(A/A_1)$. Then $A/N \cong (A/A_1)/(N/A_1)$ is semisimple Artinian and hence a completely reducible A–module. Moreover N/A_1 is nilpotent; hence $N^k \subseteq A_1$ for some least $k \geqslant 1$. Since A is prime, $A_1^2 = A_1$, and so $A_1 \subset N^i$ for all $1 \leqslant i \leqslant k$. Also

N^i/N^{i-1} is a completely reducible A/A_1–module and hence a completely reducible A–module for each $1 \leqslant i \leqslant k$. Thus the sequence $0 \subset A_1 \subset N^{k-1} \subset \cdots \subset N^2 \subset N \subset A$ is a socle sequence for A. Now assume (c) holds. We can assume that A_1 = socle of A. If $B \neq 0$ is an ideal of A then, since A is prime, $B \cap A_1 \neq 0$. If $B_1 = B \cap A_1 \neq A_1$, then $A_1 = B_1 \oplus C$ as left A–modules and so $B_1C \subseteq B_1 \cap C = 0$. Thus it must be that $A_1 \subseteq B$ for all nonzero two–sided ideals B of A. Since factor rings of semiprimary rings are semiprimary rings it suffices then to show that A/A_1 is semiprimary. But this follows easily from the fact that $A_{k-1}/A_1 = J(A/A_1)$. For $(A/A_1)/J(A/A_1) \cong A/A_{k-1}$ is completely reducible and hence a semisimple Artinian ring; also each A_i/A_1 is annihilated by A_{k-1}/A_1 and so $A_{k-1}^2 \subseteq A_{k-2}$, $A_{k-1}^3 \subseteq A_{k-3}$, $A_{k-1}^{k-1} \subseteq A_1$ thus $J(A/A_1)$ is nilpotent. This completes the proof.

For a commutative RSP–domain A, the torsion A–modules are precisely those A–modules having a socle sequence. If S denotes a simple A–module, let $\mathcal{T}_S$ denote the class of all torsion A–modules having all simple modules appearing in the socle sequence of A isomorphic to S. The class $\mathcal{T}_S$ is then a torsion class in the sense of S. E. Dickson [5], and for any torsion A–module M, $M \supseteq \oplus \Sigma\, T_S(M)$, where $T_S(M)$ denotes the unique maximal submodule of M belonging to $\mathcal{T}_S$, and S ranges over a representative set of non–isomorphic simple A–modules. Now for any ideal $I \neq 0$ of A, we have $I \supseteq M_1 \cdots M_k$ where the M_j are maximal ideals of A. Thus if $I \subseteq P$ with P maximal, then $P = M_j$ for some j and so each nonzero ideal of A, hence each nonzero element of A, lies in only a finite number of maximal ideals of A. Thus by [6, Cor. 2.7], we have that for any torsion A–module M, $M = \oplus\, \Sigma\, T_S(M)$. Thus we have shown:

Theorem 2.14. If A is an RSP–domain, then A has the primary decomposition property for torsion A–modules.

Finally, we mention that Gilmer establishes in [8, Theorem 4] that if A is an RSP–domain which is strongly integrally closed (in the sense that if $x \in K$, the quotient field of A, such that all powers of x are in a finite A–submodule of K, then $x \in A$) then A is a Dedekind domain, and so every nonzero ideal is a product of maximal ideals.

3. Some Related Restricted Conditions

In this section we consider some conditions which are related to the RM–condition.

Following Vamos [12], we say a left A–module M is *finitely–embedded* if M has a finitely generated essential socle. We say A is a *finitely–embedded ring* if ${}_AA$

is finitely–embedded. Vamos established that an A–module M is Artinian if and only if each factor module of M is finitely–embedded. Related to this we have

Proposition 3.1. Let A be a ring. Then A/I is a finitely–embedded ring for each ideal $I \neq 0$ of A if and only if A is an RM–ring.

Proof. Clearly every RM–ring A has the property that A/I is a finitely embedded ring for each ideal $I \neq 0$. Thus assume that A/I is finitely–embedded for each ideal $I \neq 0$ of A. Now each A/I has a socle sequence and so by [1, Proposition 2.1], nonzero A/I–modules have nonzero socle. If $I \neq 0$ is an ideal of A, then (A/I)/J(A/I) is a semiprime finitely–embedded ring and hence is semisimple Artinian. Thus each A/I is a right perfect ring. Suppose that K is a right perfect ring such that K and all of its factor rings are finitely–embedded. We claim that K must be left Artinian. For if N = J(K), then $N/N^2 \subseteq \text{socle}(K/N^2)$ and so N/N^2 is a finitely generated K/N, hence, K–module. By [11, Lemma 11], N is nilpotent and K is left Artinian. Combining this with the fact that each A/I is a finitely–embedded perfect ring all of whose factors are finitely–embedded, we conclude that A is an RM–ring.

It is appropriate to mention that John Beachy has also obtained the above result and it appears in [3].

A commutative subdirectly irreducible ring has essential finitely generated socle and so is finitely–embedded. Thus it might be of interest to consider rings whose factors are direct sums of subdirectly irreducible rings. For commutative domains these turn out to be Dedekind domains and this is the substance of the next result.

Proposition 3.2. For a commutative domain A the following are equivalent:

(a) A is a Dedekind domain.

(b) Every proper factor of A is a direct sum of subdirectly irreducible rings.

Proof. Assume (a); then every proper factor of A is a quasi–Frobenius principal ideal ring [7,9]. But a quasi–Frobenius principal ideal ring is a direct sum of local quasi–Frobenius principal ideal rings each of which has a unique simple submodule and so is a direct sum of subdirectly irreducible rings. Conversely assume (b) holds; then every proper factor of A is finitely embedded and so A is an RM–ring by Proposition 3.1. Hence A is also Noetherian. Since any subdirectly irreducible Artinian ring is self–injective and so quasi–Frobenius, it follows by [7] or [9] that A is a Dedekind domain, completing the proof.

Finally we should note that if A is either an RM–ring or an RSP–ring which is not semiprimary and we assume that A has enough ideals in the sense that every nonzero left ideal contains a nonzero two sided ideal then, because A is prime, A is in fact a left Ore domain. Moreover, if A is a non–simple RSP–prime ring, then A has no infinite sets of orthogonal idempotents.

References

1. J. S. Alin and E. P. Armendariz, A class of rings having all singular simple modules injective, Math. Scandinavica **23** (1968), 233-240.
2. H. Bass, Finitistic dimension and a homological generalization of semi–primary rings, Trans. Amer. Math. Soc. **95** (1960), 466-488.
3. J. Beachy, On quasi–Artinian rings, J. London Math. Soc. **46** (1971).
4. I. S. Cohen, Commutative rings with restricted minimum condition, Duke Math. J. **17** (1950), 27-42.
5. S. E. Dickson, A torsion theory for Abelian categories, Trans. Amer. Math. Soc. **12** (1966), 223-235.
6. S. E. Dickson, Decomposition of modules II: rings without chain conditions, Math. Z. **104** (1968), 349-357.
7. C. Faith, On Köthe rings, Math. Annalen **164** (1966), 207-212.
8. R. W. Gilmer, Rings in which the primary decomposition theorem holds, Proc. Amer. Math. Soc. **14** (1963), 777-781.
9. L. Levy, Commutative rings whose homomorphic images are self–injective, Pacific J. Math. **18** (1966), 149-153.
10. A. J. Ornstein, Rings with restricted minimum condition, Proc. Amer. Math. Soc. **19** (1968), 1145-1150.
11. B. L. Osofsky, A generalization of quasi–Frobenius rings, J. Algebra **4** (1966), 373-387.
12. P. Vamos, The dual of the notion of "finitely generated," J. London Math. Soc. **43** (1968), 642-646.
13. O. Zariski and P. Samuel, "Commutative Algebra I," Van Nostrand, Princeton (1958).

ALGEBRAS WITH HOCHSCHILD DIMENSION $\leqslant 1$

Goro Azumaya

Indiana University

Introduction

Let A be an algebra over a commutative ring K. Let A° be the opposite K–algebra to A, and let $A \otimes A^\circ$ be their tensor product over K, i.e., the enveloping algebra of A. Then A can be regarded as a left $A \otimes A^\circ$–module in the natural way. The homological dimension of this module is called the Hochschild dimension of A, and is denoted by K–dim A. That K–dim A = 0 means that A is $A \otimes A^\circ$–projective, i.e., A is separable over K. Such algebras are investigated by Auslander and Goldman in [1]. On the other hand, that K–dim A $\leqslant 1$ is equivalent to the following condition: if M, N are two–sided A–modules and $\varphi : M \to N$ an A–A–epimorphism then for any derivation $\epsilon : A \to N$ there exists a derivation $\delta : A \to M$ such that $\varphi \circ \delta = \epsilon$. Now, we call A a singularly segregated algebra if given two K–algebras B, C, a singular epimorphism $h : B \to C$, i.e., an algebra–epimorphism satisfying $(\ker h)^2 = 0$, and an algebra–homomorphism $g : A \to C$ there exists an algebra–homomorphism $f : A \to B$ such that $h \circ f = g$. It is proved that if A is singularly segregated then K–dim A $\leqslant 1$. The converse is not true in general, but it was actually proved by Hochschild [5] that if A is projective as a K–module then K–dim A $\leqslant 1$ implies that A is singularly segregated. The following proposition seems to suggest the difference between the two types of algebras: if A is any separable K–algebra and B any K–algebra with K–dim B $\leqslant 1$ then K–dim$(A \otimes B) \leqslant 1$, while if A is a separable K–algebra which is projective as a K–module and B a singularly segregated K–algebra then $A \otimes B$ is singularly segregated too.

We now consider the case where A is a unital subalgebra of a K–algebra B. Then B can be regarded as a left $A \otimes B^\circ$–module. The homological dimension of the module is denoted by K–dim(A, B). Then the above characterization of algebras A with K–dim A $\leqslant 1$ can be generalized as follows: that K–dim(A, B) $\leqslant 1$ is equivalent to the condition that for any two–sided A–B–modules (or equivalently, left $A \otimes B^\circ$–modules) M, N and any A–B–epimorphism $\varphi : M \to N$ every derivation $\epsilon : A \to N$ can be lifted to a derivation $\delta : A \to M$ so that $\varphi \circ \delta = \epsilon$. Also, that

K–dim$(A, B) = 0$, i.e., B is a projective $(A \otimes B^{\circ})$–module is equivalent to the condition that, for the same M, N and $\varphi : M \to N$ as above, φ maps the centralizer M^A of A in M onto the centralizer N^A of A in $N : \varphi(M^A) = N^A$. Thus it follows that if K–dim $A \leqslant 1$ then K–dim$(A, B) \leqslant 1$, while if K–dim $A = 0$ then K–dim$(A, B) = 0$. It should be mentioned, however, that the similar situation does not hold in general, or more precisely, in case $n > 1$ that K–dim $A \leqslant n$ does not always imply that K–dim$(A, B) \leqslant n$ unless A satisfies certain conditions such as projectivity over K. In particular, the case where K–dim$(A, B) = 0$ seems of interest; we call A a relatively separable subalgebra of B in this case. Indeed, it is shown that if K–dim$(A, B) = 0$ then every left B–module is (A, K)–projective (in the sense of relative homology) and every right B–module is (A, K)–injective. On the other hand, when changing the base rings the notion of relatively separable (commutative) sub–algebras plays an important role in connection with the maintenance of the property that K–dim $A \leqslant 1$ in the following sense: if L is a subalgebra of the center of A and is relatively separable in A then K–dim $A \leqslant 1$ if and only if L–dim $A \leqslant 1$.

The latter half of this paper is throughout concerned with the case where K is a Hensel ring and A is a separable algebra. Let A be a separable K–algebra which is projective as a K–module. Then A is singularly segregated by Hochschild's theorem, i.e., for any singular epimorphism $h : B \to C$ every algebra–homomorphism $g : A \to C$ can be lifted to an algebra–homomorphism $f : A \to B$ so that $h \circ f = g$. However, we can prove that the assumption that K is a Hensel ring makes it possible to replace the condition that h is a singular epimorphism by the much weaker condition that $\ker h$ is in the (Jacobson) radical of B provided both B and C are assumed to be finitely generated as K–modules. By using this fact,we can give a simplified or clearer proof to the final theorem (Generalized Wedderburn–Malcev Theorem) of the writer's previous paper [2], whose proof given there seemed somewhat complicated.

1. Relative Hochschild dimensions and relatively separable subalgebras

Let K be a commutative ring, and let A, B be K–algebras. Let B° be an opposite algebra of B, which is in an opposite isomorphism $b \to b^{\circ}$ with B. A left A– and right B–module M is called a two–sided A–B–module, or more precise–ly, a two–sided A–B–module over K, if $(ax)b = a(xb)$ and $\kappa x = x\kappa$ for $a \in A$, $b \in B$, $x \in M$, $\kappa \in K$. As is well known, every two–sided A–B–module M can be made into a left module of the tensor product K–algebra $A \otimes B^{\circ} = A \otimes_K B^{\circ}$ by setting

$$(1) \qquad (a \otimes b^{\circ})x = axb$$

for $a \in A$, $b \in B$, $x \in M$; in particular,

$$(2) \qquad (a \otimes 1)x = ax, \quad (1 \otimes b^{\circ})x = xb.$$

Conversely, every left $A \otimes B^\circ$–module M can be converted into a two–sided A–B–module by means of (2), or equivalently, by (1).

Now, we shall assume that A is a unital subalgebra of B (i.e., A is a subalgebra of B and contains the unit element 1 of B). Then B can be regarded as a two–sided A–B–module and so a left $A \otimes B^\circ$–module. The homological dimension (= projective dimension) of this module B is called the *relative Hochschild dimension* of A in B (over K) and is denoted by $K\text{–dim}(A, B)$. This is the supremum of those integers $n (\geqslant 0)$ for which $\mathrm{Ext}^n_{A\otimes B^\circ}(B, M) \neq 0$ for all left $A \otimes B^\circ$–modules M.

Consider the $A \otimes B^\circ$–epimorphism $\rho : A \otimes B^\circ \to B$ which is defined by $\rho(u) = u1$ for $u \in A \otimes B^\circ$. The kernel of ρ is denoted by $J(A, B)$. Then this is a left ideal of $A \otimes B^\circ$, and we have a short exact sequence

$$0 \to J(A, B) \to A \otimes B^\circ \to B \to 0. \tag{3}$$

The sequence yields, with any left $A \otimes B^\circ$–module M, exact sequences

$$\mathrm{Ext}^{n-1}_{A\otimes B^\circ}(A \otimes B^\circ, M) \to \mathrm{Ext}^{n-1}_{A\otimes B^\circ}(J(A, B), M) \to \mathrm{Ext}^{n}_{A\otimes B^\circ}(B, M) \to \mathrm{Ext}^{n}_{A\otimes B^\circ}(A \otimes B^\circ, M)$$

for $n \geqslant 1$. Since $A \otimes B^\circ$ is $A \otimes B^\circ$–projective, $\mathrm{Ext}^n_{A\otimes B^\circ}(A \otimes B^\circ, M) = 0$ if $n \geqslant 1$ and thus we have natural isomorphisms

$$\mathrm{Ext}^{n-1}_{A\otimes B^\circ}(J(A, B), M) \cong \mathrm{Ext}^{n}_{A\otimes B^\circ}(B, M) \tag{4}$$

for $n > 1$ as well as an exact sequence (for $n = 1$)

$$M \xrightarrow{\tau} \mathrm{Hom}_{A\otimes B^\circ}(J(A, B), M) \to \mathrm{Ext}^{1}_{A\otimes B^\circ}(B, M) \to 0 \tag{5}$$

where the first homomorphism τ is given by associating $x \in M$ with the mapping $u \to ux$ $(u \in J(A, B))$.

We now define a K–homomorphism $j : A \to J(A, B)$ by $j(a) = a \otimes 1 - 1 \otimes a^\circ$. Then we have a generalization of [3, IX, Proposition 3.1]:

Proposition 1. $J(A, B)$ is generated by the elements $j(a)$, i.e., $J(A, B) = (A \otimes B^\circ)j(A)$.

Indeed, if $u = \Sigma\, a_i \otimes b_i^\circ$ is in $J(A, B)$ then $\Sigma\, a_i b_i = u1 = 0$ and $u = \Sigma\, a_i \otimes b_i^\circ - \Sigma\, 1 \otimes (a_i b_i)^\circ = \Sigma\,(1 \otimes b_i^\circ)(a_i \otimes 1 - 1 \otimes a_i^\circ)$.

We denote by $D_K(A, M)$ the K–module of all K–derivations $\delta : A \to M$, i.e., those K–homomorphisms of A into M which satisfy $\delta(a_1 a_2) = a_1\delta(a_2) + \delta(a_1)a_2$ for $a_1, a_2 \in A$. For any $x \in M$, the mapping $a \to ax - xa$ $(a \in A)$

defines a K–derivation, called inner; the set of all such derivations forms a K–submodule $I_K(A, M)$ of $D_K(A, M)$.

The following proposition can be proved in the same way as [3, IX, Proposition 3.2].

Proposition 2. By associating with each $h \in \mathrm{Hom}_{A\otimes B^\circ}(A, M)$ the mapping $h \circ j$, we obtain a natural K–isomorphism

$$\mathrm{Hom}_{A\otimes B^\circ}(J(A, B), M) \cong D_K(A, M). \tag{6}$$

Consider the homomorphism τ in (5) and suppose $h = \tau(x)$ with some $x \in M$. Then $h(j(a)) = \tau(x)(a \otimes 1 - 1 \otimes a^\circ) = (a \otimes 1 - 1 \otimes a^\circ)x = ax - xa$, i.e., the image of $h = \tau(x)$ by the isomorphism (6) is nothing but the inner derivation $a \to ax - xa$. Thus we have an isomorphism

$$\mathrm{Ext}^1_{A\otimes B^\circ}(B, M) \cong D_K(A, M)/I_K(A, M). \tag{7}$$

Now, specializing $n = 1$, we can derive from [3, VI, Proposition 2.1] the following proposition.

Proposition 3. The following conditions are equivalent

(i) $K\text{–}\dim(A, B) \leqslant 1$,

(ii) $\mathrm{Ext}^2_{A\otimes B^\circ}(B, M) = 0$ for all two–sided A–B–modules M,

(iii) $J(A, B)$ is $A \otimes B^\circ$–projective,

(iv) $D_K(A, M)$ is a right exact functor of two–sided A–B–modules M.

Indeed, the equivalence of (i) and (iii) follows from the exact sequence (3) with projective $A \otimes B^\circ$, while the equivalence of (iii) and (iv) follows from the isomorphism (6).

For a two–sided A–B–module M, we denote by M^A the centralizer of A in M, i.e., the K–module of those elements x of M for which $ax = xa$ for all $a \in A$. We regard M as a left $A \otimes B^\circ$–module. Since $j(a)x = ax - xa$ for $a \in A$, $x \in M$ and $J(A, B)$ is generated by $j(a)$'s (Proposition 1), M^A is nothing but the annihilator of $J(A, B)$ in M. This, together with the exactness of (3), implies there there exists a natural K–isomorphism

$$\mathrm{Hom}_{A\otimes B^\circ}(B, M) \cong M^A; \tag{8}$$

in fact, the isomorphism is given by associating each $h \in \mathrm{Hom}_{A\otimes B^\circ}(B, M)$ with the image $h(1)$.

Proposition 4. The following conditions are equivalent

(i) $K\text{-dim}(A, B) = 0$,

(ii) $\mathrm{Ext}^1_{A\otimes B^\circ}(B, M) = 0$ for all two-sided A–B–modules M,

(iii) B is $A \otimes B^\circ$–projective,

(iv) $D_K(A, M) = I_K(A, M)$ for all two-sided A–B–modules M,

(v) M^A is a right exact functor of two-sided A–B–modules M.

Proof. The equivalence of (i), (ii), (iii) are clear. The equivalence of (ii) and (iv) and the equivalence of (iii) and (v) follow from the isomorphisms (7) and (8) respectively.

We shall call A *relatively* (K–)*separable* in B if they satisfy any of the equivalent conditions in Proposition 4. Clearly, a K–algebra A is (K–)separable if and only if A is relatively separable in A.

Proposition 5. Let A be relatively separable in B. Then for every two-sided A–B–module M, M^A is a direct summand of the K–module M.

Proof. Since B is $A \otimes B^\circ$–projective, it follows from the exactness of (3) that $J(A, B)$ is a direct summand left ideal of $A \otimes B^\circ$, i.e., $J(A, B)$ is generated by an idempotent e of $A \otimes B^\circ$: $J(A, B) = (A \otimes B^\circ)e$. Therefore, the annihilator M^A of $J(A, B)$ is nothing but the annihilator of e, i.e., $M^A = (1 - e)M$. Thus we have a direct decomposition $M = M^A \oplus eM$ into K–submodules.

The following proposition can be regarded as a generalization of [1, Proposition 1.4].

Proposition 6. Let A be relatively separable in B. Then for any algebra–epimorphism $B \to B'$, the image A' of A is relatively separable in B' and the image of B^A is $B'^{A'}$.

Proof. Let $h : B \to B'$ denote the given algebra–epimorphism. Then every two-sided A'–B'–module M can be converted into a two-sided A–B–module by means of h. Therefore $M^{A'} = M^A$ is a right exact functor of M, i.e., A' is relatively separable in B'. In particular, $B' = {}_{A'}B'_{B'}$ can be regarded as ${}_AB'_B$ and h becomes an epimorphism ${}_AB_B \to {}_AB'_B$. Thus h induces an epimorphism $B^A \to B'^A = B'^{A'}$.

Proposition 7. Let A be relatively separable in B. Then every left B-module is (A, K)-projective and every right B-module is (A, K)-injective.

Proof. Let M be a left B-module and N a left A-module. Then $\mathrm{Hom}_K(M, N)$ becomes a two-sided A-B-module by setting $(ah)(x) = a(h(x))$ and $(hb)(x) = h(bx)$ for $a \in A$, $b \in B$, $x \in M$, $h \in \mathrm{Hom}(M, N)$. Therefore, it follows that $\mathrm{Hom}_K(M, N)^A = \mathrm{Hom}_A(M, N)$. Let N' be another left A-module and $\varphi : N \to N'$ an A-epimorphism which is K-split. Then, by associating each $h \in \mathrm{Hom}_k(M, N)$ with $\varphi \circ h$, we have an A-B-epimorphism $\mathrm{Hom}_K(M, N) \to \mathrm{Hom}_K(M, N')$, which therefore induces an epimorphism $\mathrm{Hom}_A(M, N) \to \mathrm{Hom}_A(M, N')$ $(= \mathrm{Hom}(M, N')^A)$ because of the relative separability of A in B. Thus M is (A, K)-projective. If we next observe that for a right B-module M and a right A-module N, $\mathrm{Hom}_K(N, M)$ is a two-sided A-B-module and $\mathrm{Hom}_A(N, M) = \mathrm{Hom}(N, M)^A$, we can prove in the similar way that M is (A, K)-injective.

Finally, if we observe the fact that the modules $D_K(A, M)$ and M^A, both defined for two-sided A-B-modules M, depend only on the A-A-modules structure of M, we can derive the following immediately from Propositions 3 and 4.

Proposition 8. If K-dim A $(= \mathrm{hd}_{A \otimes A^\circ} A) \leqslant 1$ then K-dim$(A, B) \leqslant 1$ and if A is K-separable then A is relatively K-separable in B, for every algebra B which contains A as a unital subalgebra.

2. Algebras with dimension $\leqslant 1$ and singularly segregated algebras

Let A be a K-algebra, and let L be a unital subalgebra of the center of A. Then L is commutative and A can also be regarded as an L-algebra. Let M be a two-sided A-module over K. Then the centralizer M^L of L is regarded as a two-sided A-module over L.

Lemma. Let $\delta : A \to M$ be a K-derivation. Then $\delta(L) = 0$ if and only if δ is an L-derivation of A into M^A.

Proof. The "if" part follows from that $\delta(\lambda) = \lambda\delta(1)$ for all $\lambda \in L$ and $\delta(1) = 0$. Conversely assume that $\delta(L) = 0$. Then, for any $\lambda \in L$ and $a \in A$, we have $\delta(\lambda a) = \lambda\delta(a) + \delta(\lambda)a = \lambda\delta(a)$ and $\delta(a\lambda) = a\delta(\lambda) + \delta(a)\lambda = \delta(a)\lambda$. But since $\lambda a = a\lambda$, it follows $\lambda\delta(a) = \delta(a)\lambda$. Thus δ is L-linear and $\delta(A) \subset M^A$.

Proposition 9. Let A be a K-algebra and let L be a unital subalgebra of the center of A such that L is relatively K-separable in A. Then K-dim $A \leqslant 1$ if and only if L-dim $A \leqslant 1$.

Proof. Suppose K–dim A $\leqslant 1$. Let M, N be two–sided A–A–modules over L and $\varphi : {}_AM_A \to {}_AN_A$ an epimorphism. Let $\epsilon : A \to N$ be an L–derivation. Then there exists a K–derivation $\delta : A \to M$ such that $\varphi \circ \delta = \epsilon$. If we restrict δ to L, we have a K–derivation of L. Since L is relatively separable, it must be inner (Proposition 4). L is however element–wise commutative with M and so every inner derivation of L into M is 0. Thus $\delta(L) = 0$, and δ is an L–derivation by Lemma, which shows L–dim A $\leqslant 1$ according to Proposition 3.

Assume conversely L–dim A $\leqslant 1$. Let M, N be two–sided A–B–modules over K this time and $\varphi : {}_AM_B \to {}_AN_B$ an epimorphism. Let $\epsilon : A \to N$ be a K–derivation. If we restrict ϵ to L, we have a K–derivation of L, which is however an inner derivation because of the relative separability of L, i.e., there exists a $y \in N$ such that $\epsilon(\lambda) = \lambda y - y\lambda$ for all $\lambda \in L$. We define a mapping $\epsilon' : A \to N$ by $\epsilon'(a) = \epsilon(a) - (ay - ya)$ for $a \in A$. Then ϵ' is a K–derivation and satisfies $\epsilon'(\lambda) = 0$ for all $\lambda \in L$. Therefore it follows from Lemma that ϵ' is an L–derivation of A into N. Now, since L is relatively separable in A the epimorphism $\varphi : {}_AM_A \to {}_AN_A$ induces an epimorphism $M^L \to N^L$. Therefore there exists by Proposition 3 an L–derivation $\delta' : A \to M^L$ such that $\varphi \circ \delta' = \epsilon'$. We pick out an $x \in M$ such that $\varphi(x) = y$, and then define a mapping $\delta : A \to M$ by putting $\delta(a) = \delta'(a) + (ax - xa)$. It is clear that δ is a K–derivation and satisfies $\varphi \circ \delta = \epsilon$, and thus K–dim A $\leqslant 1$.

Let B, C be K–algebras. An algebra–epimorphism $h : B \to C$ is called a *singular epimorphism* if $(\ker h)^2 = 0$. A K–algebra A is called *singularly segregated* (over K) if for any B, C, any singular epimorphism. $h : B \to C$ and any algebra–homomorphism $g : A \to C$ there exists an algebra–homomorphism $f : A \to B$ such h that $h \circ f = g$.

Proposition 10. A K–algebra A is singularly segregated if and only if every singular epimorphism onto A splits.

Proof. Since the "only if" part is clear, we have only to prove the "if" part. Let there be given K–algebras B, C, a singular epimorphism $h : B \to C$ and an algebra–homomorphism $g : A \to C$. Then consider the subalgebra S of the direct sum $A \oplus B$ consisting of those elements (a, b) which satisfies $g(a) = h(b)$. If we associate $(a, b) \in S$ with $a \in A$, we have an algebra–epimorphism $\sigma : S \to A$. Furthermore, if (a, b) is in $\ker \sigma$ then $a = 0$ and $h(b) = g(0) = 0$. This shows that $\ker \sigma = 0 \oplus \ker h$ and therefore $(\ker \sigma)^2 = 0$ (because $\ker h$ satisfies the same condition). Thus σ is a singular epimorphism. Hence it must split, i.e., there exists an algebra–homomorphism $\tau : A \to S$ such that $\sigma \circ \tau = 1$. This means that for any $a \in A$ the image $\tau(a)$ is of the form (a, b). Thus, by associating a with this b, we obtain an algebra–homomorphism $f : A \to B$. Since $(a, f(a)) = \tau(a)$ is in S, it follows $g(a) = h(f(a))$, i.e., $g = h \circ f$, which proves that A is singularly segregated.

Proposition 11. If A is singularly segregated then $K\text{-dim}\, A \leqslant 1$.

Proof. Let there be given two-sided A-modules M, N, an epimorphism $\varphi : {}_AM_A \to {}_AN_A$ and a derivation $\epsilon : A \to N$. We consider the direct sum $A \oplus M$. For any two elements (a, x) and (b, y) of $A \oplus M$, we define their product by $(a, x)(b, y) = (ab, ay + xa)$. Then it is easy to prove that $A \oplus M$ forms a K-algebra with unit element $(1, 0)$ and if we identify each $x \in M$ with $(0, x)$, M becomes a two-sided ideal of $A \oplus M$ satisfying $M^2 = 0$. Similarly, $A \oplus N$ can be made into a K-algebra. Now, if we define a mapping $h : A \oplus M \to A \oplus N$ by $h(a, x) = (a, \varphi(x))$ then it is also easy to see that h is a singular epimorphism. Next, we define a mapping $g : A \to A \oplus N$ by $g(a) = (a, \epsilon(a))$. Then g is an algebra-homomorphism, because $(a, \epsilon(a))(b, \epsilon(b)) = (ab, a\epsilon(b) + \epsilon(a)b) = (ab, \epsilon(ab))$. Since A is singularly segregated, there exists an algebra-homomorphism $f : A \to A \oplus M$ such that $h \circ f = g$. Therefore $h(f(a)) = (a, \epsilon(a))$ for every $a \in A$ and thus $f(a)$ is of the form (a, z) with $z \in M$ such that $\varphi(z) = \epsilon(a)$. If we associate a with z, we have a mapping $\delta : A \to M$. By using the fact that f is an algebra-homomorphism, we can verify that δ is a derivation and satisfies $\varphi \circ \delta = \epsilon$. Thus the natural homomorphism $D_K(A, M) \to D_K(A, N)$ is an epimorphism, and therefore $K\text{-dim}\, A \leqslant 1$ by Proposition 3.

Proposition 12. If A is K-projective and $K\text{-dim}\, A \leqslant 1$ then A is singularly segregated.

This was actually proved by Hochschild [5], but we shall give a proof for completeness (cf, also [3, XIV, 2]). Let B, C be K-algebras, $h : B \to C$ a singular epimorphism and $g : A \to C$ an algebra-homomorphism. Then since A is K-projective, there exists a K-homomorphism $\varphi : A \to B$ such that $h \circ \varphi = g$. Let a_1, a_2 be any elements of A. Then $h(\varphi(a_1a_2)) = g(a_1a_2) = g(a_1)g(a_2) = h(\varphi(a_1))h(\varphi(a_2)) = h(\varphi(a_1)\varphi(a_2))$ and hence $\varphi(a_1a_2) \equiv \varphi(a_1)\varphi(a_2) \mod \ker h$, i.e., there is an element $\zeta(a_1, a_2)$ of $\ker h$ such that

$$\varphi(a_1a_2) = \varphi(a_1)\varphi(a_2) + \zeta(a_1, a_2). \tag{9}$$

We can now convert $\ker h$ into a two-sided A-module by means of φ. Namely, for any $x \in \ker h$ and $a \in A$, we define ax and xa to be $\varphi(a)x$ and $x\varphi(a)$ respectively; since $\ker h$ is a two-sided ideal of A, both ax and xa are in $\ker h$, and since $(\ker h)^2 = 0$, it follows from (9) that $(a_1a_2)x = a_1(a_2x)$, $x(a_1a_2) = (xa_1)a_2$. On the other hand, using (9) we have the following equalities: $\varphi(a_1a_2a_3) = \varphi(a_1)\varphi(a_2a_3) + \zeta(a_1, a_2a_3) = \varphi(a_1)\varphi(a_2)\varphi(a_3) + \varphi(a_1)\zeta(a_2a_3) + \zeta(a_1, a_2a_3)$, $\varphi(a_1a_2a_3) = \varphi(a_1a_2)\varphi(a_3) + \zeta(a_1a_2, a_3) = \varphi(a_1)\varphi(a_2)\varphi(a_3) + \zeta(a_1, a_2)\varphi(a_3) + \zeta(a_1a_2, a_3)$.

Thus we have

$$a_1\zeta(a_2, a_3) - \zeta(a_1a_2, a_3) + \zeta(a_1, a_2a_3) - \zeta(a_1, a_2)a_3 = 0,$$

which shows that $\zeta(a_1, a_2)$, as a function of a_1, a_2, gives a 2–dimensional cocycle of A into the two–sided A–module $\ker h$. Since K–dim $A \leqslant 1$ and hence $\mathrm{Ext}^2_{A\otimes A^\circ}(A, \ker h) = 0$ and since A is K–projective, it follows from [3, IX, 6] that every 2–dimensional cocycle is a coboundary; thus there exists a K–homomorphism $\eta : A \to \ker h$ such that

$$\zeta(a_1, a_2) = a_1\eta(a_2) - \eta(a_1a_2) + \eta(a_1)a_2 \tag{10}$$

for $a_1, a_2 \in A$. Now we put $f(a) = \varphi(a) + \eta(a)$ for each $a \in A$. Then we have a K–homomorphism $f : A \to B$. It satisfies $f(a_1a_2) = \varphi(a_1a_2) + \eta(a_1a_2) = \varphi(a_1)\varphi(a_2) + \zeta(a_1, a_2) + \eta(a_1a_2)$. On the other hand, since $\eta(a_1)\eta(a_2) = 0$, $f(a_1)f(a_2) = (\varphi(a_1) + \eta(a_1))(\varphi(a_2) + \eta(a_2)) = \varphi(a_1)\varphi(a_2) + \varphi(a_1)\eta(a_2) + \eta(a_1)\varphi(a_2) = \varphi(a_1)\varphi(a_2) + a_1\eta(a_2) + \eta(a_1)a_2$. Therefore, it follows from (10) that $f(a_1a_2) = f(a_1)f(a_2)$, i.e., f is an algebra–homomorphism. Furthermore, $h(f(a)) = h(\varphi(a)) + h(\eta(a)) = h(\varphi(a)) = g(a)$ for $a \in A$, i.e., $h \circ f = g$. Thus A is singularly segregated.

Proposition 13. Let A be a singularly segregated algebra over K and L a unital subalgebra of the center of A such that L is relatively separable in A. Then A is singularly segregated over L too.

Proof. Let B be an algebra over L and $f : B \to A$ a singular L–algebra–epimorphism. Then f splits as a K–algebra–epimorphism, i.e., there exists a K–algebra–homomorphism $g : A \to B$ such that $f \circ g = 1$. Now, L is clearly regarded as a unital subalgebra of the center of B and thus $f(\lambda) = \lambda$ for all $\lambda \in L$. Since also $f(g(\lambda)) = \lambda$ for all $\lambda \in L$, it follows that $\lambda \equiv g(\lambda) \mod \ker f$. Consider then the K–homomorphism $\delta : L \to \ker f$ defined by $\delta(\lambda) = \lambda - g(\lambda)$. Since $(\ker f)^2 = 0$, it follows $\lambda\delta(\mu) - \delta(\lambda\mu) + \delta(\lambda)\mu = \lambda(\mu - g(\mu)) - (\lambda\mu - g(\lambda)g(\mu)) + (\lambda - g(\lambda))\mu = \lambda\mu - \lambda g(\mu) + g(\lambda)g(\mu) - g(\lambda)\mu = (\lambda - g(\lambda))(\mu - g(\mu)) = 0$, and this shows that δ is a K–derivation of L into $\ker f$. Since L is relatively K–separable in A and $\ker f$ is a two–sided ideal of A, the derivation δ must be inner by Proposition 4. Since L and $\ker f$ are element–wise commutative, L has no non–zero inner derivation into $\ker f$. Thus $\delta(\lambda) = 0$, or $g(\lambda) = \lambda$ for all $\lambda \in L$, which shows that g is in fact an L–algebra–homomorphism and so A is singularly segregated over L by Proposition 10.

Proposition 14. Let L be a commutative separable algebra over K which is

K–projective and A a singularly segregated algebra over L. Then A is singularly segregated as an algebra over K too.

Proof. Let B, C be K–algebras, $h : B \to C$ a singular K–algebra–epimorphism and $g : A \to C$ a K–algebra–homomorphism. Since A is an L–algebra, we have a canonical homomorphism $p : L \to$ the center of A. Then $g \circ p : L \to C$ becomes a K–algebra–homomorphism. Since L is separable and projective over K, L is by Proposition 12 singularly K–segregated. Thus there exists a K–algebra–homomorphism $q : L \to B$ such that $h \circ q = g \circ p$. Now, $q(L)$ is, as a homomorphic image of L, separable over K. Therefore, if we put B' the centralizer of $q(L)$ in B the image $h(B')$ is the centralizer of $h(q(L)) = g(p(L))$ in C by Proposition 6. On the other hand, B' and $h(B')$ can be regarded as L–algebras by means of q and $h \circ q = g \circ p$ respectively, and thus, if we restrict h to B' we have a singular L–algebra–epimorphism : $B' \to h(B')$. Furthermore, since $p(L)$ is in the center of A, $g(A)$ is in the centralizer $h(B')$ of $g(p(L))$; thus g can actually be regarded as an L–algebra–homomorphism : $A \to h(B')$. Since A is assumed to be singularly L–segregated there must exist an L– whence K–algebra–homomorphism $f : A \to B'$ such that $h \circ f = g$. Thus A is singularly K–segregated.

Proposition 15. Let A be a separable K–algebra and B a K–algebra with K–dim $B \leqslant 1$. Then K–dim$(A \otimes B) \leqslant 1$.

Proof. Let M be a two–sided $A \otimes B$–module. Then by means of the homomorphism $a \to a \otimes 1$ $(a \in A)$ M can be converted into a two–sided A–module. Similarly, M can be regarded as a two–sided B–module, and further the centralizer M^A of A in M becomes a two–sided B–submodule. Let N be another two–sided $A \otimes B$–module and let there be given an epimorphism $\varphi : M \to N$ (as two–sided $A \otimes B$–modules). Since A is separable, φ induces an epimorphism : $M^A \to N^A$ as two–sided B–modules. Suppose moreover there is given a derivation $\epsilon : A \otimes B \to N$. Then it can be transferred to a derivation : $A \to N$, by means of the same homomorphism $A \to A \otimes B$. Since A is separable, the derivation must be inner (Proposition 4), i.e., there is a $y \in N$ such that $\epsilon(a) = ay - ya$ for all $a \in A$. Consider then a derivation $\epsilon_0 : A \otimes B \to N$ defined by $\epsilon_0(u) = \epsilon(u) - (uy - yu)$ for $u \in A \otimes B$. For any $a \in A$ and $b \in B$, $\epsilon_0(a \otimes b) = \epsilon_0((a \otimes 1)(1 \otimes b)) = a\epsilon_0(b) + \epsilon_0(a)b = a\epsilon_0(b)$ because $\epsilon_0(a) = 0$. Similarly, $\epsilon_0(a \otimes b) = \epsilon_0((1 \otimes b)(a \otimes 1)) = b\epsilon_0(a) + \epsilon_0(b)a = \epsilon_0(b)a$ and thus $a\epsilon_0(b) = \epsilon_0(b)a$. This shows that ϵ_0 can be regarded as a derivation : $B \to N^A$. Since K–dim $B \leqslant 1$, it follows that there exists a derivation $\delta_0 : B \to M^A$ such that $\varphi(\delta_0(b)) = \epsilon_0(b)$ for all $b \in B$ (Proposition 3). Now, by associating $a \otimes b$ with $a\delta_0(b)$ we have a K–homomorphism : $A \otimes B \to M$ which we shall denote again by δ_0. Then, for any $a, a' \in A$ and $b, b' \in B$, $\delta_0((a \otimes b)(a' \otimes b')) = \delta_0((aa') \otimes (bb')) = aa'\delta_0(bb') = aa'(b\delta_0(b') +$

$\delta_0(b)b') = aa'(b\delta_0(b')) + aa'\delta_0(b)b' = (aa' \otimes b)\delta_0(b') + (a\delta_0(b)a')b' = (a \otimes b)a'\delta_0(b') + a\delta_0(b)(a' \otimes b') = (a \otimes b)\delta_0(a' \otimes b') + \delta_0(a \otimes b)(a' \otimes b')$, which shows that $\delta_0 : A \otimes B \to M$ is a derivation. Furthermore, we have $\varphi(\delta_0(a \otimes b))= \varphi(a\delta_0(b)) = a\varphi(\delta_0(b)) = a\epsilon_0(b) = \epsilon_0(a \otimes b)$, i.e., $\varphi \circ \delta_0 = \epsilon_0$. Take now an $x \in M$ such that $\varphi(x) = y$, and define another derivation $\delta : A \otimes B \to M$ by $\delta(u) = \delta_0(u) + (ux - xu)$. Then clearly $\varphi \circ \delta = \epsilon$, and thus it is proved that $K\text{–dim}(A \otimes B) \leqslant 1$.

Proposition 16. Let A be a separable K–algebra which is projective over K and B a singularly segregated K–algebra. Then $A \otimes B$ is singularly K–segregated too.

Proof. Let C, C' be K–algebras and $h : C \to C'$ a singular algebra–epimorphism. Let there be given an algebra–homomorphism $g : A \otimes B \to C'$. Then there yield two algebra–homomorphisms $g_1 : A \to C'$ and $g_2 : B \to C'$ in the natural way so that $g_1 \otimes g_2 = g$. Since A is separable and projective over K, A is by Proposition 12 singularly segregated. Thus there exists an algebra–homomorphism $f_1 : A \to C$ such that $h \circ f_1 = g_1$. The homomorphic image $f_1(A)$ of A is also separable. Therefore, it follows from Proposition 6 that the centralizer D of $f_1(A)$ in C is mapped by h onto the centralizer D' of $h(f_1(A)) = g_1(A)$ in C'. The homomorphic image $g_2(B)$ of B by g_2 is clearly in D' and since B is singularly segregated, it follows that there exists an algebra–homomorphism $f_2 : B \to D$ such that $h \circ f_2 = g_2$. Since $f_1(A)$ and $f_2(B)$ are element–wise commutative, the algebra–homomorphism $f = f_1 \otimes f_2 : A \otimes B \to C$ is well defined, which satisfies $h \circ f = g$. Thus $A \otimes B$ is singularly segregated.

Remark. If L is K–separable then the "if" part of Proposition 9 follows immediately from [4, Proposition 4] (thereby, K and L should be interchanged), while if either A or B is K–flat then Proposition 15 becomes a particular case of [4, Proposition 2].

3. Separable algebras over a Hensel ring

By a finite algebra over K we mean a K–algebra which is finitely generated as a K–module. In this section, we shall throughout consider finite algebras over a Hensel ring. Let K be a Hensel ring with a unique maximal ideal P and A a finite K–algebra. Then AP is a two–sided ideal of A and the residue class ring A/AP can be regarded as a finite algebra over the residue class field K/P. The (Jacobson) radical N of A contains AP and N/AP coincides with the radical of A/AP, and therefore A/N is an Artinian semi–simple ring. Furthermore, every idempotent of A/N can be lifted to an idempotent of A [2, Theorem 24]. (In [2], "algebra" always meant "finite algebras" in our sense.) Thus A is a semi–perfect ring.

Let $p : A \to A/AP$ be the natural epimorphism. Let $a_1, a_2, \cdots, a_n$ be elements of A such that the corresponding residue classes $p(a_1), p(a_2), \cdots, p(a_n)$ form a linearly independent basis of A/AP over K/P. Then $Ka_1 + Ka_2 + \cdots + Ka_n + AP = A$. Since A is finitely generated as a K–module, it follows from [2, Theorem 1 or Theorem 5] that $Ka_1 + Ka_2 + \cdots + Ka_n = A$. Let B be another K–algebra and consider the tensor product $A \otimes B$ over K. If we denote by $q : B \to B/BP$ the natural epimorphism, then the (algebra–) epimorphism $p \otimes q : A \otimes B \to A/AP \otimes B/BP$ is well defined. Every element of $A \otimes B$ is expressed in the form $a_1 \otimes b_1 + a_2 \otimes b_2 + \cdots + a_n \otimes b_n$ with b_i in B. Suppose this element is in $\ker(p \otimes q)$, i.e., $p(a_1) \otimes q(b_1) + p(a_2) \otimes q(b_2) + \cdots + p(a_n) \otimes q(b_n) = 0$. Since $p(a_1), p(a_2), \cdots, p(a_n)$ form still a linearly independent basis of $A/AP \otimes B/BP$ over B/BP, it follows that $q(b_1) = q(b_2) = \cdots = q(b_n) = 0$. Thus $\ker(p \otimes q) = (A \otimes B)P$.

Let A° be an opposite K–algebra of A and consider the enveloping algebra $A \otimes A^\circ$. Then A is regarded as a left $A \otimes A^\circ$–module, and by associating each $u \in A \otimes A^\circ$ with $u1 \in A$ we have an $A \otimes A^\circ$–epimorphism : $A \otimes A^\circ \to A$. The kernel J of the epimorphism is a left ideal of $A \otimes A^\circ$ generated by elements of the form $a \otimes 1 - 1 \otimes a^\circ$ with $a \in A$ [3, IX, Proposition 3.1]. Since A is projective as a left A–module, J is necessarily a direct summand of $A \otimes A^\circ$ as A– whence K–modules. Therefore, it follows that $J \cap (A \otimes A^\circ)P = JP$. We again consider $\bar{A} = A/AP$ and the natural epimorphism $p : A \to \bar{A}$. Similarly, we consider $\bar{A}^\circ = A^\circ/A^\circ P$ and $p^\circ : A^\circ \to \bar{A}^\circ$. $\bar{A}^\circ$ is clearly an opposite algebra of $\bar{A}$ and so $\bar{A} \otimes \bar{A}^\circ$ is the enveloping algebra of $\bar{A}$. We have similarly the natural $\bar{A} \otimes \bar{A}^\circ$–epimorphism : $\bar{A} \otimes \bar{A}^\circ \to \bar{A}$ with the kernel $\bar{J}$, which is the left ideal of $\bar{A} \otimes \bar{A}^\circ$ generated by elements of the form $p(a) \otimes 1 - 1 \otimes p^\circ(a)$ with $a \in A$. Therefore, if we restrict $p \otimes p^\circ$ to J, its image is equal to $\bar{J}$ and furthermore its kernel is $J \cap \ker(p \otimes p^\circ) = J \cap (A \otimes A^\circ)P = JP$. Thus we have the following commutative diagram of $(A \otimes A^\circ)$–modules with exact rows and columns

$$\begin{array}{ccccccccc}
 & & 0 & & 0 & & 0 & & \\
 & & \downarrow & & \downarrow & & \downarrow & & \\
0 & \to & JP & \to & (A \otimes A^\circ)P & \to & AP & \to & 0 \\
 & & \downarrow & & \downarrow & & \downarrow & & \\
0 & \to & J & \to & A \otimes A^\circ & \to & A & \to & 0 \\
 & & \downarrow & & \downarrow & & \downarrow & & \\
0 & \to & \bar{J} & \to & \bar{A} \otimes \bar{A}^\circ & \to & \bar{A} & \to & 0 \\
 & & \downarrow & & \downarrow & & \downarrow & & \\
 & & 0 & & 0 & & 0 & &
\end{array}$$

Now, A is separable if and only if A is projective as an $A \otimes A^\circ$–module, and this is equivalent (since $A \otimes A^\circ$ is $A \otimes A^\circ$–projective) to the condition that the middle row in the above splits, i.e., J is a direct summand left ideal of $A \otimes A^\circ$, or equivalently, J is generated by an idempotent. Similarly, $\bar{A}$ is separable if and only if $\bar{J}$ is generated by an idempotent. Since $\bar{J}$ is an epimorphic image of J bv $p \otimes p^\circ$, it is clear that if J is generated by an idempotent the so is $\bar{J}$, i.e., if A is separable then $\bar{A}$ is separable (although this follows from [1, Proposition 1.4]). Conversely, suppose $\bar{J}$ is generated by an idempotent $\bar{e} : \bar{J} = (\bar{A} \otimes \bar{A}^\circ)\bar{e}$. Then, since K is Henselean and $A \otimes A^\circ$ is a finite K–algebra whence a semi–perfect ring, $\bar{e}$ can be lifted to an idempotent e in J with respect to $p \otimes p^\circ$. Since further the kernel of $p \otimes p^\circ$ restricted to J is JP, we have $J = (A \otimes A^\circ)e + JP$. J is however finitely generated as a K–module, because J is a direct summand of the finitely generated K–module $A \otimes A^\circ$. Therefore it follows $J = (A \otimes A^\circ)e$ by [2, Theorem 1]. Thus we have proved the following.

Proposition 17. Let A be a finite algebra over a Hensel ring K. Then A is separable if and only if the residue class algebra A/AP is separable, where P is the unique maximal ideal of K.

Remark. Generalizing Proposition 17, Endo and Watanabe proved in their paper, " On separable algebras over a commutative ring", Osaka J. Math. **4** (1967) that over any commutative ring K a finite algebra A is separable if and only if A/AP is separable for all maximal ideal P of K; their proof of Proposition 17 is however somewhat different from that above.

Let A be finite and separable over K. Then $\bar{A} = A/AP$ is finite and separable over the field K/P, and therefore $\bar{A}$ is semi–simple, i.e., $\bar{A}$ has radical 0. This implies AP is the radical of A. Since $\bar{A}$ is semi–simple, $\bar{A}$ is a direct sum of mutually orthogonal simple subalgebras, say, $\bar{A}_1, \bar{A}_2, \cdots, \bar{A}_r : \bar{A} = \bar{A}_1 \oplus \bar{A}_2 \oplus \cdots \oplus \bar{A}_r$. Let $1 = \bar{E}_1 + \bar{E}_2 + \cdots + \bar{E}_r$ be the corresponding decomposition of the unit element 1 of $\bar{A}$. Then each $\bar{E}_i$ is in the center $\bar{Z}$ of $\bar{A}$, and indeed $\bar{A}_i = \bar{A}\bar{E}_i (= \bar{E}_i\bar{A} = \bar{E}_i\bar{A}\bar{E}_i)$. If we put $\bar{Z}_i = \bar{Z}\bar{E}_i$ then $\bar{Z}_i$ is the center of $\bar{A}_i$ and so is regarded as a finite separable extension field of $\bar{K} = K/P$. Let Z be the center of A. Then the image of Z by the natural epimorphism $p : A \to \bar{A}$ is equal to $\bar{Z}$ by [1, Proposition 1.4]. Z is a direct summand of the Z–module A by [1, Proposition 1.2], and therefore Z is also a finite K–algebra and $Z \cap \ker p = Z \cap AP = ZP$ [1, Corollary 1.3]. Thus the orthogonal idempotents $\bar{E}_1, \bar{E}_2, \cdots, \bar{E}_r$ in $\bar{Z}$ can be lifted to orthogonal idempotents $E_1, E_2, \cdots, E_r$ in Z with respect to p. Since $E_1 + E_2 + \cdots + E_r$ is an idempotent and $1 \equiv E_1 + E_2 + \cdots + E_r \pmod{AP}$, it follows that actually $1 = E_1 + E_2 + \cdots + E_r$. If we put $A_i = AE_i$ and $Z_i = ZE_i$ for each i, then A_i is a subalgebra of A with the unit element E_i and $A = A_1 \oplus A_2 \oplus \cdots \oplus A_r$ gives an orthogonal direct decomposition of A. It follows therefore that each A_i is a finite separable K–algebra whose image by p is the simple algebra $\bar{A}_i$ and $A_i \cap \ker p = A_iP$, and besides Z_i is the center of A_i

and the image of Z_i by p is equal to the field $\bar{Z}_i$.

Proposition 18. Let K be a Hensel ring and A a finite separable K–algebra which is K–projective. Let B, C be finite K–algebras and $h : B \to C$ an algebra–epimorphism whose kernel $\ker h$ is contained in the (Jacobson) radical of B. Then for any algebra–homomorphism $g : A \to C$ there exists an algebra–homomorphism $f : A \to B$ such that $h \circ f = g$, and such an f is unique up to inner automorphisms induced by those invertible elements of B which are $\equiv 1 \pmod{\ker h}$.

Proof. For convenience, we shall, assuming the existence of f, first prove the up–to–inner–automorphisms–uniqueness of f. To do so, let $f' : A \to B$ be another algebra–homomorphism such that $h \circ f' = g$. For any $a \in A$ and $b \in B$, we define $a \cdot b = f(a)b$ and $b \cdot a = bf'(a)$. Then it is easy to see that with respect to the new multiplication $\cdot$ B becomes a two–sided A–module (over K). Similarly, C can be converted into a two–sided A–module by means of g; namely, for $a \in A$, $c \in C$, we define $a \cdot c = g(a)c$, $c \cdot a = cg(a)$. h becomes then an A–A–epimorphism $: B \to C$, because $h(a \cdot b) = h(f(a)b) = h(f(a))h(b) = g(a)h(b) = a \cdot h(b)$ and $h(b \cdot a) = h(bf'(a)) = h(b)h(f'(a)) = h(b)g(a) = h(b) \cdot a$. Since A is separable, h induces an epimorphism : $B^A \to C^A$. In particular, since 1 is in C^A, there is a $v \in B^A$ such that $h(v) = 1$. That $v \in B^A$ means, however, that $f(a)v = vf'(a)$ for all $a \in A$. On the other hand, since $h(1) = 1$, $h(v) = 1$ means $v \equiv 1$ (mod $\ker h$), which implies, since $\ker h$ is in the radical of B, that v is invertible in B. Thus $v^{-1}f(a)v = f'(a)$ for all $a \in A$, which proves our assertion.

Now, in order to prove the existence part of our proposition, it is sufficient to restrict ourselves to the case where $\ker h$ contains BP, or equivalently, $CP = 0$. To see this, suppose our proposition is proved for this case. Consider the residue class algebra C/CP and the natural epimorphism $q : C \to C/CP$. Then $q \circ h : B \to C/CP$ is an algebra–epimorphism and $\ker q \circ h = \ker h + BP$; therefore $\ker q \circ h$ contains BP and is contained in the radical of B. Thus, for the algebra–homomorphism $q \circ g : A \to C/CP$, there exists an algebra–homomorphism $f_0 : A \to B$ such that $q \circ h \circ f_0 = q \circ g$. Applying the above proved up–to–inner–automorphisms–uniqueness to $q : C \to C/CP$, $q \circ g : A \to C/CP$, $g : A \to C$, $h \circ f_0 : A \to C$ instead of $h : B \to C$, $g : A \to C$, $f : A \to B$, $f' : A \to B$, we can find a $w \in C$ such that $w \equiv 1 \pmod{\ker q}$ and $w^{-1}h(f_0(a))w = g(a)$ for all $a \in A$. We pick $v \in B$ such that $h(v) = w$. Then $v \equiv 1 \pmod{\ker q \circ h}$ and so v is invertible (because $\ker q \circ h$ is in the radical of B). We now define an algebra–homomorphism $f : A \to B$ by $f(a) = v^{-1}f_0(a)v$ $(a \in A)$. Then $h(f(a)) = h(v)^{-1}h(f_0(a))h(v) = w^{-1}h(f_0(a))w = g(a)$ for $a \in A$, i.e., $h \circ f = g$, which proves our assertion.

Thus we may and shall assume that $CP = 0$; C can therefore be regarded as a finite algebra over the field $\bar{K} = K/P$.

(a) We first treat the case where A is central over K. Generally, let M be a

K–module and consider the tensor product $M \otimes A$ (over K). Then $M \otimes A$ can be regarded as a two–sided A–module, or equivalently, a left $A \otimes A^{\circ}$–module in the natural way, and in this case $(M \otimes A)^A$ is nothing but the right annihilator of J in $M \otimes A$ (because J is generated by elements of the form $a \otimes 1 - 1 \otimes a^{\circ}$). But since A is separable J is generated by an idempotent e, and so the annihilator is equal to $(1 - e)(M \otimes A) = M \otimes (1 - e)A$. On the other hand, $(1 - e)A$ is also the right annihilator of $J = (A \otimes A^{\circ})e$ in A, which is nothing but the center K of A. Thus we have $(M \otimes A)^A = M \otimes K = M$.

We now consider three tensor products $A \otimes A^{\circ}$, $B \otimes A^{\circ}$ and $C \otimes A^{\circ}$. Then, by applying the above fact to (central separable) A° instead of A, we have $(A \otimes A^{\circ})^{A^{\circ}} = A$, $(B \otimes A^{\circ})^{A^{\circ}} = B$, and $(C \otimes A^{\circ})^{A^{\circ}} = C$. Since A is projective over the local ring K it is indeed free over K, and since $A \otimes A^{\circ}$ can be identified with the endomorphism ring of the K–module A by [1, Theorem 2.1], it follows that $A \otimes A^{\circ}$ is a matrix ring over K, i.e., there exist matrix units E_{ij} in $A \otimes A^{\circ}$ which form a free basis of $A \otimes A^{\circ}$ over $K : A \otimes A^{\circ} = \Sigma E_{ij}K$. We denote by $\bar{E}_{ij}$ the image of E_{ij} by the algebra–homomorphism $g \otimes 1 : A \otimes A^{\circ} \to C \otimes A^{\circ}$. Then $\bar{E}_{ij}$'s form matrix units in $C \otimes A^{\circ}$. We consider the algebra–epimorphism $h \otimes 1 : B \otimes A^{\circ} \to C \otimes A^{\circ}$. If we observe the fact that A° has a free basis over K and this is also a free basis of $C \otimes A^{\circ}$ over C, it is easily seen that $\ker(h \otimes 1) = (\ker h) \otimes A^{\circ}$. Since $\ker h$ is in the radical of B and $B \otimes A^{\circ}$ is finitely generated as a B–module, $(\ker h) \otimes A^{\circ}$ is small in $B \otimes A^{\circ}$ as a B–module [2, Theorem 1] and so as a one–sided ideal all the more. Thus $\ker(h \otimes 1)$ is in the radical of $B \otimes A^{\circ}$. Therefore, the matrix units $\bar{E}_{ij}$ in $C \otimes A^{\circ}$ can, by [2, Theorem 25], be lifted to matrix units E'_{ij} in $B \otimes A^{\circ}$ with respect to $h \otimes 1$. It is now clear that by associating E_{ij} with E'_{ij} we have an algebra–homomorphism $f' : A \otimes A^{\circ} \to B \otimes A^{\circ}$ which satisfies $(h \otimes 1) \circ f' = g \otimes 1$. If we restrict f' to $A = (A \otimes A^{\circ})^{A^{\circ}}$ then its image must be in $B = (B \otimes A^{\circ})^{A^{\circ}}$ and thus we have an algebra–homomorphism $f : A \to B$. On the other hand, $g : A \to C$ and $h : B \to C$ are clearly the restrictions of $g \otimes 1$ and $h \otimes 1$ to A and B respectively. Thus we have $h \circ f = g$, and f is a desired homomorphism.

(b) We next consider the case where $\bar{A} = A/AP$ is a simple algebra, i.e., AP is a maximal two–sided ideal of A. Since we assume that $CP = 0$, it follows that $g(AP) = g(A)P = 0$ and hence $\ker g = AP$. Thus we can identify $\bar{A}$ with the image $g(A)$ and assume that g induces the natural epimorphism : $A \to \bar{A}$. Let Z be the center of A. Then its image by g is the center $\bar{Z}$ of $\bar{A}$ and the kernel is ZP; thus we may identify Z/ZP with $\bar{Z}$. Now, since $\bar{Z}$ is a finite separable extension field of $\bar{K} = K/P$, $\bar{Z}$ is a simple extension of $\bar{K}$, i.e., $\bar{Z}$ is obtained by adjoining a single element $\bar{a}$ to $\bar{K} : \bar{Z} = \bar{K}[\bar{a}]$. Let $\bar{f}(t)$ be the (monic) minimal polynomial of $\bar{a}$ over $\bar{K}$. Then $\bar{f}(t) = (t - \bar{a})\bar{f}_1(t)$ with a polynomial $\bar{f}_1(t)$ in $\bar{Z}[t]$, which, since $\bar{a}$ is a separable root, is relatively prime to $t - \bar{a}$. Let $f(t)$ be a monic polynomial in $K[t]$ whose image by g is equal to $\bar{f}(t)$. Since A is finitely generated and projective as a K–module and since Z is a direct summand of

A as (Z– whence) K–modules, Z is finitely generated and projective, or equivalently, free over K; and in fact the rank of Z over K is equal to the rank (= degree) n of $\bar{Z}$ over $\bar{K}$. ZP is in the radical of Z, while $Z/ZP = \bar{Z}$ is a field, so that ZP coincides with the radical and is the only maximal ideal of Z. Thus Z is a local ring, and therefore a Hensel ring by [2, Theorem 23]. Hence $f(t)$ can be decomposed in $Z[t]$ as $f(t) = (t - a)f_1(t)$, with $a \in Z$ and $f_1(t) \in Z[t]$ whose images by g are equal to $\bar{a}$ and $\bar{f}_1(t)$ respectively. We then consider the subalgebra $K[a]$ obtained by adjoining a to K. Then $K[a] + ZP = Z$ and so we have $K[a] = Z$. Since a is a root of $f(t)$ whose degree is n (= $\deg \bar{f}(t) = [\bar{Z} : \bar{K}]$) and Z is a free K–module of rank n, it follows that the n elements $1, a, a^2, \cdots, a^{n-1}$ form a free basis of Z, and besides, by associating t with a, $K[t]/(f(t))$ is isomorphic to Z as K–algebras. We next pick $b \in B$ whose image by the epimorphism $h : B \to C$ is $\bar{a}$, and then adjoin b to K to obtain a commutative subalgebra $K[b]$ of B. Since B is a finite K–algebra, b is a root of a monic polynomial in $K[t]$ by [2, Theorem 8] and so $K[b]$ is a finite K–algebra. Furthermore, the image of $K[b]$ by h is clearly $\bar{Z} = \bar{K}[\bar{a}]$, while $K[b] \cap \ker h$ is in the radical of $K[b]$ by [2, Corollary to Theorem 9] (because $\ker h$ is in the radical of B). Thus $K[b]$ is a local ring with the only maximal ideal $K[b] \cap \ker h$ and therefore a Hensel ring by [2, Theorem 23]. Hence $f(t)$ is in $(K[b])[t]$ decomposed as $f(t) = (t - a')f_2(t)$ with $a' \in K[b]$ and $f_2(t) \in (K[b])[t]$ whose images by h are $\bar{a}$ and $\bar{f}_1(t)$ respectively. We adjoin a' to K to obtain a commutative subalgebra $Z' = K[a']$ of B. Since a' is a root of $f(t)$ and $Z = K[a] \cong K[t]/(f(t))$, we can define a K–algebra–epimorphism $f' : Z \to Z'$ by associating a with a'. Then Z' is separable over K because Z, the center of A, is separable over K. We now put $B' = B^{Z'}$, the centralizer of Z' in B. Then B' can be regarded as a Z'–algebra; but B' is indeed a finite Z'–algebra because B' is a direct summand of the K–module B by Proposition 5 and so is finitely generated even as a K–module. Similarly, we put $C' = C^{\bar{Z}}$, the centralizer of $\bar{Z}$ in C. Then C' is a finite $\bar{Z}$–algebra and contains $\bar{A} = g(A)$; thus g can be regarded as a K–algebra–homomorphism : $A \to C'$. Furthermore, it follows from Proposition 6 that the epimorphism $h : B \to C$ induces an epimorphism $B' \to C'$. We now convert B' into a (finite) Z–algebra by means of the above defined epimorphism $f' : Z \to Z'$. Similarly, if we define the epimorphism $g' : Z \to \bar{Z}$ as the restriction of g to Z, C' can be converted into a (finite) Z–algebra by means of g'. It is clear that $g : A \to C'$ becomes then a Z–algebra–homomorphism, while if we observe the fact that $h \circ f' = g'$ it is easily seen that the above epimorphism : $B' \to C'$ induced by h becomes a Z–algebra–epimorphism. Thus we can apply the case (a) above to Z–algebras A, B' and C' to obtain a Z–algebra–homomorphism $f : A \to B'$ such that $h \circ f = g$. If we regard f as a K–algebra–homomorphism $A \to B$, f is the required homomorphism.

(c) We now turn to the general case. Let $A = A_1 \oplus A_2 \oplus \cdots \oplus A_r$ be a direct decomposition of A into mutually orthogonal subalgebras such that A_i/A_iP is simple algebra for each i, as before. Let $1 = E_1 + E_2 + \cdots + E_r$ be the corresponding decomposition of 1. Then each E_i is an idempotent and is the unit element of A_i. We assume, without loss of generality, that $g(A_1) \neq 0, \cdots, g(A_s) \neq 0$ and $g(A_{s+1}) = \cdots = g(A_r) = 0$, where $1 \leqslant s \leqslant r$. Put $\bar{E}_i = g(E_i)$ for each $i = 1, \cdots, s$. Then we have a decomposition of the unit element 1 of C into orthogonal idempotents : $1 = \bar{E}_1 + \cdots + \bar{E}_s$. Further, by restricting g to A_i, we have a (unital) algebra–homomorphism $g_i : A_i \to \bar{E}_i C \bar{E}_i$ for $i = 1, \cdots, s$. Consider now the algebra–epimorphism $h : B \to C$. Since $\ker h$ is contained in the radical of B, the above decomposition can be lifted to a decomposition of the unit element 1 of B into orthogonal idempotents, say, $1 = E'_1 + \cdots E'_s$ with respect to h [2, Theorem 24], where $h(E'_i) = \bar{E}_i$ for each i. It follows then that if we restrict h to the subalgebra $E'_i B E'_i$ we have a (unital) algebra–epimorphism $h_i : E'_i B E'_i \to \bar{E}_i C \bar{E}_i$. Hence, by applying the case (b) proved above, we can find an algebra–homomorphism, $f_i : A_i \to E'_i B E'_i$ such that $h_i \circ f_i = g_i$. We are now able to compose an algebra–homomorphism $f : A \to B$ by defining $f(a) = f_1(aE_1) + \cdots + f_s(aE_s)$ for all $a \in A$; observe that $E'_1, \cdots, E'_s$ are mutually orthogonal. It is then clear that f satisfies $h \circ f = g$, and this proves our proposition completely.

From the above proposition we can now derive the following proposition,which is actually the same as [2, Theorem 33].

Proposition 19. Let B be a finite algebra over a Hensel ring K such that the residue class algebra B/N modulo the radical N of B is separable over K. Then there exists a finite separable K–subalgebra A of B such that $A + N = B$, and such a subalgebra A is unique up to inner automorphisms of B induced by those invertible elements of B which are $\equiv 1 \pmod N$.

Proof. Put $C = B/N$. Then C is semi–simple. Since $BP \subset N$ for the maximal ideal P of K, it follows that $CP = 0$ and thus C can be regarded as a finite separable algebra over the residue class algebra $\bar{K} = K/P$. We want first to show that there exists a finite separable algebra $\tilde{A}$ over K which is projective (or equivalently, free) as a K–module and such that $\tilde{A}/\tilde{A}P \cong C$. But the semi–simple algebra C is a direct sum of mutually orthogonal simple subalgebras each of which is separable over K. Therefore, it suffices to show in the case where C is a simple (and separable) algebra. Let $\bar{Z}$ be the center of C. Then $\bar{Z}$ be a finite separable extension of $\bar{K}$, of degree n, say. Let $\bar{a}$ be an element of $\bar{Z}$ such that $\bar{K}[\bar{a}] = \bar{Z}$. Let $\bar{f}(t)$ be the (monic) minimal polynomial of $\bar{a}$ over $\bar{K}$. Then $\bar{f}(t)$ be an irreducible polynomial in $\bar{K}[t]$ of degree n and $\bar{K}[t]/(\bar{f}(t)) \cong \bar{Z}$. Let $f(t)$ be a monic polynomial in $K[t]$ whose image by the natural epimorphism $K \to \bar{K}$ is equal to $\bar{f}(t)$. Consider then the residue algebra $\tilde{Z} = K[t]/(f(t))$. If we put by a the residue class of t $(\mathrm{mod}(f(t)))$ then clearly $K[a] = \tilde{Z}$ and, since $f(t)$ is a monic polynomial of degree n, the n elements $1, a, a^2, \cdots, a^{n-1}$ form a free basis of the K–module $\tilde{Z}$. Furthermore, the natural epimorphism $K[t]/(f(t)) \to \bar{K}[t]/(\bar{f}(t))$ yields clearly a K–algebra–epimorphism $\tilde{Z} \to \bar{Z}$ which maps a onto $\bar{a}$ and whose kernel is $\tilde{Z}P$. Thus $\tilde{Z}/\tilde{Z}P \cong \bar{Z}$ and therefore $\tilde{Z}$ is separable over K by Proposition 17. Also, it follows that $\tilde{Z}P$ is the only maximal ideal of $\tilde{Z}$ and hence $\tilde{Z}$ is a Hensel ring by [2, Theorem 23]. Now C is a central simple algebra over $\bar{Z}$. Then by [2, Theorem 32] there exists a central separable (= proper maximally central, since $\tilde{Z}$ is a local ring) algebra $\tilde{A}$ over $\tilde{Z}$ such that $\tilde{A}/\tilde{A}P \cong C$. (The existence of such $\tilde{A}$ is an immediate consequence of [2, Theorem 31] to the effect that the Brauer group of $\tilde{Z}$ is canonically isomorphic to the Brauer group of $\bar{Z}$; cf. also[1, Theorem 6.5].) We now consider $\tilde{A}$ as an algebra over K. Then $\tilde{A}$ is separable over K by [1, Theorem 2.3]. Since $\tilde{Z}$ is a finitely generated projective K–module and since $\tilde{A}$ is a finitely generated projective $\tilde{Z}$–module by [1, Theorem 2.1], it follows that $\tilde{A}$ is a finitely generated projective K–module too. Thus $\tilde{A}$ is a desired separable algebra. Now the isomorphism $\tilde{A}/\tilde{A}P \cong C$ yields naturally an algebra–epimorphism $g : \tilde{A} \to C$, while if we denote by h the natural epimorphism $B \to C$, then $\ker h = N$. Therefore, by applying Proposition 18 to $\tilde{A}$ instead of A, we can find an algebra–homomorphism $f : \tilde{A} \to B$ such that $h \circ f = g$. We put $A = f(\tilde{A})$. Then A is a finite separable subalgebra of B by [1, Proposition 1.4]. Furthermore, $h(A) = h(f(\tilde{A})) = g(\tilde{A}) = C$ and this shows that $A + N = B$.

Let A' be another finite separable K–subalgebra of B such that $A' + N = B$, i.e., $h(A') = C$. Let h' be the restriction of h to A'. Then we have $\ker h' = A' \cap N$, which is however contained in the radical of A' by [2, Corollary to Theorem 9]. On the other hand, the radical of the finite separable algebra A' is $A'P$, and besides $A'P \subset BP \subset N$ whence $A'P \subset A' \cap N$. Thus it follows $A'P = A' \cap N = \ker h'$. Apply again Proposition 18 to $\tilde{A}$ and $h' : A' \to C$ in place of A and $h : B \to C$ respectively. Then we know that there exists an algebra–homomorphism $f' : \tilde{A} \to A'$ such that $h' \circ f' = g'$, or what is the same, $h \circ f' = g$. It follows then $h'(f'(\tilde{A})) = g(\tilde{A}) = C$, so that $A' = f'(\tilde{A}) + \ker h' = f'(\tilde{A}) + A'P$. But since $A'P$ is a small submodule of the K–module A', it follows $A' = f'(\tilde{A})$, i.e., $f' : \tilde{A} \to A'$ is an epimorphism. Now f and f' satisfy $h \circ f = g$ and $h \circ f' = g$, and so there exists by Proposition 18 an invertible element v of B such that $v \equiv 1 \pmod{N}$ and $v^{-1}f(a)v = f'(a)$ for all $a \in \tilde{A}$; but the last equality implies, since $A = f(\tilde{A})$ and $A' = f'(\tilde{A})$, that $v^{-1}Av = A'$. Thus our proposition is proved completely.

Remark. Let A be a finite algebra over a commutative ring K . Let A satisfy the following condition, as in the case where K is Henselean and A is separable and projective over K (Proposition 18): For any finite K–algebras B, C and any algebra–epimorphism $h : B \to C$ such that $\ker h$ is in the radical of B, every algebra–homomorphism $g : A \to C$ can be lifted to an algebra–homomorphism $f : A \to B$ so that $h \circ f = g$. Then, as a particular case, A satisfies the following condition: every algebra–epimorphism onto A from a finite K–algebra splits whenever its kernel is in the radical of the algebra. However, unlike Proposition 10, it does not seem that this condition implies conversely the first condition, because the proof of the "if" part of Proposition 10 is not applicable to our case any more; indeed, with the same notation in the proof, we cannot guarantee that the subalgebra S of $A \oplus B$ is a finitely generated K–module even if both A, B are finitely generated K–modules.

References

1. M. Auslander and O. Goldman, The Brauer group of a commutative ring, Trans. Amer. Math. Soc. **97** (1960), 367-409.
2. G. Azumaya, On maximally central algebras, Nagoya Math. J. **2** (1951), 119-150.
3. H. Cartan and S. Eilenberg, "Homological algebra," Princeton, 1956.
4. S. Eilenberg, A. Rosenberg and D. Zelinsky, On the dimension of modules and algebras VIII, Nagoya Math. J. **12** (1957), 71-93.
5. G. Hochschild, On the cohomology groups of an associative algebra, Ann. of Math. **46** (1945), 58-67.

HEREDITARILY AND COHEREDITARILY PROJECTIVE MODULES

George M. Bergman*

Bedford College, Harvard University, and University of California

There is a similarity between the study of n–firs (definition recalled in §3 below) and that of right semihereditary rings. In each case one works with projective right modules P over the given ring R, with the property that the image of any homomorphism of P into a free module of finite rank is again projective. For R semihereditary, all finitely generated projective right R–modules have this property; R is an n–fir if and only if the free modules of ranks $1, \cdots, n$ satisfy this condition and certain others. We shall call a projective (right or left) module P with the above property *hereditarily* projective.

This concept leads naturally to three others, which we shall also define and study: the dual one of a cohereditarily projective module, and the two self–dual concepts of weakly and strongly hereditarily projective modules. Inversely, these last two module–theoretic concepts lead us to definitions of weakly and strongly (semi) hereditary rings, and weakly and strongly α–hereditary rings for any cardinal α.

The first half of this paper, §§1–3, is devoted to establishing the basic properties of our four classes of projective modules, with an emphasis on the finitely generated case where we have duality by $\mathrm{Hom}(\cdot, R)$. Our goal is to set up a wider context in which one can apply the ideas used in the study of n–firs [cf. 11, 18, 20, 21] and of semi–hereditary rings, and which, inversely, can give us a better understanding of results concerning such rings.

The second half is mainly concerned with the non–finitely generated cases of the above concepts. Section 4 studies infinitely generated projective modules, but raises more questions than it answers. In sections 5 and 6 we study some of the classes of *rings* referred to above. Section 5 contains a menagerie of unusual ascending and descending chain arguments! In section 6, following [7], we use the natural representation of a commutative ring C by a sheaf of rings on a Boolean space X_C (constructed from the idempotent elements of C) to characterize properties of C.

* Part of this work comes from the author's doctoral thesis, written while the author held an NSF graduate fellowship, part was done under a research grant from England's Science Research Council, and part while the author was partly supported by NSF contract GP 9152.

1. Conventions and background; duality of modules

All rings will be associative with unit and all modules unital. When a definition or result is stated for right modules, the corresponding statement for left modules will be taken for granted. When there is no chance of confusion, we shall often speak of "modules", meaning modules of either sort.

A submodule N of a module M over a ring R will be said to be a direct summand in M if there exists $N' \subseteq M$ such that $M = N \oplus N'$.

All maps will be written on the left. A module homomorphism $f : N \to M$ will thus be left invertible if and only if it is (up to isomorphism) the inclusion into M of a direct summand N, and right invertible if and only if it is the projection of N onto a direct summand M of itself.

If M is a right (left) module over a ring R, the linear functionals on M will form a left (right) R–module $\mathrm{Hom}(M, R)$, which we shall denote M^*. "$*$" gives contravariant functors from right to left R–modules, and vice versa.

If P is a finitely generated projective module over a ring R, then P^* will again be projective and finitely generated, and P^{**} is naturally isomorphic to P; in fact, it is well known that * gives an antiisomorphism between the categories of finitely generated projective right and left R–modules. (These facts can be shown by noting that they are true for finitely generated free modules, and that a projective module is one which can be written as the range of a right–invertible map from a free module, or the domain of a left–invertible map into a free module.) For P a finitely generated projective R–module, we shall identify P and P^{**}, and call P and P^* dual projective modules. Likewise, corresponding maps $f \in \mathrm{Hom}(P, Q)$ and $f^* \in \mathrm{Hom}(Q^*, P^*)$ between such modules will be called dual.

Returning to arbitrary modules, we define the *closure* of a submodule N in a module M as the intersection of the kernels of all linear functionals on M annihilating N, and call N *closed* in M if it is its own closure; that is, if it is the intersection of the kernels of some family of linear functionals on M. A submodule $N \subseteq M$ will be called *dense* if its closure is M, that is, if no nonzero linear functional on M annihilates N.

Given a map $f : N \to M$, $f(N)$ will be dense in M if and only if the map $f^* : M^* \to N^*$ is 1–1. Hence given two dual maps of finitely generated projectives, one of them will be 1–1 if and only if the other has dense image. It is also easy to see that a map of finitely generated projectives is *onto* if and only if it is right invertible, which is equivalent to left invertibility of the dual map.

Note that a direct summand of a projective module is always closed!

(Let us note one disconcerting property of our concept of closure. Though it is a closure operator in the sense of [19, Ch. 2 §1], it differs in its dependence on M from most familiar closure operators: Given $N \subseteq M' \subseteq M$, the closure of N in M' may be smaller than the intersection of M' with the closure of N in M. For instance, for the $\mathbf{Z}$-modules $\{0\} \subseteq \mathbf{Z} \subseteq \mathbf{Q}$, the closure of $\{0\}$ in $\mathbf{Z}$ is $\{0\}$, but in $\mathbf{Q}$ it is all of $\mathbf{Q}$. A submodule N of a module M may not even be dense in its closure; the reader should examine the case $R = k[x, y]/(x^2, xy)$, $M = R/(x, y^2)$, $N = \{0\}$. This is because there can be linear functionals on the closure of N that are not induced from M. However note that this cannot happen when the closure of N is a direct summand in M, which will be an important case in what follows.)

Let us prove again a very handy lemma that we have used elsewhere:

Lemma 1.1. (= [8, Lemma 1.1] $\supset$ [10, Lemma 1.1]) A projective R–module P having a finitely generated dense submodule N is finitely generated. More generally, if a projective module P has a dense submodule N generated by $< \alpha$ elements, for α some infinite cardinal, then P is generated by $< \alpha$ elements.

Proof. Write P as a direct summand of a free module F, with basis B. All members of N can be written using $< \alpha$ elements of B, i.e., N will lie in a direct summand $F_0 \subseteq F$ free of rank $< \alpha$. Since F_0 is closed in F and N is dense in P, P will also lie in F_0, hence be a direct summand in F_0, hence P can be generated by $< \alpha$ elements. ∎

2. Lemmas and definitions

Lemma 2.1. Let P be a projective module over a ring R. Then the following conditions are equivalent:

(a) For every homomorphism f of P into a finitely generated projective R–module Q, the image of f is projective; equivalently, the kernel of f is a direct summand in P.

(b) Every homomorphism f of P into a finitely generated projective R–module Q can be factored $P \overset{v}{\to} S \overset{u}{\to} Q$, where v is the projection of P onto a direct summand S, and u is an injection.

(c) If A is an arbitrary set of elements of P, and B a finite set of linear functionals on P, annihilating A, then there exists a direct sum decomposition $P = P_1 \oplus P_2$ such that all elements of A lie in P_1, and all elements of B annihilate P_1 (that is, are induced by linear functionals on P_2).

$(a^1), (b^1), (c^1)$: the same conditions with "finitely generated projective module" or "finite set of linear functionals" replaced by "free module of rank 1" and "one linear functional".

Proof. The equivalence of (a) and (b) is clear. To get (c) from (a), regard the finite family B of linear functionals as defining a map f of P into a free module of finite rank. To get (a) from (c), reduce to the case where Q is *free* of finite rank (for any finitely generated projective module is a direct summand in such a free module), regard the map f as a finite family B of linear functionals on P, and take for A the whole kernel of f.

By exactly the same arguments, conditions $(a^1), (b^1), (c^1)$ are equivalent to one another. Clearly (a)-(c) imply (a^1)-(c^1); to get the converse, suppose that condition (a) holds for all maps of our given module P into two particular modules, Q and Q'. Let f be a map of P into $Q \oplus Q'$. Composing with the projection of $Q \oplus Q'$ onto Q', we get a map of P into Q', hence we can write $P = P_1 \oplus P_2$, where P_1 is the kernel of this map. Under f, P_1 goes into Q, and its image there can be regarded as a homomorphic image of P (since P_1 is the image of a projection map on P), hence this image is projective, hence $\operatorname{Ker} f = \operatorname{Ker} f \mid P_1$ is a direct summand in P_1 and so in P.

By this argument, we can get from the case of Q free of rank 1 to any free Q of finite rank, and hence to any finitely generated projective Q. ∎

Definition 2.2. A projective module P satisfying the above equivalent conditions will be called *hereditarily* projective.

Note that a ring R will be right *(semi)hereditary* if and only if all (finitely generated) projective right R–modules are hereditarily projective.

Dually, we have:

Lemma 2.3. Let P be a projective module over a ring R. Then the following conditions are equivalent:

(a) The closure of any finitely generated submodule of P is a direct summand in P.

(b) Every homomorphism f from a finitely generated projective R–module Q into P can be factored $Q \overset{v}{\to} S \overset{u}{\to} P$, where $v(Q)$ is dense in S, and u is the inclusion of a direct summand S in P.

(c) If A is a finite family of elements of P, and B an arbitrary family of linear functionals on P annihilating A, then there exists a direct sum decomposition $P = P_1 \oplus P_2$ such that all elements of A lie in P_1, and all elements of B annihilate P_1.

$(a_1), (b_1), (c_1)$: the same conditions with "finitely many" etc. replaced by "one" etc.

Proof. Analogous to the preceding. In proving (a_1) implies (a), we suppose that the result holds for the closure of any homomorphic image in P of either of two given finitely generated modules Q and Q′. Now suppose we have an image $f(Q \oplus Q')$ in P. We write $P = P_1 \oplus P_2$ where P_2 is the closure of the image of Q′. Let s be a projection of P onto P_1. Then the closure of $sf(Q)$ will be a direct summand P_3 in P, and will lie in P_1, hence we get a decomposition $P_1 = P_4 \oplus P_3$. It is not hard to verify that $P_2 \oplus P_3$ is the closure of $f(Q \oplus Q')$ in P. ∎

Definition 2.4. A projective module P satisfying the above equivalent conditions will be called *cohereditarily* projective.

(Note: Mohan S. Shrikhande defines in [30] concepts of *hereditary* and *co–hereditary* modules, but these appear to be essentially opposite to our pair of concepts. E.g., his hereditary modules are modules all submodules of which are projective; cf. §7 below.)

Conditions (c) of the two preceding lemmas were mirror images of one another. They suggest two "symmetrized" conditions, one strong and one weak. To formulate the strong one, let us define a free power module as a module of the form R^S, S a set; i.e., a direct product of copies of the right or left module R. A module M is embeddable in a free power module if and only if elements of M can be "separated" by linear functionals; equivalently, if and only if the submodule $\{0\}$ is closed in M. Chase [14] calls such an M "torsionless".

Lemma 2.5. Let P be a projective module over a ring R. Then the following conditions are equivalent:

(a) Any homomorphic image of P in a free power module – i.e., any quotient of P torsionless in the sense of Chase – is projective.

(b) Every closed submodule of P is a direct summand in P.

(c) If A is an arbitrary family of elements of P, and B an arbitrary family of linear functionals on P annihilating A, then there exists a direct sum decomposition $P = P_1 \oplus P_2$ such that all elements of A lie in P_1, and all elements of B annihilate P_1.

Proof. Immediate. ∎

Definition 2.6. A projective module P satisfying the above equivalent conditions will be called *strongly hereditarily* projective.

In establishing our earlier Lemmas 2.1 and 2.3, to get condition (a) from (a^1), respectively (a_1), we used an argument that proved on the way that a direct summand in a module satisfying the condition in question again satisfies it. Perhaps the essence of that argument can be seen in terms of conditions (c), (c_1) and (c^1). Let P satisfy one of these conditions from one of those Lemmas, let $P = P' \oplus P''$, and say we wish to prove P' satisfies the same condition. Given sets A and B of elements and linear functionals on P', as in the condition in question, we consider them as elements and functionals of P. In each of the conditions mentioned, one of the sets A, B, has no restriction as to size. Hence we may either throw into A *all* elements of P'', or throw into B all linear functionals on P''. Then, applying to P the condition assumed, we get a direct sum decomposition, in which one of the factors will contain P'' or be contained in P'. This induces a decomposition of P', which will have the required properties with respect to our original A and B, proving that P' does satisfy our condition.

The same argument works, clearly, for condition (c) of Lemma 2.5 (though we did not need to use it there). On the other hand, in the weak property which we shall define next, both A and B will be required to be finite, and the above argument will not work unless P is finitely generated. Rather, we shall want to impose "carrying over to direct summands" as an additional condition.

Precisely, if (m) is a module–theoretic condition, let us say P satisfies (m+) if P and every direct summand in P satisfies condition (m). We can now state:

Lemma 2.7. Let P be a projective module over a ring R, and consider the obviously equivalent conditions:

(a) Given finitely generated projective R–modules Q, Q', and maps $Q \xrightarrow{f} P \xrightarrow{f'} Q'$ which compose to zero, there will exist a direct summand P_1 of P containing f(Q) and contained in Ker f'.

(c) Given a finite set A of elements of P, and a finite set B of linear functionals on P, there exists a direct sum decomposition $P = P_1 \oplus P_2$ such that $A \subseteq P_1$, and B annihilates P_1,

and likewise the pairs of conditions $(a_1) \Leftrightarrow (c_1)$, $(a^1) \Leftrightarrow (c^1)$, $(a_1^1) \Leftrightarrow (c_1^1)$ obtained by replacing "finitely many" by "one" in the conditions on Q and A, resp. Q' and B, resp. both of these.

Then the strengthened conditions (a+), (c+), (a_1+), (c_1+), (a^1+), (c^1+), (a_1^1+), (c_1^1+) are all equivalent. If P is finitely generated, these are also equivalent to the original conditions (a), ⋯, (c_1^1).

Proof. Straightforward, using the preceding observations. ∎

Definition 2.8. A projective module P over a ring R will be called *weakly hereditarily* projective if it satisfies the equivalent conditions (a+), (c+), ⋯, (c_1^1+) of the preceding Lemma.

(We do not know any examples of modules satisfying (c_1^1) but not (c), or (c) but not (c+). Proposition 4.4, especially the implications (i) ⇒ (ii), (iii) ⇒ (ii), and part (b), will indicate why it should be difficult to find such examples.)

Some of the conditions we have defined can be given characterizations of somewhat different sorts. Let us call a partially ordered set *directed,* or *upward directed* if every pair of elements has an upper bound, and *downward directed* if every pair of elements has a lower bound. Families of submodules of a module will be given the partial ordering by inclusion if the contrary is not stated.

Lemma 2.9. A finitely generated projective module P is weakly hereditarily projective if and only if the kernel of any homomorphism of P into a finitely generated projective module Q is the union of a directed family of direct summands.

An arbitrary projective module P is cohereditarily projective if and only if every closed submodule of P is the union of a directed family of direct summands.

Proof. The first claim is straightforward, as is the implication "⇒" in the second. To get the reverse implication, let Q be a finitely generated submodule of P. If its closure is the union of a directed family of direct summands, one of these summands will contain Q, and, since it is closed in P, must equal the closure of Q, hence the closure of Q is indeed a summand. ∎

Let us call a partially ordered set S semicomplete if given any upward directed family $U \subseteq S$, and downward directed family V, with all elements of U less than all elements of V, there exists $s \in S$, which is an upper bound for U and a lower bound for V. (This is in general weaker than either every ascending chain having an l.u.b., *or* every descending chain having a g.l.b.; but if S is a lattice, all these conditions are equivalent to completeness.)

Lemma 2.10. Let P be a projective module over a ring R. Then the following conditions are equivalent:

(i) P is strongly hereditarily projective.

(ii) P is hereditarily and cohereditarily projective, and the partially ordered set of direct summands in P is semicomplete.

(iii) P is hereditarily projective, and the class of direct summands in P is closed under taking intersections of descending chains.

(iv) P is cohereditarily projective, and the class of direct summands in P is closed under taking closures of unions of ascending chains.

(v) P is weakly hereditarily projective, and the class of direct summands in P is closed under both of the above operations.

Proof. Clearly (i) implies all the other conditions. To get the reverse implication in case (ii), let K be a closed submodule of P, and compare the upward directed system of closures of finitely generated submodules of K with the downward directed system of kernels of finite families of linear functionals annihilating K The arguments for (iii) and (iv) are straightforward, using Lemma 2.9 in the latter case. We reduce (v) to (iv): suppose Q is a finitely generated submodule of P. Then using the first of our closure conditions, we can find a minimal direct summand $P' \subseteq P$ containing Q. P' is again weakly hereditarily projective, and no proper direct summand contains Q, hence Q is dense in P', hence P' is the closure of Q, so P is cohereditarily projective, and thus satisfies (iv). ∎

With respect to case (ii), note that in a *finitely generated* hereditarily and cohereditarily projective module, the partially ordered set of direct summands will be a lattice, so semicompleteness = lattice completeness.

There are various "naturally occurring" conditions which imply the chain conditions on direct summands occurring in the above lemma. The simplest of these is clearly the one described by the lemma to follow.

While a partial ordering on a set S is usually described as a *reflexive transitive antisymmetric* relation "$\geqslant$", it can equally well be described by giving the *transitive antireflexive* relation "$>$". Let us call such a relation a "$>$-type" partial ordering on S.

Lemma 2.11. Let M be an element of an abelian category. Then the following conditions are equivalent:

(i) The class of direct summands of M has ascending chain condition.

(i′) The class of direct summands of M has descending chain condition.

(ii) For any chain of decompositions:

$$M = N_1 \oplus M_1 = N_1 \oplus N_2 \oplus M_2 = \cdots \qquad (M_i = N_{i+1} \oplus M_{i+1})$$

only finitely many of the N_i are nonzero.

(iii) The endomorphism ring of M does not contain an infinite family of mutally orthogonal idempotents.

(iii′) The endomorphism ring of M does not contain an infinite family of distinct commuting idempotents.

(iv) If we define on the set of isomorphism classes of direct summands of M a relation $A > B$ meaning that (a representative of) A has (a representative of) B as a proper direct summand, then this is a $>$-type partial ordering, and has descending chain condition.

Proof. The equivalence of (i)-(iii′) is straightforward. To show (ii) equivalent to (iv), note that for $>$ not to be a partial ordering would here mean $A > A$ for some A, and this is so if and only if we can find a chain contradicting (ii) with infinitely many M_i isomorphic to each other; while $>$ will satisfy descending chain condition if and only if we cannot find such a chain with infinitely many *non*isomorphic terms. ∎

Definition 2.12. An object of an abelian category, satisfying the above equivalent conditions, will be said to have *chain condition on direct summands,* abbreviated $ACC_{\oplus}$.

Corollary 2.13. For a projective module P over a ring R:

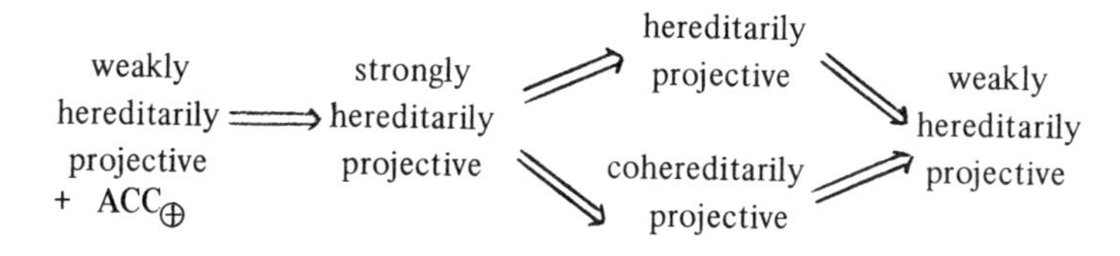

Proof. First step by Lemma 2.10 (v) ⇒ (i); others obvious. ∎

Let us note down two results inherent in what has gone above:

Lemma 2.14. Let P be a projective module over a ring R. If P is strongly hereditarily projective, resp. hereditarily projective, cohereditarily projective, or weakly hereditarily projective, or has $ACC_{\oplus}$, then the same holds for any direct summand of P. ∎

Lemma 2.15. Let P be a finitely generated projective module over a ring R. Then P is hereditarily projective if and only if the dual module P^* is cohereditarily projective, while P is strongly or weakly hereditarily projective, or has $ACC_{\oplus}$, if and only if P^* has the *same* property. ∎

One can also prove without difficulty:

Lemma 2.16. Let $(R_i)_{i \in I}$ be a family of rings, and $R = \Pi R_i$ their direct product. Then an R–module M will be finitely generated, say by n elements, if and only if it can be written ΠM_i where each M_i is an R_i–module generated by n elements; and such a finitely generated M will be (strongly, weakly, cohereditarily) projective if and only if every M_i is so. ∎

(The corresponding statements for infinitely generated modules are false in general, but true if I is finite.)

3. Applications and examples

Recall that a ring R is called right (semi) hereditary if every (finitely generated) right ideal is projective as a module. As we previously observed, R will be right (semi)-hereditary if and only if every all free right modules (of finite rank) are hereditarily projective; equivalently, if and only if all (finitely generated) projective right R–modules are hereditarily projective. Generally, we define R to be right α–hereditary (α any cardinal) if and only if every right ideal generated by $\leqslant \alpha$ elements is projective; this is then equivalent to the condition that every projective right R–module generated by $\leqslant \alpha$ elements be hereditarily projective; equivalently, that the free right R–module of rank α be hereditarily projective.

Using Lemma 2.15 (duality) we see that R will be *left* semihereditary if and only if every finitely generated projective right R–module is *co*hereditarily projective. (Because duality by * applies only to finitely generated projective modules, we cannot similarly characterize left hereditary rings in terms of right modules. More on this in Proposition 4.5 below.)

Now Corollary 2.13 tells us that for projective modules with $ACC_{\oplus}$ the conditions of being hereditarily projective and cohereditarily projective are equivalent. Using the characterization of $ACC_{\oplus}$ in terms of idempotents in endomorphism rings, and recalling that the endomorphism ring of the free right R–module of rank n is the $n \times n$ matrix ring over R, we recover a result of Lance Small's:

Corollary 3.1. (Small [32, Thm. 3]). A ring R such that for all positive integers n, the $n \times n$ matrix ring over R has no infinite set of mutually orthogonal idempotents is right semihereditary if and only if it is left semihereditary. ∎

Recall [18] that a right *fir* (free ideal ring) is defined as a ring in which all right ideals are free as right modules, and all free modules have unique rank (= cardinality of a basis. This uniqueness is automatic except for the finite ranks.) More generally, for any cardinal α, a right α–fir is defined as a ring in which right ideals generated by $\leqslant \alpha$ elements are free, and free modules of ranks $\leqslant \alpha$ have unique rank.

In particular, for n a positive integer, we see that a right n–fir can be characterized as a ring R such that (a) the right R–module R^n is hereditarily projective, and (b) the map from the set $\{0, \cdots, n\}$ with a partially defined operation +, into the set of isomorphism classes of right R–modules embeddable as direct summands in R^n, with partial operation $\oplus$, given by $m \mapsto R^m$, is an isomorphism.

Applying * we see that condition (b) is equivalent to its right–left dual. Further, this condition implies that R^n satisfies $ACC_{\oplus}$ (see Lemma 2.11(iv)!). Therefore by Corollary 2.13 (a) is equivalent in the presence of (b) to the condition that R^n be *strongly* (or *weakly*) hereditarily projective, which is also *–invariant. Hence:

Corollary 3.2. (Cohn [11]). The condition of being a right n–fir and of being a left n–fir are equivalent. ∎

(Hence one simply says n–fir. A ring which is an n–fir for all n is called a semifir.)

Let us now give some examples that distinguish among the five properties considered in Corollary 2.13. All modules used will be finitely generated, and where possible, the base ring will be commutative.

Example 3.3. *A finitely generated strongly hereditarily projective module not having* $ACC_{\oplus}$: A free module of rank 1 over any infinite direct product of fields (by Lemma 2.16).

Example 3.4. *A hereditarily and cohereditarily projective module that is not strongly hereditarily projective:* Let R be the ring of all functions from the 1–point compactification of the positive integers into a field k, that are continuous in the discrete topology on k. R is easily shown to be hereditary, hence the free R–module of rank 1 is hereditarily and cohereditarily projective. But the submodule thereof consisting of functions zero at all even integers is easily seen to be closed, but not a direct summand.

Example 3.5. *A hereditarily but not cohereditarily projective module:* Chase [15] has constructed an example of a right– but not left–semihereditary ring R. (Small [31] shows that R can even be made right hereditary.) A free right module of sufficiently large rank (in the example in question, rank 1) over such a ring will be hereditarily, but not cohereditarily projective. We do not know whether there is an example over a commutative ring; we suspect not. (Cf. end of §6).

Example 3.6. *A weakly hereditarily projective module that is neither hereditarily nor cohereditarily projective:* We could be lazy, and take for R the direct product of a right–but–not–left semihereditary ring and a left–but–not–right semihereditary ring, and let P be a free module of appropriate rank over R. But we can do better, and get a commutative example. Let $f: A \to B$ be a non–1–1 homomorphism of commutative integral domains. Let R be the ring of all sequences $(a_0, a_1, \cdots, a_\infty)$ with $a_i \in B$ $(i < \infty)$ and $a_\infty \in A$, such that for all sufficient large i, $a_i = f(a_\infty)$. We leave it to the reader to verify that the free module of rank 1 over R is weakly hereditarily projective, but that if $a \in A$ is a nonzero member of the kernel of f, then the linear functional on that module given by multiplication by $(0, 0, \cdots, a)$ has kernel which is not finitely generated, hence not a direct summand. Hence this module is not hereditarily projective, and by duality, not cohereditarily projective either.

The ring–theoretic concepts of hereditary, α–hereditary and semihereditary rings suggested the definitions of §2. Let us now return the favor.

Definition 3.7. A ring R will be called *strongly right hereditary* if all free right R–modules are strongly hereditarily projective. For any cardinal α, R will be called strongly right α–hereditary if the free right R–module of rank α is strongly hereditarily projective. For α finite, the qualifier "right" (or "left") can be dropped (by Lemma 2.15). R will be called *strongly semihereditary* if it is strongly n–hereditary for all integers n.

Definition 3.8. For any positive integer n, a ring R will be called *weakly n–hereditary* if the free right R–module of rank n (equivalently, if the free left R–module of rank n) is weakly hereditarily porjective. R will be called *weakly semihereditary* if it is weakly n–hereditary for all n.

It is clear from Lemma 2.15 why we did not have to define right "co–n–hereditary" (and "co–semi–hereditary") rings. The reason we have not defined right co(–α–)hereditary, or right weakly (α–)hereditary rings (for infinite cardinals α), will be seen in the next section.

From Corollary 2.13, we get obvious implications among the above conditions, and the standard ones (right and left semihereditary, etc.). From Lemma 2.16 we see:

Corollary 3.9. A direct product $R = \Pi_{i \in I} R_i$ of rings is *weakly*, resp. *strongly, right*, or *left* n–hereditary ($n < \infty$) or semihereditary, if and only if each R_i has the same property. ∎

4. Some problems and some results on infinitely generated modules

It is known that if R is a right hereditary ring, any submodule of any free right R–module is projective. This suggests the generalization that if P is a hereditarily projective module over a ring R, then a homomorphic image of P in any free right R–module is projective, i.e., that conditions (a) and (b) of Lemma 2.1 are equivalent to the same conditions with the qualifier "finitely generated" deleted from before "projective R–module Q". We do not know whether this is true; let us call a projective module P satisfying this strengthened property *absolutely hereditarily projective.*

It is immediate that all finitely generated hereditarily projective modules, all strongly hereditarily projective modules, and any direct summand in an absolutely hereditarily projective module will be absolutely hereditarily projective. Aside from this, the best results we have been able to get are the following Lemma and Corollaries:

Lemma 4.1. Let P be a projective module over a ring R, which is the union of a family of submodules $(P_i)_{i\in I}$, each of which is absolutely hereditarily projective, and whose direct sum $\oplus P_i$ is hereditarily projective. Then P is absolutely hereditarily projective.

Proof. Let $f: P \to F$ be a homomorphism of P into a free module, let S be a well-ordered basis for F, and let $p_s : F \to R$ $(s \in S)$ be the coordinate functions. For every $s \in S$, let $F_{\leqslant s}$ and $F_{>s}$ denote the free submodules of F spanned by the basis elements $\leqslant s$ and $>s$ respectively, and let $p_{\leqslant s}$ and $p_{>s}$ be the projections of F onto these summands.

Now fix $s \in S$. For all $i \in I$, P_i is absolutely hereditarily projective, hence $\mathrm{Ker}(p_{>s}\, f) \mid P_i$ is a direct summand in P_i, hence its image under f, $(F_{\leqslant s}) \cap f(P_i)$, is a homomorphic image of P_i, hence so in turn is $p_s((F_{\leqslant s}) \cap f(P_i)) \subseteq R$. Therefore $p_s((F_{\leqslant s}) \cap f(P)) = \cup_i\, p_s((F_{\leqslant s}) \cap f(P_i))$ will be a homomorphic image of $\oplus_I P_i$. (If in fact we could say at this point that it was a homomorphic image of P, we could strengthen our result!) Since $\oplus_I P_i$ is hereditarily projective, $p_s((F_{\leqslant s}) \cap f(P))$ is projective. Hence by splitting an exact sequence, we get a submodule $M_s \subseteq (F_{\leqslant s}) \cap f(P)$ mapped isomorphically onto this image by p_s. It is now easy to show that $f(P) = \oplus M_s$, so $f(P)$ is projective. ∎

Corollary 4.2. Let P_0 be a projective module over a ring R, such that for every positive integer n, P_0^n is absolutely hereditarily projective, and let P be a direct sum of copies of P_0. Then if P is hereditarily projective, it is absolutely hereditarily projective.

Proof. Let P be the direct sum of a family of α copies of P_0, where we can assume that α is an infinite cardinal; and let (P_i) be the set of all the submodules of P generated by finite subsets of this family. Clearly, $\oplus P_i$ will again be the direct sum of precisely α copies of P_0, hence isomorphic to P, hence hereditarily projective, so we can apply the preceding lemma to get our conclusion. ∎

In the above corollary, if P_0 is finitely generated, the hypothesis on P_0^n can be deleted, by our earlier observations. Taking for P_0 the free right module of rank 1, we get:

Corollary 4.3. If a free module over a ring R is hereditarily projective, it is absolutely hereditarily projective. I.e., if R is right α–hereditary, for some cardinal α, then any submodule generated by $\leqslant\alpha$ elements of any free right R–module is projective. ∎

We remark that if the conditions of hereditary and absolutely hereditary projectivity turn out to be distinct, yet both of interest, then a still weaker condition might also be of interest; we could call a projective module P *loosely* hereditarily (or "semihereditarily"?) projective if any *finitely generated* homomorphic image of P in a free module is projective. This is equivalent to the other two conditions for finitely generated P, but *not* so for arbitrary P: clearly any projective module over a right semihereditarily ring will be loosely hereditarily projective, but not necessarily hereditarily projective!

Since the definition of weakly hereditarily projective modules also involves a map into a free module of finite rank, one can similarly replace this by an arbitrary free module, and get a concept of "weakly absolutely hereditarily projective" module. This is, in fact, equivalent to a number of other interesting conditions:

Proposition 4.4. (a) Let P be a projective right module over a ring R. Then the following conditions are equivalent:

(i) If a finitely generated submodule $A \subseteq P$ is annihilated by a homomorphism f of P into a projective module Q, then A is contained in a direct summand of P annihilated by f.

(ii) P is weakly hereditarily projective, and every finitely generated submodule of P is contained in a finitely generated direct summand.

(iii) Every finitely generated submodule of P is contained in a finitely generated weakly hereditarily projective direct summand.

(iv) P can be written as a direct sum $\oplus_I P_i$ of finitely generated projective modules, such that every finite subsum of this sum is weakly hereditarily projective.

(b) Any direct summand in a projective module P with the above properties again has them, and any finitely generated weakly hereditarily projective module has them.

(c) Any projective right module over a weakly semihereditary ring has these properties.

Proof. Let us first observe that any direct summand in a projective module P satisfying condition (i) will again satisfy it, by the same type of argument used in §2: if $P = P' \oplus P''$, and $A \subseteq P'$, $f : P' \to Q$ annihilating A, then we extend f to a map of $P = P' \oplus P''$ into $Q \oplus P''$, and apply our condition on P. Since (i) is formally stronger than the definition of "weakly hereditarily projective", except for lacking the condition of going over to direct summands, we now see that (i) implies P weakly hereditarily projective. To get the second part of (ii), write P as a direct summand in a free module F, and let A be any finitely generated submodule of P. Then we can write $F = F' \oplus F''$, where $A \subseteq F'$ and F' is finitely generated. Applying condition (i) to the submodule A, and the projection of F onto F'', we see that A will lie in a direct summand $P' \subseteq P$ which is contained in F'. Then P' will be a direct summand in F', hence will be finitely generated, establishing (ii).

Clearly (ii) ⇒ (iii), and (iii) ⇒ (i) is straightforward because any finitely generated weakly hereditarily projective module will certainly satisfy (i). So (i)–(iii) are equivalent. Clearly also (iv) ⇒ (iii). To show that (i)–(iii) imply (iv), we first want to show that P is a direct sum of finitely generated submodules. By a result of Kaplansky's [23, Thm. 3] P will be a direct sum of countably generated projective modules, each of which we know will again satisfy (i), so we are reduced to the case of P countably generated. But then, using (iii), it is easy to see that P can be written as the union of an ascending chain of finitely generated direct summands. As each will be a direct summand in the next, we get an expression for P as a direct sum of finitely generated submodules.

To establish the last clause of (iv), we note that *any* subsum of the P_i's is a direct summand in P, which is weakly hereditarily projective by (ii), hence will indeed be weakly hereditarily projective.

We have established (b) in the course of the above argument. To get (c), note that any projective module is a direct summand in a free module, i.e., a direct sum of copies of R, which will clearly satisfy (iv) if R is weakly semihereditary. ∎

Assertion (c) above generalizes a result of Kaplansky [23] and its two generalizations by Albrecht [1] and Bass [5], saying that over a *commutative* semihereditary, resp. a *right* semihereditary, resp. a *left* semihereditary ring R, every projective right module is a direct sum of finitely generated ones.

Suppose we call a projective module P satisfying the equivalent conditions of the above proposition *weakly absolutely hereditarily projective,* and, on the other hand, define a *weakly loosely hereditarily projective* module by analogy with the definition we suggested for "loosely hereditarily projective", we wonder whether any or all of the following implications are reversible:

$$\text{a.h.p.} \Rightarrow (\text{w.a.h.p.} + \text{h.p.}) \Rightarrow \text{h.p.}$$

$$\text{w.a.h.p.} \Rightarrow \text{w.h.p.} \Rightarrow \text{w.l.h.p.}$$

None of these problems arise for the class of cohereditarily projective modules! It is easily seen that such a module will satisfy condition (i) of the preceding proposition, and one deduces without difficulty:

Proposition 4.5. Let P be a projective module over a ring R. Then the following conditions are equivalent:

(i) P is cohereditarily projective.

(ii) Every finitely generated submodule of P is contained in a finitely generated

generated cohereditarily projective direct summand.

(iii) P can be written as a direct sum $\oplus_I P_i$ of finitely generated submodules, such that every finite subsum is cohereditarily projective.

Further, any projective right module over a left semihereditary ring is cohereditarily projective. ■

(If the reader wishes to prove this directly, observe, to get (i) ⇒ (ii), that the closure of a finitely generated submodule of P will be a direct summand by (i), and hence finitely generated by Lemma 1.1.)

The last assertions of the above two propositions are the reason for the apparent incompleteness of the set of definitions we gave at the end of the last section. The condition on a ring that we would have called being "weakly right hereditary" is simply that of being weakly semihereditary, and the condition we would have called "right cohereditary" is equivalent to left semihereditary.

It is worth recalling, in connection with the above propositions, that there *do* exist projective modules that are not direct sums of finitely generated submodules: The standard example takes for R the ring of continuous real–valued functions on the unit interval [0,1], and for P the ideal consisting of elements zero in the neighborhood of the point 0. (We have a variant of this example, which we hope to describe elsewhere, with the additional property that w.gl.dim R = 1.)

There are dual questions to the ones with which we began this note, which might possibly be of interest: in the definitions of *co*hereditarily and weakly hereditarily projective modules, can the finitely generated projective module which we map into P be replaced by an arbitrary free power module R^S?

(In this connection, recall E. Specker's unusual result [33, Satz 3], that any linear functional on the free countable–power module over the integers, $\mathbf{Z}^{\aleph_0}$, is a linear combination of finitely many of the coordinate functions, from which it can be deduced that any map from this power module into a free module can be factored through a finitely generated module. Specker's result can be established for right modules over any ring R satisfying: (i) The "noncompleteness" condition, for any sequence $a_1, a_2, \cdots$, of nonzero elements of R, there exists a chain $I_1 \supseteq I_2 \supseteq \cdots$ of finitely generated left ideals of R, and elements $b_i \in I_i$, such that the (I_i)–adic Cauchy sequence $\Sigma^n a_i b_i$ does not converge in R, i.e., such that $\cap_n (I_{n+1} + \Sigma^n a_i b_i) = \phi$. This allows us to prove that any linear functional on $R^{\aleph_0}$ must annihilate all but finitely many of the "Kronecker–delta" elements. (ii) The "nonlocalness" property: there exist two chains of finitely generated left ideals $I_1 \supseteq \cdots$ and $J_1 \supseteq \cdots$ such that $\cap I_i = \cap J_i = \{0\}$, but $\forall i,\ I_i + J_i = R$. From this one argues that if a functional h on $R^{\aleph_0}$ annihilates all the "delta functions", then $h(R^{\aleph_0}) = h(\Pi I_i) + h(\Pi J_i) \subseteq \cap I_i + \cap J_i = \{0\}$.)

In view of the number of variants to our original conditions discussed above, it may well be that the classes of modules defined in §2 will not all turn out to be as natural or as useful as some others (though it seems that for finitely generated modules they are the "right" ones). In any case, the terminology of this paper should be thought of only as a suggestion, subject to revision.

The subsidiary concepts introduced in this section will not be used in the remaining two sections, which will be concerned mainly with strongly hereditarily projective modules.

5. Strongly (α–) hereditary rings

By Corollary 2.13, a right semihereditary ring R will be strongly semihereditary if all free (right or left) R–modules of finite rank have $ACC_\oplus$. This makes the class of strongly semihereditary rings quite large. Let us prove one useful criterion for these $ACC_\oplus$–conditions to hold:

If R is a subring of S, and P a projective right R–module, then $P_S =_{def.} P \otimes_R S$ will be a projective S–module, and it is clear that $End_R(P)$ embeds in $End_S(P_S)$; hence that the partially ordered set of direct summands of P embeds in the corresponding set for P_S. Hence:

Proposition 5.1. If $R \subseteq S$ are rings and P a projective R–module such that P_S has $ACC_\oplus$, then P has $ACC_\oplus$. ∎

Corollary 5.2. A ring R that can be embedded in a ring S such that all finitely generated projective S–modules have $ACC_\oplus$, has the same property. In particular, any commutative integral domain R, more generally any ring R that can be embedded in a sfield, or still more generally, any ring R that can be embedded in a ring which, modulo its radical, is right or left Noetherian, has this property. ∎

A. A. Klein studies some stronger consequences of embeddability in a sfield, of a similar sort, in [25, 26].

Examples of strongly semihereditary rings not satisfying this $ACC_\oplus$ condition can be gotten by taking direct products, as in Example 3.3. More interesting examples can be gotten from the following considerations:

If M is an object of an abelian category, the ring R = End(M) will be von Neumann regular (see [29] for definition) if and only if the kernel and image of every endomorphism of M are direct summands in M. R will then be semihereditary, and its finitely generated right ideals will all be principal, and will be in natural 1–1 correspondence with the direct summands in M. The class of such ideals or modules will be closed under sums and intersections, and it follows from Lemma 2.10 (ii) that R will be strongly 1–hereditary if and only if they form a *complete* lattice. More generally, R will be strongly n–hereditary if and only if the same is true of direct summands in M^n. (An example in §6 will show us that a commutative von Neumann regular ring can be strongly 1–hereditary but not strongly 2–hereditary!) Let us here merely note a simple consequence of the above observations:

Proposition 5.3. Let V be a vector space over a field k. Then the endomorphism ring $\mathrm{End}_k(V)$ is strongly semihereditary. On the other hand, if V is infinite–dimensional, the quotient of $\mathrm{End}_k(V)$ by the subring of endomorphisms of V of finite rank is not strongly 1–hereditary. ∎

To study $\aleph_0$–hereditary and strongly $\aleph_0$–hereditary rings, let us now prove:

Proposition 5.4 Let R be a left semihereditary ring, and M a right R–module every finitely generated submodule of which is projective. Then the following conditions are equivalent:

(1) Every countably generated submodule of M is projective.

(2) Every chain $A_1 \subseteq A_2 \subseteq \cdots$ of finitely generated submodules of M, each of which is dense in the next, is eventually constant.

(3) If $A \subseteq N \subseteq M$, where A is a finitely generated submodule of M, and N an arbitrary submodule, then there exists a finitely generated submodule $B \subseteq N$, in which A is dense, and such that B is a direct summand in any finitely generated submodule of N containing it. (Such a B will be unique.)

Proof. (1) ⇒ (2): Given a chain as in (2), its union will be projective by (1), and A_1 is dense therein. Hence by Lemma 1.1, this union is finitely generated, i.e., the given chain terminates.

(2) ⇒ (3): The class of finitely generated submodules of N containing A as a dense submodule is closed under finite sums, hence directed as a partially ordered set; and by (2) it has ascending chain condition, hence it has a maximal member B. If C is a finitely generated submodule of N containing B, then because R is left semihereditary, C is cohereditarily projective, and the closure B' of B in C is a direct summand. B is dense in B' so by the maximality assumption $B' = B$. So B is a direct summand in C as desired. (Uniqueness is immediate.)

(3) ⇒ (1): Let the countably generated submodule $N \subseteq M$ be the union of the chain $A_1 \subseteq A_2 \subseteq \cdots$ of finitely generated submodules. Applying (3) to each $A_i \subseteq N$, we get a chain $B_1 \subseteq B_2 \subseteq \cdots$ of finitely generated (hence projective) submodules of N, containing the respective A_i. Each will be a direct summand in the next: $B_i = B_{i-1} \oplus C_i$ (with $B_0 = \{0\}$), hence $N = C_1 \oplus C_2 \oplus \cdots$, so N is projective. ∎

This immediately gives:

Theorem 5.5. A left and right semihereditary ring R will be right $\aleph_0$-hereditary if and only if every free right R–module of finite rank, M, satisfies the chain condition (2) of the preceding proposition. ■

See [10, §§1–4] and [8] for applications of this chain condition (called ACC_{dense}) to the study of $\aleph_0$–hereditary rings. Note also that in the proof of Proposition 5.4, the implications (3) ⇒ (1) ⇒ (2) did not require the assumption that R is left semihereditary.

The next result was discovered by looking for the "convex hull" of the two corollaries and one proposition which follow it, though in fact it was not possible to make it encompass the last of these. I am indebted to G. Sabbagh for supplying two of these clues.

Theorem 5.6. Any strongly right $\aleph_0$–hereditary ring R will satisfy the following two conditions, which are equivalent for arbitrary R:

(a) If $P_1 \supseteq P_2 \supseteq \cdots$ is a descending chain of finitely generated projective *left* R–modules, then it is true for all but finitely many i, that any finitely generated projective submodule $Q \subseteq P_i$, such that $Q \cap P_{i+1} = \{0\}$, is a direct summand in P_i.

(a*) Given families of finitely generated projective *right* R–modules $A_1, A_2, \cdots$; $B_1, B_2, \cdots$, and maps $(f_i, g_i) : A_i \to A_{i+1} \oplus B_i$ with dense image, the g_i will be surjective for all but finitely many i.

Conversely, any right $\aleph_0$–hereditary ring R satisfying the above conditions, *and* such that all free modules of finite rank have $ACC_{\oplus}$, will be strongly right $\aleph_0$–hereditary.

Proof. To see the equivalence of (a) and (a*), note that (a) is certainly equivalent to (a′): given P_i as in (a), and $Q_i \subseteq P_i$ such that $Q_i \cap P_{i+1} = \{0\}$, all but finitely many of the Q_i's will be direct summands in the corresponding P_i's. Abstractly, this situation can be described by a family of embeddings of the direct sums $Q_i \oplus P_{i+1}$ in the modules P_i, and the conclusion says that almost all of these embeddings, restricted to the Q–component, give left–invertible maps. Applying the functor *, and recalling that "embedding" is dual to "map with dense image", we get statement (a*). We shall find statement (a) more useful for applications, but (a*) what we need to complete this proof.

Given strongly right $\aleph_0$–hereditary R, to obtain condition (a*), let A_i, B_i, f_i, g_i be as in that condition. Then we get a chain of maps with dense images:

$$A_1 \to B_1 \oplus A_2 \to B_1 \oplus B_2 \oplus A_3 \to \cdots .$$

These modules map "in the limit" into $\Pi_i B_i$. (Explicitly, we map each term $B_1 \oplus \cdots \oplus B_{i-1} \oplus A_i$ into this product by sending each B_j $(j < i)$ to the factor B_j identically, and sending A_i into $\Pi_{j \leqslant i} B_j$ by the map $(g_i, g_{i+1}f_i, g_{i+2}f_{i+1}f_i, \cdots)$.) Now ΠB_i is a submodule of a free power module, hence by the hypothesis that R is strongly right $\aleph_0$-hereditary, every countably generated submodule thereof is projective; so by Proposition 5.3, the chain of images of the $B_1 \oplus \cdots \oplus B_{i-1} \oplus A_i$ will eventually be constant. It follows that the g_i must all eventually be surjective, as desired.

Now assume R satisfies the hypotheses of our converse statement. Then R will be strongly semihereditary, and in particular, left semihereditary (Corollary 2.13). To prove R strongly right $\aleph_0$-hereditary it will suffice, according to Proposition 5.4, to show for any set X that any ascending chain $U_1 \subseteq U_2 \subseteq \cdots$ of finitely generated submodules (necessarily projective) of the right module R^X, each dense in the next, is eventually constant. Let U denote the union of such a chain.

Using the fact that each U_i has $ACC_{\oplus}$, we easily conclude that for each i there will exist a finite subset $X_i \subseteq X$ such that the projection $p_{X_i} : R^X \to R^{X_i}$ will be 1–1 on U_i. Clearly we can take $X_i \subseteq X_{i+1}$; let us also write $X_0 = \phi$.

Since R^{X_i} is free of finite rank, and R is right $\aleph_0$-hereditary, Proposition 5.4 tells us that for each i, the chain $p_{X_i}(U_i) \subseteq p_{X_i}(U_{i+1}) \subseteq \cdots$ eventually becomes constant. Going to a subsequence of our chain of U_i if necessary, we can assume without loss of generality that this constancy always occurs after one step: $p_{X_i}(U_{i+1}) = p_{X_i}(U)$.

Since $p_{X_i}(U_{i+1})$ is projective, we can write $U_{i+1} = C_i \oplus A_{i+1}$, where $A_{i+1} = \ker(p_{X_i} | U_{i+1})$. Because $p_{X_i}(U_{i+1}) = p_{X_i}(U)$, C_i will also be a direct summand, complementing $\ker(p_{X_i} | U)$ in U, and using this fact we can arrange to have each C_i contained in, and hence a direct summand in, C_{i+1}; say $C_{i+1} = C_i \oplus B_{i+1}$. Thus $C_i = B_1 \oplus \cdots \oplus B_i$, and our chain of dense inclusions of U_i becomes:

$$A_1 \subseteq B_1 \oplus A_2 \subseteq B_1 \oplus B_2 \oplus A_3 \subseteq \cdots .$$

By our hypotheses (a*), for all large i the map from A_i to B_i is surjective. This means the map from U_i into the summand C_i of U_{i+1}, by projection along A_{i+1}, is surjective. But U_i does not meet $A_{i+1} = \ker p_{X_i}$, so we get $U_{i+1} = U_i \oplus A_{i+1}$. Since U_i is dense in U_{i+1}, this says $A_{i+1} = \{0\}$, and this clearly means that our chain of dense inclusions must terminate, QED. ■

Corollary 5.7. An integral domain (ring without zero-divisors) R is strongly right $\aleph_0$-hereditary if and only if it is right $\aleph_0$-hereditary and left Ore.

Proof. An integral domain R which is not left Ore will contain two left linearly independent elements x and y, and the chain of left ideals $R \supseteq Rx \oplus Ry \supseteq Rx \oplus Rxy \oplus Ry^2 \supseteq \cdots$ will contradict condition (a).

Conversely, if R is left Ore, it will be embeddable in a sfield so that free R–modules of finite rank have $ACC_{\oplus}$; and R will satisfy condition (a) because it is easy to show that the left module R^n can contain no more than n linearly independent elements. Applying Theorem 5.6, R will be strongly right $\aleph_0$–hereditary. ∎

It is well known that if k is a (skew) field and $\alpha : k \to k$ a nonsurjective endomorphism, then the skew polynomial ring $k[x, \alpha]$ defined by the relations $cx = xc^{\alpha}$ $(c \in k)$ will be right by not left Ore, and will be a right and left fir, hence right and left hereditary. It follows from the above corollary that such a ring will be strongly left $\aleph_0$–hereditary but not strongly right $\aleph_0$–hereditary.

The method of the second half of the proof of Corollary 5.7 easily extends to give the next corollary. The key point to note is that if $S^{-1}R$ is a ring of left quotients of R with respect to a multiplicative semigroup S of non–zero–divisors, and A, B submodules of a free left R–module F with $A \cap B = \{0\}$, then $S^{-1}A \cap S^{-1}B = \{0\}$ in $S^{-1}F$.

Corollary 5.8. A right $\aleph_0$–hereditary ring will be strongly right $\aleph_0$–hereditary if it is left Noetherian (G. Sabbagh), or more generally, if it has a left Noetherian ring of left fractions $S^{-1}R$ with respect to some multiplicative semigroup S of non–zero–divisors. ∎

A result we cannot get from Theorem 5.6 is:

Proposition 5.9. (G. Sabbagh). Any strongly semihereditary von Neumann regular ring is strongly right (and left) $\aleph_0$–hereditary.

Proof. Over a von Neumann regular ring, any finitely generated submodule B of a finitely generated projective module A is a direct summand, hence if B is dense in A, B = A. With this observation, the result follows trivially from Proposition 5.4. ∎

For the same reasons, every von Neumann regular ring, being right and left semihereditary, is also right and left $\aleph_0$–hereditary (Kaplansky [24]).

Onward, now, to higher cardinals! Here's a nice result:

Proposition 5.10. Any strongly hereditarily projective finitely generated right or left module P over a right $\aleph_1$–hereditary ring R has $ACC_{\oplus}$.

Proof. By Lemma 2.14 (duality) it will suffice to consider right modules. Supposing the result false, let $P = Q_1 \oplus P_1 = Q_1 \oplus Q_2 \oplus P_2 = \dots$, ($P_i = Q_{i+1} \oplus P_{i+1}$, with all $Q_i \neq \{0\}$.) Then $Q = \oplus Q_i$ will be a submodule of P which (i) is countably but not finitely generated, and (ii) has the property that every finitely generated submodule is contained in a submodule which is a direct summand in P. Let us consider the class C of $Q' \subseteq P$ which have these properties, and (iii) contain Q as a dense submodule. Clearly C is closed under taking unions of countable chains. If we have a chain of elements of C indexed by the first uncountable ordinal, its union will be projective, hence by Lemma 1.1, countably generated, so the chain must be eventually constant. It can be deduced that C is closed under unions of arbitrary chains, hence by Zorn's Lemma, C has a maximal member U.

Clearly, using conditions (i) and (ii), we can write $P = U_1 \oplus P'_1 = U_1 \oplus U_2 \oplus P'_2 = \dots$, ($P'_i = U_{i+1} \oplus P'_{i+1}$, no $U_i = \{0\}$), so that $U = \oplus U_i$. Let $V = \oplus U_{2n-1}$, $W = \oplus U_{2n}$, and let $\bar{V}$ denote the closure of V in P. Since P is strongly hereditarily projective, $\bar{V}$ will be a direct summand therein, and in particular, finitely generated and so strictly larger than V. We claim that $\bar{V} + W$ satisfies properties (i)–(iii) above, and that the sum is direct, making the module strictly larger than U. This will contradict the maximality assumption on U, completing our proof.

Clearly, $\bar{V} + W$ is countably generated, and contains Q as a dense submodule. The remainder of our assertions will follow immediately if we show that for all n, the sum $\bar{V} + (U_2 \oplus \dots \oplus U_{2n})$ is direct, and forms a direct summand in P. But indeed, if we let f denote the projection of P onto $U_2 \oplus \dots \oplus U_{2n}$ obtained from our chain of decompositions, V lies in $\operatorname{Ker} f$, hence so does $\bar{V}$, hence, being a direct summand in P, $\bar{V}$ will also be one in $\operatorname{Ker} f$, from which the desired direct–sum statements follow. ∎

Corollary 5.11. If a ring R is right $\aleph_1$–hereditary and strongly semihereditary, then every finitely generated projective (right or left) R–module has $ACC_\oplus$. ∎

Corollary 5.12. The only strongly 1–hereditary and right $\aleph_1$–hereditary von Neumann regular rings are the semisimple Artinian rings, that is, the finite products of finite matrix rings over sfields. ∎

The ring of integers is an example of a ring satisfying the conditions of Corollary 5.11. In fact, it is left and right hereditary, and strongly $\aleph_0$–hereditary. But we shall soon see that it is not strongly $\aleph_1$–hereditary!

The doctoral thesis of S. U. Chase [13] contains a characterization of the class of strongly right hereditary rings – in his terminology, rings over which every torsionless right module is projective. Theorem 5.14 below is a strengthening of Chase's result. I shall use much of Chase's original proof (unpublished), but also show that strongly $\aleph_1$–hereditary = strongly hereditary, and obtain a more precise description of the rings in question.

We shall need the following silly:

Lemma 5.13. Let $A_1 \supset A_2 \supset \cdots$ be a strictly descending chain of finitely generated strongly hereditarily projective left modules, all having $ACC_{\oplus}$, over a ring R. Then there exists a strictly descending chain $B_1 \supset B_2 \supset \cdots$ where each B_i is isomorphic to a direct summand of some A_{n_i} $(n_1 < n_2 < \cdots)$, and such that no B_i is a direct sum of proper submodules.

If R is right $\aleph_0$–hereditary, then $\cap B_i = \{0\}$.

If the A_i are principal left ideals of R, the B_i may also be taken to be such, and can be written $Rx_1 \supset Rx_2x_1 \supset Rx_3x_2x_1 \supset \cdots$, such that for every i, the right multiplication map $Rx_i \to Rx_ix_{i-1} \cdots x_1$ is 1–1. Hence if $\cap B_i = \{0\}$, then for all i we have $\cap_{j \geqslant i} Rx_jx_{j-1} \cdots x_i = \{0\}$.

Proof. Suppose $U \subseteq V$ are submodules of a direct sum of modules, $M = M_1 \oplus M_2$. Write $U_1 \subseteq V_1$ for the intersections of these modules with M_1, $U_2 \subseteq V_2$ for their *projections* in M_2. If the original inclusion was strict, it is clear that one of these inclusions must be. Also note that if M is projective, and U, V strongly hereditarily projective, then U_1, U_2 will be isomorphic to direct summands of U, and V_1, V_2 to direct summands in V.

Let us call a projective R–module "long" if it contains an infinite strictly descending chain of finitely generated strongly hereditarily projective submodules with $ACC_{\oplus}$. It is clear from the above observations that if a direct sum of finitely many projective modules is long, at least one of the summands will be.

Now let us be given $A_1 \supset \cdots$ as in the hypothesis. Clearly A_1 is long. By $ACC_{\oplus}$, A_1 can be written as a direct sum of finitely many indecomposable summands. At least one of these will be long; call it B_1. By the definition of "long", the long module B_1 must contain a finitely generated strongly hereditarily projective *long* proper submodule; call it $\bar{A}_2$, and treat it the same way$\cdots$. We thus get a sequence of indecomposable B_i, and it is easy to see that these arguments can be sharpened to yield the conclusion that each B_i is isomorphic to a direct summand of some A_{n_i}.

Now suppose R were right $\aleph_0$–hereditary ,but $\cap B_i \ni x \neq 0$. The closure of Rx in each B_i will be a direct summand, hence since B_i is indecomposable, it will be all of B_i, i.e., Rx is dense in B_i. The chain of dense inclusions $B_1 \supset B_2 \supset \cdots \supset Rx$ dualizes to an ascending chain of dense inclusions of finitely generated projective right R–modules, $B_1^* \subset B_2^* \subset \cdots \subset P$, which is impossible by Proposition 5.4 (1) $\Rightarrow$ (2)! So $\cap B_i = \{0\}$.

If the A_i are all principal left ideals, it is clear that the B_i can also be taken to be so. We can write any two successive B's as $Rv \supset Ruv$. Since the map right–multiplication–by–v: $Ru \to Ruv$ has projective image, this will be the 1–1 image of a direct summand of Ru, which we shall call Rx_i and we can rewrite our inclusion as $Rv \supset Rx_i v$. By induction we then get the desired representation of our chain. The last sentence is clear. ∎

We shall also need an observation about rates of growth of integer–valued functions: Let $\mathbf{Z}_+$ denote the set of positive integers, let $\mathbf{S} = \mathbf{Z}_+^{\mathbf{Z}_+}$, and for $f, g \in \mathbf{S}$, let us write $f > g$ if for all but finitely many i, $f(i) > g(i)$. Then "$>$" is clearly a pseudo–partial–ordering on $\mathbf{S}$. It is easy to show by a Cantor diagonal argument that given any countable family of elements of $\mathbf{S}$, there exists an element strictly greater than all of these, under this relation. Hence one can find a strictly increasing sequence of elements of $\mathbf{S}$ indexed by the first uncountable ordinal.

We are now ready to characterize strongly hereditary rings. We shall refer the reader at one key point to an argument in Chase's paper [15], and shall make use of results of Chase [14], Auslander [2], and Bass [4]. We also remark that the proof of (2) ⇒ (3) below belongs to a long tradition of results to the effect that infinite direct product–modules behave quite differently from, say, free modules, unless things are fairly special. Cf. R. Baer [3, Thm. 12.4], E. Specker [33, Sätze 2, 3] (and parenthetical comments at the end of §4 above), S. U. Chase [14, Thm. 3.1]; [16], [17] and B. L. Osofsky [27, 28].

Recall that a ring R is called right perfect if every descending chain of principal *left* ideals in R is eventually constant. This concept was introduced by Bass, who gives several equivalent conditions in [4, Thm. P].

Theorem 5.14. (after Chase). For any ring R, the following conditions are equivalent:

(1) R is strongly right hereditary,

(2) R is strongly right $\aleph_1$–hereditary,

(3) R is right perfect, and w. gl. dim. $R \leqslant 1$,

(4) In R, 1 can be written as a sum of orthogonal idempotents $e_1, \cdots, e_n$ such that:

(i) the rings $R_i = e_i R e_i$ are all simple Artinian (= matrix rings over sfields),

(ii) (triangularity) for all $i < j$, $e_j R e_i = \{0\}$, and

(iii) for all $i < k$, the kernel of the natural map $\oplus_{i<j<k} (e_i R e_j \otimes_{R_j} e_j R e_k) \to e_i R e_k$ (induced by multiplication in R) is generated only by elements of the form $(xy) \otimes z - x \otimes (yz)$ $(x \in e_i R e_{j_1},\ y \in e_{j_1} R e_{j_2},\ z \in e_{j_2} R e_k;\ i < j_1 < j_2 < k.)$

(1*)–(4*) The left–right duals of (1)–(4).

Proof. We shall show (1) ⇒ (2) ⇒ (3) ⇒ (1), (3) ⇒ (4) ⇒ (3 + 3*).

(1) $\Rightarrow$ (2) is trivial.

(2) $\Rightarrow$ (3): Clearly w. gl. dim. R $\leqslant$ 1. Suppose R not right perfect. Let $Rx_1 \supset Rx_2x_1 \supset \cdots$ be a strictly descending chain of principal left ideals. By Lemma 5.13 and Proposition 5.10 we can assume that for all i, $\cap_{j>i} Rx_j \cdots x_i = \{0\}$.

Let M be the right R–module $R^{Z_+ \times Z_+}$. Given any $f \in S$ (defined above) and $a_1, a_2, \cdots, \in R$, we shall write $(a_1, a_2, \cdots)^f$ for the element of M having value a_i at $(i, f(i)) \in Z_+ \times Z_+$ for all i, and zero everywhere else.

Let (f_α) be a strictly increasing sequence of elements of S, indexed by the first uncountable ordinal. Let M_0 be the submodule of M consisting of elements with finite support in $Z_+ \times Z_+$, and let N be the submodule of M generated by M_0 and the elements:

$$u_{\alpha i} = (0, \cdots, 0, x_i, \cdots, x_j \cdots x_i, \cdots)^{f_\alpha} \quad \text{(0 in the first } i{-}1 \text{ places)}.$$

It is easy to see from the assumptions on the f_α's that N is generated by $\aleph_1$ but not by fewer elements.

Modulo M_0, we have $u_{\alpha i} \equiv u_{\alpha j} x_{j-1} \cdots x_i$ $(j \geqslant i)$. Hence for any linear functional $h : N \to R$ annihilating M_0 we will have $h(u_{\alpha i}) \in \cap_j Rx_j \cdots x_i = \{0\}$. Therefore the countably generated module M_0 is dense in N. By Lemma 1.1, N is not projective, contradicting (2). So R must be right perfect.

(3) $\Rightarrow$ (1) and (4): By Bass [4, Thm. P] (quoted in [14, Thm. 3.2]) the condition that R is right perfect is equivalent to saying that every flat right R–module is projective. Hence our condition w. gl. dim. R $\leqslant$ 1 makes r. gl. dim. R $\leqslant$ 1. I.e., R is right hereditary.

The result of Bass quoted also asserts that the quotient of R by its Jacobson radical J will be semisimple Artinian. Hence every finitely generated right R/J–module has $ACC_\oplus$; hence using Nakayama's Lemma so will every finitely generated right R–module. In particular, R will also be left semihereditary. But by [14, Thm. 4.1], left semihereditary is equivalent to every torsionless right module being flat. Hence by the result of Bass quoted above, every torsionless right module will be projective, establishing (1).

Now using the right perfectness condition, we can find $x \neq 0$ in R minimizing Rx. Hence Rx will be a simple (and by the above observations) projective left module. The method of proof of Chase [15, Thm. 4.2] can now be applied to establish (by induction) conditions (i) and (ii); that is, to write R as an upper triangular "matrix ring";

$$\begin{pmatrix} R_1 & \cdots & e_1Re_n \\ \vdots & & \vdots \\ 0 & \cdots & R_n \end{pmatrix}$$

where the R_i are simple Artinian. It is easily seen from this representation that R has Jacobson radical $J = \Sigma_{i<j} e_i R e_j$.

We now wish to prove condition (iii). If we take the direct sum of the maps referred to in this condition, over i and k, we see that (iii) is equivalent to the statement that the kernel of the single map $\oplus_j (Je_j \otimes_{R_j} e_j J) \to J$ is generated by elements $(xy) \otimes z - x \otimes (yz)$ $(x \in Je_{j_1},\ y \in e_{j_1} R e_{j_2},\ z \in e_{j_2} R,\ j_1 < j_2)$. But it is easy to show from the construction for tensor products that these elements generate the kernel of the natural map $\oplus (Je_j \otimes_{R_j} e_j J) \to J \otimes_R J$. By right flatness of J, $J \otimes_R J$ embeds in J, so this will also be the kernel of the map into J, establishing (iii).

(4) ⇒ (3), (3*): A ring R satisfying (4) will be semiprimary (i.e., J will be nilpotent and R/J Artinian semisimple), hence in particular, by one of the criteria in the theorem of Bass cited above, right and left perfect. Also, by Auslander [2, Cor. 12] a semiprimary ring will have left and right global dimension $\leqslant 1$ if $\mathrm{Tor}_2^R(R/J, R/J) = \{0\}$. This *Tor* can be described as the kernel of the map $J \otimes J \to J$, and as we just observed, condition (iii) of (4) is equivalent to the statement that this is zero, establishing conditions (3) and (3*). ∎

So $\aleph_0$ is the only cardinal α for which strongly right α–hereditary ≠ strongly left α–hereditary!

Note that condition (4) of the above theorem gives us a prescription for constructing all strongly hereditary rings R! Namely:

Step 0: Choose a finite sequence of simple Artinian rings, $R_1, \cdots, R_n$.

Step d $(0 < d < n)$: Assume we have already constructed for every i, k with $0 < k-i < d$ an R_i–R_k–bimodule R_{ik}, with R_j–bilinear "multiplications": $R_{ij} \times R_{jk} \to R_{ik}$ for $i < j < k$. Form, for every i, k with $k-i = d$, the quotient Q_{ik} of the direct sum $\oplus_j (R_{ij} \otimes_{R_j} R_{jk})$ by the sub–bimodule spanned by elements $xy \otimes z - x \otimes yz$ (x, y, z as in condition (iii)); and choose for R_{ik} any R_i–R_k–bimodule *containing* Q_{ik}.

Final step: Take R to be the matrix ring:

$$\begin{pmatrix} R_1 & \cdots & R_{1n} \\ \vdots & & \vdots \\ 0 & \cdots & R_n \end{pmatrix}.$$

(This construction has a surprising formal similarity to one we shall give in [9] for the most general ring with weak algorithm, in the sense of Cohn!)

As a simple example, let k be a field, V an infinite–dimensional k–vector–space, considered as a k–k bimodule, and R the "matrix ring" $\begin{pmatrix} k & V \\ 0 & k \end{pmatrix}$. As Chase observes in [15], this strongly hereditary ring is not right or left Noetherian. It is also an example of a ring satisfying the conditions (a), (a*) of Theorem 5.6, and for which all free modules of finite rank have $ACC_{\oplus}$, but which does not have ACC on right, left, or even 2–sided ideals, and does not even have a right or left ring of fractions with such a chain condition.

Another example which we shall give without proof: Let k be a field, A and B k–vector–spaces, and M a nonzero subspace of $A \otimes_k B$. Let r(M) denote the minimum of the ranks, as tensors, of nonzero elements of M (cf. [6]). Then the ring

$$R = \begin{pmatrix} k & A & A\otimes B/M \\ 0 & k & B \\ 0 & 0 & k \end{pmatrix}$$

will be (r(M)−1)–hereditary, but not r(M)–hereditary. In fact, if we write $R = U \oplus V \oplus W$, where U, V and W are the three rows, then every finitely generated projective right R–module can be written uniquely $U^p \oplus V^q \oplus W^r$, and this will be (strongly = weakly = co– =) hereditarily projective if and only if $q < r(M)$. Also, gl. dim R = w. gl. dim. R = 2.

6. Commutative rings

If C is a commutative ring, B(C) the set of idempotent elements of C made into a Boolean ring in the well–known manner, and X_C the prime spectrum of B(C), then C has a natural representation as the ring of global sections of a sheaf of rings on X_C. This sheaf representation of C is uniquely characterized by the properties that the base space X_C is compact Hausdorff totally disconnected, and the stalks C_x ($x \in X_C$) have no idempotent elements other than 0 and 1. This is discussed by Pierce in [29] and by me in [7], where I establish necessary and sufficient conditions, in terms of this representation, for C to be 1–hereditary, semihereditary, or (α–)–hereditary. We shall here examine in the same fashion the "weak" and "strong" conditions.

We observed that the stalks in our sheaf representation of a commutative ring C have no nontrivial idempotents. Let us begin by characterizing our conditions for rings with this property:

Lemma 6.1. Let R be a ring containing no idempotent elements other than 0 and 1. Then the following conditions are equivalent: R is weakly 1–hereditary, R is right 1–hereditary, R is left 1–hereditary, R is strongly 1–hereditary, R is an integral domain (ring without zero–divisors).

Proof. Straightforward. ∎

Lemma 6.2. Let C be a commutative integral domain. Then the following conditions are equivalent: (i) C is weakly 2–hereditary, (ii) C is 2–hereditary, (iii) C is strongly 2–hereditary, (iv) C is weakly semihereditary, (v) C is semihereditary, (vi) C is strongly semihereditary, (vii) C is a Prüfer domain.

Proof. Using Corollary 5.2 and Corollary 2.13, we see that conditions (i)–(iii) are equivalent, and similarly (iv)–(vi). But (ii), (v) and (vii) are known to be equivalent.

(The proof of (ii) ⇒ (v) given in [22, Lemma 2, p. 27] can be reduced to the following observation: If I, J and K are ideals such that I + J, I + K, J + K are invertible, then so is I + J + K, the inverse being given by $I(I+J)^{-1}(I+K)^{-1} + J(I+J)^{-1}(J+K)^{-1}$.) ∎

To study rings that may not be integral domains, let us note that the condition on a ring R that the free right R–module of rank n be weakly hereditarily projective can be expressed by a family of statements of the form $\forall\ u\ \exists\ v,\ p(u) \Rightarrow q(u, v)$, where u and v are finite families of variables representing elements of R, and p and q are finite families of equations. Namely, u represents the coordinates of the elements of A and B as in Lemma 2.7, and v the coordinates of the projection map of P onto P_1. It is not hard to show that such a condition will hold for global sections of a sheaf of algebras on a totally disconnected compact Hausdorff space if it holds at each stalk. (This is essentially Proposition 3.4 of [29], except that p is there taken vacuous.) The converse is also true, so long as $\exists\ u,\ p(u)$ is satisfied by the global algebra, so that any solution of p(u) at a stalk yields a global solution. This applies to the conditions we are interested in, making it very easy to characterize the weakly 1–hereditary and weakly semihereditary rings:

(In the propositions to follow, conditions will be numbered so as to extend the system we used in [7], except that the condition there called (a) will be replaced by the equivalent *pair* of conditions (a′) + (b_{domain}), as in Proposition 6.3 below – cf. [7, Lemma 3.1 (ii)].)

Proposition 6.2. A commutative ring C is weakly 1–hereditary if and only if it satisfies the condition (b_{domain}): for every $x \in C$, the stalk C_x is an integral domain.

C is weakly 2–hereditary, equivalently, weakly semihereditary, if and only if it satisfies the condition ($b_{Prüfer}$): for every $x \in X_C$, the stalk C_x is a Prüfer domain. ∎

(In [22, Thm. 1, p. 77] it is shown that a commutative integral domain C is Prüfer if and only if it satisfies the condition (ν): the localization of C at every maximal ideal, equivalently, at every prime ideal, is a valuation ring. Hence the above condition for an arbitrary commutative ring to be weakly semihereditary, ($b_{\text{Prüfer}}$), is equivalent to (b_{domain}) + (ν). Condition (ν) alone characterizes commutative rings of weak gl. dim. $\leqslant 1$, that is, rings all of whose ideals are flat.)

Let us recall, for comparison, these results of [7]:

Proposition 6.3. [= 7: parts of 3.1, 4.1, 4.4, 4.5]. A commutative ring C will be 1–hereditary if and only if it satisfies:

(a′) The support in X_C of any element of C is open–closed, and

(b_{domain}) For every $x \in C$, the stalk C_x is an integral domain.

C will be 2–hereditary, equivalently, semihereditary, if and only if it satisfies (a′) and:

($b_{\text{Prüfer}}$) For every $x \in C$, the stalk C_x is a Prüfer domain.

For C to be $\aleph_0$–hereditary, we must replace ($b_{\text{Prüfer}}$) by (b_{Dedekind}) (defined in the obvious manner) and add:

(c) Every non–zero–divisor in C is invertible at all but finitely many points of X_C.

Finally, to make C (α–)hereditary, ($\alpha > \aleph_0$) we must add the condition (vacuous for $\aleph_0$!)

(d) (resp. d_α). The Boolean ring $B(C)$ is (α–)hereditary. ∎

Example 3.6 can now be seen to be an example of a ring satisfying (b_{domain}), but not (a′).

To study the "strong" conditions, recall that a Boolean ring B is called *complete* if it is complete as a lattice. The corresponding totally disconnected compact Hausdorff spaces are said to be *extremally disconnected,* and characterized by the property that the closure of any open set is open. Such spaces are used by analysts. From Lemma 2.10 (ii), we easily get:

Proposition 6.4. A commutative ring C is strongly 1–hereditary if and only if:

(a′) The support in X_C of any element of C is closed,

(b_{domain}) For all $x \in X_C$, C_x is an integral domain, and

(e) The Boolean ring of idempotents $B(C)$ is complete. ∎

In particular, a Boolean ring B is strongly 1–hereditary if and only if it is complete.

(We remark that Example 3.6 can be modified by replacing the space $\{0, 1, \cdots, \infty\}$ by an infinite extremally disconnected compact Hausdorff space (e.g., the Stone–Czech compactification of the integers) and the distinguished point ∞ by any nonisolated point thereof, to get a ring C satisfying (b_{domain}) and (c), but not (a′). The free C–module of rank 1 will be weakly hereditarily projective, and have the property that its direct summands form a complete lattice, but will not be strongly hereditarily projective – contrast Lemma 2.10, (ii) and (v)!)

We would expect, by analogy with Propositions 6.2 and 6.3, that to characterize commutative strongly semihereditary rings, we should only have to replace (b_{domain}) in the above proposition by ($b_{\text{Prüfer}}$). But in fact this is not enough. For let X be any infinite extremally disconnected compact Hausdorff space, k an infinite field, and C the ring of k–valued functions on X, continuous in the discrete topology on k. Then we get $X_C = X$ (whence $B(C)$ is complete) and $C = k$ for all $x \in C$; so C satisfies (a′), ($b_{\text{Prüfer}}$), (e). However, let $U_1, U_2, \cdots$ be a sequence of disjoint open–closed subsets of X, let $e_1, e_2, \cdots$ be their characteristic functions, considered as idempotent elements of C, let $\alpha_1, \alpha_2, \cdots$ be a sequence of distinct elements of k, and let M be the submodule of the free C–module of rank 2, defined as the intersection of the kernels of the functionals $(a, b) \mapsto (a - \alpha_i b)e_i$ $(i = 1, 2, \cdots)$. M is not finitely generated, since every member of it is zero on all but finitely many of the sets X_i, hence it is not a direct summand in C^2, so C is not strongly 2–hereditary. (This means, also, that strongly 1–hereditary + von Neumann regular $\not\Rightarrow$ strongly semihereditary.)

On the other hand, by Corollary 3.9, any direct product of Prüfer rings will be strongly semihereditary. Another example is the ring C of all *bounded* real–valued functions on a set S. A general closed submodule M of C^n is easily seen to be given by assigning to each $s \in S$ a subspace $M_s \subseteq \mathbf{R}^n$. If p_s is a projection of $\mathbf{R}^n$ onto M_s, written as an $n \times n$ matrix, we can "put these matrices together" to get an $n \times n$ matrix of real–valued functions on S. This will give a projection of C^n onto M – *if* the p_s's have bounded entries. And, indeed, we can choose the p_s's to have entries all of absolute value $\leqslant 1$ by using the *orthogonal* projections onto the M_s's for our p_s's! (Essentially the same argument applies to the ring $\mathcal{L}^\infty$ of bounded Borel functions on the real line, modulo functions with supports in null sets.)

We do not know whether every strongly 2–hereditary commutative ring is strongly semihereditary! We remark that the condition 'C is strongly n–hereditary' can be expressed as (a′) + ($b_{\text{Prüfer}}$) + (e) + the condition that certain "Grassmannian sheaves" on X_C be *flabby* (cf. [12, p. 34] for this term).

Proposition 6.5. A commutative ring C is strongly $\aleph_0$–hereditary if and only if it is the direct product of a strongly semihereditary commutative von Neumann regular ring, and a finite number of Dedekind domains (commutative hereditary integral domains.)

Proof. By Corollary 5.7, any Dedekind domain is strongly $\aleph_0$–hereditary, by Proposition 5.9, any strongly semihereditary von Neumann regular ring is strongly $\aleph_0$–hereditary, and it is clear that the class of strongly $\aleph_0$–hereditary rings is closed under finite direct products. This establishes the sufficiency of the conditions given.

Conversely, let C be strongly $\aleph_0$–hereditary. In particular, it will be $\aleph_0$–hereditary. Examining the description of $\aleph_0$–hereditary rings in Proposition 6.3, we see that the additional conditions we need to prove are (1) that for all but finitely many $x \in X_C$, C_x is a field, and (2) that those points $x \in X_C$ for which C_x is not a field are all isolated.

Suppose (1) failed. Then we can easily find a family $U_1, U_2, \cdots$ of mutually disjoint open–closed subsets of X_C, each of which contains at least one point at which the stalk of C is not a field. Let $e_1, e_2, \cdots$ be the (idempotent) characteristic functions of the U_i, and for each i let a_i be an element of C which at some point of U_i is neither zero or invertible. Note that a_ie_iR will *not* be a direct summand in R.

For $i = 1, 2, \cdots$, let $f_i = 1 - e_1 - \cdots - e_{i-1}$. Then we have the descending chain $R = f_1R \supseteq f_2R \supseteq \cdots$. For each i, e_ia_iR will be a submodule of f_iR disjoint from $f_{i+1}R$, but not a direct summand in f_iR. This contradicts Theorem 5.6; so (1) must hold.

Now suppose (2) failed: let x be a nonisolated point of X_C, such that C_x is not a field. Let E be the set of all idempotent elements of C whose supports do not contain x, and define $f : C \to C^E$ by $f(c) = (ce)_{e \in E}$ $(c \in C)$. Because x is not isolated, this map will be $1–1$ (in view of condition (a$'$)).

Because C_x is not a field, we can choose an element $a \in C$ whose stalk at x is neither zero nor invertible. Using condition (c) of Proposition 6.3, and elementary sheaf theory, we see that we can take a to be invertible at every point of X_C except x. This means that though there is no "a^{-1}" in C, "$a^{-1}e$" can be defined for all $e \in E$. (I.e., an element b satisfying $ab = e$, $be = b$.) Hence we can construct maps $a^{-n}f : C \to C^E$ for all n. The images of these maps will form an infinite strictly ascending chain of finitely generated submodules of C^E, each dense in the next, contradicting Proposition 5.4. Hence (2) must also hold. ∎

As an example, suppose C is a strongly $\aleph_0$–hereditary commutative von Neumann regular ring, $q_x : C \to C_x$ the map to the stalk at one point x, and $A \subseteq C_x$ a Dedekind domain which is not a field. Then $C' = q_x^{-1}(A)$ will be $\aleph_0$–hereditary if and only if $x \in X_K$ is isolated. In the contrary case, C' will be an $\aleph_0$–hereditary ring which satisfies the equivalent conditions (a) and (a*) of Theorem 5.6, but is not strongly $\aleph_0$–hereditary. Indeed, it was in trying to prove the converse statement of that theorem without the $ACC_{\oplus}$ hypothesis that I was led to this example, and to the preceding result.

(The above proposition tells us, in effect, that every commutative strongly $\aleph_0$–hereditary ring is the direct product of a ring to which the second half of Theorem 5.6 applies and a ring to which Proposition 5.9 applies. Conceivably this might even be true in the noncommutative case: Every strongly right $\aleph_0$–hereditary ring might be a direct product of a von Neumann regular one, and one all of whose free modules of finite rank have $ACC_{\oplus}$. A weaker conjecture would be that a strongly right $\aleph_0$–hereditary ring R can be decomposed $R = V \oplus A$ as a right module over itself, so that every finitely generated submodule of V is a direct summand, and every finite direct sum of copies of A has $ACC_{\oplus}$. A first test ring for such questions might be the ring of endomorphisms of a group $\mathbf{Z} \oplus U$, where U is a vector space over the rational numbers.)

Finally, the $\aleph_1$–conditions. Propositions 5.10 and 6.3 yield:

Corollary 6.7. An $\aleph_1$–hereditary and strongly 1–hereditary commutative ring C will be a finite direct product of Dedekind domains. Any such direct product is in fact hereditary and strongly $\aleph_0$–hereditary. ∎

And from Theorem 5.14 we have:

Corollary 6.8. The only strongly $\aleph_1$–hereditary commutative rings are the finite direct products of fields. These are in fact strongly hereditary. ∎

It is probably not hard to strengthen many of the results of this section to criteria for an arbitrary projective module over a commutative ring C to be hereditarily, cohereditarily, weakly hereditarily or strongly hereditarily projective. To take the easiest case: If P is a finitely generated projective module over a commutative ring C it defines a continuous integer–valued "rank" function on Spec C, hence on X_C. It appears that P should be hereditarily projective if and only if the conditions (a′) and (b_{domain}) are satisfied "on" the open–closed subset of X_C where rank $P > 0$, and ($b_{\text{Prüfer}}$) as well where rank $P > 1$. (A point to note is: If P has rank n at $x \in X_C$, then on localizing at any point of Spec C lying over x, P becomes *free* of rank n.) If this is so, it follows that P is hereditarily projective if and only if P^* is, i.e., if and only if P is cohereditarily projective, and hence there will be no finitely generated example like Example 3.5 over a commutative ring.

For non–finitely–generated P things will not be so simple. The rank function of P may be bounded but not continuous, everywhere finite, but unbounded, or infinite in some places. We have not investigated what our various conditions might look like for such P; in particular, we don't know whether a hereditarily projective P need be cohereditarily projective. Note, however, that this is no longer equivalent to the question of whether a cohereditarily projective P need be hereditarily projective; and in fact, the answer to *this* is no: A free module P of infinite rank over a Prüfer ring C that is not a Dedekind domain will be cohereditarily projective by Proposition 4.5 (last assertion), but not hereditarily projective.

7. **Further notes**

M. S. Shrikhande [30] calls a module M *(semi)hereditary* if every (finitely generated) submodule of M is projective. Clearly, this and our concept of a hereditarily projective module could both be subsummed in a study of pairs of modules (M, N) with the property that any homomorphic image of M in N is projective. By arguments given in §2 we see that if (M, N) and (M, N′) both have this property, so will (M, N ⊕ N′).

Another direction in which the ideas of this paper might be modified would be to see what happens if we put flatness in place of projectivity. One might consider flat modules whose homomorphic images in free modules of finite rank are all flat, whose homomorphic images in flat modules are all flat etc. (Note, in this context,Chase's result [14, Thm. 4.1]: R is semihereditary if and only if every torsionless right R–module is flat.)

References

1. F. Albrecht, On projective modules over semihereditary rings, Proc. Am. Math. Soc. **12** (1961), 638-639.
2. M. Auslander, On the dimension of modules and algebras III, Nagoya Math. J. **9** (1955), 65-67.
3. R. Baer, Abelian groups without elements of finite order, Duke Math. J. **3** (1937), 68-122.
4. H. Bass, Finitistic dimension and a homological generalization of semiprimary rings, Trans. Am. Math. Soc. **95** (1960), 466-488.
5. H. Bass, Projective modules over free groups are free, J. Alg. **1** (1964), 367-373.
6. G. M. Bergman, Ranks of tensors and change of base field, J. Alg. **11** (1969), 613-621.
7. G. M. Bergman, Hereditary commutative rings and centers of hereditary rings, Proc. London Math. Soc. 23(1971), 214-236.
8. G. M. Bergman, Infinite multiplication of ideals in $\aleph_0$–hereditary rings, J. Alg. (to appear, 1972).
9. G. M. Bergman, Construction of the general ring with weak algorithm (to appear, title uncertain).

10. G. M. Bergman and P. M. Cohn, The centres of 2–firs and hereditary rings, Proc. London Math. Soc. 23 (1971), 83-98.
11. G. M. Bergman and P. M. Cohn, Dependence in rings I: n–firs (to appear) .
12. G. Bredon, "Sheaf Theory", McGraw-Hill, 1967.
13. S. U. Chase, Module–theoretic characterizations of rings, Thesis, U. of Chicago, (1960).
14. S. U. Chase, Direct products of modules, Trans. Amer. Math. Soc. **97** (1960), 457-473.
15. S. U. Chase, A generalization of the ring of upper triangular matrices, Nagoya Math. J. **18** (1961), 13-25.
16. S. U. Chase, A remark on direct products of modules, Proc. Amer. Math. Soc. **13** (1962), 214-216.
17. S. U. Chase, On direct sums and products of modules, Pacific J. Math. **12** (1962), 847-854.
18. P. M. Cohn, Free ideal rings, J. Alg. **1** (1964), 47-69. (Note: §4 contains some errors. See J. Alg. **8** (1968), 376-383 and **10** (1968), 123.)
19. P. M. Cohn, "Universal Algebra", Harper and Row (1965), New York.
20. P. M. Cohn, Torsion modules over free ideal rings, Proc. London Math. Soc. , (3) **17** (1967), 577-599.
21. P. M. Cohn, The embedding of firs in skew fields (to appear).
22. P. Jaffard, "Les systèmes d'ideaux", Dunod, Paris, 1960.
23. I. Kaplansky, Projective modules, Annals of Math. **68** (1958), 372-377.
24. I. Kaplansky, On the dimension of modules and algebras X. A right hereditary ring which is not left hereditary, Nagoya Math. J. **13** (1958), 85-88.
25. A. A. Klein, Matrix rings of finite degree of nilpotency, Pacific J. Math. (to appear).
26. A. A. Klein, A remark concerning embedding of rings in fields, J. Alg. (to appear).
27. B. L. Osofsky, Noninjective cyclic modules, Proc. Amer. Math. Soc. **19** (1968), 1383-1384.
28. B. L. Osofsky, Homological dimension and cardinality, Trans. Amer. Math. Soc. **151** (1970), 641-649.
29. R. S. Pierce, Modules over commutative regular rings, Mem. Amer. Math. Soc. No. 70, 1967.
30. M. S. Shrikhande, Hereditary and cohereditary modules (to appear).
31. L. W. Small, Hereditary rings, Proc. Nat. Acad. Sci. U.S.A. **55** (1966), 25-27.
32. L. W. Small, Semihereditary rings, Bull. Amer. Math. Soc. **73** (1967), 656-658.
33. E. Specker, Additive Gruppen von Folgen ganzer Zahlen, Portugaliae Mathematica **9** (1950), 131-140.

LIFTING MODULES AND A THEOREM

ON FINITE FREE RESOLUTIONS*

David A. Buchsbaum

and

David Eisenbud

Brandeis University

Summary

The purpose of this paper is to expound some results connected with the lifting problem of Grothendieck, and to record a new theorem about finite free resolutions over commutative noetherian rings which has proved useful in connection with the lifting problem. The result on free resolutions, a statement of which can be found at the end of section 1, has some other interesting applications; it yields Burch's theorem on the structure of cyclic modules of homological dimension two, and also the main results of [7]. In particular, it yields a new proof that any regular local ring is a unique factorization domain.

1. Discussion of the lifting problem, and statement of the main result

Throughout this paper, rings will be commutative and will have units.

The lifting problem of Grothendieck poses the following question. Suppose R is a regular local ring with maximal ideal M, and $x \in M - M^2$. Set $S = R/(x)$, and let B be a finitely generated S–module. Does there exist an R–module A such that

(1) The element x is not a zero divisor on A; and

(2) $A/xA \cong B$?

We shall call any module A satisfying these conditions a *lifting* of B to R.

* The work on this paper was done while the authors were partially supported by NSF GP 23119.

The reason for Grothendieck's interest in the lifting problem was that an affirmative answer to the question it raises would yield a proof of a conjecture of Serre on multiplicities [9, p. 2]. Serre's conjecture is that if B and C are modules over a regular local ring S such that $\mathrm{Tor}_i^S(B, C)$ has finite length for all i, then $\Sigma_{i=0}^{\infty} (-1)^i \ell_S(\mathrm{Tor}_i^S(B, C)) \geqslant 0$, where $\ell_S(\mathrm{Tor}_i^S(B, C))$ is the length of an S–composition series for $\mathrm{Tor}_i^S(B, C)$. In [9], the problem is reduced to the case where S is complete, and B and C are cyclic. Also, the conjecture is proved in the case where S is equicharacteristic or unramified.

A positive solution to the lifting problem would allow one to reduce the general (complete regular local) case of Serre's conjecture to the unramified case. For, if S is a complete regular local ring, then by the Cohen structure theory [8, p. 106], there is an unramified regular local ring R with maximal ideal M and an element $x \in M - M^2$ such that $S \cong R/(x)$. If A is a lifting to R of the S–module B, then

$$\mathrm{Tor}_i^R(A, C) \cong \mathrm{Tor}_i^S(B, C).$$

To see this, note that the fact that x is a non–zero divisor on A implies that $\mathrm{Tor}_i^R(A, S) = 0$ for all $i > 0$. Thus if $P \to A$ is an R–free resolution of A, then $S \otimes_R P \to S \otimes_R A \cong B$ is an S–free resolution of B, so the two sides of the desired formula are the homology, respectively, of $P \otimes_R (S \otimes_S C)$ and $(P \otimes_R S) \otimes_S C$.

Thus $\Sigma_{i=0}^{\infty} (-1)^i \ell_S(\mathrm{Tor}_i^S(B, C)) = \Sigma_{i=0}^{\infty} (-1)^i \ell_R(\mathrm{Tor}_i^R(A, C)) \geqslant 0$ by Serre's result for the unramified case.

We note in passing that because of Serre's reduction of the conjecture to the cyclic case, it would be sufficient to be able to lift cyclic modules.

We now sketch a few of the known results on lifting. Let R be a ring, $S = R/(x)$, where x is a non–zero divisor in R. Most theorems about lifting take advantage of the well–known result (lemma 3.1 of this paper) which says that lifting an S–module B to R is the same thing as "lifting" an S–projective resolution of B to an R–projective complex. From this it follows easily that any S–module of free dimension less than or equal to 1 can be lifted, and that any cyclic module $S/(s_1 \cdots s_n)$, where $s_1 \cdots s_n$ forms an R–sequence can be lifted; see section 3.

We now specialize to the case where R and S are local, and the module to be lifted is cyclic, say of the form S/I. Then if S/I has homological dimension 2, S/I is liftable. Moreover, if I is generated by at most 3 elements, then S/I is liftable. The first of these results is a simple consequence of a structure theorem for cyclic modules of homological dimension 2 which is due (in this generality) to Burch [3, Theorem], though a special case of it goes back to Hilbert [4, pp. 239-240 in the Ges. Abh., V. 2]. (See [5, p. 148, ex. 8] for a slick, short proof of Burch's theorem that works even in the non–noetherian case.) We will now describe Burch's theorem.

Let S be a local ring, $I = (s_1 \cdots s_n)$ an ideal of S such that S/I has homological dimension 1. The free resolution of S/I will have the form

$$0 \longrightarrow S^{n-1} \xrightarrow{\phi_2} S^n \xrightarrow{\phi_1} S \longrightarrow S/I \longrightarrow 0.$$

Suppose that with respect to the canonical bases of S^{n-1}, S^n, and S, ϕ_2 has the form (ϕ_{ij}) and ϕ_1 has the form $(s_1,\cdots,s_n)$. We will write $[\phi_2;i]$ for the minor obtained from the matrix (ϕ_{ij}) by omitting the ith row. We are now ready to state Burch's theorem.

Theorem. There exists a non-zero divisor $s \in S$ such that $s\cdot[\phi_2\,;i] = s_i$ for $i = 1,\cdots,n$.

The result on lifting cyclic modules S/I where S/I has finite homological dimension and I is generated by three elements follows from a broad generalization of Burch's theorem which gives information about the form of a finite free resolution of any length.

We need some notation. We will say that a map of finitely generated free S-modules $\phi : F \to G$ has rank r if $\Lambda^r \phi \neq 0$ and $\Lambda^{r+1} \phi = 0$. Equivalently, ϕ has rank r if ϕ has a non-zero minor of order r, but every minor of order $r + 1$ is 0. If we choose a basis for F indexed by a set μ and a basis for G indexed by υ, we will write $\phi : S^\mu \to S^\upsilon$, and we will identify ϕ and its matrix with respect to the canonical bases of S^μ and S^υ.

For the minor of ϕ with columns $\mu' \subset \mu$ and rows $\upsilon' \subset \upsilon$, we will write $[\phi\,;\mu', \upsilon']$. Of course this notation only makes sense when $\#\mu'$ (the number of elements in μ') is the same as $\#\upsilon'$, and in this case, $[\phi\,;\mu', \upsilon'] \in S$. If $\mu \supseteq \mu'$, we write $\mu - \mu'$ for the complement of μ' in μ.

Recall that the depth of an ideal $I \subseteq S$ is defined to be the length of a maximal S-sequence contained in I.

Theorem 1. Let S be a noetherian ring, and let

$$0 \longrightarrow S^{\mu(n)} \xrightarrow[\phi_n]{} S^{\mu(n-1)} \xrightarrow[\phi_{n-1}]{} \cdots \xrightarrow[\phi_1]{} S^{\mu(0)}$$

be an exact sequence of finitely generated free S-modules with chosen bases. Then, for each $0 \leqslant k \leqslant n-1$ and $\mu \subset \mu_k$ with $\#\mu = \text{rank}\,\phi_k$, there is an element $\alpha(k, \mu)$ such that

(1) if $\mu \subseteq \mu(k)$ and $\upsilon \subseteq \mu(k-1)$ with $\#\mu = \#\upsilon = \text{rank}\,\phi_k$, then, $[\phi_k : \mu, \upsilon] = \alpha(k, \mu)\alpha(k-1, \mu(k-1)-\upsilon)$,

(2) if $\mu \subseteq \mu(n-1)$ and $\#\mu = \text{rank } \phi_{n-1}$, then $\alpha(n-1, \mu) = [\phi_n ; \mu(n), \mu(n-1)-\mu]$, and

(3) for each k, the depth of the ideal generated by $\{\alpha(k, \mu) | \mu \subseteq \mu(k)$ and $\#\mu = \text{rank } \phi_k\}$ is at least $k + 1$.

Remarks.

(1) Note that $\text{rank } \phi_k + \text{rank } \phi_{k-1} = \#\mu_{k-1}$, so parts (1) and (2) of the theorem make sense.

(2) There is also an "intrinsic" version of the theorem: if we let $F_k = R^{\mu(k)}$, its statement depends only on a choice of basis for $\Lambda^{\mu(k)} F_k$. Its conclusion asserts, instead of the existence of elements $\alpha(k, \mu)$, the existence of maps α_k making the following diagrams commute

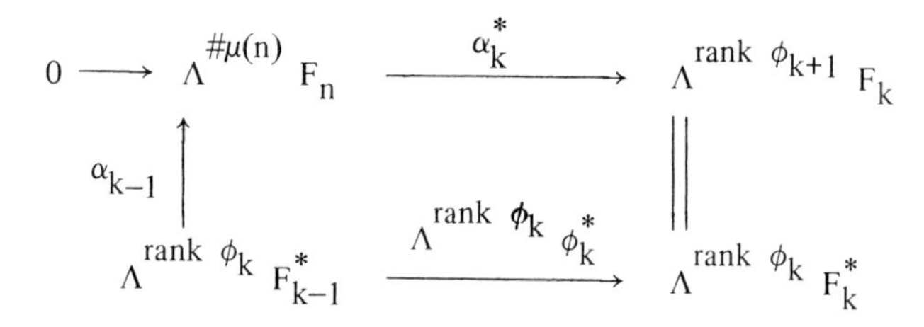

where the isomorphism on the right is that given by the exterior product with a chosen basis of $\Lambda^\lambda F_k$ where $\lambda = \text{rank } F_k$; see [1, A, III, §11].

(3) Burch's Theorem is the special case of Theorem 1 in which $n = 2$ and $\#\mu(0) = 1$.

(4) It is easy to see that if $\text{Cok } \phi_1$ is annihilated by a non–zero divisor of S, then $\text{rank } \phi_1 = \#\mu(0)$, so there will be a non–zero divisor $\alpha(0, \mu(0)) \in S$. This is precisely MacRae's $G(\text{Cok } \phi_1)$ [7, p. 159]. Our result thus gives a new (and more direct) derivation of many of the results in [7]. In particular, we obtain a new proof of the well–known theorem that a regular local ring is a unique factorization domain. This proof is detailed in the next section of this paper.

(5) The proof of Theorem 1, along with the proof that any cyclic module S/I such that I is generated by three elements and S/I has finite homological dimension can be lifted, will appear elsewhere. In this paper we will do a special case of the liftability theorem – the case where S/I has homological dimension 3 and I has only 3 relations. This is contained in section four of this paper; the lifting technique to be used there is discussed in section three.

2. Some consequences of Theorem 1

Our first application is to the proof of a result proved for domains in MacRae's paper [6, Cor. 4.4] and extended, although not explicitly, to all noetherian rings in [7]. We abbreviate projective dimension to pd.

Theorem 2. Let S be a noetherian ring, $I = (s_1, s_2)$ an ideal generated by 2 elements. If $\text{pd } S/I < \infty$, then $\text{pd } S/I \leqslant 2$.

Proof. It suffices to prove that $\text{pd } S/I$ is locally at most 2, so we may assume that S is local, so that S/I has a finite free resolution of the form

$$0 \to S^{\mu(n)} \to S^{\mu(n-1)} \to \dots \to S^2 \xrightarrow[\phi_1]{} S \to S/I \to 0$$

where 2 is to be thought of as the index set with elements $\{1,2\}$, and where the matrix representing ϕ_1 is (s_1, s_2).

Of course we may assume that $\text{pd } S/I \geqslant 2$, and that the above resolution is minimal.

We have $[\phi_1 ; i, 1] = s_i$, $i = 1,2$, so (2) of Theorem 1 tells us that there are elements $\alpha = \alpha(0, 1)$ and $\alpha_i = \alpha(1, i)$, such that $\alpha\alpha_i = s_i$, $i = 1,2$. Moreover, (3) of Theorem 1 tells us that the depth of the ideal (α_1, α_2) is at least 2, so (α_1, α_2) forms an S–sequence (if one of the α_i were a unit, s_1, s_2 would not form a minimal generating set for I). Part (3) of Theorem 1 also tells us that α is a non–zero divisor, so $(\alpha_1, \alpha_2) \cong (s_1, s_2)$ as modules. Since α_1, α_2 is an S–sequence, $\text{pd } (\alpha_1, \alpha_2) = 1$. Consequently $\text{pd } (s_1, s_2) = 1$, and $\text{pd } S/I = 2$, as required.

Corollary 1. Every regular local ring is a unique factorization domain.

Proof. Let S be a regular local ring. We will show that the intersection of principal ideals (s_1) and (s_2) in S is again a principal ideal; from this, unique factorization follows at once. Since S is local, projective ideals are principal. It is easy to check that the obvious sequence $0 \to (s_1) \cap (s_2) \to (s_1) \oplus (s_2) \to (s_1, s_2) \to 0$ is exact. But by Theorem 2, $\text{pd}(s_1, s_2) = \max(0, \text{pd } S/(s_1, s_2) - 1) \leqslant 1$, so $(s_1) \cap (s_2)$ is projective.

We will now state the special case of Theorem 1 that we will need for the lifting result to be proved in the next section.

Lemma 1. Suppose

$$0 \to S^1 \xrightarrow[\phi_3]{} S^\mu \xrightarrow[\phi_2]{} S^\mu \xrightarrow[\phi_1]{} S$$

is exact, where $\mu = \{1,2,3\}$. Suppose that the matrices of ϕ_3, ϕ_2, ϕ_1 with respect to the given bases are

$$\phi_3 = \begin{pmatrix} \sigma_1 \\ \sigma_2 \\ \sigma_3 \end{pmatrix},$$

$$\phi_2 = \begin{pmatrix} \phi_{11} & \phi_{12} & \phi_{13} \\ \phi_{21} & \phi_{22} & \phi_{23} \\ \phi_{31} & \phi_{32} & \phi_{33} \end{pmatrix},$$

$$\phi_1 = (s_1 \ \ s_2 \ \ s_3);$$

then there exist $\alpha, \alpha_1, \alpha_2, \alpha_3$ such that α is a non–zero divisor, depth $(\alpha_1, \alpha_2, \alpha_3) \geqslant 2$, $s_i = \alpha\alpha_i$, $i = 1,2,3$, and $[\phi\,; \mu - i, \mu - j] = \sigma_i\alpha_j$, $i,j = 1,2,3$. Moreover, since depth$(\sigma_1, \sigma_2, \sigma_3)$ is at least 3, $\sigma_1, \sigma_2, \sigma_3$ form an S–sequence.

3. A technique for lifting

Our approach to lifting is based on the following lemma.

Lemma 2. Let R be a ring, $x \in R$, and $S = R/(x)$. Let B be an S–module, and let

$$\mathfrak{F} : F_2 \xrightarrow{\phi_2} F_1 \xrightarrow{\phi_1} F_0$$

be an exact sequence of S–modules with $\operatorname{cok} \phi_1 \cong B$. Suppose that

$$\Gamma : G_2 \xrightarrow{\psi_2} G_1 \xrightarrow{\psi_1} G_0$$

is a complex of R–modules such that

(1) The element x is a non–zero divisor on each G_i,

(2) $G_i \otimes_R S = F_i$, and

(3) $\psi_i \otimes_R S = \phi_i$.

Then, coker ψ_1 is a lifting of B to R.

Proof. Conditions (1) and (2) imply that for each i the sequence $0 \to G_i \xrightarrow{x} G_i \to F_i \to 0$ is exact. Using condition (3) we obtain a commutative diagram with exact columns:

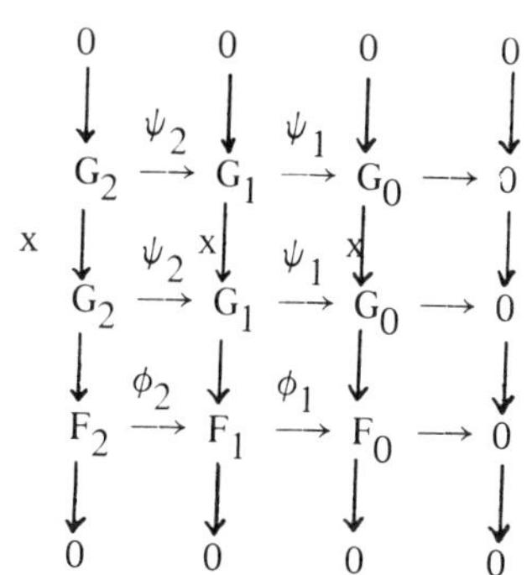

Thus we obtain an exact sequence in homology $0 = H_1(\mathfrak{F}) \to H_0(\Gamma) \xrightarrow{x} H_0(\Gamma) \to H_0(\mathfrak{F}) \cong B \to 0$. But $H_0(\Gamma) = \mathrm{cok}(\psi_1)$, so $\mathrm{cok}(\psi_1)$ is a lifting of B as required.

Remarks.

(1) If we assume $x \in \mathrm{Rad}\ R$ and take $\mathfrak{F}'$ to be a complete resolution of B by S–modules, and Γ' a complex of R modules which reduces, modulo x, to $\mathfrak{F}'$, then the same argument, together with Nakayama's Lemma, shows that Γ' is exact.

(2) A sort of converse to Lemma 2 is also easy: Suppose x is a non–zero divisor in R, $S = R/(x)$. If A is a lifting to R of the S–module B, and if Γ is a projective R–resolution of A, then $\Gamma \otimes_R S$ is a projective resolution of B. Thus the problem of lifting a module from S to R is equivalent to the problem of finding an R–projective complex which reduces to a given S–projective resolution.

(3) Lemma 2 suggests a stronger form of the lifting problem: If ϕ_1 and ϕ_2 are matrices over S such that $\phi_1 \cdot \phi_2 = 0$, are there matrices over R which reduce modulo x to ϕ_1 and ϕ_2 and whose composite is 0? Of course, the answer is in general "no", but if S is a discrete valuation ring, every matrix over S is diagonalizable, so the answer is "yes". The question seems at least reasonable if R and S are regular local rings.

We now turn to two easy applications of Lemma 2. Recall that a sequence $s_1 \cdots s_n$ of elements of S is said to be a generalized S–sequence if the Koszul complex

$$K(S;s_1,\cdots,s_n) : \ldots \to \Lambda^3 R^n \to \Lambda^2 R^n \to R^n \xrightarrow{(s_1 \cdots s_n)} R$$

is exact.

Corollary 2. With the notation of Lemma 2, suppose x is a non–zero divisor on R and $B = S/(s_1,\cdots,s_n)$, where $s_1,\cdots,s_n$ form a generalized S–sequence. Then B can be lifted to R.

Proof. Lift the s_i arbitrarily to elements r_i of R. The Koszul complex $K(R;r_1,\cdots,r_n)$ reduces modulo x to $K(S;s_1,\cdots,s_n)$. By Lemma 2, $R/(r_1,\cdots,r_n)$ is a lifting.

Another class of modules which can be lifted for a similar reason is the class of cokernels of matrices whose minors generate an ideal of "large" depth. Explicitly, suppose that S is a noetherian ring, $\phi : S^m \to S^n$ a map of rank n of free S–modules. Suppose that the $n \times n$ minors of ϕ generate an ideal of depth $m - n + 1$ (which is the maximum possible depth). Then cok ϕ can be lifted. An equivalent condition to the largeness of depth of the ideal above is that $\mathrm{Ext}^k(\mathrm{cok}\ \phi, S) = 0$, $0 \leqslant k \leqslant m - n$. The reason that this works is that under these hypotheses one can write down a free resolution for cok ϕ in terms of the matrix ϕ itself, much as in the case of the Koszul complex. This resolution is called a generalized Koszul complex (see [2, Theorem 2.4] for details). Since there is a similar complex resolving cok $\Lambda^n \phi$, modules of this form can also be lifted.

As a last result in this direction, we mention that since one can also explicitly write down a complex resolving modules of the form S/I where I is generated by any set of monomials in an S–sequence, these modules can also be lifted. For details of this complex, see [10, section IV].

Finally, Lemma 2 shows that any module of projective dimension 1 can be lifted.

Corollary 3. With notation as in Lemma 2, suppose x is a non–zero divisor on R and $\mathrm{pd}_S B = 1$. Then B can be lifted to R.

Proof. Let $0 \to F_1 \xrightarrow{\phi} F_0 \to B \to 0$ be an S–free resolution of B. Lift F_1 and F_0 to free R–modules G_1 and G_0. Since the canonical map $\mathrm{Hom}_R(G_1, G_0) \to \mathrm{Hom}_S(F_1, F_0)$ is onto, we may choose a map $G_1 \xrightarrow{\psi} G_0$ which reduces to ϕ modulo x. Lemma 2 now applies to the complex $0 \to G_1 \xrightarrow{\psi} G_0$, showing that coker ϕ is a lifting of B.

4. Ideals with 3 generators and 3 relations

As we mentioned in the introduction, Theorem 1 gives enough information about the structure of finite free resolutions to allow the lifting of any cyclic modules S/I where I is generated by three elements $I = (s_1, s_2, s_3)$. In this section we will exhibit a simple special case of the techniques involved, namely, the case in which there are just three relations on the elements (s_1, s_2, s_3).

Theorem 3. Suppose R is a noetherian ring, $x \in R$ a non-zero divisor, $S = R/(x)$. Suppose that $I = (s_1, s_2, s_3)$ is an ideal of S, and that S/I has a free resolution of the form

$$(*) \qquad 0 \to S \xrightarrow{\phi_3} S^{\mu} \xrightarrow{\phi_2} S^{\mu} \xrightarrow{\phi_1} S \to S/I$$

where $\mu = \{1,2,3\}$. Then S/I can be lifted to R.

Proof. We will adopt the notations of Lemma 1 of Section 2. Because coker(ϕ_2) is torsion free, the sequence

$$0 \to S \xrightarrow[\phi_3]{} S^{\mu} \xrightarrow[\phi_2]{} S^{\mu} \xrightarrow[(\alpha_1, \alpha_2, \alpha_3)]{} S$$

is exact. If this sequence could be lifted to a complex of free R-modules

$$0 \to R \xrightarrow[\psi_3]{} R^{\mu} \xrightarrow[\psi_2]{} R^{\mu} \xrightarrow[(\beta_1, \beta_2, \beta_3)]{} R,$$

then if β is any element of R which reduces modulo x to α, it is clear that the sequence

$$0 \to R \xrightarrow[\psi_3]{} R^{\mu} \xrightarrow[\psi_2]{} R^{\mu} \xrightarrow[(\beta\beta_1, \beta\beta_2, \beta\beta_3)]{} R$$

is a lifting of (*). Thus it suffices to treat the case $\alpha = 1$, $\alpha_i = s_i$, $i = 1,2,3$, so that $\phi_1 = (\alpha_1, \alpha_2, \alpha_3)$.

Since the Koszul complex associated with ϕ_1 is a free complex, and since (*) is exact, there is a map $\gamma : \Lambda^2 S^{\mu} \to S^{\mu}$ making the following diagram commutative:

$$\begin{array}{ccccccc} \Lambda^2 S^\mu & \xrightarrow{\delta} & S^\mu & \xrightarrow{\phi_1} & S & \to & S/I \\ \gamma \downarrow & & & & & & \\ S^\mu & \xrightarrow[\phi_2]{} & S^\mu & \xrightarrow{\phi_1} & S & \to & S/I \end{array},$$

where δ is the boundary map of the Koszul complex associated to ϕ_1. If we take $\epsilon_1, \epsilon_2, \epsilon_3$ to be the canonical basis of S^μ, then we may write

$$\gamma(\epsilon_{i_1} \wedge \epsilon_{i_2}) = \sum_j \gamma_j^{i_1 i_2} \epsilon_j,$$

and the commutativity yields

$$\begin{aligned} \phi_2(\gamma(\epsilon_{i_1} \wedge \epsilon_{i_2})) &= \sum_k (\sum_j \gamma_j^{i_1 i_2} \phi_{kj}) \epsilon_k \\ &= \delta(\epsilon_{i_1} \wedge \epsilon_{i_2}) \\ &= \alpha_{i_1} \epsilon_{i_2} - \alpha_{i_2} \epsilon_{i_1} \end{aligned}$$

for all $i_1 < i_2$.

On the other hand, by virtue of Lemma 2,

$$\begin{aligned} \sigma_j \alpha_{i_1} \epsilon_{i_2} - \sigma_j \alpha_{i_2} \epsilon_{i_1} &= [\phi_2 ; \mu - j, \mu - i_1] \epsilon_{i_2} - [\phi_2 ; \mu - j, \mu - i_2] \epsilon_{i_1} \\ &= \phi_2 (\sum_{k \neq j} \pm [\phi_2 ; \omega, \lambda] \epsilon_k) \end{aligned}$$

where $\omega = \mu - \{ j, k \}$ and $\lambda = \mu - \{ i_1, i_2 \}$. Thus $\sigma_j \gamma(\epsilon_{i_1} \wedge \epsilon_{i_2}) - \sum_{k \neq j} \pm [\phi_2 ; \omega, \lambda] \epsilon_k$ is in $\ker \phi_2$.

Since (*) is exact, we may write

$$\sigma_j \gamma(\epsilon_{i_1} \wedge \epsilon_{i_2}) - \sum_{k \neq j} \pm [\phi_2 ; \omega, \lambda] \epsilon_k = \phi_2(s_j^{i_1 i_2}) = \sum_k s_j^{i_1 i_2} \sigma_k \epsilon_k ,$$

for some $s_j^{i_1 i_2} \in S$.

Now the coefficient of ϵ_j in $\sum_{k \neq j} \pm [\phi_2 ; \omega, \lambda] \epsilon_k$ is 0, so

$$\sigma_j \gamma_j^{i_1 i_2} = \sigma_j s_j^{i_1 i_2} .$$

Since by Lemma 2, σ_j is part of an S–sequence, it is a non–zero divisor, so

$$\gamma_j^{i_1 i_2} = s_j^{i_1 i_2} .$$

This gives

$$[\phi_2 ; \omega,\lambda] = \phi_{\omega,\lambda} = \gamma_k^{i_1 i_2} \sigma_j - \gamma_j^{i_1 i_2} \sigma_k .$$

We have now expressed the elements ϕ_{ij} of ϕ_2 in terms of ϕ_3 and γ. But the elements α_i of ϕ_1 are expressible in terms of ϕ_2 and γ, since $\delta = \phi_2\gamma$. Thus both ϕ_1 and ϕ_2 may be expressed in terms of ϕ_3 and γ. In fact, if we form the matrix

$$(**) \qquad \begin{pmatrix} \sigma_1 & \sigma_2 & \sigma_3 \\ \gamma_1^{12} & \gamma_2^{12} & \gamma_3^{12} \\ \gamma_1^{13} & \gamma_2^{13} & \gamma_3^{13} \\ \gamma_1^{23} & \gamma_2^{23} & \gamma_3^{23} \end{pmatrix},$$

then it is easy to verify that the element ϕ_{ij} is $\pm$ the 2×2 minor of $(**)$ obtained by omitting the ith column and the rows involving $\gamma_k^{i_1 i_2}$ with $i_1 = j$ or $i_2 = j$. The element α_i is $\pm$ the 3×3 minor of $(**)$ obtained by omitting the row involving $a_k^{i_1 i_2}$ with $i \neq i_1$ or i_2.

It is now easy to lift ϕ_2 and ϕ_1 to matrices $\widetilde{\phi}_2 = (\widetilde{\phi}_{ij})$ and $\widetilde{\phi}_1 = (\widetilde{\alpha}_1, \widetilde{\alpha}_2, \widetilde{\alpha}_3)$ such that $\widetilde{\phi}_1\widetilde{\phi}_2 = 0$; this will complete the lifting, by virtue of Lemma 2.

Simply find elements $\widetilde{\sigma}_i$ and $\widetilde{\gamma}_k^{ij}$ in R which reduce modulo x to σ_i and γ_k^{ij}. Form the matrix

$$(**)^{\sim} \qquad \begin{pmatrix} \widetilde{\sigma}_1 & \widetilde{\sigma}_2 & \widetilde{\sigma}_3 \\ \widetilde{\gamma}_1^{12} & \widetilde{\gamma}_2^{12} & \widetilde{\gamma}_3^{12} \\ \widetilde{\gamma}_1^{13} & \widetilde{\gamma}_2^{13} & \widetilde{\gamma}_3^{13} \\ \widetilde{\gamma}_1^{23} & \widetilde{\gamma}_2^{23} & \widetilde{\gamma}_3^{23} \end{pmatrix}$$

Now define the elements $\widetilde{\phi}_{ij}$ and $\widetilde{\alpha}_i$ of $\widetilde{\phi}_2$ and $\widetilde{\phi}_1$ to be the minors formed from $(**)^{\sim}$ in the same way that we formed minors of $(**)$ to represent ϕ_{ij} and α_i, so that $\widetilde{\phi}_{ij}$ and $\widetilde{\alpha}_i$ will reduce, modulo x, to ϕ_{ij} and α_i respectively.

It is now easy to check by direct computation that $\widetilde{\phi}_1\widetilde{\phi}_2 = 0$.

Note Added in Proof. The computation involved in the proof of Theorem 3 can be considerably shortened by comparing (*) with the dual of the Koszul complex associated to ϕ_3^* and applying Theorem 1 to get α's for this Koszul complex. This trick eliminates the use of the Koszul complex of ϕ_1.

References

1. N. Bourbaki, Éléments de Mathématique, "Algebra," Ch. III, Hermann, 1970.
2. David A. Buchsbaum and Dock Sang Rim, A generalized Koszul Complex, II: Depth and multiplicity, Trans. Amer. Math. Soc. **111** (1964), 197-224.
3. Lindsay Burch, On ideals of finite homological dimension in local rings, Proc. Cam. Phil. Soc. **64** (1968), 941-946.
4. David Hilbert, Über die Theorie der Algebraischen Formen, Math. Ann. **36** (1890), 473-534.
5. Irving Kaplansky, "Commutative Rings," Allyn and Bacon, 1970.
6. R. E. MacRae, On the homological dimension of certain ideals, Proc. Amer. Math. Soc. **14** (1963), 746-750.
7. R. E. MacRae, On an application of Fitting invariants, J. Alg. **2** (1965), 153-169.
8. Masayoshi Nagata, "Local Rings," John Wiley and Sons, 1962.
9. J. -P. Serre, "Algébre Locale–Multiplicités," Springer Lecture Notes in Mathematics **11**, 1965.
10. Diana Taylor, Ideals generated by monomials in an R–sequence, Thesis, University of Chicago, 1966.

ON THE AUTOMORPHISM SCHEME
OF A PURELY INSEPARABLE FIELD EXTENSION

Stephen U. Chase[1]

Cornell University

Let k be a field, and K be a finite extension of k. For every commutative k–algebra T, we can consider the group $\mathbf{Aut}_{K/k}(T)$ of T–algebra automorphisms of $K \otimes T = K \otimes_k T$; this defines a functor from k–algebras to groups. It is well known, and easy to see [1, pp. 90-91], that this functor is representable by a commutative k–algebra A(K/k); that is, for each T as above there is an isomorphism of sets

$$\mathrm{Alg}_k(A(K/k), T) \xrightarrow{\approx} \mathbf{Aut}_{K/k}(T)$$

and these isomorphisms are natural in T. Since $\mathbf{Aut}_{K/k}$ is a group–valued functor, A(K/k) becomes — via the usual categorical arguments — a cogroup object in the category of commutative k–algebras, which is simply a commutative Hopf algebra with antipode. That is to say, Spec(A(K/k)) is an affine group scheme over k, called the *automorphism scheme* of the extension K/k.

The automorphism scheme of a finite field extension, and its significance in field theory, have recently been studied by several authors; see, e.g., [1, 6]. We refer the reader to these papers, as well as to [3, II, §1, 2.6-2.7, pp. 152-154] for a discussion of general properties of the automorphism scheme, and to [10] for an explanation of its connection with other approaches to field theory, such as the cocommutative Hopf algebras of [8] and [2].

In this paper we investigate the automorphism scheme of K/k for the special case in which k has characteristic $p \neq 0$ and K is purely inseparable and modular over k (i.e., isomorphic to a tensor product over k of primitive extensions [9]). Section 1 catalogues a number of elementary and essentially well–known facts about automorphism schemes, and provides generators and relations for A(K/k), with K/k modular. In Section 2 we prove that, if K is a tensor product over k of primitive purely inseparable extensions of k of *equal* degree (e.g., if the exponent of K/k is one), then A(K/k) is an integral domain, and its quotient field is isomorphic to a pure function field over K. In particular, the automorphism scheme of K/k is, in

[1] Supported in part by NSF GP 25600.

this case, irreducible and reduced. This is not true for an arbitrary purely inseparable modular extension, however; Section 3 exhibits such an extension K/k for which $A(K/k)$ possesses non–trivial nilpotent elements.

If K is a finite purely inseparable extension of k and $\bar{k}$ is an algebraic closure of k, then it is not difficult to show that $\bar{k} \otimes A(K/k)$ always has non–trivial nilpotent elements (see, e.g., [1, Proposition 12, p. 105]). Hence, if K/k satisfies the conditions of the preceding paragraph, then the automorphism scheme of K/k is reduced, but its base extension to a group scheme over $\bar{k}$ is not. It is well known that this phenomenon cannot happen for group schemes over a perfect field; this is, for example, an easy consequence of [7, Theorem 3, p. 521]. These observations illustrate a pleasant feature of automorphism schemes of purely inseparable extensions; namely, they provide a class of examples of group schemes with properties markedly different from those of group schemes over perfect fields.

R. Rasala [5] has recently developed a theory of splitting fields of purely inseparable extensions in which the modular ones assume the role played by the normal separable extensions in classical field theory, and there is accumulating evidence (see, e.g., [8], [6], [2]) that a purely inseparable Galois theory can be developed with similar features. To this end, we introduce in Section 4 a certain finite subgroup scheme G_t of the automorphism scheme G of a purely inseparable modular extension K/k and establish a weak Galois correspondence between a class of subgroup schemes of G_t and certain intermediate fields in the extension K/k. The group scheme G_t – which we call the *truncated automorphism scheme* of K/k – has rank $[K : k]^{[K : k]}$ over k, is represented by a commutative Hopf k–algebra $P(K/k)$ with a particularly simple algebra structure, and in general appears to be a good candidate for an inseparable analogue of a Galois group. Further properties of G_t, and a more complete Galois correspondence for modular extensions, are announced in [2] in the language of cocommutative Hopf algebras (the Hopf k–algebra $H(K/k)$ of that note being simply the linear dual of $P(K/k)$).

In Section 5 we prove that the quotient scheme G/G_t is isomorphic to an open subscheme of affine space over k of suitable dimension.

Our treatment of the automorphism scheme is largely self–contained, although in a few places we quote theorems of Begueri [1] (these could, however, be deduced as easy consequences of certain results proved here). The material of Section 1 is closely related to that of [1, pp. 102-105], and Theorem 5.6 can be viewed as a stronger form of [1, Proposition 10, p. 104] in the modular case. We shall rely heavily on the language and general theory of group schemes as set forth, for example, in [3], and shall also make cavalier use of the more basic notions of category theory. Throughout the paper k will be a field of characteristic $p \neq 0$, and if V and W are k–spaces we shall write $V \otimes W$ for $V \otimes_k W$. Recall that the *exponent* of a purely inseparable field extension K/k is the smallest natural number e such that $K^{p^e} \subseteq k$.

We are indebted to many colleagues – in particular, to S. Lichtenbaum, S. Shatz, M. E. Sweedler, and W. Waterhouse – for countless stimulating discussions, during the past several years, of matters related to the material of this paper.

1. The Hom scheme and representability

We first introduce a construction closely related to the automorphism scheme. Let K and L be finite extensions of our ground field k, and define a functor $\mathbf{Hom}_k(K, L)$ from commutative k–algebras to sets by the condition–

$$\mathrm{Hom}_k(K, L)(T) = \mathrm{Alg}_T(K \otimes T, L \otimes T) \approx \mathrm{Alg}_k(K, L \otimes T)$$

the latter isomorphism being a familiar adjointness formula. It is not difficult to show that the functor $\mathbf{Hom}_k(K, L)$ is representable by a commutative k–algebra $A(K, L)$, which will play an important role in our discussion. The affine k–scheme $\mathrm{Spec}(A(K, L))$ will be called the *homomorphism* (or *hom*) *scheme* from K to L, and its coordinate k–algebra $A(K, L)$ will be called its *representing algebra.*

We shall compute $A(K, L)$ for the cases of primary interest to us. Recall that K/k is called *modular* [9] if there exist $\alpha_1, \cdots, \alpha_s$ in K such that the k–algebra homomorphism–

$$k(\alpha_1) \otimes \cdots \otimes k(\alpha_s) \to K \tag{1.1}$$

induced by multiplication, is an isomorphism. In this situation we shall usually say, for brevity, that $K = k(\alpha_1, \cdots, \alpha_s)$ is modular over k, with the tacit understanding that (1.1) holds. If K/k is purely inseparable and α is in K, then the *exponent* of α over k is the smallest natural e such that α^{p^e} is in k. If K/k is modular and the elements $\alpha_1, \cdots, \alpha_s$ of (1.1) all have *equal* exponent e, then we shall say that K/k is *equi-exponential of exponent* e.

The following notation and remarks will be useful in our discussion. Let $\beta_1, \cdots, \beta_n$ be a basis of L/k, and e be a natural number. We can then find $n(e) < n$ such that $\beta_1^{p^e}, \cdots, \beta_{n(e)}^{p^e}$ is a basis of kL^{p^e}/k, in which case we may write, for each $i > n(e)$–

$$\beta_i^{p^e} = \sum_{\upsilon=1}^{n(e)} c_{i\upsilon}^{(e)} \beta_\upsilon^{p^e} \tag{1.2}$$

for some uniquely determined elements $\{c_{i\upsilon}^{(e)}\} \subseteq k$. A routine computation then establishes the lemma below.

Lemma 1.3. Let V be a k space and $v_1, \cdots, v_n$ be in V. Then

$$\sum_{i=1}^{n} \beta_i^{p^e} \otimes v_i = 0$$

in $K \otimes V$ if and only if the equations

$$v_\upsilon + \sum_{i=n(e)+1}^{n} c_{i\upsilon}^{(e)} v_i = 0$$

hold for all $\upsilon \leqslant n(e)$.

Proposition 1.4. Let $k \subseteq K \subseteq L$, where $K = k(\alpha_1, \cdots, \alpha_s)$ is modular over k, and $\alpha_j^{p^{e_j}} = a_j$ in k, with e_j the exponent of α_j over k. Let $\beta_1, \cdots, \beta_n$ be a basis of L/k, and for each $j \leqslant s$ let $n(j) = n(e_j)$ and $c_{i\upsilon}^{(j)} = c_{i\upsilon}^{(e_j)}$ ($\upsilon \leqslant n(j) < i$) be as in (1.2). Then $A(K, L) \approx B/I$, where $B = k[t_{ij}]$ is the k–algebra of polynomials in the indeterminates t_{ij} for $i \leqslant n, j \leqslant s$, and I is the ideal of B generated by the elements

$$t_{\upsilon j}^{p^{e_j}} + \sum_{i=n(j)+1}^{n} c_{i\upsilon}^{(j)} t_{ij}^{p^{e_j}} \quad (j \leqslant s, \upsilon \leqslant n(j)).$$

Proof. Let $A = B/I$, and x_{ij} be the image in A of t_{ij}, in which case

$$x_{\upsilon j}^{p^{e_j}} + \sum_{i=n(j)+1}^{n} c_{i\upsilon}^{(j)} x_{ij}^{p^{e_j}} = 0 \quad (j \leqslant s, \upsilon \leqslant n(j)). \tag{1.5}$$

Now define elements $\theta_{(K, L)}(\alpha_j)$ in $L \otimes A$ ($j \leqslant s$) by the formulae

$$\theta_{(K, L)}(\alpha_j) = \alpha_j \otimes 1 + \sum_{i=1}^{n} \beta_i \otimes x_{ij}. \tag{1.6}$$

It follows easily from (1.5) and Lemma 1.3 that

$$\sum_{i=1}^{n} \beta_i^{p^{e_j}} \otimes x_{ij}^{p^{e_j}} = 0$$

and therefore $\theta_{(K, L)}(\alpha_j)^{p^{e_j}} = a_j \otimes 1$. Our hypotheses on K/k then guarantee the existence of a unique k–algebra homomorphism $\theta_{(K, L)} : K \to L \otimes A$ whose value on each α_j is as in (1.6).

We next show that the pair $(A, \theta_{(K, L)})$ possesses the universal property that characterizes $A(K, L)$. Let T be a commutative k–algebra, and $\theta : K \to L \otimes T$ be a k–algebra homomorphism. Since $\beta_1, \cdots, \beta_n$ is a basis of L/k, we can find f_{ij} in T such that

$$\theta(\alpha_j) = \alpha_j \otimes 1 + \sum_{i=1}^{n} \beta_i \otimes f_{ij} \quad (j \leqslant s).$$

The equation $\alpha_j^{p^{e_j}} = a_j$ then yields easily that

$$\sum_{i=1}^{n} \beta_i^{p^{e_j}} \otimes f_{ij}^{p^{e_j}} = 0 \quad (j \leqslant s)$$

and therefore, by Lemma 1.3

$$f_{\upsilon j}^{p^{e_j}} + \sum_{i=n(j)+1}^{n} c_{i\upsilon}^{(j)} f_{i\upsilon}^{p^{e_j}} = 0 \quad (j \leqslant s, \upsilon \leqslant n(j)).$$

Therefore, by definition of the ideal I, the mapping $B \rightarrow T$ $(t_{ij} \rightarrow f_{ij})$ factors through A to yield a k–algebra homomorphism $\varphi : A \rightarrow T$ such that $\varphi(x_{ij}) = f_{ij}$. It is then clear that $(1_L \otimes \varphi)\theta_{(K, L)} = \theta$ as maps from K to $L \otimes T$. Moreover, since $\beta_1, \cdots, \beta_n$ is a basis of L/k, φ is uniquely determined by this property.

Familiar categorical considerations can now be applied to show that A represents the functor $\mathbf{Hom}_k(K, L)$, and the uniqueness of representing objects then yields that $A(K, L) \approx A = B/I$. We shall omit the details of this argument, except for the following remarks. If $\varphi : A \rightarrow T$ is a homomorphism of commutative k–algebras, then the k–algebra map $\theta : K \rightarrow L \otimes T$, corresponding to φ under the desired natural isomorphism $\mathrm{Alg}_k(A, T) \overset{\approx}{\rightarrow} \mathrm{Alg}_k(K, L \otimes T)$, is given by the formula $\theta = (1_L \otimes \varphi)\theta_{(K, L)}$. Conversely, if θ is as above, then our earlier remarks guarantee the existence of a map $\varphi : A \rightarrow T$ related to θ via the formula just given. This completes the proof.

Another way of stating the above proposition is to say that $A(K, L)$ is generated as a k–algebra by elements $\{ x_{ij} | i \leqslant n, j \leqslant s \}$ subject only to the relations (1.5). Moreover, as is implicit in the preceding discussion, the fact that $A(K, L)$ represents the functor $\mathbf{Hom}_k(K, L)$ is equivalent to the following universal property of the pair $(A(K, L), \theta_{(K, L)})$:

(1.7) Given any commutative k–algebra T and k–algebra homomorphism $\theta : K \rightarrow L \otimes T$, there is a unique k–algebra homomorphism $\varphi : A(K, L) \rightarrow T$ such that the diagram below commutes

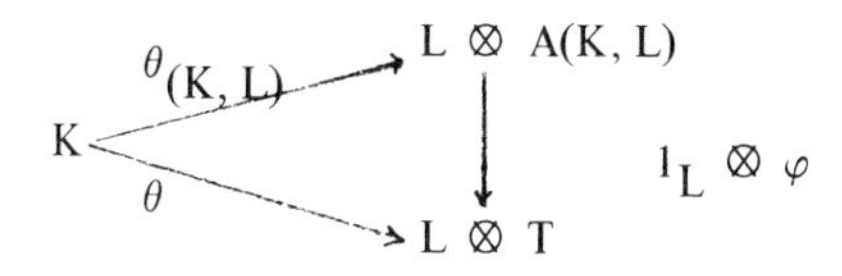

where $\theta_{(K,L)}$ is the unique k–algebra homomorphism satisfying the equations (1.6). $\varphi(x_{ij}) = f_{ij}$, where $\{ f_{ij} \mid i \leq n, j \leq s \}$ are the unique elements of T such that

$$\theta(\alpha_j) = \alpha_j \otimes 1 + \sum_{i=1}^{n} \beta_i \otimes f_{ij} \quad (j \leq s).$$

It is now an easy matter to describe the representing algebra of the automorphism scheme of a modular extension.

Proposition 1.8. Let K/k be modular. Set $A = A(K, K)$, and let $\overline{\theta}_{(K,K)} : K \otimes A \to K \otimes A$ be the A–algebra endomorphism of $K \otimes A$ defined by the formula $\overline{\theta}_{(K,K)}(\alpha \otimes u) = \theta_{(K,K)}(\alpha)(1 \otimes u)$ for α in K, u in A, with $\theta_{(K,K)} : K \to K \otimes A$ as in (1.6) for the special case in which L = K. Viewing $K \otimes A$ as a free A–module of rank [K : k], let S be the multiplicatively closed subset of A consisting of powers of the element $\delta = \det(\overline{\theta}_{(K,K)})$ of A. Then $A(K/k) \approx A_S$, the localization of A at S. If $\varphi : A(K/k) \to T$ is a homomorphism of commutative k–algebras, then the T–algebra automorphism $\overline{\theta}$ of $K \otimes T$, corresponding to φ under the natural isomorphism

$$\mathrm{Alg}_k(A(K/k), T) \overset{\approx}{\to} \mathbf{Aut}_{K/k}(T)$$

is defined by the equation $\overline{\theta}(\alpha \otimes t) = \theta(\alpha)(1 \otimes t)$ (α in K, t in T), where $\theta = (1_K \otimes \varphi)\theta_{K/k} : K \to K \otimes T$. Here $\theta_{K/k} : K \to K \otimes A(K/k)$ is the composite

$$K \xrightarrow{\theta_{(K,K)}} K \otimes A \xrightarrow{1_K \otimes \sigma} K \otimes A(K/k)$$

with $\sigma : A \to A_S \approx A(K/k)$ the map which sends u in A to u/1 in A_S.

Proof. The proposition follows easily from the universal property (1.7) of the pair $(A(K,K), \theta_{(K,K)})$ via arguments similar to those of the proof of Proposition 1.4. One uses, of course, the fact that a T–algebra endomorphism $\overline{\theta} : K \otimes T \to K \otimes T$, as above, is an automorphism if and only if $\det(\overline{\theta})$ is a unit in T. We omit the details; see also [1, Proposition 10, p. 104].

The result below, which provides a simpler form of the equations (1.5) for an important special case, can be derived easily from Proposition 1.4 (or, perhaps more directly, from a repetition of the argument used in its proof).

Proposition 1.9. Let $k \subseteq K \subseteq L$, where $K = k(\alpha_1, \cdots, \alpha_s)$ is equi–exponential, over k, of exponent e, and $L^{p^e} \subseteq k$. Let $\beta_1, \cdots, \beta_n$ be a basis of L/k, with $\beta_1 = 1$, and set $b_i = \beta_i^{p^e}$ in k for $i \leq n$. Then $A(K, L) \approx B/I$, where $B = k[t_{ij}]$ is the k–algebra of polynomials in the indeterminates t_{ij} for $i \leq n, j \leq s$, and I is the ideal of B generated by the elements

$$t_{1j}^{p^e} + \sum_{i=2}^{n} b_i t_{ij}^{p^e} \quad (j \leqslant s).$$

Hence, if x_{ij} is the image in $A(K, L)$ of t_{ij}, then $A(K, L)$ is generated as a k–algebra by the elements $\{x_{ij} \mid i \leqslant n, j \leqslant s\}$ subject only to the relations

$$x_{1j}^{p^e} + \sum_{i=2}^{n} b_i x_{ij}^{p^e} = 0 \quad (j \leqslant s). \tag{1.10}$$

Moreover, if $\theta_{(K, L)} : K \to L \otimes A(K, L)$ is the k–algebra homomorphism of (1.6), then the pair $(A(K, L), \theta_{(K, L)})$ and the elements x_{ij} satisfy (1.7).

At this point we record two observations which will be useful in the computations of the next section. The first of these is an obvious restatement of Proposition 1.9.

Lemma 1.11. Let $k \subseteq K \subseteq L$, etc., be as in Proposition 1.9, and set $K_0 = k(\alpha_1, \cdots, \alpha_{s-1}) \subseteq K$, $A_0 = A(K_0, L)$. Then $A(K, L) \approx A_0[t_1, \cdots, t_n]/(t_1^{p^e} + b_2 t_2^{p^e} + \cdots + b_n t_n^{p^e})$, where $A_0[t_1, \cdots, t_n]$ is the A_0–algebra of polynomials in the indeterminates $t_1, \cdots, t_n$.

Proposition 1.12. Let $k \subseteq K \subseteq L$, etc., be as in Lemma 1.11. Assume that $A(K, L)$ is an integral domain with quotient field F, and let $k(x)$ be the pure function field over k in the (algebraically independent) indeterminates $x = \{x_{ij} \mid 1 < i \leqslant n, j \leqslant s\}$. Then $F \approx k(x)(x_{11}, x_{12}, \cdots, x_{1s}) \subseteq L(x)$, where

$$x_{1j} = -\sum_{i=2}^{n} \beta_i x_{ij} \quad (j \leqslant s).$$

In particular, F has transcendence degree $s(n-1)$ over k. Moreover, $A(K, L) \approx k[x_{ij}] \subseteq L(x)$, the k–subalgebra of $L(x)$ generated by all x_{ij} for $i \leqslant n, j \leqslant s$.

Proof. Let K_0, A_0, etc., be as in Lemma 1.11. Since the ideal of $A_0[t_1, \cdots, t_n]$ generated by $t_1^{p^e} + b_2 t_2^{p^e} + \cdots + b_n t_n^{p^e}$ clearly has trivial intersection with $A' = A_0[t_2, \cdots, t_n]$, we see that $A(K, L)$ is an integral extension of A'. In particular, A_0 is an integral domain; moreover, [4, Theorem 48, p. 32] and an easy induction argument on s then yield that the Krull dimension of $A(K, L)$ is $s(n-1)$.

Now let $B = k[t_{ij}]$ $(i \leqslant n, j \leqslant s)$ be as in Proposition 1.9, and define a homomorphism $\psi : B \to L(x)$ by the formulae $\psi(t_{ij}) = x_{ij}$, with x and x_{ij} as above. Since $\beta_i^{p^e} = b_i$, we have easily that $\psi(I) = 0$, with I the ideal of B exhibited in Proposition 1.9. Hence ψ gives rise to a k–algebra homomorphism $\bar{\psi} : A(K, L) \to L(x)$, the image of which is clearly the subalgebra $k[x_{ij}]$ of $L(x)$ generated by all x_{ij} for $1 \leqslant i \leqslant n, j \leqslant s$. This subalgebra has Krull dimension

$s(n-1)$ by [11, VII, §7, Theorem 20, p. 193], since its quotient field $k(x_{ij})$ has transcendence degree $s(n-1)$ over k. A comparison of Krull dimensions then yields that $\bar{\psi}$ is an isomorphism, and the proposition follows easily.

In Section 2 we shall whenever permissible and convenient, identify $A(K, L)$ and F with their images in $L(x)$, as above.

We end this section with some remarks on the construction of the automorphism scheme of a finite field extension K/k as an affine group scheme. Since $A(K/k)$ represents the functor $\mathbf{Aut}_{K/k} : \mathbf{A} \to$ Groups (with $\mathbf{A}$ the category of commutative k–algebras), it follows that $A(K/k)$ is a cogroup object in the category $\mathbf{A}$. Since $\otimes = \otimes_k$ gives a coproduct in $\mathbf{A}$, this means simply that $A(K/k)$ is a Hopf k–algebra with antipode (or, in the language of [3], a "bigèbre" with involution); the structure maps are uniquely determined by the group structure on the values of the functor $\mathbf{Aut}_{K/k}$. We summarize the relevant considerations in the proposition below, the proof of which is routine and will be omitted. The reader may consult [3, II, §1, no. 1, pp. 139-147] for a general discussion of these matters.

Proposition 1.13. $A(K/k)$ is a commutative Hopf k–algebra with antipode. The diagonal map $\Delta_{K/k} : A(K/k) \to A(K/k) \otimes A(K/k)$ and augmentation $\epsilon_{K/k} : A(K/k) \to k$ satisfy the (equivalent) conditions (a) and (b) below, and are uniquely determined by either of them. In particular, the automorphism scheme $\mathrm{Spec}(A(K/k))$ is an affine group scheme over k.

(a) If T is any commutative k–algebra, then the following diagrams commute

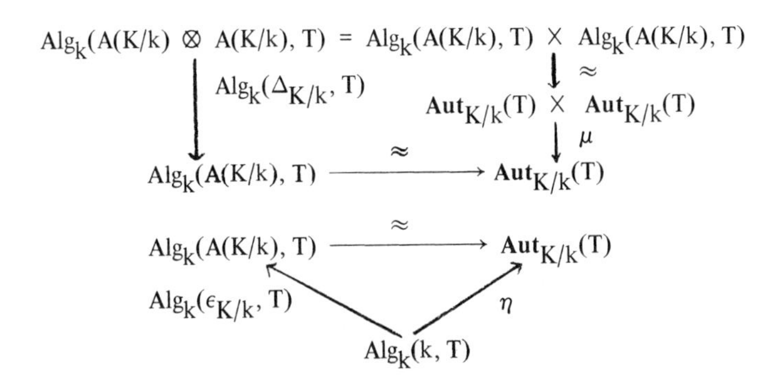

In the first diagram, μ is the multiplication map for the group $\mathbf{Aut}_{K/k}(T)$, the horizontal arrows represent the natural isomorphisms of Proposition 1.8, and the equality denotes the natural isomorphism which expresses the fact that $\otimes$ gives a coproduct in the category $\mathbf{A}$. In the second diagram, η maps the one–point set $\mathrm{Alg}_k(k, T)$ to the identity element of $\mathbf{Aut}_{K/k}(T)$.

(b) The diagrams below commute

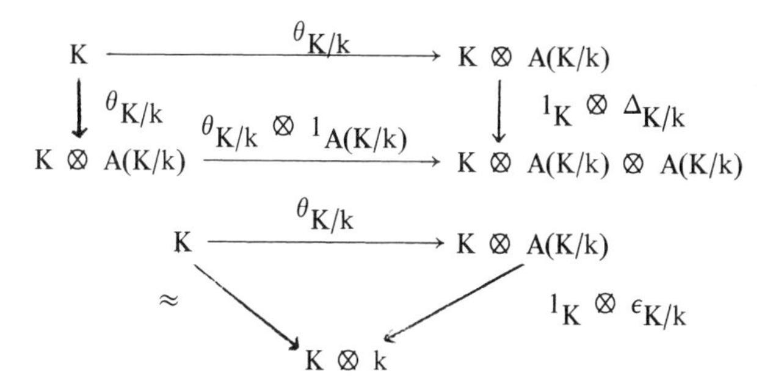

The antipode $\lambda_{K/k} : A(K/k) \to A(K/k)$ is related to the inverse operation in the group $\mathbf{Aut}_{K/k}(T)$ via a diagram analogous to those of (a) above, but we shall not reproduce it here; see [3, II, §1, 1.8(c), p. 147].

2. The equi–exponential case

We begin with an easy but important lemma.

Lemma 2.1. Let R be an integral domain with quotient field F of characteristic $p \neq 0$. Let a be in R, but not in F^p. Then, for any $e \geqslant 1$, $t^{p^e} - a$ generates a prime ideal in the polynomial ring $R[t]$.

Proof. We must show that $S = R[t]/(t^{p^e} - a)$ is an integral domain. Since a is not in F^p, we have that $F \otimes_R S \approx F[t]/(t^{p^e} - a)$ is a field. Moreover, since $t^{p^e} - a$ is monic of degree p^e, S is a free R–module of rank p^e, and since R is an integral domain, we obtain that no non–zero element of R is a zero divisor in S. Since $F \otimes_R S$ is naturally isomorphic to the localization of S at the multiplicatively closed subset consisting of all non–zero elements of R, it then follows from standard properties of localization that the map $S \to F \otimes_R S$, where $x \to 1 \otimes x$, is injective. The lemma follows.

The next several lemmas and remarks will treat the following data.

(2.2) Let $K = k(\alpha_1, \cdots, \alpha_s) \subseteq L$, and assume that

(a) K/k is equi–exponential of exponent e and $\alpha_j^{p^e} = a_j$ in k for $j \leqslant s$.

(b) L/k is equi–exponential of exponent e and $[L : k] = p^{re} = n$.

(c) $A_0 = A(K_0, L)$ is an integral domain with quotient field F_0, where $K_0 = k(\alpha_1, \cdots, \alpha_{s-1})$. In particular, by Proposition 1.12, $F_0 \approx k(x_0)(x_{11}, x_{12}, \cdots, x_{1, s-1}) \subseteq L(x_0)$, where $x_0 = \{ x_{ij} \mid 1 < i \leqslant n, j < s \}$ are algebraically inde–

pendent over k and $x_{1j} = -\Sigma_{i=2}^{n} \beta_i x_{ij}$ with $\beta_1, \cdots, \beta_n$ a basis of L/k and $\beta_1 = 1$. Also, $A_0 \approx k[x_{ij}]$, the k–subalgebra of $L(x_0)$ generated by all x_{ij} for $i \leqslant n, j < s$. We shall identify A_0 and F_0 with their images in $L(x_0)$.

(d) $L^{p^{e-1}} \not\subset F_0$.

(e) $[F_0 : k(x_0)] = p^{(s-1)e} = [K_0 : k]$.

Note that $x_{1j}^{p^e}$ is in $k(x_0)$ for $j < s$, and so, in any case, $[F_0 : k(x_0)] \leqslant p^{(s-1)e}$.

Lemma 2.3. Let $k \subseteq K \subseteq L$, etc., be as in (2.2). Then $A = A(K, L)$ is an integral domain.

Proof. As above, let $\beta_1, \cdots, \beta_n$ be a basis of L/k, with $\beta_1 = 1$. Then, by Lemma 1.11, $A \approx R[t_1]/(t_1^{p^e} - a)$, where $R = A_0[t_2, \cdots, t_n]$, the ring of polynomials, over the integral domain A_0, in the indeterminates $t_2, \cdots, t_n$, and

$$a = -\sum_{i=2}^{n} b_i t_i^{p^e}$$

an element of R, with $b_i = \beta_i^{p^e}$ in k. The quotient field of R is then the pure function field $F_0(t_2, \cdots, t_n)$.

Suppose now that a is in $F_0(t_2, \cdots, t_n)^p = F_0^p(t_2^p, \cdots, t_n^p)$. Then, since a^p is in $R^p = A_0^p[t_2^p, \cdots, t_n^p]$, a must be in the integral closure of R^p in $F_0(t_2, \cdots, t_n)^p$. But $R^p \subseteq F_0^p[t_2^p, \cdots, t_n^p]$, and since the latter is integrally closed in $F_0^p(t_2^p, \cdots, t_n^p)$ it follows that a is in $F_0^p[t_2^p, \cdots, t_n^p]$, from which we see easily that $b_2, \cdots, b_n$ are in F_0^p. But then $\beta_2^{p^{e-1}}, \cdots, \beta_n^{p^{e-1}}$ are in F_0, and so, since $\{1, \beta_2, \cdots, \beta_n\}$ is a basis of L/k, we obtain that $L^{p^{e-1}} \subseteq F_0$, contradicting (2.2d). Therefore a is not in $F_0(t_2, \cdots, t_n)^p$, and so we may apply Lemma 2.1 to conclude that $t_1^{p^e} - a$ generates a prime ideal in $R[t_1]$. Thus A is an integral domain, and the proof is complete.

Lemma 2.4. Let $k \subseteq K \subseteq L$, etc., be as in (2.2), in which case, by Lemma 2.3, $A = A(K, L)$ is an integral domain. Let $F = k(x)(x_{11}, \cdots, x_{1s})$ be the quotient field of A, with $x = \{x_{ij} | 1 < i \leqslant n, j \leqslant s\}$ and x_{ij} $(j \leqslant s)$ as in Proposition 1.12. Write $x = x_0 \cup x_s$, with x_0 as in (2.2c) and $x_s = \{x_{is} | 1 < i \leqslant n\}$. Then

(a) The set x_s is algebraically independent over F_0,

(b) $F_0(x_s) \subseteq F \subseteq L(x)$,

(c) $[F : k(x)] = p^{se}$ and $[L(x) : F] = p^{(r-s)e}$, where $[L : k] = p^{re}$ as in (2.2b).

Proof. That x_s is algebraically independent over F_0 follows from the facts that x is algebraically independent over k and F_0 is, by (2.2e), an algebraic extension of $k(x_0)$. That $F_0(x_s) \subseteq F \subseteq L(x)$ (or, more precisely, $F_0(x_s)$ can be identified in an obvious way with a subfield of F) follows from comparison of (2.2c) and Proposition 1.12. Observe, moreover, that $F = F_0(x_s)(x_{1s})$, and $x_{1s}^{p^e} = -b_2x_{2s}^{p^e} - \cdots - b_nx_{ns}^{p^e}$ is in $F_0(x_s)$, by Proposition 1.12.

Now suppose that $x_{1s}^{p^{e-1}} = -\beta_2^{p^{e-1}}x_{2s}^{p^{e-1}} - \cdots - \beta_n^{p^{e-1}}x_{ns}^{p^{e-1}}$ is in $F_0(x_s)$. Then $\beta_2^{p^{e-1}}, \cdots, \beta_n^{p^{e-1}}$ is in F_0, and hence $L^{p^{e-1}} \subseteq F_0$, since $\{1, \beta_2, \cdots, \beta_n\}$ is a basis of L/k. This contradicts (2.2d), and so we may conclude that $x_{1s}^{p^{e-1}}$ is not in F_0. It then follows that $[F : F_0(x_s)] = p^e$, and therefore

$$[F : k(x)] = [F : F_0(x_s)]\,[F_0(x_s) : k(x)] = p^e[F_0 : k(x_0)] = p^ep^{(s-1)e} = p^{se}$$

the second and third inequalities following from the identity $x = x_0 \cup x_s$ and (2.2e), respectively. But then

$$p^{re} = [L : k] = [L(x) : k(x)] = [L(x) : F]\,[F : k(x)] = [L(x) : F]\,p^{se}$$

whence $[L(x) : F] = p^{(r-s)e}$, completing the proof of the lemma.

Lemma 2.5. Let $k \subseteq K \subseteq L$ and A, F, etc., be as in Lemma 2.4. If $K \neq L$, then $L^{p^{e-1}} \not\subseteq F$.

Proof. Setting $E = k(x)$, we have from the description of F given in Lemma 2.4 that the map

$$E(x_{11}) \otimes_E \cdots \otimes_E E(x_{1s}) \to F$$

induced by the inclusions $E(x_{1j}) \subseteq F$, is surjective. Since $x_{1j}^{p^e}$ is in E for $j \leqslant s$, it then follows from Lemma 2.4(c) that the map is bijective, whence $F = E(x_{11}, \cdots, x_{1s})$ is equi-exponential over E of exponent e and, of course, degree p^{se}. Then $F \cap E^{1/p} = E(x_{11}^{p^{e-1}}, \cdots, x_{1s}^{p^{e-1}})$, and so $[F \cap E^{1/p} : E] = p^s$. Let $D = kL^{p^{e-1}} \cap F$; then $D(x) \subseteq F \cap E^{1/p}$, and so $[D : k] = [D(x) : k(x)] \leqslant [F \cap E^{1/p} : E] = p^s < p^r$, the last inequality holding because $K \neq L$. Moreover, since L/k is equi-exponential of exponent e and degree p^{re}, $[kL^{p^{e-1}} : k] = p^r$. Therefore $[D : k] < [kL^{p^{e-1}} : k]$, whence $D \neq kL^{p^{e-1}}$ and $L^{p^{e-1}} \not\subseteq F$, completing the proof.

We are now ready to prove the principal result of this section.

Theorem 2.6. Let $k \subseteq K \subseteq L$, with K/k and L/k equi–exponential of exponent e. Then

(a) $A(K, L)$ is an integral domain.

Now let F be the quotient field of $A(K, L)$.

(b) $k(x) \subseteq F \subseteq L(x)$ and $[F : k(x)] = [K : k]$, with x as in Proposition 1.12.

(c) If $K \neq L$, then $L^{p^{e-1}} \not\subseteq F$.

(d) If $K = L$, then $F = L(x)$.

Proof. Let $[L : k] = p^{re}$, and write $K = k(\alpha_1, \cdots, \alpha_s)$ as in (1.1), with $\alpha_j^{p^e} = a_j$ in k. Then $[K : k] = p^{se}$ and $s \leqslant r$. Set $K_i = k(\alpha_1, \cdots, \alpha_i)$ for $1 \leqslant i \leqslant s$.

Observe now that the conditions of (2.2) trivially hold, with k and K_1 playing the roles of K_0 and K, respectively. Then, by Lemmas 2.3-2.5, the tower $k \subseteq K_1 \subseteq L$ satisfies (a)-(c) above. Proceeding by induction on i, we may assume that $k \subseteq K_{i-1} \subseteq L$ satisfy (a)-(c) above, and therefore the conditions of (2.2) again hold with K_{i-1}, K_i playing the roles of K_0 and K_1, respectively. Then Lemmas 2.3-2.5 again apply to yield that K_i satisfies (a)-(c).

We may then conclude that $K = K_s$ satisfies (a) and (b) above, and also (c) if $K \neq L$. If, however, $K = L$, then, by (b), $[F : k(x)] = [K : k] = [L : k] = [L(x) : k(x)]$, whence $F = L(x)$. This establishes (d) and completes the proof of the theorem.

By Proposition 1.8, $A(K/k)$ is a localization of $A(K, K)$ at some multiplicatively closed subset, and therefore is likewise an integral domain with the same quotient field as $A(K, K)$. Therefore

Corollary 2.7. Let $K = k(\alpha_1, \cdots, \alpha_s)$ be equi–exponential over k of exponent e. Then $A(K/k)$ is an integral domain with quotient field $K(x)$, the field of rational functions over K in the set

$$x = \{ x_{ij} \mid 1 < i \leqslant n, j \leqslant s \} \quad \text{of} \quad s(n-1)$$

indeterminates, where $[K : k] = n = p^{se}$. In particular, the automorphism scheme $\mathrm{Spec}(A(K/k))$ of the extension K/k is irreducible and reduced.

3. An example

In this section we shall exhibit a modular extension K/k with the property that

$A(K/k)$ has non–trivial nilpotent elements.

Let $\mathbf{F}_2$ be the field of two elements, and $k = \mathbf{F}_2(a, b)$, the field of rational functions over $\mathbf{F}_2$ in the indeterminates a and b. We set $K = k(\alpha, \beta)$, with $\alpha^4 = a$ and $\beta^2 = b$. Then clearly $[K : k] = 8$, whence the multiplication map $k(\alpha) \otimes k(\beta) \to K$ is an isomorphism and K/k is modular.

We show first that $A = A(K, K)$ is not an integral domain. Since the set $\{1, \alpha, \alpha^2, \alpha^3, \beta, \alpha\beta, \alpha^2\beta, \alpha^3\beta\}$ is a basis of K/k, we may apply Proposition 1.4 and the remarks following its proof to obtain that $A = k[t_0, t_1, \cdots, t_7, u_0, u_1, \cdots, u_7]$, subject only to the relations

$$t_0^4 + at_1^4 + a^2t_2^4 + a^3t_3^4 + b^2t_4^4 + ab^2t_5^4 + a^2b^2t_6^4 + a^3b^2t_7^4 = 0$$

$$u_0^2 + au_2^2 + bu_4^2 + abu_6^2 = 0 \tag{3.1}$$

$$u_1^2 + au_3^2 + bu_5^2 + abu_7^2 = 0$$

The k–algebra homomorphism $\theta_{(K, K)} : K \to K \otimes A$ of (1.6) is then given by the formulae

$$\theta_{(K, K)}(\alpha) = \alpha \otimes 1 + 1 \otimes t_0 + \alpha \otimes t_1 + \alpha^2 \otimes t_2 + \alpha^3 \otimes t_3 +$$

$$\beta \otimes t_4 + \alpha\beta \otimes t_5 + \alpha^2\beta \otimes t_6 + \alpha^3\beta \otimes t_7 \tag{3.2}$$

$$\theta_{(K, K)}(\beta) = \beta \otimes 1 + 1 \otimes u_0 + \alpha \otimes u_1 + \alpha^2 \otimes u_2 + \alpha^3 \otimes u_3 +$$

$$\beta \otimes u_4 + \alpha\beta \otimes u_5 + \alpha^2\beta \otimes u_6 + \alpha^3\beta \otimes u_7$$

Moreover, by [1, Proposition 11, p. 104], or an easy direct argument, we see that

(3.3) A has Krull dimension 13.

Now let $B = K[x_1, \cdots, x_7, y_2, y_4, y_6]$ be the polynomial ring over K in the ten indeterminates $x_1, \cdots, x_7, y_2, y_4, y_6$, and let $B' = B[z_1, z_3, z_5, z_7]$, subject only to the relation $z_1^2 + az_3^2 + bz_5^2 + abz_7^2 = 0$. It is then easy to see that the following conditions hold

(3.4a) Krull $\dim(B') = 13$.

(3.4b) Every non–zero element of B is a non–zero divisor in B'.

(3.4c) If $w = z_1 + \alpha^2 z_3 + \beta z_5 + \alpha^2\beta z_7$ in B', then $w^2 = 0$ but $w \neq 0$.

Moreover, (3.1) yields easily the existence of a unique k–algebra homomorphism $\varphi : A \to B'$ such that $\varphi(t_i) = x_i$ for $0 \leqslant i \leqslant 7$, $\varphi(u_j) = y_j$ for $j = 0, 2, 4, 6$, and $\varphi(u_j) = z_j$ for $j = 1, 3, 5, 7$, where

$$(3.5) \quad x_0 = \alpha x_1 + \beta x_4 + \alpha\beta x_5 + \alpha^2(x_2 + \alpha x_3 + \beta x_6 + \alpha\beta x_7)$$

$$y_0 = \beta y_4 + \alpha^2(y_2 + \beta y_6)$$

Let $C' = \mathrm{Im}(\varphi) = k[x_0, x_1, \cdots, x_7, y_0, y_2, y_4, y_6, z_1, z_3, z_5, z_7] \subseteq B'$. Then $B' = C'[\alpha, \beta]$ is integral over C', and so (3.4a) and [4, Theorem 48, p. 32] guarantee that

$$(3.6) \quad \text{Krull dim } (C') = 13.$$

Now, $C = k[x_0, x_1, \cdots, x_7, y_0, y_2, y_4, y_6] \subseteq B \cap C'$ is an integral domain, since B is. Let F and E be the quotient fields of B and C, respectively; then $F_0 = k(x_1, \cdots, x_7, y_2, y_4, y_6) \subseteq E = k(x_0, x_1 \cdots, x_7, y_0, y_2, y_4, y_6) \subseteq F = K(x_1, \cdots, x_7, y_2, y_4, y_6) = F_0(\alpha, \beta)$. Moreover, $[F : F_0] = [K : k] = 8$.

By (3.5), x_0^4 is in F_0; however, since α^2 is not in k, it follows from an argument similar to that of Lemma 2.3 that x_0^2 is not in F_0. Hence $[E : F_0] \geqslant 4$, and so $[F : E] \leqslant 2$, from which we may conclude that α^2 is in E. But then, by (3.5)

$$\begin{aligned}\alpha^2 &= (y_2 + \beta y_6)^{-1}(y_0 + \beta y_4)\\ &= (y_2 + \beta y_6)^{-2}(y_2 + \beta y_6)(y_0 + \beta y_4)\\ &= (y_2^2 + b y_6^2)^{-1}[(y_2 y_0 + b y_4 y_6) + \beta(y_0 y_6 + y_2 y_4)]\end{aligned}$$

is in E. Since β is not in k, y_0 is not in F_0, and so $y_0 y_6 + y_2 y_4 \neq 0$, whence β is in E.

We can then find non–zero elements f_1, f_2, g_1, g_2 of C such that $f_1\alpha^2 = f_2$, $g_1\beta = g_2$. Since C is an integral domain and is contained in $B \cap C'$, $f_1 g_1$ is a non–zero element of $B \cap C'$, whence, by (3.4b) and (3.4c), $f_1 g_1 w$ is a non–zero element of C', with w as in (3.4c). But, again by (3.4c), $(f_1 g_1 w)^2 = 0$, and so C' possesses non–trivial nilpotent elements. But, since φ maps A surjectively onto C', and, by (3.3) and (3.6), Krull dim(A) = Krull dim(C'), it then follows easily that A cannot be an integral domain. That A has non–trivial nilpotent elements is then an immediate consequence of the general result below.

Proposition 3.7. Let K/k be a finite purely inseparable extension, and A be either $A(K, K)$ or $A(K/k)$. Then the set of nilpotent elements of A is a prime ideal of A. In particular, the automorphism scheme $\mathrm{Spec}(A(K/k))$ is irreducible.

Proof. By [1, Proposition 10, p. 104] there is a subring A_0 of A which is an integral domain such that, for any a in A, a^{p^n} is in A_0 for large n. Let a, b be elements of A such that ab is nilpotent; say $(ab)^m = 0$. Then we may select n so large that a^{p^n} and b^{p^n} are in A_0. Since $a^{mp^n}b^{mp^n} = 0$ and A_0 is an integral domain, either $a^{mp^n} = 0$ or $b^{mp^n} = 0$, and so either a or b is a nilpotent element of A. Since the set of nilpotent elements of any commutative ring is an ideal, the proposition follows.

We wish to show that, in the special case outlined above, $A(K/k)$ also has non-trivial nilpotent elements. To this end we first note a useful property of the automorphism scheme of an arbitrary modular extension.

Proposition 3.8. Let K/k be any finite purely inseparable modular extension. Then the element $\delta = \det \bar{\theta}_{(K, K)}$ of Proposition 1.8 is a non-zero divisor of $A(K, K)$, and hence the k-algebra homomorphism $\sigma : A(K, K) \to A(K/k)$ of that proposition is injective.

Proof. In the notation of Proposition 1.4, the relations (1.5) are clearly homogeneous in the variables $\{x_{ij} \mid i \leqslant n, j \leqslant s\}$, and so $A(K, K)$ is a graded k-algebra, with the monomials, in these variables, of total degree m spanning the k-subspace of homogeneous elements of degree m. Moreover, if the endomorphism $\bar{\theta}_{(K, K)} : K \otimes A(K, K) \to K \otimes A(K, K)$ is as in Proposition 1.8 (i.e., $\bar{\theta}_{(K,K)}(\gamma \otimes u) = \theta_{(K,K)}(\gamma)(1 \otimes u)$ for γ in K, u in $A(K, K)$, with $\theta_{(K, K)}$ as in (1.6)), then an explicit computation shows that $\delta = \det \bar{\theta}_{(K, K)} = 1 + z$, with z in $A(K, K)$ a sum of homogeneous elements of degree greater than zero.

Suppose now that $u \neq 0$ is in $A(K, K)$ and $\delta u = 0$. We may write $u = u_1 + u_2$, with $u_1 \neq 0$ homogeneous of degree r, say, and u_2 a sum of homogeneous elements of degree greater than r. Then $0 = u_1 + (u_2 + zu_1 + zu_2)$, and therefore $u_1 = 0$, since the term in parentheses is a sum of homogeneous elements of degree greater than r. This is a contradiction, and the proposition follows.

The result below is then an immediate consequence of Proposition 3.8 and earlier discussion.

Proposition 3.9. Let K/k be the modular extension introduced at the beginning of this section. Then $A(K/k)$ possesses non-trivial nilpotent elements, and thus the automorphism scheme of K/k is not reduced.

4. The truncated automorphism scheme

Let K/k be a finite purely inseparable modular extension. In this section we shall show that the functor $\mathbf{Aut}_{K/k}$ is representable on a certain full subcategory of the category $\mathbf{A}$ of commutative k–algebras, and shall analyze the representing object of this functor.

Definition and Remarks 4.1. A k–algebra T will be called a *truncated polynomial algebra* if it is isomorphic to $k[t_1, \cdots, t_r]/I$, where I is the ideal of the polynomial ring $k[t_1, \cdots, t_r]$ generated by $t_1^{\upsilon_1}, \cdots, t_r^{\upsilon_r}$, for some $r > 0$ and $\upsilon_1, \cdots, \upsilon_r > 0$ which depend upon T. Note that a truncated polynomial algebra is local with residue field k, and has finite k–dimension. We shall denote by $\mathbf{T}$ the full subcategory of $\mathbf{A}$ of which the objects are the truncated polynomial k–algebras.

Truncated polynomial algebras seem to be ubiquitous in the theory of purely inseparable field extensions; see, e.g., [5], where they are called "STP algebras".

Lemma 4.2. Let $k \subseteq K$ be fields, and T be a truncated polynomial k–algebra. Let $f_1, \cdots, f_n$ be in T, and $\beta_1, \cdots, \beta_n$ in K be linearly independent over k. If, for some $e \geqslant 0$

$$\beta_1^{p^e} \otimes f_1^{p^e} + \cdots + \beta_n^{p^e} \otimes f_n^{p^e} = 0 \tag{4.3}$$

in $K \otimes T$, then $f_1^{p^e} = \cdots = f_n^{p^e} = 0$.

Proof. Representing T as a residue class algebra of a polynomial algebra $k[t_1, \cdots, t_r]$, as in Definition 4.1, we see that there is a collection of monomials in $t_1, \cdots, t_r$ whose images in T form a k–basis X of T. Moreover, if x is in X and $x^s \neq 0$ (s a natural number) then x^s is likewise in X.

Now let x be in X, and c_i in k be the coefficient of x in f_i. Then $c_i^{p^e}$ is the coefficient of x^{p^e} in $f_i^{p^e}$, and

$$c_1^{p^e}\beta_1^{p^e} + \cdots + c_n^{p^e}\beta_n^{p^e}$$

is the coefficient of $1 \otimes x^{p^e}$ in the expression (4.3). Since $\{1 \otimes y | y \text{ in } X\}$ is a K–basis of $K \otimes T$, we see that if $x^{p^e} \neq 0$ then

$$(c_1\beta_1 + \cdots + c_n\beta_n)^{p^e} = c_1^{p^e}\beta_1^{p^e} + \cdots + c_n^{p^e}\beta_n^{p^e} = 0$$

whence $c_1\beta_1 + \cdots + c_n\beta_n = 0$. But then $c_1 = \cdots = c_n = 0$, since $\beta_1, \cdots, \beta_n$ are linearly independent over k.

We may conclude that, in any case, $(c_i x)^{p^e} = 0$ for all $i \leqslant n$. This holds for all x in X, whence $f_1^{p^e} = \cdots = f_n^{p^e} = 0$.

Lemma 4.4. Let K/k be a finite extension, T be a truncated polynomial k–algebra, and $\bar{\theta} : K \otimes T \to K \otimes T$ be a T–algebra endomorphism of $K \otimes T$. Then $\bar{\theta}$ is an automorphism.

Proof. We have noted that T is a local k–algebra, of finite dimension, with residue field k. Let $\pi: T \to k$ be the canonical map, and $\sigma : K \to K$ be the composite

$$K \xrightarrow{\theta} K \otimes T \xrightarrow{1_K \otimes \pi} K \otimes k = K$$

where $\theta(\alpha) = \bar{\theta}(\alpha \otimes 1)$ for α in K. σ is a k–algebra endomorphism of K, and hence is an isomorphism since K/k is finite. If $m = \mathrm{Ker}(\pi)$ is the maximal ideal of T, then $m(K \otimes T) = \mathrm{Ker}(1_K \otimes \pi)$, and we then have that $\mathrm{Im}(\theta) + m(K \otimes T) = K \otimes T$, whence $\mathrm{Im}(\bar{\theta}) + m(K \otimes T) = K \otimes T$. Since $K \otimes T$ is a finitely generated T–module and $\mathrm{Im}(\bar{\theta})$ is a T–submodule, it follows from Nakayama's Lemma that $\mathrm{Im}(\bar{\theta}) = K \otimes T$; i.e., $\bar{\theta}$ is surjective. A trivial dimension argument then yields the desired result.

We are now ready to prove the representation theorem promised at the beginning of the section. With applications to inseparable Galois theory in mind, we shall work with a somewhat more general class of functors than the automorphism functor discussed earlier. Namely, let $k \subseteq F \subseteq K$ be fields, and define a group–valued functor $\mathbf{Aut}_{K/F/k}$ on **T** by the condition that

(4.5) $\mathbf{Aut}_{K/F/k}(T)$ is the group of $F \otimes T$–algebra automorphisms of $K \otimes T$.

Of course, $\mathbf{Aut}_{K/k/k}$ is simply the functor $\mathbf{Aut}_{K/k}$ (or, more precisely, its restriction to **T**) which was defined in the introduction and discussed in Proposition 1.8.

Theorem 4.6. Let $k \subseteq F \subseteq K$ be fields, with $[K : k] = n < \infty$ and $K = F(\alpha_1, \cdots, \alpha_s)$ modular over F. Let $\alpha_j^{p^{e_j}} = a_j$ in F, with e_j the exponent of α_j over F, and let $P(K/F/k) = k[t_{ij}]$ be the truncated polynomial k–algebra with relations $t_{ij}^{p^{e_j}} = 0$, where $i \leqslant n$ and $j \leqslant s$. If $\beta_1, \cdots, \beta_n$ is a basis of K/k, then

(a) There exists a unique F–algebra homomorphism $\theta_{K/F/k} : K \to K \otimes P(K/F/k)$ such that

$$\theta_{K/F/k}(\alpha_j) = \alpha_j \otimes 1 + \sum_{i=1}^{n} \beta_i \otimes t_{ij}$$

for all $j \leqslant s$.

(b) If $\theta : K \to K \otimes T$ is any F–algebra homomorphism, with T a truncated polynomial k–algebra, then there is a unique k–algebra homomorphism $\varphi : P(K/F/k) \to T$ such that the diagram below commutes

$$\begin{array}{ccc} & \xrightarrow{\theta_{K/F/k}} & K \otimes P(K/F/k) \\ K & & \downarrow 1_K \otimes \varphi \\ & \xrightarrow{\theta} & K \otimes T \end{array}$$

$\varphi(t_{ij}) = f_{ij}$, where $\{ f_{ij} | i \leqslant n, j \leqslant s \}$ are the unique elements of T satisfying the equations

$$\theta(\alpha_j) = \alpha_j \otimes 1 + \sum_{i=1}^{n} \beta_i \otimes f_{ij}$$

for all $j \leqslant s$.

(c) There exist isomorphisms

$$\mathrm{Alg}_k(P(K/F/k), T) \xrightarrow{\approx} \mathbf{Aut}_{K/F/k}(T)$$

for T a truncated polynomial k–algebra, which are natural in T and satisfy the following conditions. If $\varphi : P(K/F/k) \to T$ is a k–algebra homomorphism, then the corresponding $F \otimes T$–algebra automorphism $\overline{\theta}$ of $K \otimes T$ is defined by the equation $\overline{\theta}(\alpha \otimes f) = \theta(\alpha)(1 \otimes f)$, with α in K, f in T, and $\theta = (1_K \otimes \varphi)\theta_{K/F/k} : K \to K \otimes T$. Conversely, if $\overline{\theta}$ is in $\mathbf{Aut}_{K/F/k}(T)$, then the corresponding k–algebra homomorphism $\varphi : P(K/F/k) \to T$ is obtained via (b) from the map $\theta : K \to K \otimes T$, where $\theta(\alpha) = \overline{\theta}(\alpha \otimes 1)$ for α in K.

(d) $[P(K/F/k) : k] = [K : F]^{[K : k]}$.

Proof. Assertions (a) and (b) follow from an argument entirely similar to that of the proof of Porposition 1.4, with Lemma 4.2 being used in place of Lemma 1.3. (c) is an easy consequence of (b) and Lemma 4.4, via reasoning analogous to that of the last part of the proof of Proposition 1.4; we omit the details. (d) is obtained from a trivial counting argument.

Definition and Remarks 4.7. Let $K = k(\alpha_1, \cdots, \alpha_s)$ be finite, purely inseparable, and modular over k. We shall write $P(K/k) = P(K/k/k)$ and $\theta_{K/k} = \theta_{K/k/k}$: $K \to K \otimes P(K/k)$ (note the possibility of confusion with the map $\theta_{K/k} : K \to K \otimes A(K/k)$ of Proposition 1.8). If $P(K/k) = k[t_{ij}]$ is as in Theorem 4.5 (with $F = k$), then (1.7) yields immediately the existence of a unique k–algebra homomorphism $A(K, K) \to P(K/k)$ such that $x_{ij} \to t_{ij}$. By Lemma 4.4 and Proposition 1.8, this map factors through $A(K/k)$; i.e., we obtain a commutative diagram

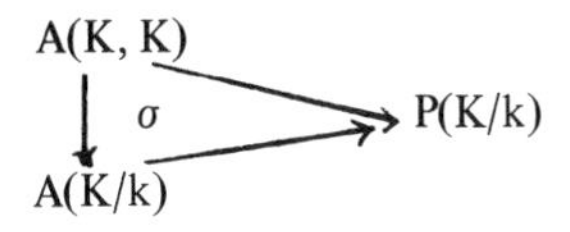

with σ as in Proposition 1.8. We shall call the two unlabeled arrows the *canonical maps*. Both are surjective, since $P(K/k) = k[t_{ij}]$.

We turn now to the additional structure on $P(K/F/k)$ arising from the fact that it represents a group-valued functor.

Proposition 4.8. Let $k \subseteq F \subseteq K$ be fields, with K/k finite and K/F purely inseparable and modular.

(a) $P(K/F/k)$ is a commutative Hopf k-algebra with antipode. The diagonal map $\Delta_{K/F/k} : P(K/F/k) \to P(K/F/k) \otimes P(K/F/k)$ and augmentation $\epsilon_{K/F/k} : P(K/F/k) \to k$ satisfy the two (equivalent) conditions which are the analogues of (a) and (b) of Proposition 1.13, and are uniquely determined by either of these conditions. In particular, $\mathrm{Spec}(P(K/F/k))$ is a finite group scheme over k ("finite" meaning simply that it is an affine group scheme represented by a k-algebra of finite dimension).

(b) If K/k is also purely inseparable and modular, then the canonical map $A(K/k) \to P(K/k)$ is a surjection of Hopf k-algebras.

Proof. (a) follows immediately from the same considerations as are involved in the proof of Proposition 1.13, but applied to the category **T** rather than to the full category **A** of commutative k-algebras; one need only note that **T** is closed under $\otimes$, which then yields a coproduct in **T**. (b) is an easy consequence of Proposition 1.13(b) (and its analogue for $P(K/k)$), and the universal property of $A(K/k)$.

Definition and Remarks 4.9. Let K/k be finite, purely inseparable, and modular. We shall write $\Delta_{K/k} = \Delta_{K/k/k}$, $\epsilon_{K/k} = \epsilon_{K/k/k}$, etc., for the structure maps of $P(K/k)$ (these are, of course, the same symbols used to denote the structure maps of $A(K/k)$; however, in practice no confusion will arise). The finite group scheme $\mathrm{Spec}(P(K/k))$ will be called the *truncated automorphism scheme* of the extension K/k; by Theorem 4.6(d), it is a finite group scheme of rank $[K:k]^{[K:k]}$ over k. Finally, by Proposition 4.8(b), the canonical surjection $A(K/k) \to P(K/k)$ induces a closed immersion of group schemes $\mathrm{Spec}(P(K/k)) \to \mathrm{Spec}(A(K/k))$, by means of which we may identify $\mathrm{Spec}(P(K/k))$ with a closed subgroup scheme of $\mathrm{Spec}(A(K/k))$.

Next we explore the connections between the various group schemes introduced above which arise from a tower of fields.

Proposition 4.10. Let $k \subseteq E \subseteq F \subseteq K$ be fields, with K/k finite and K/F, K/E purely inseparable and modular. Then there exists a unique k–algebra homomorphism $\varphi_{F/E} : P(K/E/k) \to P(K/F/k)$ such that $\theta_{K/F/k} = (1_K \otimes \varphi_{F/E})\theta_{K/E/k}$. $\varphi_{F/E}$ is a surjection of Hopf k–algebras. Moreover, the following statements are equivalent

(a) $F = E$.

(b) $\theta_{K/E/k} : K \to K \otimes P(K/E/k)$ is an F–algebra homomorphism.

(c) $\varphi_{F/E}$ is an isomorphism.

Proof. Since $E \subseteq F$, $\theta_{K/F/k} : K \to K \otimes P(K/F/k)$ is an E–algebra homomorphism. The existence of a unique k–algebra map $\varphi_{F/E}$, with $\theta_{K/F/k} = (1_K \otimes \varphi_{F/E})\theta_{K/E/k}$, then follows from Theorem 4.6(b).

Now let $K = F(\alpha_1, \cdots, \alpha_s)$ and $P(K/F/k) = k[t_{ij}]$ $(i \leqslant n = [K : k], j \leqslant s)$ as in Theorem 4.6, with $\beta_1, \cdots, \beta_n$ a basis of K/k. Then, since $\beta_1 \otimes 1, \cdots, \beta_n \otimes 1$ is a basis of the free $P(K/E/k)$–module $K \otimes P(K/E/k)$, we may write

$$\theta_{K/E/k}(\alpha_j) = \alpha_j \otimes 1 + \sum_{i=1}^{n} \beta_i \otimes u_{ij}$$

for certain u_{ij} in $P(K/E/k)$ $(1 \leqslant n, j \leqslant s)$. Since

$$\theta_{K/F/k}(\alpha_j) = \alpha_j \otimes 1 + \sum_{i=1}^{n} \beta_i \otimes t_{ij}$$

the equation $\theta_{K/F/k} = (1_K \otimes \varphi_{F/E})\theta_{K/E/k}$ then yields immediately that $\varphi_{F/E}(u_{ij}) = t_{ij}$ for all i, j, whence $\varphi_{F/E}$ is surjective. That $\varphi_{F/E}$ is a Hopf k–algebra map follows easily from the universal property of $P(K/E/k)$, via Proposition 4.8(a) and a standard argument.

We turn now to the equivalence of conditions (a)-(c) above. If (a) holds,then the definition of $\theta_{K/E/k}$ guarantees that (b) holds also. If (b) is true, then Theorem 4.6(b) provides a unique k–algebra map $\varphi : P(K/F/k) \to P(K/E/k)$ such that $\theta_{K/E/k} = (1_K \otimes \varphi)\theta_{K/F/k}$. The uniqueness part of Theorem 4.6(b) then guarantees easily that $\varphi \circ \varphi_{F/E} = 1_{P(K/E/k)} = \varphi_{F/E} \circ \varphi$, whence (c) holds. But if (c) is true, then $[K : E] = [K : F]$ by Theorem 4.6(d), and so (a) is likewise true. This completes the proof of the proposition.

Remark 4.11. It perhaps goes without saying that, if T is a truncated polynomial k–algebra, then the map $\mathbf{Aut}_{K/F/k}(T) \to \mathbf{Aut}_{K/E/k}(T)$, induced by the homomorphism $\varphi_{F/E}$ just constructed, is simply the natural injection (i.e., view $F \otimes T$–algebra automorphisms as $E \otimes T$–algebra automorphisms). This observation provides an alternate method of establishing the properties of $\varphi_{F/E}$ noted above.

Corollary 4.12. Let $k \subseteq F_3 \subseteq F_2 \subseteq F_1 \subseteq K$ be fields, with K/k finite and K/F_i purely inseparable and modular for $i \leqslant 3$. Then $\varphi_{F_1/F_3} = \varphi_{F_1/F_2} \circ \varphi_{F_2/F_3}$.

Proof. This is an immediate consequence of the uniqueness of φ_{F_1/F_3} noted in Proposition 4.10.

We turn now to a weak Galois correspondence for a finite purely inseparable modular extension K/k; it has the same form as that of [6], but uses the truncated automorphism scheme rather than the full automorphism scheme. The analogy with classical Galois theory will become more apparent if we employ the language of group schemes rather than that of Hopf algebras. If $k \subseteq F \subseteq K$ with K/k finite and K/F purely inseparable and modular, we shall write $G_t(K/F/k) = \mathrm{Spec}(P(K/F/k))$. If $k \subseteq E \subseteq F \subseteq K$, with K/E also purely inseparable and modular, then the Hopf algebra surjection $\varphi_{F/E} : P(K/E/k) \to P(K/F/k)$ of Proposition 4.10 yields a closed immersion $\mathrm{Spec}(\varphi_{F/E}) : G_t(K/F/k) \to G_t(K/E/k)$, and we shall identify $G(K/F/k)$ with the corresponding closed subgroup scheme of $G_t(K/E/k)$ (Corollary 4.12 guarantees that all such identifications are compatible with one another.)

Now let $k \subseteq F \subseteq K$ be as above, and let $X = \mathrm{Spec}(A)$ be any closed affine subscheme of $G_t(K/F/k)$; the closed immersion $X \rightarrowtail G_t(K/F/k)$ corresponds to a surjection $\varphi : P(K/F/k) \to A$ of k–algebras, which in turn induces an injection

$$X(T) = \mathrm{Alg}_k(A, T) \to \mathrm{Alg}_k(P(K/F/k), T) = \mathbf{Aut}_{K/F/k}(T)$$

for any commutative k–algebra T (the right–most equality denoting the natural isomorphism of Theorem 4.6(c)). We shall identify $X(T)$ with its image in $\mathbf{Aut}_{K/F/k}(T)$. $X = \mathrm{Spec}(A)$ will be called a *truncated* scheme if A is a truncated polynomial k–algebra.

Lemma 4.13. Let $k \subseteq F \subseteq K$ and $X = \mathrm{Spec}(A)$ be as in the preceding discussion, with A a truncated polynomial k–algebra. Then the following are equivalent for any α in K

(a) Given any truncated polynomial k–algebra T and σ in $X(T) \subseteq \mathbf{Aut}_{K/F/k}(T)$, $\sigma(\alpha) = \alpha$ in $K \subseteq K \otimes T$.

(b) $\theta(\alpha) = \alpha \otimes 1$, where $\theta = (1_K \otimes \varphi)\theta_{K/F/k} : K \to K \otimes A$.

The proof is a routine consequence of the universal property of $\theta_{K/F/k}$ as noted in Theorem 4.6(b), and will be omitted.

Definition and Remarks 4.14. Let $k \subseteq F \subseteq K$ be fields, with K/k finite and K/F purely inseparable and modular, and $X = \mathrm{Spec}(A)$ be a closed truncated subscheme of $G_t(K/F/k)$. We let K^X be the set of all elements α in K satisfying the conditions of Lemma 4.13. Lemma 4.13(a) guarantees immediately that K^X is a subfield of K; it will be called the *fixed field* of X. One checks easily that K^X is independent of F in the following sense: If $k \subseteq E \subseteq F \subseteq K$ with K/E also purely inseparable and modular, then $X \subseteq G_t(K/F/k) \subseteq G_t(K/E/k)$, and we obtain the same fixed field K^X whether we view X as a subscheme of $G_t(K/F/k)$ or of $G_t(K/E/k)$.

Theorem 4.15. Let $k \subseteq F \subseteq K$ be fields, with K/k finite, and K/F purely inseparable and modular.

(a) $K^{G_t(K/F/k)} = F$.

(b) If X is a closed truncated subscheme of $G_t(K/F/k)$, then K/K^X is modular.

(c) Let $k \subseteq E \subseteq F$, with K/E also purely inseparable and modular, and Y be a closed truncated subscheme of $G_t(K/E/k)$. Then $F \subseteq K^Y$ if and only if $Y \subseteq G_t(K/F/k)$. In particular, $G_t(K/F/k)$ contains every closed truncated subscheme of $G_t(K/E/k)$ with fixed field F.

Proof. (c): By Lemma 4.13, $K^Y \subseteq G_t(K/F/k)$ if and only if, for any truncated polynomial k–algebra T, $Y(T) \subseteq \mathbf{Aut}_{K/F/k}(T)$ ($\subseteq \mathbf{Aut}_{K/E/k}(T)$). Passing from these functors to the corresponding schemes, we see that this condition holds if and only if the inclusion $Y \to G_t(K/E/k)$ factors through $G_t(K/F/k)$; i.e., $Y \subseteq G_t(K/F/k)$.

(b): If $X = \mathrm{Spec}(A)$, then, by Lemma 4.13(b), $K^X = \{\alpha$ in $K \mid \theta(\alpha) = \alpha \otimes 1\}$, where $\theta = (1_K \otimes \varphi)\theta_{K/F/k} : K \to K \otimes A$ and $\varphi : P(K/F/k) \to A$ is the k–algebra surjection corresponding to the inclusion $X \to G_t(K/F/k)$. Since A is a truncated polynomial k–algebra, we may apply an important theorem of M. E. Sweedler [9, Theorem 1, p. 403], in a form given by R. Rasala [5, Chapter 1, §5, Theorem 4], to conclude that K/K^X is modular.

(a): Let $F_1 = K^{G_t(K/F/k)}$; then K/F_1 is modular, by (b). Moreover, by Lemma 4.13(b) and Proposition 4.10, $F_1 = \{\alpha$ in $K \mid \theta_{K/F/k}(\alpha) = \alpha \otimes 1\}$, with $\theta_{K/F/k} : K \to K \otimes P(K/F/k)$ as in Theorem 4.6(b). In particular, $\theta_{K/F/k}$ is an F_1–algebra homomorphism. We may then apply Proposition 4.10(b) to conclude that $F_1 = F$, completing the proof.

In particular, if K/k is finite, purely inseparable, and modular,then Theorem 4.15 yields a many–to–one correspondence between the truncated subgroup schemes G of $G_t(K/k)$ and the subfields F of K containing k for which K/F is modular, the subfield corresponding to G being K^G. In general, if F is as above, then there are infinitely many truncated subgroup schemes of $G_t(K/k)$ whose fixed field is F;

however, by Theorem 4.15(b), $G_t(K/F/k)$ is the unique maximal such. There remains the problem of providing an intrinsic characterization of those truncated subgroup schemes of $G_t(K/k)$ of the form $G_t(K/F/k)$ for some F. A solution to this problem is announced in [2] (albeit in the language of cocommutative Hopf algebras), and we shall say no more about it here.

One technical point needs to be clarified. Namely, let $k \subseteq F \subseteq K$ be fields, with K/k finite and K/F purely inseparable and modular. What relation, if any, is there between the group schemes $G_t(K/F/k)$ and $G_t(K/F)$ $(=G_t(K/F/F))$? We treat the equivalent question regarding the relation between the Hopf k–algebra $P(K/F/k)$ and the Hopf F–algebra $P(K/F)$. The connection is made via the so–called restriction functor on truncated polynomial algebras, which we now introduce.

Definition 4.16. Let F/k be a finite field extension, and A be a truncated polynomial F–algebra. We define the truncated polynomial k–algebra $\rho_{F/k}(A)$ by the condition that it represent the set–valued functor

$$T \rightsquigarrow \mathrm{Alg}_F(A, F \otimes T)$$

on the category $\mathbf{T}$ of truncated polynomial k–algebras. That is, there exists an isomorphism

$$\mathrm{Alg}_k(\rho_{F/k}(A), T) \overset{\approx}{\to} \mathrm{Alg}_F(A, F \otimes T)$$

for each truncated polynomial k–algebra T, and these isomorphisms are natural in T. $\rho_{F/k}(A)$ will be called the *restriction of A to k.*

Remarks 4.17. (a) Note that, if $\rho_{F/k}(A)$ exists, then it is unquely determined , up to k–algebra isomorphism, by the condition that it represent the functor introduced above.

(b) It is not difficult to show, using the techniques of the proof of Theorem 4.6, that the above functor is always representable, and hence $\rho_{F/k}(A)$ exists for any truncated polynomial F–algebra A. In fact, we obtain a functor $\rho_{F/k} : \mathbf{T}_F \to \mathbf{T}$ (with $\mathbf{T}_F$ the category of truncated polynomial F–algebras) which is left adjoint to the base extension functor $F \otimes (-) : \mathbf{T} \to \mathbf{T}_F$ However, we omit the proofs of these results, as they will not be needed in our work.

Lemma 4.18. Let $k \subseteq E \subseteq F$ be fields, with F/k finite. If A is a truncated polynomial F–algebra, and $\rho_{F/k}(A)$ and $\rho_{F/E}(A)$ exist, then $\rho_{E/k}(\rho_{F/E}(A))$ likewise exists, and is isomorphic to $\rho_{F/k}(A)$ as k–algebras.

Proof. In view of Definition 4.16, we need only show that $\rho_{F/k}(A)$ represents the functor $\mathrm{Alg}_E(\rho_{F/E}(A), E \otimes (-)) : \mathbf{T} \to \mathrm{Sets}$. But, if T is a truncated polynomial k–algebra, Definition 4.16 yields isomorphisms

$$\mathrm{Alg}_E(\rho_{F/E}(A), E \otimes T) \approx \mathrm{Alg}_F(A, F \otimes_E (E \otimes T))$$

$$\approx \mathrm{Alg}_F(A, F \otimes T) \approx \mathrm{Alg}_k(\rho_{F/k}(A), T)$$

which are natural in T, and we are done.

Theorem 4.19. Let $k \subseteq F \subseteq K$ be fields, with K/k finite and K/F purely inseparable and modular. Then $\rho_{K/k}(K \otimes_F K)$ and $\rho_{F/k}(P(K/F))$ exist, and both are isomorphic to $P(K/F/k)$ as k–algebras.

Proof. If T is a truncated polynomial k–algebra, we have isomorphisms

$$\mathrm{Alg}_k(P(K/F/k), T) \approx \mathbf{Aut}_{K/F/k}(T) \approx \mathrm{Alg}_{F \otimes T}(K \otimes T, K \otimes T)$$

$$\approx \mathrm{Alg}_F(K, K \otimes T) \approx \mathrm{Alg}_K(K \otimes_F K, K \otimes T)$$

which are natural in T. Here the first isomorphism is the definition of $P(K/F/k)$, the second is Lemma 4.4, and the others are familiar adjointness formulae. It follows that $P(K/F/k)$ represents the functor $\mathrm{Alg}_K(K \otimes_F K, K \otimes (-)) : \mathbf{T} \to \mathrm{Sets}$, whence, by Definition 4.16, $\rho_{K/k}(K \otimes_F K)$ exists and is isomorphic to $P(K/F/k)$ as k–algebras.

A similar argument yields that $\rho_{K/F}(K \otimes_F K)$ exists and is isomorphic to $P(K/F)$ as F–algebras, whence, by Lemma 4.18, $\rho_{F/k}(P(K/F))$ exists and is isomorphic to $\rho_{K/k}(K \otimes_F K) \approx P(K/F/k)$ as k–algebras. This completes the proof of the theorem.

As remarked in (4.17(b)), the restriction functor $\rho_{F/k} : \mathbf{T}_F \to \mathbf{T}$ is a left adjoint, hence preserves coproducts, and therefore carries Hopf F–algebras to Hopf k–algebras. In particular, $\rho_{F/k}(P(K/F))$ is a Hopf k–algebra. It can then be shown without difficulty that the isomorphism $P(K/F/k) \approx \rho_{F/k}(P(K/F))$ established above is an isomorphism of Hopf k–algebras, but we shall not do this here.

5. The Coset space by the truncated Aut scheme

Throughout this section we shall deal with a fixed purely inseparable modular extension K/k. We shall denote the automorphism scheme and truncated automorphism scheme of K/k by G and G_t, respectively; i.e., $G = \mathrm{Spec}(A(K/k))$ and $G_t = \mathrm{Spec}(P(K/k))$, **with** $A(K/k)$ and $P(K/k)$ the Hopf k–algebras introduced in Sections 1 and 4. As noted in (4.9), the canonical map $A(K/k) \to P(K/k)$ of

Proposition 4.8 yields a closed immersion $G_t \to G$, by means of which we may (and shall) identify G_t with a closed subgroup scheme of G.

In order to clarify the relation between the group schemes G and G_t, we shall analyze the left coset space (or quotient scheme) G/G_t. Our principal result is, in essence, that G/G_t is an open subscheme of affine m–space for suitable m.

There exist modular extensions K/k for which G_t is not a normal subgroup scheme of G. However, if K/k is equi–exponential, then G_t is normal in G; in fact, G_t is then easily seen to be the kernel of the e'th power of the Frobenius map of G, where e is the exponent of K/k. But we shall not prove these assertions here.

We first introduce the data and notation with which we shall work throughout this section, and at the same time recall certain facts established earlier.

(5.1). (a) $K = k(\alpha_1, \cdots, \alpha_s)$ is finite, purely inseparable, and modular over k, where $\alpha_j^{p^{e_j}} = a_j$ in k and e_j is the exponent of α_j over k.

(b) $\beta_1, \cdots, \beta_n$ is a basis of K/k, where $n = [K:k] = p^{e_1 + \cdots + e_s}$. Then, given $j \leqslant s$, there exists a unique $n(j) < n$ such that $\beta_1, \cdots, \beta_{n(j)}$ is a basis of $kK^{p^{e_j}}/k$, in which case, for each $i > n(j)$

$$\beta_i^{p^{e_j}} = \sum_{\upsilon=1}^{n(j)} c_{i\upsilon}^{(j)} \beta_\upsilon^{p^{e_j}}$$

for uniquely determined elements $\{ c_{i\upsilon}^{(j)} \} \subseteq k$ (see (1.2)).

(c) $A(K, K) = k[x_{ij}]$ $(i \leqslant n, j \leqslant s)$, subject only to the relations

$$x_{\upsilon j}^{p^{e_j}} + \sum_{i=n(j)+1}^{n} c_{i\upsilon}^{(j)} x_{ij}^{p^{e_j}} = 0$$

for all $j \leqslant s$ and $\upsilon \leqslant n(j)$. The k–algebra homomorphism $\theta_{(K,K)} : K \to K \otimes A(K, K)$ satisfies the equations

$$\theta_{(K,K)}(\alpha_j) = \alpha_j \otimes 1 + \sum_{i=1}^{n} \beta_i \otimes x_{ij} \quad (j \leqslant s)$$

(see Proposition 1.4 and (1.5)).

(d) $P(K/k) = k[t_{ij}]$ $(i \leqslant n, j \leqslant s)$ with relations $t_{ij}^{p^{e_j}} = 0$. The k–algebra homomorphism $\theta_{K/k} : K \to K \otimes P(K/k)$ satisfies the equations

$$\theta_{K/k}(\alpha_j) = \alpha_j \otimes 1 + \sum_{i=1}^{n} \beta_i \otimes t_{ij}$$

(see Theorem 4.6).

(e) $A(K/k) \approx A(K, K)_S$, the localization of $A(K, K)$ at the multi–

plicatively closed subset S consisting of powers of $\delta = \det(\bar{\theta}_{(K,K)})$, where $\bar{\theta}_{(K,K)}$ is the $A(K,K)$–algebra endomorphism of $K \otimes A(K,K)$ described in Proposition 1.8. Viewing this isomorphism as an identification, we then have the commutative diagram below

$$\begin{array}{ccc} A(K,K) & & \\ \downarrow \sigma & \searrow & P(K/k) \\ A(K/k) & \nearrow & \end{array}$$

where $\sigma(u) = u/1$ and the unlabeled arrows denote the canonical maps of (4.7), the map $A(K,K) \to P(K/k)$ sending x_{ij} to t_{ij}. σ is injective by Lemma 3.8, and we shall identify $A(K,K)$ with its image in $A(K/k)$. Finally, the k-algebra homomorphism $\theta_{K/k}: K \to K \otimes A(K/k)$ is determined by the requirement that the diagram below commute

$$\begin{array}{ccc} & \overset{\theta_{(K,K)}}{\nearrow} & K \otimes A(K,K) \\ K & & \downarrow 1 \otimes \sigma \\ & \underset{\theta_{K/k}}{\searrow} & K \otimes A(K/k) \end{array}$$

(see Proposition 1.8).

In order to describe the coset space G/G_t, we must scrutinize carefully the subalgebras of $A(K,K)$ and $A(K/k)$ defined below.

Definition 5.2. Given K/k, $A(K,K) = k[x_{ij}]$, etc., as in (5.1). Set $z_{ij} = x_{ij}^{p^{e_j}}$, and let $B(K,K) = k[z_{ij}]$, the k–subalgebra of $A(K,K)$ generated by all z_{ij} for $i \leqslant n, j \leqslant s$. Define, moreover, $B(K/k) = B(K,K)[\delta^{-p^e}]$, a k–subalgebra of $A(K/k)$, where δ is as in (5.1(e)) and e is the exponent of K/k.

We shall need to establish that $A(K/k)$ is a faithfully flat $B(K/k)$–algebra. To this end, we first record a trivial observation on extensions of rings.

Lemma 5.3. Let R be a commutative ring, and $f_1(t), \cdots, f_r(t)$ be monic polynomials in $R[t]$. Let $S = R[t_1, \cdots, t_r]/J$, where J is the ideal of $R[t_1, \cdots, t_r]$ generated by $f_1(t_1), \cdots, f_r(t_r)$. Then S is a finitely generated free R–module.

Proof. The assertion is proved by a routine induction argument on r, the step $r = 1$ being obvious. We omit the details.

The lemma below establishes the properties of $B(K/k)$ which we shall need in the proof of our main theorem.

Lemma 5.4. Given K/k, etc., as in (5.1), and $B(K,K)$, $B(K/k)$ as in Definition 5.2.

(a) Let $k[Y]$ be the k–algebra of polynomials in the indeterminates $Y = \{y_{ij} \mid j \leq s, n(j) < i \leq n\}$. Then the map $\varphi : k[Y] \to B(K,K)$, where $\varphi(y_{ij}) = z_{ij}$, is a k–algebra isomorphism, and induces an isomorphism of $B(K/k)$ with the localization of $k[Y]$ at a multiplicatively closed subset consisting of powers of a single polynomial in $k[Y]$.

(b) $A(K/k)$ is a finitely generated free $B(K/k)$–module.

(c) The diagram below is a coequalizer diagram of commutative k–algebras

$$B(K/k) \underset{\epsilon}{\overset{i}{\rightrightarrows}} A(K/k) \to P(K/k)$$

where the unlabeled arrow denotes the canonical map, i is the inclusion, and ϵ is the composite

$$B(K/k) \xrightarrow{i} A(K/k) \xrightarrow{\epsilon_{K/k}} k \to A(K/k)$$

(d) $\Delta_{K/k}(B(K/k)) \subseteq A(K/k) \otimes B(K/k)$, where $\Delta_{K/k} : A(K/k) \to A(K/k) \otimes A(K/k)$ is the diagonal map.

Proof. Let $X = \{x_{ij} \mid j = 1, \cdots, s; n(j) < i \leq n\} \subseteq A(K,K)$. Inspection of the relations (5.1(c)) yields easily that the elements of X are algebraically independent over k, and hence the subalgebra $k[X] \subseteq A(K\ K)$ may be viewed as a polynomial k–algebra with these elements as indeterminates. Now let $Z = \{z_{ij} \mid j = 1, \cdots, s; n(j) < i \leq n\} \subseteq k[X]$. Since, by (5.1(c))

$$z_{\upsilon j} + \sum_{i=n(j)+1}^{n} c_{i\upsilon}^{(j)} z_{ij} = 0$$

for all $j \leq s$ and $\upsilon \leq n(j)$, we see that $B(K,K) = k[Z] \subseteq k[X]$. That φ is an isomorphism then follows immediately from the structure of $k[X]$, as noted above, and the fact that $z_{ij} = x_{ij}^{p^{e_j}}$. Moreover, we may apply Lemma 5.3 with $k[X]$ and $A(K,K)$ playing the roles of R and S, respectively, and r equal to the number of relations in (5.1(c)) to obtain that $A(K\ K)$ is a finitely generated free $k[X]$–module. A similar argument shows that $k[X]$ is a finitely generated free $B(K,K)$–module, whence $A(K,K)$ is likewise a finitely generated free $B(K,K)$–module.

Observe now that $A(K,K)^{p^e} \subseteq B(K,K)$, where e is the exponent of K/k. Hence, if δ is as in (5.1(e)), then the powers of δ^{p^e} constitute a multiplicatively closed subset Σ of $B(K,K)$. Standard properties of localization then yield that the inclusion maps $A(K, K) \to A(K/k)$ and $B(K K) \to B(K/k)$ induce isomorphisms $A(K, K)_\Sigma \overset{\approx}{\to} A(K/k)$ and $B(K, K)_\Sigma \overset{\approx}{\to} B(K/k)$. This, together with the preceding discussion, establishes (b), and (a) follows from the k–algebra isomorphisms $k[Y]_{\varphi^{-1}(\Sigma)} \approx B(K, K)_\Sigma \approx B(K/k)$, the first being induced by φ.

In order to prove (c), we must show that the kernel I of the canonical map $A(K/k) \to P(K/k)$ is identical with the ideal J of $A(K/k)$ generated by all elements of the form $b-\epsilon(b)$, with b in $B(K/k)$. Now, ϵ is determined by the requirement that $\epsilon(z_{ij}) = 0$ for all $i \leqslant n, j \leqslant s$. Since, by (5.1(e)), each z_{ij} is mapped to zero by the canonical map, it follows that $b-\epsilon(b)$ is in I for all b in $B(K, K)$. Moreover, since $P(K/k)^{p^e} \subseteq k$ by (5.1(d)), $\delta^{-p^e}-\epsilon(\delta^{-p^e})$ is also in I, and so $b-\epsilon(b)$ is in I for all b in $B(K/k)$; i.e., $J \subseteq I$.

Viewing the isomorphisms $A(K, K)_\Sigma \overset{\approx}{\to} A(K/k)$ and $B(K, K)_\Sigma \overset{\approx}{\to} B(K/k)$ as identifications, we have that $I = I_0 A(K/k)$, where $I_0 = I \cap A(K, K)$. Hence, in order to prove that $I \subseteq J$ (and therefore $I = J$), it is sufficient to show that $I_0 \subseteq J_0$ where J_0 is the ideal of $A(K, K)$ generated by all $b-\epsilon(b)$ with B in $B(K, K)$. But it follows immediately from the structure of the truncated polynomial k–algebra $P(K/k)$ (as reviewed in (5.1(d))) that even $I_0 = J_0$ and the proof of (c) is complete.

We turn finally to (d). As before, we identify $A(K, K)$ with its image in $A(K/k)$ under the injection $\sigma : A(K, K) \to A(K/k)$; then (5.1(c)) and the commutativity of the last diagram of (5.1(e)) guarantee that $\theta_{K/k} : K \to K \otimes A(K/k)$ satisfies the equations

$$\theta_{K/k}(\alpha_j) = \alpha_j \otimes 1 + \sum_{i=1}^{n} \beta_i \otimes x_{ij} \quad (j \leqslant s) .$$

Moreover, since $\beta_1 \otimes 1, \cdots, \beta_n \otimes 1$ is a basis of the free $A(K/k)$–module $K \otimes A(K/k)$, we may write

$$\theta_{K/k}(\beta_i) = \beta_i \otimes 1 + \sum_{\mu=1}^{n} \beta_\mu \otimes w_{\mu i}$$

for some $w_{\mu i}$ in $A(K/k)$. Then, applying to each α_j both sides of the equality

$$(1_K \otimes \Delta_{K/k})\theta_{K/k} = (\theta_{K/k} \otimes 1_{A(K/k)})\theta_{K/k} : \; K \to K \otimes A(K/k) \otimes A(K/k)$$

of Proposition 1.13(b), we obtain after a little computation that

$$\sum_{i=1}^{n} \beta_i \otimes \Delta_{K/k}(x_{ij}) = \sum_{i=1}^{n} \beta_i \otimes (x_{ij} \otimes 1 + 1 \otimes x_{ij}) + \sum_{i,\mu=1}^{n} \beta_\mu \otimes w_{\mu i} \otimes x_{ij}$$

for all $j \leqslant s$. Equating coefficients of $\beta_i \otimes 1 \otimes 1$ for $i \leqslant n$, it follows that

$$\Delta_{K/k}(x_{ij}) = x_{ij} \otimes 1 + 1 \otimes x_{ij} + \sum_{\mu=1}^{n} w_{i\mu} \otimes x_{\mu j} \quad (i \leqslant n, j \leqslant s).$$

Raising both sides of this equation to the p^{e_j}–th power, we obtain finally that

$$\Delta_{K/k}(z_{ij}) = z_{ij} \otimes 1 + 1 \otimes z_{ij} + \sum_{\mu=1}^{n} w_{i\mu}^{p^{e_j}} \otimes z_{\mu j}$$

is in $A(K/k) \otimes B(K/k)$ for all $i \leqslant n, j \leqslant s$, and thus $\Delta_{K/k}(B(K,K)) \subseteq A(K/k) \otimes B(K/k)$. Also, as noted earlier, $A(K,K)^{p^e} \subseteq B(K,K)$, and then $A(K/k)^{p^e} = (A(K,K)_\Sigma)^{p^e} \subseteq B(K,K)_\Sigma = B(K/k)$. Therefore $\Delta_{K/k}(\delta^{-p^e})$ is in $A(K/k)^{p^e} \otimes A(K/k)^{p^e} \subseteq B(K/k) \otimes B(K/k)$, and since $B(K/k) = B(K,K)[\delta^{-p^e}]$ we at last obtain that $\Delta_{K/k}(B(K/k)) \subseteq A(K/k) \otimes B(K/k)$. This establishes (d) and completes the proof of the lemma.

We note, in passing, an easy consequence of the lemma just proved.

Remark 5.5. Let K/k, etc., be as in (5.1), and m_A, m_P be the augmentation ideals of $A(K/k)$ and $P(K/k)$, respectively; i.e., $m_A = \mathrm{Ker}(\epsilon_{K/k} : A(K/k) \to k)$, etc. Then the canonical map $A(K/k) \to P(K/k)$ induces an isomorphism $m_A/m_A^2 \stackrel{\approx}{\to} m_P/m_P^2$. Hence, applying the functor $(-)^* = \mathrm{Hom}_k(-,k)$, we obtain that the Lie p–algebra map $(m_P/m_P^2)^* \stackrel{\approx}{\to} (m_A/m_A^2)^*$ of tangent spaces at the origin, induced by the closed immersion $G_t \to G$, is an isomorphism.

Proof. For the relevant material on Lie p–algebras over k, and tangent spaces of group schemes over a field of characteristic p, see [3, II, §7, no. 3, pp. 274-281] and [3, II, §4, no. 4, pp. 209-216].

In the notation of (5.1), m_A is generated by $\{x_{ij} | i \leqslant n, j \leqslant s\}$ and m_P is generated by $\{t_{ij} | i \leqslant n, j \leqslant s\}$. Since the canonical map sends x_{ij} to t_{ij}, we at least obtain a surjection $m_A/m_A^2 \to m_P/m_P^2$. But, by Lemma 5.4(c) (or its proof), the kernel of the map $A(K/k) \to P(K/k)$ is generated by the elements $\{z_{ij} = x_{ij}^{p^{e_j}} | i \leqslant n, j \leqslant s\}$, and hence is contained in m_A^2. The desired result follows easily.

Recall now that, if $X = \mathrm{Spec}(A)$ is an affine scheme over k, then the corresponding set–valued functor $T \rightsquigarrow X(T) = \mathrm{Alg}_k(A,T)$ is a sheaf on the category of commutative k–algebras with the faithfully flat topology [3, III, §1, 1.3, p. 285]. We shall denote this sheaf by the same letter X. $X(T)$ is what we have earlier called the set of T–valued points of X. A map $\varphi : X \to Y = \mathrm{Spec}(B)$ of affine k–schemes arises from a unique k–algebra map $\psi : B \to A$; this in turn yields a corresponding sheaf map $\mathrm{Alg}_k(\psi, -) : X \to Y$ which we shall also denote by φ. The passage from an affine k–scheme to its corresponding sheaf is functorial.

With K/k, G, G_t as before, we define the quotient sheaf G/G_t by the following coequalizer diagram in the category of sheaves

$$G \times G_t \underset{\pi}{\overset{\mu}{\rightrightarrows}} G \rightarrow G/G_t$$

where μ is the restriction to $G \times G_t$ of the multiplication map $G \times G \rightarrow G$. and π is the projection on the left factor [3, II, §3, 1.4, p. 324]. We shall show that G/G_t "is" an open subscheme of affine k–space, in the sense described below.

Theorem 5.6. Let $K = k(\alpha_1, \cdots, \alpha_s)$ be a finite purely inseparable modular extension of k, and $G = \mathrm{Spec}(A(K/k))$ and $G_t = \mathrm{Spec}(P(K/k))$ be the automorphism and truncated automorphism schemes of K/k, respectively. Then the sheaf G/G_t is isomorphic to that corresponding to an open subscheme of affine space over k of dimension m, where

$$m = \sum_{j=1}^{s} (n - n(j))$$

$n = [K : k]$, and $n(j) < n$ is as in (5.1(b)).

Proof. We use all of the notation and conventions of the preceding discussion. Let $F = \mathrm{Spec}(B(K/k))$; then the inclusion map $B(K/k) \rightarrow A(K/k)$ yields a homomorphism of k–schemes $p : G \rightarrow F$. By Lemma 5.4(a), F is isomorphic to an open subscheme of affine k–space of dimension m. Hence it will be sufficient to establish a sheaf isomorphism $F \approx G/G_t$.

Translation of Lemma 5.4(d) gives to F an action of G from the left which is preserved by p; i.e., there is a k–scheme map $G \times F \rightarrow F$, which is associative and unitary in the obvious sense, such that the diagram below commutes

$$\begin{array}{ccc} G \times G & \xrightarrow{1_G \times p} & G \times F \\ \downarrow & & \downarrow \\ G & \xrightarrow{p} & F \end{array}$$

the map $G \times G \rightarrow G$ being multiplication. Moreover, Lemma 5.4(c) gives rise to the following equalizer diagram in the category of k–schemes

$$G_t \rightarrow G \underset{\eta}{\overset{p}{\rightrightarrows}} F$$

with $\eta = \mathrm{Spec}(\epsilon)$. In particular, if T is a commutative k–algebra, then we have the following diagram of sets

$$G_t(T) \rightarrow G(T) \underset{\eta(T)}{\overset{p(T)}{\rightrightarrows}} F(T) \tag{5.7}$$

$F(T)$ is a set with left action by the group $G(T)$, $p(T)$ preserves this action, and one checks easily that

$$\eta(T)(g) = p(T)(1_T)$$

for all g in $G(T)$, with 1_T the identity element of $G(T)$. Since passage from a k–scheme to its set of T–valued points preserves limits, (5.7) is an equalizer diagram in the category of sets, which, in view of the preceding remarks, means simply that

$$(5.8) \qquad G_t(T) = \{ g \text{ in } G(T) | p(T)(g) = p(T)(1_T) \} = \{ g \text{ in } G(T) | g x_T = x_T \}$$

with $x_T = p(T)(1_T)$ in $F(T)$.

Now, by Lemma 5.4(b), $A(K/k)$ is a faithfully flat $B(K/k)$–algebra of finite type, and so $p: G \to F$ is an epimorphism of sheaves [3, III, §1, 2.10, p. 295]. Hence, if x is the point of F given by the composite map

$$\mathrm{Spec}(k) \xrightarrow{\mathrm{Spec}(\epsilon_{K/k})} G \xrightarrow{p} F$$

then F is the orbit of x under the action of G, in the sense of [3, III, §3, 1.6, p. 325]. This reference, together with (5.8), then yields a sheaf isomorphism $\rho : G/G_t \overset{\approx}{\to} F$ such that the diagram below commutes

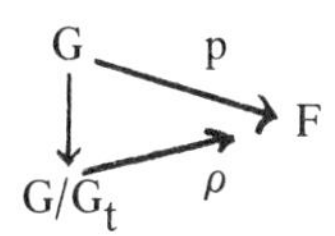

This completes the proof of the theorem.

References

1. L. Begueri, Schema d'automorphismes et application a l'etude d'extensions finies radicielles, Bull. Sc. Math. **93**, (1969), 89-111.
2. S. Chase, On inseparable Galois theory, Bull. Amer. Math. Soc. **77** (1971), 413-417.
3. M. Demazure and P. Gabriel, "Groupes algebrique", Tome I, North-Holland Publishing Co., Amsterdam, 1970.
4. I. Kaplansky, "Commutative rings", Allyn and Bacon, Inc., Boston, 1970.
5. R. Rasala, Inseparable splitting theory, (to appear in Trans. Amer. Math. Soc.).
6. S. Shatz, Galois theory, Lecture notes in mathematics vol. 86 (category theory, homology theory and their applications I, Battelle Institute Conference 1968),

Springer-Verlag 1969.

7. M. E. Sweedler, Hopf algebras with one grouplike element, Trans. Amer. Math. Soc. **127** (1967), 515-526.
8. M. E. Sweedler, The Hopf algebra of an algebra applied to field theory, J. Algebra **8** (1968), 262-276.
9. M. E. Sweedler, Structure of inseparable extensions, Annals of Math. **87** (1968), 401-410.
10. W. Waterhouse, Automorphism schemes and forms of Witt Lie algebras, J. Algebra **17**, 34-40.
11. O. Zariski and P. Samuel, "Commutative algebra," Vol. II, D. van Nostrand Co., Inc., Princeton, 1960.

GENERALIZED RATIONAL IDENTITIES

P. M. Cohn

Tulane University and Bedford College

1. Introduction

In what follows, all rings will be algebras over a commutative field k. The free k–algebra on a set X will be denoted by $k\langle X\rangle$. If A is any algebra, then each mapping $X \to A$ extends to a unique homomorphism $k\langle X\rangle \to A$. A non–zero element p of $k\langle X\rangle$ that is mapped to zero by all such homomorphisms is called a *polynomial identity for* A and we say: A *satisfies* $p = 0$. E.g., any commutative algebra satisfies $xy - yx = 0$.

The foundation of the subject was laid by Kaplansky who proved in [15]: If A is a primitive ring with a polynomial identity, then A is a simple algebra, finite–dimensional over its centre C, in fact $(A : C) \leqslant (d/2)^2$, where d is the degree of the identity holding in A.

Amitsur [1] generalized this in 1965 by allowing non–central coefficients. If A is any k–algebra, define

$$A_k\langle X\rangle = A \underset{k}{*} k\langle X\rangle,$$

where $\underset{k}{*}$ denotes the free product, taken over k (cf. [5]). This algebra may also be described as the tensor–ring over A on $|X|$ copies of $A \otimes_k A$. By a *generalized polynomial identity* (g.p.i.) in A one means a non–zero element of $A_k\langle X\rangle$ which vanishes under all mappings $X \to A$. In [1] Amitsur showed that a primitive ring with a g.p.i. is a dense ring of linear transformations over a skew field. We shall be interested in the following special case of this result: A skew field satisfies a generalized polynomial identity if and only if it is finite–dimensional over its centre.

A little later Amitsur turned to rational identities. Here there is a difficulty in that not all substitutions can be allowed (e.g., $x \to 0$ in x^{-1}) and also that there are some rational identities universally valid, e.g., $(xy)^{-1} = y^{-1}x^{-1}$ or $y^{-1}(x^{-1} + y^{-1})^{-1}x^{-1} = (x + y)^{-1}$ or the identity used in Hua's proof of the fundamental theorem of projective geometry:

$$xyx = x - [x^{-1} + (y^{-1} - x)^{-1}]^{-1}.$$

Amitsur's basic results in [2] were: (i) all skew fields of infinite dimension over a given infinite centre satisfy the same rational identities, and (ii) the rational identities in a skew field of finite degree over the centre depend only on the degree.

Amitsur's proof was simplified by Bergman [4] and at the same time generalized to the case of identities with non–central coefficients; further he allowed the centre to vary but it had to satisfy a comparability condition and be infinite. My aim here is to outline a simple proof of the Amitsur–Bergman result, with a somewhat different condition on the centres (which are still assumed infinite). Although this proof requires some preparation, it is conceptually quite simple. The main tool is the notion of a *universal skew field of fractions*. This was constructed by Amitsur in the special case of free algebras, using precisely his results on rational identities. By contrast, I shall use the explicit construction given recently [11] for universal skew fields of fractions to prove the result on rational identities. The details are briefly recalled in §2 and in §3 we generalize Bergman's inertia theorem [3] to the form used here. The proof of this occupies §4.

2. Universal skew fields of fractions

Henceforth we shall use the term field to mean "not necessarily commutative division ring" and occasionally use the prefix "skew" for emphasis.

Let R be any ring; by a *field of fractions* of R one understands a field K with an embedding $R \to K$ such that K is the field generated by the image. In the commutative case such a K exists if and only if R is an integral domain, and it is then unique. In general no necessary and sufficient conditions are known[1] for a field of fractions to exist, and even when it does exist it need not be unique. Given two fields of fractions K, L of R, let us define a *specialization* $K \to L$ as a homomorphism $f: K_0 \to L$ where K_0 is a subring of K containing R and such that $xf \neq 0 \Rightarrow x^{-1} \in K_0$. Then the non–units in K_0 form an ideal, namely ker f, and so K_0 is a local ring, with residue–class field $K_0/\ker f$. This is a subfield of L containing R, hence equal to L, thus every specialization of fields of fractions of R is necessarily surjective.

$$\begin{array}{ccc} & R & \\ \swarrow & & \searrow \\ K & \to & L \end{array}$$

For every ring R one obtains in this way a category whose objects are the fields of fractions of R and whose maps are specializations. This category may of course be empty (e.g., if R has zero–divisors) or consist of a single object (when R is a commutative integral domain), but in general there will be several fields of fractions. An initial object U in this category (if it exists) is called a *universal field of fractions*

[1] Added May 10, 1971: I have now found such conditions. They will be described in the lecture notes of the Tulane Symposium, to appear in the Springer lecture note series.

of R; such a field U is then characterized by the property that for every field of fractions K of R there is a unique specialization from U to K.

The main difficulty in constructing fields of fractions is that there is no convenient way of expressing the elements in the general case. When R is a commutative domain, or more generally an Ore domain, all elements of the field of fractions can be obtained as solutions of

$$(1) \qquad ax + b = 0, \quad a, b \in R, \ a \neq 0.$$

But in general we may have to perform repeated inversions (e.g., to get elements like $(ab^{-1}c + de^{-1}f)^{-1}$). Nevertheless there is a method for getting any field element by a single inversion, as in (1). We need only replace the element a in (1) by a *matrix*. We therefore consider the effect of inverting matrices.

If Σ is any set of square matrices over R and $f : R \to S$ is a homomorphism, we say f is *Σ-inverting* if for each $A \in \Sigma$ the image A^f is invertible over S. Given such a Σ-inverting map, the set R' of all components of inverses of matrices in the image of Σ under f is called the *Σ-rational closure* of R in S under f. When R is commutative, nothing is gained by adjoining inverses of matrices instead of elements, for inverting a matrix A simply amounts to inverting det A. In the commutative case det A is a polynomial in the elements of A, whereas in general it is at best a rational function (cf. Dieudonné [12]).

Given a homomorphism $f : R \to S$, if Σ is the set of all matrices inverted by f, the rational closure is a subring. In general it need not be, but it is not hard to find sufficient conditions for this to hold. Let us call Σ *admissible* if

(i) $1 \in \Sigma$ (where 1 is the unit matrix).

(ii) Σ admits elementary row and column operations.

(iii) If $A, B \in \Sigma$, then $\begin{pmatrix} A & C \\ 0 & B \end{pmatrix} \in \Sigma$ for any C of appropriate size.

Using this notion we obtain the following description of the rational closure (based on the Schützenberger–Nivat criterion [16] for the rationality of power series):

Theorem 1. Let R be any ring, Σ any admissible set of matrices over R and $f : R \to S$ a Σ-inverting homomorphism. Then the rational closure R' is a subring of S containing R^f and for any $x \in S$ the following are equivalent:

(a) $x \in R'$,

(b) $x = u_1$ is the first component of the solution of an equation

$$(2) \qquad Au + a = 0$$

where $A \in \Sigma^f$ and a is a column over R^f,

(c) $x = u_1$ is the first component of the solution of an equation

$$Au + e_1 = 0 \tag{3}$$

where $A \in \Sigma^f$ and $e_1 = (1, 0, \cdots, 0)^T$ (T indicates transpose).

Sketch of proof (cf. [11]). Clearly (c) $\Rightarrow$ (a) $\Rightarrow$ (b). From (b) we get

$$\begin{pmatrix} 0 & -1 \\ A & a \end{pmatrix}\begin{pmatrix} u \\ 1 \end{pmatrix} + \begin{pmatrix} 1 \\ 0 \end{pmatrix} = 0$$

which is of the form (3) (because Σ is admissible).

To show that R' is a ring we use (b). Any $c \in R^f$ satisfies $1 \cdot u - c = 0$, so $R' \supseteq R^f$. Next if u_1 is the first component of the solution u of (2), while v_1 is the first component of a column v satisfying $Bv + b = 0$, then $u_1 - v_1$ satisfies

$$\left(\begin{array}{c|cc|cc} 1 & -1 & 0 & 1 & 0 \\ \hline 0 & A & & 0 & \\ 0 & 0 & & B & \end{array}\right)\begin{pmatrix} u_1 - v_1 \\ u \\ v \end{pmatrix} + \begin{pmatrix} 0 \\ a \\ b \end{pmatrix} = 0,$$

and $u_1 v_1$ satisfies

$$\left(\begin{array}{c|cc} A & a & 0 \\ \hline 0 & B & \end{array}\right)\begin{pmatrix} uv_1 \\ v \end{pmatrix} + \begin{pmatrix} 0 \\ b \end{pmatrix} = 0.$$

This shows R' to be a subring containing R^f, as claimed.

Given any set Σ of square matrices over a ring R, we can always find a Σ-inverting homomorphism $\lambda : R \to R_\Sigma$ which is *universal* in the sense that any Σ-inverting homomorphism can be uniquely factored by λ. We simply adjoin to R, for every $n \times n$ matrix $A \in \Sigma$, a set of n^2 indeterminates, written in matrix form as A', with defining relations $AA' = A'A = I$. It may of course happen that $R_\Sigma = 0$; we are concerned with finding conditions under which $R \to R_\Sigma$ is injective. Let us recall a definition and a result from [10]. A matrix A is said to be *full* if it is square say $n \times n$, and *not* of the form $A = PQ$ where P is $n \times r$ and Q is $r \times n$, with $r < n$. Any matrix inverted under a homomorphism into a field is necessarily full, and in [10] it is shown that for the class of firs (free ideal rings) the universal Φ-inverting mapping, where Φ is the set of all full matrices, is an embedding. But when

$R \to R_\Phi$ is an embedding, R_Φ is actually a field, necessarily the universal field of fractions because the greatest set of matrices is inverted here. (For a detailed proof, see [11].)

An example of a fir is the ring $D_k\langle X\rangle = D * k\langle X\rangle$, where D is any field with k as central subfield, and X is any set. Thus we see that $D_k\langle X\rangle$ has a universal field of fractions; moreover, any automorphism of $D_k\langle X\rangle$ extends to a unique automorphism of the universal field of fractions. This follows essentially because an automorphism maps every full matrix to a full matrix (it is an 'honest' mapping).

3. Inertia

If R is any ring, we denote by R^n the set of all rows of length n over R, and by ${}^nR$ the set of all columns of length n. Note that R^n has a natural (R, R_n)–bimodule structure, and ${}^nR$ a natural (R_n, R)–bimodule structure, where R_n is the ring of all $n \times n$ matrices over R.

Let R be a ring and X a subring not necessarily containing the 1 of R. Then the product ab of a row and a column $a \in R^n$, $b \in {}^nR$ is said to lie *trivially* in X if for each $i = 1, \cdots, n$, *either* a_i and b_i lie in X, *or* $a_i = 0$, *or* $b_i = 0$. Suppose that X is such that for any families (a_λ) of rows in R^n and (b_μ) of columns in ${}^nR$, such that $a_\lambda b_\mu \in X$ for all λ, μ, there exists $P \in GL_n(R)$ such that each product $a_\lambda P \cdot P^{-1} b_\mu$ lies trivially in X; then X is called n–*inert* in R. If X is n–inert for all n we say X is *totally inert* in R. E.g., an n–fir (defined in [9]) may be defined as a ring in which $\{0\}$ is ν–inert for $1 \leqslant \nu \leqslant n$. Let X be any n–inert subring of R; then it is not har to see that any full $n \times n$ matrix over X remains full over R.

Many results on the relation between two rings can be stated as inertia theorems (cf. e.g., [3, 17, 8, 14, 13, 18]). We shall be particularly concerned with Bergman's theorem. Let $R = k\langle X\rangle$ be a free associative k–algebra and $\hat{R} = k\langle\langle X\rangle\rangle$ its power series completion, then Bergman's inertia theorem [3] states: $k\langle X\rangle$ is 1–inert in $k\langle\langle X\rangle\rangle$. He proves this result for any graded ring R with inverse weak algorithm (cf. [6, 3]) and its completion $\hat{R}$. In fact we actually have total inertia:

Theorem 2. Let R be a graded ring with inverse weak algorithm and $\hat{R}$ its power series completion; then R is totally inert in $\hat{R}$.

The proof is an adaptation of Bergman's proof of his result in [3] and will not be given here. The theorem can be further generalized as follows: Let E be a field and D a subfield, while h is a central subfield of E such that D and h are linearly disjoint over $k = D \cap h$. Then $D_k\langle X\rangle$ is totally inert in $E_h\langle X\rangle$.

4. Generalized rational identities

Let D be any field (with k as central subfield) and consider the free algebra $R = D_k\langle X\rangle \equiv D * k\langle X\rangle$. As we saw in §2, R has a universal field of fractions which we shall denote by $D(\langle X\rangle)$ or simply by $\bar{R}$. If E is a field containing D, then by a *generalized rational identity in* E one understands a non–zero element $p \in \bar{R}$ such that p maps to 0 under any substitution $X \to E$ for which p is defined. Equivalently, we can say: p^{-1} is an element of $\bar{R}$ which is nowhere defined on E, i.e., *degenerate*. Our aim will be to find conditions under which no element of $\bar{R}$ is degenerate on E.

Theorem 3. Let D be a field (with k as central subfield) and let E be any field containing D, with centre C, such that

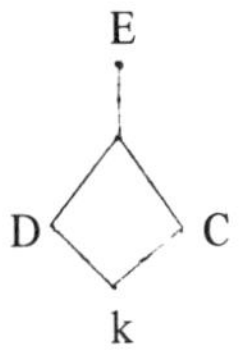

(i) $(E : C) = \infty$,

(ii) C is infinite,

(iii) C and D are linearly disjoint over k,

E

D C

k

then every element of $D(\langle X\rangle)$ is non–degenerate on E.

For the proof we observe that any element $p \in D(\langle X\rangle)$ can be obtained as the solution $p = u_1$ of an equation

$$Au + e_1 = 0,$$

where A is a full matrix over $D_k\langle X\rangle$, and p is non–degenerate on E provided that A goes over into a non–singular matrix under some substitution. Thus we must prove: if a square matrix A over $D\langle X\rangle$ is singular for all values in E, then A is not full.

We proceed by a series of lemmas:

Lemma 1. Let A be a square matrix over a field, of the form $A = \begin{pmatrix} B_1 & B_2 \\ B_3 & B_4 \end{pmatrix}$, where B_1 is an invertible $r \times r$ matrix. Then rank $A \geqslant r$, with equality if and only if $B_4 = B_3 B_1^{-1} B_2$, and then

$$A = \begin{pmatrix} B_1 \\ B_3 \end{pmatrix} \begin{pmatrix} I & B_1^{-1}B_2 \end{pmatrix}.$$

For, by applying elementary transformations to A we get $\begin{pmatrix} B_1 & B_2 \\ 0 & B_4 - B_3B_1^{-1}B_2 \end{pmatrix}$.
Here the first r columns are linearly independent, and there will be more linearly independent columns, unless $B_4 = B_3B_1^{-1}B_2$, and in that case,

$$A = \begin{pmatrix} B_1 & B_2 \\ B_3 & B_3B_1^{-1}B_2 \end{pmatrix} = \begin{pmatrix} B_1 \\ B_3 \end{pmatrix}\begin{pmatrix} I & B_1^{-1}B_2 \end{pmatrix}.$$

Lemma 2. Let E be a field with infinite centre C and $A(t)$ a square matrix over the polynomial ring $E[t]$. Then the rank of $A(t)$ over the quotient field $E(t)$ equals the sup of the ranks of $A(\tau)$ for $\tau \in C$; in fact, this sup is assumed for all but a finite number of values of τ.

In fact we obtain a diagonal form for $A(t)$ by PAQ–reduction (over the principal ideal domain $E[t]$), and the product of the diagonal terms gives us a polynomial for t whose zeros in C are the only points of C at which rank $A(t)$ falls short of its maximum.

The remaining fact needed is best expressed as a

Proposition. Let E be a field with infinite centre C. Then any full matrix over $E_C\langle X\rangle$ is non–singular for some set of values of X in E.

Proof. Let $A(X)$ be an $n \times n$ matrix over $E_C\langle X\rangle$ and r the sup of the ranks of its values on E^X. By a change of variables we may assume that this maximum rank is assumed at the point $0 \in E^X$, and by elementary operations we may take the principal $r \times r$ minor of $A(0)$ to be invertible. Thus if

$$A(X) = \begin{pmatrix} B_1(X) & B_2(X) \\ B_3(X) & B_4(X) \end{pmatrix},$$

then $B_1(0)$ is invertible.

Given any $a \in E^X$ and any $\tau \in C$ we have rank $A(\tau a) \leqslant r$, hence by Lemma 2, the rank of $A(ta)$ (over $E(t)$) is $\leqslant r$, and so the same holds over $E((t))$. Now $B_1(ta)$ is a polynomial in t, with matrix coefficients, and constant term $B_1(0)$, a unit, hence $B_1(ta)$ is a unit in $E((t))$. By Lemma 1, the equation $B_4(ta) = B_3(ta)B_1^{-1}(ta)B_2(ta)$ holds in $E((t))$, and hence in $E[[t]]$, for all $a \in E^X$. But the coefficient of every power of t is a polynomial in the components of a, hence by Amitsur's result on generalized polynomial identities, we must have

$$B_4(tX) = B_3(tX)B_1^{-1}(tX)B_2(tX) \tag{4}$$

in the completion of $R = E_C\langle X\rangle[t]$ with respect to the ideal generated by t. Hence the same equation holds in the completion with respect to the ideal generated by t and the X's, and also by the X's alone; for in each case the matrix $B_1(tX)$ has constant term $B_1(0)$ and so is invertible. Thus we may set $t = 1$ in (4) and find $B_4(X) = B_3(X)B_1^{-1}(X)B_2(X)$ in $E_C\langle\langle X\rangle\rangle$, i.e., $A(X)$ is non–full over $E_C\langle\langle X\rangle\rangle$, and by Theorem 2, over $E_C\langle X\rangle$.

Now Theorem 3 is an easy consequence.

We conclude with some remarks on the hypotheses of Theorem 3. (i) is clearly necessary since there are even polynomial identities when $(E : C)$ is finite, but (ii) may not be needed. Bergman conjectures that (ii) may be replaced by the assumption that E be infinite (which is clearly necessary). It is used in proving Lemma 2, and could be omitted if the following (very plausible) conjecture could be proved: Let E be an infinite field with finite centre. Then for any square matrix A over E there exists $c \in E$ such that the matrix $cI - A$ is non–singular. (iii) seems unduly strong, and it would be interesting if this could be shown to be either necessary or unnecessary.

I am indebted to W. S. Martindale and W. Dicks for pointing out an error in a first draft, and to G. M. Bergman for suggestions for simplifying the proof of Theorem 3.

References

1. S. A. Amitsur, Generalized polynomial identities, Trans. Amer. Math. Soc. **114** (1965), 210-226.
2. S. A. Amitsur, Rational identities and applications to algebra and geometry, J. Algebra **3** (1966), 304-359.
3. G. M. Bergman, Centralizers in free algebras and related topics in ring theory, thesis, Harvard University, 1967.
4. G. M. Bergman, Skew fields of noncommutative rational functions, after Amitsur, Séminaire Schützenberger–Lentin–Nivat, Année 1967/70, No. 16 Paris 1970.
5. P. M. Cohn, On the free product of associative rings I, Math. Zeit. **71** (1959), 380-398.
6. P. M. Cohn, Rings with a weak algorithm, Trans. Amer. Math. Soc. **109** (1963), 332-356.
7. P. M. Cohn, "Universal Algebra", New York, London, Tokyo, 1965.
8. P. M. Cohn, Bezout rings and their subrings, Proc. Cambridge Phil. Soc. **64** (1968), 251-264.

9. P. M. Cohn, Dependence in rings II, Trans. Amer. Math. Soc. **135** (1969),267-279.
10. P. M. Cohn, The embedding of firs in skew fields, Proc. London Math. Soc. (3) **23** (1971), 193-213.
11. P. M. Cohn, Universal skew fields of fractions, Symposia Math. vol. VIII.
12. J. Dieudonné, Les déterminants sur un corps non–commutatif, Bull. Soc. Math. France **71** (1943), 27-45.
13. M. Fliess, Inertie et rigidité des séries rationnelles et algébriques, C. R. Acad. Sci. Paris, Ser. A., **270** (1970), 221-223.
14. R. Gilmer, An embedding theorem for HCF–rings, Proc. Cambridge Phil. Soc. **68** (1970), 583-587.
15. I. Kaplansky, Rings with a polynomial identity, Bull. Amer. Math. Soc. **54** (1948), 575-580.
16. M. Nivat, Series rationnelles et algebriques en variables non–commutatives, Cours DEA (unpublished).
17. B. V. Tarasov, On free associative algebras (Russian), Algebra i Logika **6**, **4** (1967), 93-105.
18. A. Zaks and A. Evyatar, Rings of polynomials, Proc. Amer. Math. Soc. **25** (1970), 559-562.

K_2 OF POLYNOMIAL RINGS AND OF FREE ALGEBRAS

P. M. Cohn

Tulane University and Bedford College

The K_2 functor of Milnor has been computed for certain fields by Bass and Tate, and it turns out that some deep number theory is required in the process. By contrast, K_2 of polynomial rings can be reduced to K_2 of the ground field, using only a presentation of GL_n for these polynomial rings. Such a presentation was recently found by John R. Silvester [9], who also worked out the application to K_2, and it is his work that I want to describe briefly.

If R is a ring (always with 1), we denote the group of units by $U(R)$, by R_n the ring of all $n \times n$ matrices over R and define the general linear group over R as $GL_n(R) = U(R_n)$. As usual, the matrix units are written e_{ij}. For $i \neq j$ and $a \in R$ we put $B_{ij}(a) = I + ae_{ij}$ and introduce the following notation for diagonal matrices: $[\alpha]_i = I + (\alpha - 1)e_{ii}$, $[\alpha, \beta]_{ij} = [\alpha]_i [\beta]_j$, $D_{ij}(\alpha) = [\alpha, \alpha^{-1}]_{ij}$ and $[\alpha_1, \cdots, \alpha_n] = [\alpha_1]_1 [\alpha_2]_2 \cdots [\alpha_n]_n$. We generally use the convention of reserving greek letters to denote units.

When R is a field, or more generally a Euclidean ring, $GL_n(R)$ is generated by all $B_{ij}(a)$ and $[\alpha_1, \cdots, \alpha_n]$. In a general ring R these matrices generate a subgroup, $GE_n(R)$. If $GE_n(R) = GL_n(R)$ for all n, we call R a *generalized Euclidean ring* (*GE–ring* for short).

Now consider the following relations, which hold over any ring:

(1) $B_{ij}(a)B_{ij}(b) = B_{ij}(a + b)$

(2) $B_{ij}(a)B_{k\ell}(b) = B_{k\ell}(b)B_{ij}(a), \quad i \neq \ell,\ j \neq k$

(3) $B_{ij}(a)B_{jk}(b) = B_{jk}(b)B_{ij}(a)B_{ik}(ab), \quad i \neq k$

(4) $B_{ij}(\alpha - 1)B_{ji}(1) = D_{ij}(\alpha)B_{ji}(\alpha)B_{ij}(1 - \alpha^{-1})$

(5) $B_{ij}(a)[\alpha_1, \cdots, \alpha_n] = [\alpha_1, \cdots, \alpha_n] B_{ij}(\alpha_i^{-1} a \alpha_j)$

(6) $[\alpha_1, \cdots, \alpha_n][\beta_1, \cdots, \beta_n] = [\alpha_1\beta_1, \cdots, \alpha_n\beta_n]$.

For $n = 2$, (3) does not occur; in that case one usually adds the relation

$$(7)\quad B_{ij}(a) = B_{ji}(1)B_{ij}(-1)B_{ji}(-a)B_{ij}(1)B_{ji}(-1)$$

which for $n \geqslant 3$ is a consequence of (1) – (6). In fact the case $n = 2$ is quite different (see [2]), and we shall therefore exclude it here; there is all the more reason for doing this since we shall mainly be interested in the stable linear group.

We shall call (1) – (6) the *universal relations*; (1) – (3) are the well–known Steinberg relations, (4) can be used to define D_{ij} in terms of the B's, while (5), (6) are relations between diagonal matrices and B's. If (1) – (6) is a complete set of defining relations for $GE_n(R)$, we call R *universal for* GE_n. In particular a GE–ring universal for GE_n is also called a *universal* GE_n*–ring.*

For example, **Z** is a universal GE_n–ring for all n. This was proved for $n = 3$ by Nielsen [8], and extended to $n > 3$ by Magnus [5]. In essence, the proof goes as follows: In any expression of a matrix $M \in GL_n(\mathbf{Z})$ as a product of elementary matrices: $M = P_1P_2 \cdots P_r$, consider the last row of $P_1 \cdots P_i$. Using (1) – (6) we can modify this sequence so that the sum of the absolute values in the last row of $P_1 \cdots P_i$ is a monotone increasing function of i. If we start with a relation $P_1 \cdots P_r = I$ and apply this process, we reach the trivial relation $I = I$ using only (1) – (6). So everything is a consequence of (1) – (6). In [9], Silvester analysed (and somewhat simplified) Nielsen's proof, and then used the same method to prove that the polynomial ring $k[x]$ in one indeterminate x over a field k is a universal GE_n–ring for $n \geqslant 3$ (for $n = 2$ this was shown in [2]). In a later version of [9], Silvester proved the same more generally for free algebras $k\langle x_1, x_2, \cdots\rangle$ in any number of (non–commuting) indeterminates. Meanwhile this result had also been obtained independently by R. G. Swan [10], while the result for polynomial rings has now been proved by another method by K. Dennis [3].

To sum up: In order to show that R is a GE–ring one has to express every invertible matrix as a product of elementary matrices. To prove that R is universal for GE_n one has to show that every product of elementary matrices can be brought to a certain standard form using only (1) – (6), and the only product in this standard form equal to I is the empty product. We now outline Silvester's method of reducing products to such a standard form.

Let R be a ring with a weak algorithm relative to a degree function. Then R is a GE–ring (see [1]); if we can show it to be universal, this will certainly cover the case of free k–algebras. In essence one needs to make an induction on the degree; but it will be necessary to order in some way the elements of a given degree. For this reason it is better to start with a more general set–up.

We take a totally ordered semigroup Σ, with $1 \leqslant s$ for all $s \in \Sigma$, and satisfying the descending chain condition. E.g., if X is any set, the free semigroup X^* on X satisfies these conditions if we order X^* by degree, and elements of the same degree lexicographically. Returning to the general Σ, we suppose that a norm is defined on our ring R, with values in $\Sigma \cup \{0\}$ (where $0 < 1$) satisfying

1. $|a| = 0 \Leftrightarrow a = 0$,
2. $|a| = 1 \Leftrightarrow a \in U(R)$,
3. $|a + b| \leqslant \max\{|a|, |b|\}$,
4. $|ab| = |a|\,|b|$,
5. If $|a| = |b| \neq 0$, $\exists \lambda \in U(R)$ such that $|a + b\lambda| < |a|$,
6. If $|a| \leqslant |b|$ and $\exists c, d \in R$ such that $|ad - bc| < |bc|$, then $\exists e \in R$ such that $|b - ae| < |b|$.

For example, if $R = k\langle X\rangle$ where X is any set, and $\Sigma = X^*$ is the free semigroup on X, ordered as before, then for any $a \in R$, $a \neq 0$, we can define $|a|$ as the highest monomial occurring with non–zero coefficient, and we can easily verify that 1–6 hold. In particular this includes the case $k[x]$ (when $X = \{x\}$ consists of a single element). More generally, all rings with weak algorithm can be treated in this way, as the characterization of such rings in [1] shows.

Theorem 1. Any ring R with weak algorithm is a universal GE_n–ring for all n.

Sketch of proof. We know from [2] that R is a GE–ring. Now we define, for any $M \in GL_n(R)$, $\sigma(M)$ as

$$\sigma(M) = (|a_1|, |a_2|, \cdots, |a_n|) \in \Sigma^n,$$

where $(a_1, \cdots, a_n)$ is the last row of M, rearranged so that $|a_1| \geqslant \cdots \geqslant |a_n|$. We observe that $\sigma(M) = \sigma(MQ)$, where Q is any diagonal or permutation matrix. Now M has an expression of the form

$$M = P_1 P_2 \cdots P_r Q \tag{8}$$

where each P_ρ is some $B_{ij}(a)$ and Q is a product of diagonal and permutation matrices. We put $\sigma_i = \sigma(P_1 \cdots P_i)$, thus in particular, $\sigma_0 = (1, 0, \cdots, 0)$, $\sigma_r = \sigma(M)$. The sequence $(\sigma_0, \sigma_1, \cdots, \sigma_r)$ will be called the *diagram* of M (more precisely, the diagram of the expression (8) for M). Now by modifying the expression (8) for M, using only the relations (1) – (6), we can make the diagram monotone increasing (just as in the case $R = \mathbf{Z}$; note that r may increase in this process). This requires a number of case distinctions that I won't reproduce (see [9]). Most are straightforward except one or two where 5–6 play an essential role. It follows that any relation $P_1 \cdots P_r Q = I$ is a consequence of (1) – (6), i.e., R is universal.

This theorem then supplies us with a presentation of $GL_n(k\langle X\rangle)$; it also shows that every field (even skew) is universal for all GE_n, a fact which is not entirely obvious.

In order to define K_2 we need to consider the stable linear group $GL(R)$ of a ring R. This is defined as the union, or more precisely, the direct limit of all the $GL_n(R)$,

$$GL(R) = \lim_{\rightarrow} GL_n(R),$$

where $GL_n(R)$ is embedded in $GL_{n+1}(R)$ by writing each matrix in the top left–hand corner and bordering it by 1 on the diagonal and zeros elsewhere. Let $E(R)$ be the subgroup of $GL(R)$ generated by all the $B_{ij}(a)$.

Of the universal relations, (1) – (3) involve only the B_{ij}. We define the Steinberg group $St(R)$ as the group with generators $b_{ij}(a)$ $(a \in R)$ and defining relations (1) – (3) with $B_{ij}(a)$ replaced by $b_{ij}(a)$. There is a natural mapping $St(R) \rightarrow E(R)$ given by $b_{ij}(a) \mapsto B_{ij}(a)$; it is surjective, and the kernel is defined to be $K_2(R)$, so that we have the exact sequence

$$0 \rightarrow K_2(R) \rightarrow St(R) \rightarrow E(R) \rightarrow 0.$$

Any ring–homomorphism $f : R \rightarrow S$ induces a group–homomorphism $f^* : St(R) \rightarrow St(S)$. Suppose R is a retraction of S, i.e., there are homomorphisms $f : R \rightarrow S$, $g : S \rightarrow R$ such that $gf : S \rightarrow S$ is 1 (read left to right). Then $g^*f^* = 1$, hence g^* is an injection $St(S) \rightarrow St(R)$. We claim that g^* also induces an injection $K_2(S) \rightarrow K_2(R)$. This follows by a simple diagram chase (or even more simply by observing that K_2 is a functor and hence preserves retractions):

$$\begin{array}{ccc} & R & \\ g\nearrow & & \searrow f \\ S & \xrightarrow{1} & S \end{array}$$

$$\begin{array}{ccccccccc} & & 0 & & 0 & & 0 & & \\ & & \downarrow & & \downarrow & & \downarrow & & \\ 0 & \rightarrow & K_2(S) & \rightarrow & St(S) & \rightarrow & E(S) & \rightarrow & 0 \\ & & \downarrow & & \downarrow & & \downarrow & & \\ 0 & \rightarrow & K_2(R) & \rightarrow & St(R) & \rightarrow & E(R) & \rightarrow & 0 \end{array}$$

To be able to compute $K_2(R)$ we need a presentation for $E(R)$:

Theorem 2. If R is universal for GE_n for infinitely many n, then $E(R)$ has defining relations (1) – (3) and

(4)′ $D_{ij}(\alpha) = B_{ij}(\alpha - 1)B_{ji}(1)B_{ij}(\alpha^{-1} - 1)B_{ji}(-\alpha)$, $\alpha \in U(R)$,

(5)′ The relations between the D_{ij}'s.

The proof consists in taking a relation $P_1 \cdots P_r = I$ in $E(R)$. This follows from (1) – (6). Now $E(R)$ has relations (1) – (6), where in (5), (6) we replace $[\alpha_1, \cdots, \alpha_n]$ by $[\alpha_1, \cdots, \alpha_n, \alpha]$ with $\alpha_1 \alpha_2 \cdots \alpha_n \alpha = 1$ (to bring it within $E(R)$). Now (4)′ is merely a rearrangement of (4), so it remains to prove that (5)′ ⇒ (5) – (6). We omit the details (see [9]); note that Silvester assumes $U(R)$ abelian; this enables one to write down explicitly the relations between the D_{ij}'s, but it is not essential for the proof.

Theorem 2 allows us to give the following prescription for obtaining $K_2(R)$. In $St(R)$, let us define, for each $\alpha \in U(R)$, and any i, j, an element $d_{ij}(\alpha)$ by

$$d_{ij}(\alpha) = b_{ij}(\alpha - 1) b_{ji}(1) b_{ij}(\alpha^{-1} - 1) b_{ji}(-\alpha).$$

This corresponds of course to the relation (4)′. Now $K_2(R)$ is the subgroup of $St(R)$ consisting of all products $\Pi\, d_{ij}(\alpha)$ which reduce to I in $E(R)$, on replacing $d_{ij}(\alpha)$ by $D_{ij}(\alpha)$.

Using this description of $K_2(R)$ we can without difficulty prove

Theorem 3. Let S be a retraction of R and assume (i) g maps $U(S)$ onto $U(R)$, (ii) R is universal for GE_n, for infinitely many n. Then $K_2(R) \cong K_2(S)$.

$$\begin{array}{ccc} & R & \\ g\nearrow & & \searrow f \\ S & \xrightarrow{1} & S \end{array}$$

Proof. We already have an injection $K_2(S) \to K_2(R)$; if we identify S with its image in R, we may regard $K_2(S)$ as a subgroup of $K_2(R)$. Now $K_2(R)$ consists of the products $\Pi\, d_{ij}(\alpha)$ that reduce to I in $E(R)$. Each $d_{ij}(\alpha)$ already lies in $St(S)$, by (i), and any relation is a product of universal relations in R: If $\varphi \in K_2(R)$, then $\varphi = \Pi\, u_\upsilon^{-1} \theta_\upsilon u_\upsilon$ where θ_υ is a universal relation in R. Now $f : R \to S$ is a homomorphism which reduces to the identity mapping on S. Hence $\varphi = \varphi^f = \Pi\, (u_\upsilon^f)^{-1} \theta_\upsilon^f (u_\upsilon^f)$, which shows $\varphi \in K_2(S)$. Thus $K_2(R) = K_2(S)$.

Corollary. $K_2(k\langle X\rangle) \cong K_2(k)$ for any field k.

In the situation of this theorem it would be interesting to take S but not R universal and regard the quotient $K_2(R)/K_2(S)$ as a measure of non–universality of R. This has been done by Keith Dennis, who pointed out in the discussion following this talk, that by his results in [3], there is a converse to Theorem 3, namely if S is universal and $K_2(R) = K_2(S)$, then R is universal (for GE_n for all n). In fact, he obtains more precise information on the unstabilized K_2–functors there; the connexion between these and the stabilized form is given in [4]. A simple proof of Dennis' converse to Theorem 3 has also been obtained by R. G. Swan (private communication).

Silvester has made a second application of Theorem 1, to express $GL_n(R)$ as a generalized free product. Let G be any group, and (H_λ) a family of subgroups of G, then G is said to be the *generalized free product* of the H_λ if G has a presentation which is obtained by taking suitable presentations of the subgroups H_λ. Here "suitable" means essentially that the presentation of H_λ includes a presentation of each intersection $H_\lambda \cap H_\mu$ and the isomorphism between the two presentations of $H_\lambda \cap H_\mu$ (once as subgroup of H_λ and once as subgroup of H_μ) has been specified. A detailed account is given by B. H. Neumann [7], but in constructions of free products practically the only case considered is that where any two distinct subgroups H_λ, H_μ have the same intersection. The analysis of $GL_n(R)$ given below is one of the few cases of a free product not falling under this heading (i.e., the intersections $H_\lambda \cap H_\mu$ vary with λ and μ).

For any ring R, denote by $H_n(R)$ the subgroup of $GE_n(R)$ generated by the diagonal matrices and all $B_{ij}(a)$ with $i < n$. These are the matrices of the form shown here: $\begin{pmatrix} & & * \\ & & * \\ & * & \vdots \\ & & * \\ 0 & \cdots 0 & \lambda \end{pmatrix}$. The group $H_n(R)$ has as a complete set of defining relations those of $GE_{n-1}(R)$ together with those of (1) – (6) that involve its generators. This is proved by showing that any expression in $H_n(R)$ can be brought to the form $\Pi_i B_{in}(c_i)A_1[\alpha]_n$, where $A_1 \in GE_{n-1}(R)$, using (1) – (6). If this expression is I, we must have all $c_i = 0$, $\alpha = 1$ and $A_1 = I$, as is easily checked.

Similarly we write $F_n(R)$ for the subgroup of $GE_n(R)$ generated by the diagonal matrices and all $B_{ij}(a)$ with $j > 1$. This group has a presentation consisting of a presentation for $GE_{n-1}(R)$ and those of (1) – (6) that involve its generators. $\begin{pmatrix} \alpha & * \cdots * \\ 0 & \\ \vdots & * \\ 0 & \end{pmatrix}$

Theorem 4. Given a ring R with a subring S such that $U(S) = U(R)$, assume R, S universal for GE_n and R universal for GE_{n-1}. Then $GE_n(R)$ is the generalized free product of $GE_n(S)$, $H_n(R)$ and $F_n(R)$.

For the proof we take a presentation of $GE_n(R)$ and show that every relation follows from one holding in one or more of the given subgroups. Note that the generators of $GE_n(R)$ all lie in at least one of the subgroups, except $B_{n1}(a)$, but we can write $B_{n1}(a) = P_{n1}B_{1n}(a)P_{n1}$, where P_{n1} is the permutation matrix with rows 1 and n interchanged.

Examples. Let us represent elements of S by $\cdot$ and elements of R by $*$, then for $n = 2$, $GE_2(R)$ can be written as generalized free product of $\begin{pmatrix} \cdot & \cdot \\ \cdot & \cdot \end{pmatrix}$ and

$\begin{pmatrix} . & * \\ 0 & . \end{pmatrix}$. Thus taking $S = k$, $R = k[x]$, we obtain Nagao's theorem [6]:

$$GL_2(k[x]) \cong GL_2(k) * H_2(k[x]).$$

For $n = 3$, we find that $GE_3(R)$ is the free product of the matrix groups of type $\begin{pmatrix} \cdot & \cdot & \cdot \\ \cdot & \cdot & \cdot \\ \cdot & \cdot & \cdot \end{pmatrix}$, $\begin{pmatrix} * & * & * \\ * & * & * \\ 0 & 0 & \cdot \end{pmatrix}$ and $\begin{pmatrix} . & * & * \\ 0 & * & * \\ 0 & * & * \end{pmatrix}$. Now in $\begin{pmatrix} * & * & * \\ * & * & * \\ 0 & 0 & \cdot \end{pmatrix}$ the top left-hand 2×2 corner is the general element of $GE_2(R)$, and this is the free product of $\begin{pmatrix} \cdot & \cdot & * \\ \cdot & \cdot & * \\ 0 & 0 & \cdot \end{pmatrix}$ and $\begin{pmatrix} . & * & * \\ 0 & \cdot & * \\ 0 & 0 & \cdot \end{pmatrix}$. Similarly, $\begin{pmatrix} . & * & * \\ 0 & * & * \\ 0 & * & * \end{pmatrix}$ is the free product of $\begin{pmatrix} . & * & * \\ 0 & \cdot & \cdot \\ 0 & \cdot & \cdot \end{pmatrix}$ and $\begin{pmatrix} . & * & * \\ 0 & \cdot & * \\ 0 & 0 & \cdot \end{pmatrix}$. Hence $GE_3(R)$ is the free product of the types $\begin{pmatrix} \cdot & \cdot & \cdot \\ \cdot & \cdot & \cdot \\ \cdot & \cdot & \cdot \end{pmatrix}$, $\begin{pmatrix} \cdot & \cdot & * \\ \cdot & \cdot & * \\ 0 & 0 & \cdot \end{pmatrix}$, $\begin{pmatrix} . & * & * \\ 0 & \cdot & \cdot \\ 0 & \cdot & \cdot \end{pmatrix}$ and $\begin{pmatrix} . & * & * \\ 0 & \cdot & * \\ 0 & 0 & \cdot \end{pmatrix}$. For the case $S = k$, $R = k[x]$ this establishes a conjecture made by Serre on the basis of some function–theoretic evidence (fundamental domains).

References

1. P. M. Cohn, Rings with a weak algorithm, Trans. Amer. Math. Soc. **109** (1963), 332-356.
2. P. M. Cohn, On the structure of the GL_2 of a ring, Publs. Math. IHES (Paris) No. 30 (1966), 3-54.
3. K. Dennis, Universal GE_n–rings and the functor K_2, (to appear).
4. K. Dennis, K_2 and the stable range condition, (to appear).
5. W. Magnus, Über n–dimensionale Gittertransformationen, Acta Math. **64** (1935), 353-367.
6. H. Nagao, On $GL(2, K[x])$, J. Inst. Polytech. Osaka City Univ. Ser A, **10** (1959), 117-121.
7. B. H. Neumann, An essay on free products of groups with amalgamations, Phil. Trans. Roy. Soc. London Ser. A, **246** (1954), 503-554.
8. J. Nielsen, Die Gruppe der dreidimensionalen Gittertransformationen, D. K. Danske Vid. Sel., Mat–fys. Med. V (1924), 1-29.
9. J. R. Silvester, On the K_2 of a free associative algebre, to appear.
10. R. G. Swan, unpublished.

TRIVIAL EXTENSIONS OF ABELIAN CATEGORIES AND APPLICATIONS TO RINGS: AN EXPOSITORY ACCOUNT

R. Fossum, P. Griffith, I. Reiten*

University of Illinois

Introduction

During the Fall term of 1970, at the University of Illinois, in a course and seminar concerning applications of category theory to the study of rings and algebras, Professor M. Auslander asked whether the triangular matrix ring $\begin{pmatrix} \Lambda & 0 \\ \Lambda & \Lambda \end{pmatrix}$ has finitistic dimension bounded by a function of the finitistic dimension of Λ. He also asked whether this triangular matrix ring has any Gorenstein property if the ring Λ does. Auslander also defined the representation dimension of a finite dimensional algebra, and asked whether the triangular matrix ring inherited a finite representation dimension.

This last notion, a generalization of an algebra which has a finite number of isomorphism classes of indecomposable modules, was central to Auslander's discussions. We have established that the representation dimension of $\begin{pmatrix} \Lambda & 0 \\ \Lambda & \Lambda \end{pmatrix}$ increases by at most two over the representation dimension of Λ. Janusz has given an example of a finite dimensional algebra Λ which has a finite number of indecomposable modules,

* The first and third authors have been supported in part by the National Science Foundation. The third author has also been supported in part by a Norwegian Research Council (NAVF) research stipend.

but such that $\begin{pmatrix} \Lambda & 0 \\ \Lambda & \Lambda \end{pmatrix}$ does not have a finite number. So Auslander's representation dimension seems to be a good measure of how far away a ring is from having a finite number of indecomposable modules. Results on this subject can be found in Section 6.

In order to handle these ideas in a more concise manner, it is desirable to consider the category of maps between objects FA and B where $F : \mathfrak{A} \to \mathfrak{B}$ is a functor from an abelian category $\mathfrak{A}$ to an abelian category $\mathfrak{B}$. This category is well known, appearing already in Grothendieck's famous Tohoku paper, but discussed further in Mitchell [11, 12, 13] and Harada [7].

We want, in particular, to know which morphisms are projective or injective, and to know the injective envelopes of morphisms. We indicate, in Section 1, the categorical foundations of this theory with particular attention paid to the projective and injective objects and the homology of the change of category functors. We give several applica – tions of this theory which seem to be new. In particular, we obtain a theorem, in categorical language, which yields the corollary that the trivial extension of a ring by a two-sided module is perfect if and only if the ring is perfect.

Many of our applications concern coherent rings. We could use Roos' results [21] to show that the triangular matrix ring

$$\begin{pmatrix} R & 0 \\ {}_SM_R & S \end{pmatrix}$$

is left–coherent provided certain conditions on the derived functors $\mathrm{Tor}^R(M, -)$ are satisfied. We have been able to generalize Roos' theorems [21, 22] to a categorical setting, obtaining more straightforward proofs, and the results are announced in Section 2.

In Section 3 we discuss the homological dimension of a map $f : FA \to B$ in terms of the homological dimensions (projective dimension or injective dimension) of A, B and the derived functors L_iFA. This enables us to show that the category of morphisms has finite finitistic dimension if both $\mathfrak{A}$ and $\mathfrak{B}$ have finite finitistic dimension, irrespective of the right exact functor F.

We consider pseudo–duality in Section 4. The setting for a pseudo–duality is established, and then a k–Gorenstein category is defined. We give the relation between the property of being k–Gorenstein and the minimal injective resolution of projective objects. This was first proved for rings by Auslander. We can give conditions on a functor which entail that the category of morphisms has a pseudo–duality. However we are unable to give general conditions on F that imply the category is k–Gorenstein. On the contrary, many nice functors destroy any Gorenstein properties. However we can show that Λ is (k–) Gorenstein if and only if

$$\begin{pmatrix} \Lambda & 0 \\ \Lambda & \Lambda \end{pmatrix}$$

is (k–) Gorenstein. This answers one of the original questions.

Section 5 deals with applications of several of the formal categorical notions studied in earlier sections. In particular, starting with a given finite dimensional algebra Λ, a general method is given for constructing finite dimensional algebras of arbitrarily large (finite) dominant dimension. In addition, the algebra Λ is a particularly nice ring homomorphic image of the algebras so constructed. This construction allows us to characterize the category of finitely generated reflexive modules for a certain class of finite dimensional algebras of dominant dimension $\geqslant 2$.

Finally we wish to thank Professor Maurice Auslander for his stimulating discussions and constant encouragement. He will recognize many of his ideas and methods in this report. It is difficult for us to distinguish, in some of this theory, between results which he announced and our own findings. Sometimes we have rephrased his results in our own language, other results have been taken directly from his lectures without proper reference. His comments and criticism have greatly improved our presentation. Also, our colleague, Professor G. Janusz has contributed several examples which have helped in the formulation of some of these concepts. We take this opportunity to thank him as well.

1. Foundations

A. The categories

Let $\mathfrak{A}$ be an abelian category. For an additive functor $F : \mathfrak{A} \to \mathfrak{A}$ consider the category, which we denote by $(F, \mathfrak{A})$, whose objects are pairs (f, A) where f is a morphism $f : FA \to A$ in $\mathfrak{A}$, such that $f \cdot Ff = 0$, and where $\alpha : (f, A) \to (f', A')$ is a morphism in $(F, \mathfrak{A})$ when $\alpha : A \to A'$ is a morphism in $\mathfrak{A}$ such that the diagram

$$\begin{array}{ccc} FA & \xrightarrow{F\alpha} & FA' \\ \downarrow f & & \downarrow f' \\ A & \xrightarrow{\alpha} & A' \end{array}$$

commutes. Composition in $(F, \mathfrak{A})$ is obvious. We will have need to consider also the category $(\mathfrak{A}, G)$, where $G : \mathfrak{A} \to \mathfrak{A}$ is an additive functor; $(\mathfrak{A}, G)$ has as objects pairs (A, g), where g is a morphism $g : A \to GA$ such that $Gg \cdot g = 0$, and has the obvious morphisms and composition. The first result we note gives sufficient conditions on F and G to insure that the categories are abelian.

Theorem 1.1. Let $F, G : \mathcal{A} \to \mathcal{A}$ be endofunctors. Let $(F, \mathcal{A})$ and $(\mathcal{A}, G)$ be the categories defined above.

(a) If $\alpha : (f, A) \to (f', A')$ and if $i : A_0 \to A$ is the kernel of α in $\mathcal{A}$, then there is a unique $f_0 : FA_0 \to A_0$ such that (i) $f_0 \cdot Ff_0 = 0$ (so (f_0, A_0) is an object in $(F, \mathcal{A})$), (ii) $i : (f_0, A_0) \to (f, A)$ is a morphism in $(F, \mathcal{A})$ and (iii) i is the kernel of α.

(a′) If $\alpha : (A, g) \to (A', g')$ and if $\pi : A' \to A_1$ is the cokernel of α, then there is a unique $g_1 : A_1 \to GA_1$ such that (i) $Gg_1 \cdot g_1 = 0$, (ii) π is a morphism in $(\mathcal{A}, G)$ and (iii) π is the cokernel of α.

(b) If F is right exact, $\alpha : (f, A) \to (f', A')$ and $\pi : A' \to A_1$ is the cokernel of α, then there is a unique $f_1 : FA_1 \to A_1$ such that (i) $f_1 \cdot Ff_1 = 0$, (ii) π is a morphism in $(F, \mathcal{A})$ and (iii) π is the cokernel of α.

(b′) If G is left exact, then any morphism in $(\mathcal{A}, G)$ has a kernel.

(c) If F is right exact, then $(F, \mathcal{A})$ is an abelian category. Furthermore, a sequence in $(F, \mathcal{A})$

$$(f', A') \xrightarrow{\alpha} (f, A) \xrightarrow{\beta} (f'', A'')$$

is exact if and only if $A' \xrightarrow{\alpha} A \xrightarrow{\beta} A''$ is exact in $\mathcal{A}$.

(c′) If G is left exact, then $(\mathcal{A}, G)$ is an abelian category. A sequence

$$(A', g') \xrightarrow{\alpha} (A, g) \xrightarrow{\beta} (A'', g'')$$

is exact in $(\mathcal{A}, G)$ if and only if $A' \xrightarrow{\alpha} A \xrightarrow{\beta} A''$ is exact in $\mathcal{A}$.

The reason for considering both these categories is that we can consider projective objects in $(F, \mathcal{A})$ and injective objects in $(\mathcal{A}, G)$. When the categories are equivalent, then we can get injective and projective objects in either category (provided there are such objects in $\mathcal{A}$).

We give an outline of a proof for the next result.

Theorem 1.2. If F is a left adjoint of G, then the adjoint isomorphism $\psi_{A,B} : \mathcal{A}(FA, B) \cong \mathcal{A}(A, GB)$ induces an isomorphism of categories $\psi : (F, \mathcal{A}) \to (\mathcal{A}, G)$.

Proof. If $(f, A) \in (F, \mathcal{A})$, then $\psi(f, A)$ is given by $(A, \psi_{A,A} f)$. The only problem here is to check that $G(\psi f) \cdot \psi f = 0$ when $f \cdot Ff = 0$. This is routine. Now $\psi\alpha = \alpha$. The remainder of the proof consists in straightforward diagram chasing.

Before continuing, we need a bit more notation. If $A \in \mathcal{A}$ and $F : \mathcal{A} \to \mathcal{A}$ we can define a *standard object* $TA = \left(\begin{pmatrix} 0 & 0 \\ 1 & 0 \end{pmatrix}, A \oplus FA\right)$. So a standard object is is the morphism

$$\begin{pmatrix} 0 & 0 \\ 1 & 0 \end{pmatrix} : FA \oplus FFA \to A \oplus FA.$$

If $\alpha : A \to B$ in $\mathcal{A}$, we can define $T\alpha : TA \to TB$ by $T\alpha = \begin{pmatrix} \alpha & 0 \\ 0 & F\alpha \end{pmatrix}$ where $\begin{pmatrix} \alpha & 0 \\ 0 & F\alpha \end{pmatrix} : A \oplus FA \to B \oplus FB$. Thus $T : \mathcal{A} \to (F, \mathcal{A})$ is a functor. If we consider the codomain functor $\text{cod} : (F, \mathcal{A}) \to \mathcal{A}$, then it is easy to see that T is left adjoint to cod.

For $(\mathcal{A}, G)$ we define the standard object functor $H : \mathcal{A} \to (\mathcal{A}, G)$ by $HA = \left(\begin{pmatrix} 0 & 0 \\ 1 & 0 \end{pmatrix}, GA \oplus A\right)$ and $H\alpha = \begin{pmatrix} G\alpha & 0 \\ 0 & \alpha \end{pmatrix}$. Likewise consider $\text{dom} : (\mathcal{A}, G) \to \mathcal{A}$ as the domain functor. Then G is right adjoint to dom.

When F is a left adjoint of G there are relations between the standard objects. Let $\sigma : I_{\mathcal{A}} \to GF$ and $\tau : FG \to I_{\mathcal{A}}$ be the natural transformations associated to the adjointness isomorphism ψ. Now TA is an object in $(F, \mathcal{A})$. To it corresponds the object $\psi TA = \left(A \oplus FA, \begin{pmatrix} 0 & 0 \\ \sigma_A & 0 \end{pmatrix}\right)$; i.e., the morphism

$$\begin{pmatrix} 0 & 0 \\ \sigma_A & 0 \end{pmatrix} : A \oplus FA \to GA \oplus GFA.$$

The object HA in $(\mathcal{A}, G)$ has corresponding to it the object $\psi^{-1}HA = \left(\begin{pmatrix} 0 & 0 \\ \tau_A & 0 \end{pmatrix}, GA \oplus A\right)$.

B. Projective and injective objects, resolutions and homology

In the remainder of the paper we suppose always that F is right exact and G is left exact. However we do *not* always assume them to be adjoint.

In this section we discuss projective and injective objects.

Theorem 1.3. Let F be right exact and G left exact.

(a) An object $(f, A) \in (F, \mathcal{A})$ is projective if and only if (i) $\text{coker}\, f$ is projective in $\mathfrak{A}$ and (ii) (f, A) is isomorphic to T(coker f).

(a′) An object $(A, g) \in (\mathfrak{A}, G)$ is injective in $(\mathfrak{A}, G)$ if and only if (i) ker g is injective in $\mathfrak{A}$ and (ii) $(A, g) \cong H(\ker g)$.

(b) If A is nonzero, then the canonical morphism $(f, A) \to (0, \operatorname{coker} f)$ is a minimal epimorphism.

(b′) If A is nonzero, then the canonical morphism $(\ker g, 0) \to (A, g)$ is an essential monomorphism.

As corollaries we can read off results about $(F, \mathfrak{A})$ and $(\mathfrak{A}, G)$.

Corollary 1.4. (1) If $\mathfrak{A}$ is perfect, i.e., each object in $\mathfrak{A}$ has a projective cover, then $(F, \mathfrak{A})$ is perfect.

In fact, if $P \to \operatorname{coker} f$ is a projective cover of coker f in $\mathfrak{A}$, then $TP \to (f, A)$ is a projective cover of (f, A).

(2) If $\mathfrak{A}$ has enough injectives, then so does $(\mathfrak{A}, G)$. If $\ker g \to I$ is an injective envelope of ker g, then $(A, g) \to H(I)$ is an injective envelope of (A, g).

We now compute the derived functors of the standard object functors in terms of the derived functors of F and G. Actually we do this only for T, leaving the statements and details for H to the reader. It is possible to work out details in order to compute Yoneda's Ext in $(F, \mathfrak{A})$, for example, but they become exceedingly messy. We prefer to have available enough projectives (injectives in the category $(\mathfrak{A}, G)$) and consider derived functors with respect to projective resolutions. Thus our result.

Theorem 1.5. Let A be an object in $\mathfrak{A}$. Then $(L_iT)A \cong (0, (L_iF)A)$ for $i > 0$. (L_i is the left derived functor of the right exact functor.)

C. Abstract application

Let $\mathfrak{A}$, $\mathfrak{B}$ be abelian categories. If $F : \mathfrak{A} \to \mathfrak{B}$, we can consider the functor $\begin{pmatrix} 0 & 0 \\ F & 0 \end{pmatrix} : \mathfrak{A} \times \mathfrak{B} \to \mathfrak{A} \times \mathfrak{B}$ where $A \times B \to 0 \times FA$ and $\alpha \times \beta \to 0 \times F\alpha$. $\begin{pmatrix} 0 & 0 \\ F & 0 \end{pmatrix}$ is right exact if and only if F is right exact. When F is right exact, we get the abelian category $\left(\begin{pmatrix} 0 & 0 \\ F & 0 \end{pmatrix}, \mathfrak{A} \times \mathfrak{B}\right)$.

There is another way of looking at this category. In particular, let $\operatorname{Map}(F\mathfrak{A}, \mathfrak{B})$ (write: $(F\mathfrak{A}, \mathfrak{B})$) denote the category whose objects are triples (A, f, B) where $A \in \mathfrak{A}$, $B \in \mathfrak{B}$ and $f : FA \to B$. A morphism is a pair $\alpha/\beta : (A, f, B) \to (A', f', B')$

where $\alpha : A \to A'$, $\beta : B \to B'$ and the diagram

$$\begin{array}{ccc} FA & \xrightarrow{F\alpha} & FA' \\ f \downarrow & & \downarrow f' \\ B & \xrightarrow{\beta} & B' \end{array}$$

is commutative. However (A, f, B) is seen to be $(0 \times f, A \times B) \in (\begin{pmatrix} 0 & 0 \\ F & 0 \end{pmatrix}$, $\mathfrak{A} \times \mathfrak{B})$. And α/β is just $\alpha \times \beta$ (where $\begin{pmatrix} 0 & 0 \\ F & 0 \end{pmatrix} \alpha \times \beta = 0 \times F\alpha$). Since $\begin{pmatrix} 0 & 0 \\ F & 0 \end{pmatrix}\begin{pmatrix} 0 & 0 \\ F & 0 \end{pmatrix} = 0$, the condition $\psi \cdot \begin{pmatrix} 0 & 0 \\ F & 0 \end{pmatrix} \psi = 0$ is vacuous. We see that $\mathrm{Map}(F\mathfrak{A}, \mathfrak{B}) = (\begin{pmatrix} 0 & 0 \\ F & 0 \end{pmatrix}, \mathfrak{A} \times \mathfrak{B})$. In the next theorem, we read off the statements of the theorems in the previous sections for the category $\mathrm{Map}(F\mathfrak{A}, \mathfrak{B})$ when F is right exact.

Of course for $G : \mathfrak{B} \to \mathfrak{A}$, we get a category $(\mathfrak{B} \times \mathfrak{A}, \begin{pmatrix} 0 & 0 \\ G & 0 \end{pmatrix}) = \mathrm{Map}(\mathfrak{A}, G\mathfrak{B}) = (\mathfrak{A}, G\mathfrak{B})$. This has the interpretation as the category whose objects are triples (A, g, B), $g : A \to GB$ with the obvious morphisms and composition.

The standard object functor T induces a functor $S : \mathfrak{A} \to (F\mathfrak{A}, \mathfrak{B})$ by $SA = T(A \times 0)$. It is seen that $SA = (A, id_{FA}, FA)$. Also H induces $K : \mathfrak{B} \to (\mathfrak{A}, G\mathfrak{B})$ given by $K\mathfrak{B} = H(B \times 0) = (GB, id_{GB}, B)$.

Theorem 1.6. Let $F : \mathfrak{A} \to \mathfrak{B}$, $G : \mathfrak{B} \to \mathfrak{A}$ be right and left exact respectively. Then $\mathrm{Map}(F\mathfrak{A}, \mathfrak{B})$ and $\mathrm{Map}(\mathfrak{A}, G\mathfrak{B})$ are abelian categories.

(a) An object (A, f, B) is projective in $\mathrm{Map}(F\mathfrak{A}, \mathfrak{B})$ if and only if A is projective, f is a monomorphism and $\mathrm{coker}\, f$ is projective. So $(A, f, B) \cong SA \oplus (0, 0, B)$ where A and B are projective.

(b) The dual statement for injectives holds in $\mathrm{Map}(\mathfrak{A}, G\mathfrak{B})$.

(c) The left derived functors of S, L_iS, can be computed in $\mathfrak{B}$. In particular $(L_iS)(A) = (0, 0, L_iFA)$.

D. Concrete examples

Let R be a ring, M an R–bimodule. The *trivial extension* of R by M is the ring whose underlying set is $R \times M$ with ring multiplication given by $(r, m)(s, n) = (rs, ms + rn)$. If $\mathfrak{A} = {}_R\mathrm{Mod}$ and $F = M \otimes_R \cdot$, then $(F, \mathfrak{A})$ is isomorphic to the category ${}_{R\times M}\mathrm{Mod}$. We have the converse of this remark.

Theorem 1.7. Let $\mathfrak{A} = {}_R\mathrm{Mod}$ and $F : \mathfrak{A} \to \mathfrak{A}$ a right exact functor. The following are equivalent.

(a) $(F, \mathfrak{A})$ is a module category.

(b) F preserves coproducts.

(c) $F = M \otimes_R \cdot$ for an R–bimodule M.

(d) $(F, \mathfrak{A}) \cong {}_{R\times M}\mathrm{Mod}$.

If R, S are rings and M is an S–R–bimodule, then the module M becomes an $R \times S$–bimodule. $(R \times S) \times M$ is the ring of matrices $\Omega = \begin{pmatrix} R & 0 \\ M & S \end{pmatrix}$ with ordinary matrix multiplication. If $F = M \otimes_R \cdot : {}_R\mathrm{Mod} \to {}_S\mathrm{Mod}$, then ${}_\Omega\mathrm{Mod} \cong \mathrm{Map}(F_R\mathrm{Mod}, {}_S\mathrm{Mod})$.

We give here an interesting application of Theorem 1.3.

Theorem 1.8. Let R be a ring. Let **P**(R) denote the additive full subcategory of projective left R–modules of finite type. Let M be a bi–R–module. Then the functor $(R \times M) \otimes_R \cdot : {}_R\mathrm{Mod} \to {}_{R\times M}\mathrm{Mod}$ induces an isomorphism $\mathbf{P}(R) \to \mathbf{P}(R \times M)$. Consequently, $K_0(R) \cong K_0(R \times M)$, where $K_0(R)$ denotes the Grothendieck group of **P**(R).

Further applications involving the Picard group will yield an isomorphism $\mathrm{Pic}(R) \cong \mathrm{Pic}(R \times M)$.

One can also look at injectives over $R \times M$. It is possible to use the change of rings arguments directly. (In fact the functors T, H, ker, coker are exactly the change of rings functors $R \to R \times M$ and $R \times M \to R \times M/M \cong R$. These can be used to give the results above for projectives.)

Theorem 1.9. Let R be a ring, M a bimodule, A the right annihilator of M in R.

(a) $R \times M$ is an essential extension of the left ideal $A \times M$.

(b) An injective envelope of the left $R \times M$–module $R \times M$ is the left $R \times M$–module $\mathrm{Hom}_R(R \times M, E_R(A \times M)) \cong \mathrm{Hom}_R(M, E_R(A \times M)) \times E_R(A \times M)$.

(c) $R \times M$ is left self–injective if and only if

(i) A and M are left injective

(ii) $R \to \mathrm{Hom}_R(M, M)$ is an epimorphism (the map is right multiplication).

and (iii) $\mathrm{Hom}_R(M, A) = 0$.

(d) When $R \times M$ is self–injective, let $A = Re$, e an idempotent and $f = 1 - e$. Then $eRf = 0$, $fRf \cong \mathrm{Hom}_R(M, M)$ and so R is the triangular matrix ring

$$\begin{pmatrix} eRe & 0 \\ fRe & fRf \end{pmatrix}.$$

2. Coherence

Let $\mathcal{A}$ be an abelian category which admits a small full additive subcategory $\mathcal{P}$ whose objects are projective in $\mathcal{A}$. Recall that an object $A \in \mathcal{A}$ is (a) of *finite* $\mathcal{P}$*–type* if there is an epimorphism $P \to A$ for some $P \in \mathcal{P}$; (b) of *finite* $\mathcal{P}$*–presentation* if there is an exact sequence $P_1 \to P_0 \to A \to 0$ in $\mathcal{A}$ with $P_0, P_1 \in \mathcal{P}$; (c) $\mathcal{P}$*–coherent* if A is of finite $\mathcal{P}$–type and each subobject of finite $\mathcal{P}$–type is of finite $\mathcal{P}$–presentation; (d) *pseudo–*$\mathcal{P}$*–coherent* if each subobject of finite $\mathcal{P}$–type is coherent. Let $\mathrm{Coh}_{\mathcal{P}}\mathcal{A}$ be the full additive subcategory of $\mathcal{P}$–coherent objects. $\mathrm{Coh}_{\mathcal{P}}\mathcal{A}$ is abelian. We say $\mathcal{P}$ is *coherent* if $\mathcal{P} \subseteq \mathrm{Coh}_{\mathcal{P}}\mathcal{A}$. The next result, whose statement is due to Roos [22], gives conditions on extension of coherence.

Theorem 2.1. Let $\mathcal{P}$ be a small additive full subcategory of projective objects in $\mathcal{A}$. Let $F : \mathcal{A} \to \mathcal{A}$ be right exact. Let $T(\mathcal{P})$ denote the small additive full subcategory of standard objects over $\mathcal{P}$ in $(F, \mathcal{A})$. Then $T(\mathcal{P})$ is coherent if and only if $\mathcal{P}$ satisfies the conditions:

(i) $\mathcal{P}$ is coherent.

(ii) If A is $\mathcal{P}$–coherent, then FA is pseudo–$\mathcal{P}$–coherent and L_iFA is coherent for $i > 0$.

(iii) If A is $\mathcal{P}$–coherent and if B is a subobject of FA of finite $\mathcal{P}$–type, then FB is of finite $\mathcal{P}$–type (but it need not be coherent).

When this is applied to the more special case of a functor $F : \mathcal{A} \to \mathcal{B}$, we get the corollary.

Corollary 2.2. Let $F : \mathcal{A} \to \mathcal{B}$ be right exact. Let $\mathcal{P}$, $\mathcal{Q}$ be small full additive subcategories of projective objects in $\mathcal{A}$ and $\mathcal{B}$ respectively. Let $T(\mathcal{P} \times \mathcal{Q})$ be the associated category in $\mathrm{Map}(F\mathcal{A}, \mathcal{B})$. Then $T(\mathcal{P} \times \mathcal{Q})$ is coherent if and only if

(a) $\mathcal{P}$ and $\mathcal{Q}$ are coherent.

(b) If A is $\mathcal{P}$–coherent, then FA is pseudo–$\mathcal{Q}$–coherent and $L_i FA$ is $\mathcal{Q}$–coherent for $i > 0$.

When the conditions of Theorem 2.1 or Corollary 2.2 are satisfied, we say F is a *coherent preserving* functor.

Recall that in Section 1 we considered the trivial extension $R \times M$, R a ring and M a bimodule. This theorem and its corollary yield conditions on M that insure that $R \times M$ is left coherent when, say, R is left coherent.

Corollary 2.3. $R \times M$, the trivial extension of R by the bimodule M, is left coherent if and only if (a) R is left coherent, (b) if A is a coherent left R–module, then $M \otimes_R A$ is a pseudo coherent left R–module and $\mathrm{Tor}_i^R(M, A)$ is coherent for $i > 0$ and (c) if A is a coherent left R–module and B is a submodule of $M \otimes_R A$, then $M \otimes_R B$ is of finite type as a left R–module.

A more concrete example of an application of this theorem can be given.

Corollary 2.4. Let R be a noetherian local ring with residue field k. Let E be the injective envelope of k. Then $R \times E$ is coherent if and only if $\dim R = 0$ or $\dim R = 1$ and the total ring of quotients of R is injective as an R–module. [Hence the total ring of quotients is self–injective. This condition is equivalent to many other conditions, for example, R is 1–Gorenstein.]

Another application is to generalized matrix rings. Let R, S be rings, ${}_R M_S$ and ${}_S N_R$ bimodules with the indicated actions. We form the ring $(R \times S) \times (M \times N)$ where $M \times N$ is a right $R \times S$–module by the action of R on the right of N and S on the right of N, etc. A more suggestive notation, due to Roos [22], is $\begin{pmatrix} R & M \\ N & S \end{pmatrix}$ where the multiplication is ordinary matrix multiplication, with the proviso that $nm = 0$, $mn = 0$ for $n \in N$, $m \in M$.

Corollary 2.5 (Roos [22]). $\Gamma = \begin{pmatrix} R & M \\ N & S \end{pmatrix}$ is left coherent if and only if (a) R and S are left coherent; (b) (i) if A is a coherent left R–module then $N \otimes_R A$ is pseudo–coherent left S–module and $\mathrm{Tor}_i^R(N, A)$ is coherent for each $i > 0$, (ii) if B is a coherent left S–module, then $M \otimes_S B$ is a pseudo coherent left R–module and $\mathrm{Tor}_i^S(M, B)$ is a coherent left R–module for $i > 0$; (c) (i) if A is a coherent left R–module and B is a finitely generated S–submodule of $N \otimes_R A$, then $M \otimes_S B$ is a finitely generated left R–module, (ii) if C is a coherent left S–module and D is a finitely generated R–submodule of $M \otimes_S C$, then $N \otimes_R D$ is a finitely generated left S–module.

This is the original theorem of Roos. It can be obtained by applying our Theorem 2.1 to the special case.

3. Homological dimension in Map($F\mathfrak{A}$, $\mathfrak{B}$)

With the general properties of Map($F\mathfrak{A}$, $\mathfrak{B}$) having been provided in Section 1, we now consider various homological dimensions in Map($F\mathfrak{A}$, $\mathfrak{B}$) = $\mathfrak{M}$ via a "change of category" procedure for computing the bifunctor $\mathrm{Ext}^i_{\mathfrak{M}}(\cdot, \cdot)$. Our attention is generally focused on finitistic homological dimension, as defined by Bass [3], since results on global dimension follow as corollaries. It should be mentioned that Chase [4], Eilenberg–Rosenberg–Zelinsky [5], Harada [7], Mitchell [11, 12, 13] and Fields [F] have obtained results related to ours.

At the end of this section as well as in subsequent sections (see Sections 5 and 6) we provide several applications of the results developed in this section. In addition, some typical and pathological examples are discussed.

Before describing our results, a few definitions concerning $\mathfrak{M}$ = Map($F\mathfrak{A}$, $\mathfrak{B}$) are needed. Recall from Section 1, an object $M \in \mathfrak{M}$ corresponds to a triple (A, f_M, B) where $A \in \mathfrak{A}$, $B \in \mathfrak{B}$ and $f_M : FA \to B$ (called the map associated with M). This correspondence yields two exact functors

$$\mathbf{a} : \mathfrak{M} \to \mathfrak{A} \quad \text{defined by} \quad \mathbf{a}(M) = A, \quad \text{and}$$

$$\mathbf{b} : \mathfrak{M} \to \mathfrak{B} \quad \text{defined by} \quad \mathbf{b}(M) = B,$$

with corresponding actions on morphisms. In addition, the functors **a** and **b** give rise to two exact functors $\mathbf{a}^* : \mathfrak{A} \to \mathfrak{M}$ and $\mathbf{b}^* : \mathfrak{B} \to \mathfrak{M}$, respectively, such that

(i) $\mathbf{a}^*$ is a right adjoint of **a**,

(ii) $\mathbf{b}^*$ is a left adjoint of **b**,

(iii) $\mathbf{a}\mathbf{a}^* = 1_{\mathfrak{A}}$, $\mathbf{b}\mathbf{b}^* = 1_{\mathfrak{B}}$.

(iv) There is an exact sequence of functors

$$0 \to \mathbf{b}^*\mathbf{b} \to 1_{\mathfrak{M}} \to \mathbf{a}^*\mathbf{a} \to 0.$$

If $A \in \mathfrak{A}$, then $\mathbf{a}^*A$ corresponds to the triple $(A, 0, 0)$ in $\mathfrak{M}$ and, if $B \in \mathfrak{B}$, $\mathbf{b}^*B$ corresponds to the triple $(0, 0, B)$ in $\mathfrak{M}$. Our final definition in this section is that of a functor $S : \mathfrak{A} \to \mathfrak{M}$ which sends an object $A \in \mathfrak{A}$ to the object in $\mathfrak{M}$ corresponding to the triple $(A, 1_{FA}, FA)$. In terms of the terminology of Section 1,

S corresponds to the functor $A \to T(A \times 0)$. We remark that $S : \mathcal{A} \to \mathcal{M}$ is a right exact functor since $F : \mathcal{A} \to \mathcal{B}$ is a right exact functor. However, the most crucial property possessed by the functor S is the natural equivalence of functors

$$L_iS \cong b^*L_iF, \quad \text{for } i \geqslant 1$$

(established in Section 1). It is this natural equivalence together with properties of the aforementioned functors **a, a*, b** and **b*** which enable us to reduce homological problems in $\mathcal{M} = \mathrm{Map}(F\mathcal{A}, \mathcal{B})$ to corresponding ones in $\mathcal{A}$ and $\mathcal{B}$. Finally, one can derive the following two statements from Section 1.

(v) The functors **a, b*** and S are projective preserving functors.

(vi) An object in $\mathcal{M}$ is projective if and only if it is isomorphic to $SP \oplus b^*Q$, for some projectives P and Q in $\mathcal{A}$ and $\mathcal{B}$, respectively. We shall assume, throughout the remaining portions of this section, that the abelian categories $\mathcal{A}$ and $\mathcal{B}$ have enough projectives; hence $\mathcal{M} = \mathrm{Map}(F\mathcal{A}, \mathcal{B})$ also has enough projectives (see Section 1).

Our first, somewhat amusing, result on change of categories is the following theorem.

Theorem 3.1. Let $\mathcal{M} = \mathrm{Map}(F\mathcal{A}, \mathcal{B})$ and let $A \in \mathcal{A}$ and $B \in \mathcal{B}$.

(a) There is a natural isomorphism

$$\mathrm{Ext}^1_{\mathcal{M}}(a^*A, b^*B) \cong \mathrm{Hom}_{\mathcal{B}}(FA, B).$$

(b) If F is an exact functor, there are natural isomorphisms

$$\mathrm{Ext}^{i+1}_{\mathcal{M}}(a^*A, b^*B) \cong \mathrm{Ext}^i_{\mathcal{B}}(FA, B) \quad \text{for } i \geqslant 0.$$

Under the assumption that F is an exact functor, the following exact sequences are useful for computing $\mathrm{Ext}^i_{\mathcal{M}}(M, N)$, where M and N are arbitrary objects in $\mathcal{M}$.

(i) $\cdots \to \mathrm{Ext}^i_{\mathcal{M}}(M, b^*bN) \to \mathrm{Ext}^i_{\mathcal{M}}(M,N) \to$
$\mathrm{Ext}^i_{\mathcal{A}}(aM, aN) \to \mathrm{Ext}^{i+1}_{\mathcal{M}}(M,b^*bN) \to \cdots$

(ii) $\mathrm{Ext}^{i-1}_{\mathcal{B}}(FaM, bN) \to \mathrm{Ext}^i_{\mathcal{M}}(M, b^*bN) \to \mathrm{Ext}^i_{\mathcal{B}}(bM, bN).$

If f_M is the map associated with M, there is a natural map $M \to b^*(\mathrm{coker}\, f_M)$. If M_e denotes the kernel of this map, then M_e has the property that its associated map is an epimorphism and furthermore

(iii) $\mathrm{Ext}^i_{\mathfrak{B}}(\ker f_M, B) \cong \mathrm{Ext}^{i+1}_{\mathfrak{M}}(M_e, b^*B)$.

An application of (i), (ii) and (iii) provides us with our best result concerning "change of categories".

Theorem 3.2. Let $\mathfrak{M} = \mathrm{Map}(F\mathfrak{A}, \mathfrak{B})$ with F an exact functor. Let $M \in \mathfrak{M}$ with the property that its associated map f_M is monic in $\mathfrak{B}$ and let N be an arbitrary object in $\mathfrak{M}$. Then, for $i \geq 0$, the following sequence is exact.
$\cdots \to \mathrm{Ext}^i_{\mathfrak{B}}(\mathrm{coker}\, f_M, bN) \to \mathrm{Ext}^i_{\mathfrak{M}}(M, N) \to \mathrm{Ext}^i_{\mathfrak{A}}(aM, aN) \to \mathrm{Ext}^{i+1}_{\mathfrak{B}}(\mathrm{coker}\, f_M, bN) \to \cdots$

In the remainder of this section, we turn our attention to projective dimension in $\mathrm{Map}(F\mathfrak{A}, \mathfrak{B})$ (F need no longer be exact). Our first proposition follows trivially from the fact that the functors **a** and **b*** are exact projective preserving functors.

Proposition 3.3. Let $M \in \mathfrak{M} = \mathrm{Map}(F\mathfrak{A}, \mathfrak{B})$ and let $B \in \mathfrak{B}$. Then

(i) $\mathrm{pd}_{\mathfrak{M}} b^*B = \mathrm{pd}_{\mathfrak{B}} B$.
(ii) $\mathrm{pd}_{\mathfrak{M}} M \geq \mathrm{pd}_{\mathfrak{A}} aM$.

As a consequence of Proposition 3.3 and the additional fact that L_iS is naturally equivalent to b^*L_iF for $i \geq 1$, we obtain a fundamental result on projective dimension in $\mathrm{Map}(F\mathfrak{A}, \mathfrak{B})$. Define $d = \sup\{\mathrm{pd}_{\mathfrak{B}} FP : P \text{ is projective in } \mathfrak{A}\}$.

Theorem 3.4. Let $M \in \mathfrak{M} = \mathrm{Map}(F\mathfrak{A}, \mathfrak{B})$. If $\mathrm{pd}_{\mathfrak{M}} M < \infty$, then there is some object $B \in \mathfrak{B}$ such that $\mathrm{pd}_{\mathfrak{M}} M \leq 1 + \mathrm{pd}_{\mathfrak{A}} aM + \mathrm{pd}_{\mathfrak{B}} B < \infty$ and if $d < \infty$ then $\mathrm{pd}_{\mathfrak{M}} M \leq \sup\{\mathrm{pd}_{\mathfrak{B}} B, 1 + d + \mathrm{pd}_{\mathfrak{A}} aM\}$.

One makes use of $L_iS = b^*L_iF$ in the proof of 3.4 in the following way. Let $P. \to aM$ be a projective resolution of aM in $\mathfrak{A}$. This gives a complex $SP. \to M$ in $\mathfrak{M}$ whose homology lies in the image of the functor **b***, since $L_iS(aM) = b^*L_iF(aM)$. Hence, one is able to repair the inexactness of $SP. \to M$ with projectives Q_i in $\mathfrak{B}$ and obtain a projective resolution of M in $\mathfrak{M}$ of the form $SP. \oplus Q. \to M$.

Corollary 3.5. Let $\mathfrak{M} = \mathrm{Map}(F\mathfrak{A}, \mathfrak{B})$. Then the following inequalities hold.

(a) sup(finite gl. dim. $\mathfrak{A}$, finite gl. dim. $\mathfrak{B}$) $\leq$ finite gl. dim. $\mathfrak{M} \leq 1 +$ finite gl. dim. $\mathfrak{A}$ + finite gl. dim. $\mathfrak{B}$.

(b) sup(dim. $\mathfrak{A}$, dim. $\mathfrak{B}$, $1 + d$) $\leq$ dim $\mathfrak{M} \leq$ sup($1 + d +$ dim. $\mathfrak{A}$, dim. $\mathfrak{B}$) when d is finite, and dim. is either finite gl. dim. or gl. dim.

This has the obvious application to triangular matrix rings, in which case the upper bound in (b) agrees with Fields' result [F], and the lower bound agrees with a result of Ingegerd Palmer (private communication). The bounds for finite gl. dim. seem to be new, even for triangular matrix rings.

Corollary 3.6. Let R and S be rings and let M be an S–R–bimodule. Let $T = \begin{pmatrix} R & 0 \\ M & S \end{pmatrix}$ with the usual matrix multiplication. Then, for left or right homological dimension,

(a) sup(finite gl. dim. R, finite gl. dim. S) $\leqslant$ finite gl. dim. T $\leqslant 1 +$ finite gl. dim. R + finite gl. dim. S.

(b) sup(dim. R, dim. S, $1 + d$) $\leqslant$ gl. dim. T $\leqslant$ sup(dim. R + 1 + d, dim. S) when d is finite and (as above) dim denotes either gl. dim. or finite gl. dim.

Corollary 3.6 has the obvious translation into the following setting. Suppose R is a ring and e is an idempotent of R with either $eR(1 - e) = 0$ or $(1 - e)Re = 0$. Then, in either case (for right or left global dimension), sup(finite gl. dim. eRe, finite gl. dim. $(1 - e)R(1 - e)$) $\leqslant$ finite gl. dim. R $\leqslant 1 +$ finite gl. dim. eRe + finite gl. dim. $(1 - e)R(1 - e)$. The same inequalities hold for global dimension.

One further observation in regard to Corollary 3.6 should be made. Namely, the finiteness of the global or finitistic global dimension of T is independent of the algebraic or homological properties of the S–R–bimodule M. However, as (b) indicates, the bimodule M does affect the value of finite gl. dim. T and gl. dim. T between the upper and lower extreme values of the inequalities in (a) and (b).

We return to the general setting $\mathfrak{M} = \mathrm{Map}(F\mathfrak{A}, \mathfrak{B})$. If $M \in \mathfrak{M}$, one always has the exact sequence

$$0 \to \mathbf{b^*b}M \to M \to \mathbf{a^*a}M \to 0.$$

Since $\mathrm{pd}_{\mathfrak{M}}\mathbf{b^*b}M = \mathrm{pd}_{\mathfrak{B}}\mathbf{b}M$ (Proposition 3.3 (i)), one would expect more accurate calculations of finite gl. dim. $\mathfrak{M}$ and gl. dim. $\mathfrak{M}$ within the inequalities of Corollary 3.5 if one could give a more precise calculation of $\mathrm{pd}_{\mathfrak{M}}\mathbf{a^*a}M$. This will be accomplished in Theorem 3.11 under the assumption that F is an exact functor. However, we are able to provide some information even when F is not exact. This somewhat crude estimate of $\mathrm{pd}_{\mathfrak{M}}\mathbf{a^*a}M$ will follow as a corollary of our next theorem.

Theorem 3.7. Let $\mathfrak{C}$ and $\mathfrak{D}$ be abelian categories with enough projectives and let $G : \mathfrak{C} \to \mathfrak{D}$ be a right exact functor. Let $d = \sup\{\mathrm{pd}_{\mathfrak{D}}GP : P$ projective in $\mathfrak{C}\}$. Then, for any object $C \in \mathfrak{C}$,

$$\mathrm{pd}_{\mathcal{D}} GC \leqslant d + \mathrm{pd}_{\mathcal{C}} C + \sup\{1 + \mathrm{pd}_{\mathcal{D}} L_i G(C) : i \geqslant 1,\ L_i G(C) \neq 0\}.$$

(We agree sup(null set) = 0.)

Corollary 3.8 Let $\mathcal{M} = \mathrm{Map}(F\mathcal{A}, \mathcal{B})$ and let $A \in \mathcal{A}$. Then

(a) $\mathrm{pd}_{\mathcal{A}} A \leqslant \mathrm{pd}_{\mathcal{M}} SA \leqslant \mathrm{pd}_{\mathcal{A}} A + \sup\{1 + \mathrm{pd}_{\mathcal{B}} L_i FA : i \geqslant 1 \text{ and } L_i FA \neq 0\}$.

(b) $\mathrm{pd}_{\mathcal{A}} A \leqslant \mathrm{pd}_{\mathcal{M}} \mathbf{a}^*A \leqslant \mathrm{pd}_{\mathcal{A}} A + \sup\{1 + \mathrm{pd}_{\mathcal{B}} L_i FA : i \geqslant 0 \text{ and } L_i FA \neq 0\}$.

The left hand inequalities in 3.8 follow from Proposition 3.3 (ii). The right hand inequality in 3.8 (a) is a consequence of Theorem 3.7 with $G = S$ and the fact S preserves projectives. Then 3.8 (b) follows from 3.3 (a) and the exact sequence $0 \to \mathbf{b}^*FA \to SA \to \mathbf{a}^*A \to 0$.

Corollary 3.9. If $\mathcal{M} = \mathrm{Map}(F\mathcal{A}, \mathcal{B})$ with F exact and if $A \in \mathcal{A}$, then $\mathrm{pd}_{\mathcal{M}} SA = \mathrm{pd}_{\mathcal{A}} A$.

Remark. The general setup in Theorem 3.7 is applicable to several well–known change of rings theorems, since the left derived functors $L_i G$ of G, in these instances, possess some distinguished property. For example, if R is a commutative domain and if I is an invertible ideal of R, then one has $\text{gl. dim. } R \geqslant \text{gl. dim.}(R/I) + 1$ if $\text{gl. dim.}(R/I) < \infty$. In this case, $G. = (R/I) \otimes_R \cdot$ is right exact and projective preserving, $L_i G = 0$ for $i \geqslant 2$ and $L_1 G. = \mathrm{Tor}_1^R(R/I, \cdot)$ has the property that $\mathrm{pd}_{R/I}(L_1 GB) = \mathrm{pd}_{R/I} GB$ whenever $IB = 0$.

Our determination of $\mathrm{pd}_{\mathcal{M}} \mathbf{a}^*A$ makes use of the following lemma which is a consequence of Theorem 3.2.

Lemma 3.10. Let $\mathcal{M} = \mathrm{Map}(F\mathcal{A}, \mathcal{B})$ with F exact. If $M \in \mathcal{M}$ and $\mathrm{pd}_{\mathcal{M}} M = \mathrm{pd}_{\mathcal{A}} \mathbf{a}M < \infty$ and if the map f_M associated with M is monic in $\mathcal{B}$, then $\mathrm{pd}_{\mathcal{B}}(\mathrm{coker}\, f_M) \leqslant \mathrm{pd}_{\mathcal{A}} \mathbf{a}M$.

Theorem 3.11. Let $\mathcal{M} = \mathrm{Map}(F\mathcal{A}, \mathcal{B})$ with F exact and let $A \in \mathcal{A}$. Then

$$\mathrm{pd}_{\mathcal{M}} \mathbf{a}^*A = \sup\{1 + \mathrm{pd}_{\mathcal{B}} FA,\ \mathrm{pd}_{\mathcal{A}} A\}.$$

Corollary 3.12. Let $\mathcal{M} = \mathrm{Map}(F\mathcal{A}, \mathcal{B})$ with F exact. If $\mathrm{pd}_{\mathcal{B}} FA <$ finite gl. dim. $\mathcal{B}$ whenever $\mathrm{pd}_{\mathcal{M}} \mathbf{a}^*A < \infty$ then

$$\text{finite gl. dim. } \mathcal{M} = \sup\{\text{finite gl. dim. } \mathcal{B},\ \text{finite gl. dim. } \mathcal{A}\};$$

otherwise,

$$\text{finite gl. dim. } \mathcal{M} = \sup\{1 + \text{finite gl. dim. } \mathcal{B},\ \text{finite gl. dim. } \mathcal{A}\}.$$

We conclude this section with a number of consequences and examples of the preceding theory on homological dimension in $\mathcal{M} = \mathrm{Map}(F\mathcal{A}, \mathcal{B})$. The following corollary was first obtained by Eilenberg, Rosenberg, and Zelinsky [5] for global dimension of the $n \times n$ lower triangular matrices over a ring. The proof makes use of Corollary 3.12 as well as elementary induction argument.

Corollary 3.13. Let $T_n(R)$ denote the $n \times n$ lower triangular matrices over a ring R. Then, for $n \geqslant 2$,

$$\text{left finite gl. dim. } T_n(R) = 1 + \text{left finite gl. dim. } R.$$

The statement holds for left global dimension as well as the corresponding right homological dimensions.

The following example shows that the estimates on projective dimension in Corollary 3.5, Theorem 3.7, and Corollary 3.8 cannot, in general, be improved.

Example 3.14. Let $\mathbf{Z}$ denote the ring of integers, $\mathbf{Z}_p$ the integers modulo $p > 1$ and let R be the ring

$$R = \begin{pmatrix} \mathbf{Z} & 0 \\ \mathbf{Z}_p & \mathbf{Z} \end{pmatrix}.$$

Then, left gl. dim. R = right gl. dim. $R = 1 + \text{gl. dim. } \mathbf{Z} + \text{gl. dim. } \mathbf{Z} = 3$. Also $\mathrm{pd}_R(S\mathbf{Z}_p) = 3 = \mathrm{pd}_{\mathbf{Z}}\mathbf{Z}_p + (1 + \mathrm{pd}_{\mathbf{Z}}\mathrm{Tor}_1^{\mathbf{Z}}(\mathbf{Z}_p, \mathbf{Z}_p))$, where S is the functor defined at the beginning of this section.

In [1] M. Auslander proved, if R is a ring, left gl. dim. $R = \sup\{\mathrm{pd}_R C : C$ is a cyclic left R-module$\}$. Our next example shows, for R and S rings, $\mathcal{M} = \mathrm{Map}(F_R\mathrm{Mod}, {}_S\mathrm{Mod})$ need not inherit the above property concerning global dimension even though $\mathcal{M}$ has a small projective generator. Indeed, the example suggests the sort of pathology which may arise when the functor F does not commute with coproducts.

Example 3.15. Let $\mathbf{Z}$ denote the ring of integers, $\mathbf{Q}$ the field of rational numbers and let $\mathcal{M} = \mathrm{Map}(F\ \mathrm{Mod}\ \mathbf{Z}, \mathrm{Mod}\ \mathbf{Q})$, where $F. = \mathrm{Ext}_{\mathbf{Z}}^1(\mathbf{Q}, \cdot)$. Then $\mathcal{M}$ is a coherent abelian category with a small projective generator $P = SZ \oplus b^*Q$ such that

$$\sup\{pd_{\mathcal{M}} M : M \text{ of finite P–type}\} = 1, \text{ but gl. dim. } \mathcal{M} = 2.$$

The crucial observations needed in the above example are that F is exact on finitely generated Z–modules and M is of finite P–type if and only if aM is a finitely generated Z–module and coker f_M (associated map of M) is a finitely generated Q–module.

Our final example of this section is a sort of generalization of an example of Small [22]. It shows that the upper estimate of gl. dim. $\mathcal{M}$ in Corollary 3.5 cannot, in general, be sharpened even when F is an exact functor. We note that the lower estimate of gl. dim. $\mathcal{M}$ in Corollary 3.5 occurs whenever F is the zero functor.

Example 3.16. Let k be a countable field $R = k[x_1, \cdots, x_n]$, $n \geq 2$, and let Q be the field of quotients of the integral domain $R/(x_1, \cdots, x_{n-1})$. Set

$$T = \begin{pmatrix} R & 0 \\ Q & Q \end{pmatrix}.$$

Then left gl. dim. $T = n$ and right gl. dim. $T = 1 + n$.

Remark. Theorems dual to those in this section hold for injective dimension in the abelian category Map($\mathcal{A}$, G$\mathcal{B}$) of "maps over G" where G is a left exact functor from $\mathcal{A}$ to $\mathcal{B}$ (see Section 1). Furthermore, if $\mathcal{A} = {}_R\text{Mod}$, $\mathcal{B} = {}_S\text{Mod}$ and $F : \mathcal{A} \to \mathcal{B}$ is right exact and commutes with coproducts (i.e., by Watt's Theorem [5], $F. = M \otimes_R \cdot$ for some S–R–bimodule M), then results similar to the preceding may be established with respect to flat dim Map(F$\mathcal{A}$, $\mathcal{B}$).

4. Pseudo–duality

A ring R which is coherent (left coherent and right coherent) is said to be k–Gorenstein if flat. dim $E_i(R) \leq i$ for $0 \leq i \leq k-1$, where $E_i(R)$ is the ith term in a minimal injective resolution of the left module R. Auslander has shown this to be left–right independent. The main purpose of this section is to show that R is k–Gorenstein if and only if $T_2(R)$ is k–Gorenstein. We could proceed directly to this result. However we want to include a discussion of pseudo–duality, which includes a new result about the Gorenstein dimension of a category (a result anticipated by Auslander). Also several interesting problems arise which we cannot solve, and we take this opportunity to mention them.

Let $\mathcal{P}$ be a small additive category. **Ab** denotes the category of abelian groups. By $[\mathcal{P}, \mathbf{Ab}]$ we denote the category of additive covariant functors from $\mathcal{P}$ to **Ab**, $h^P = \mathrm{Hom}_{\mathcal{P}}(P, \cdot)$ is the P–representable object in $[\mathcal{P}, \mathbf{Ab}]$, $\mathrm{fp}[\mathcal{P}, \mathbf{Ab}]$ denotes the full subcategory of finitely $\mathcal{P}$–presented functors and $\mathrm{Coh}[\mathcal{P}, \mathbf{Ab}]$ the category of $\mathcal{P}$–coherent functors. Similar notations will prevail for the contravariant functors.

There is a tensor product

$$\otimes : [\mathcal{P}^{op}, \mathbf{Ab}] \times [\mathcal{P}, \mathbf{Ab}] \to \mathbf{Ab}$$

which represents the functor

$$(R, L)_A \to \mathrm{n.t.}[R, \mathrm{Hom}_Z(L\cdot, A)]$$

for any abelian group A. For the properties we refer to Oberst and Röhrl [20]. The tensor product has derived functors $\mathrm{Tor}_i^{\mathcal{P}}(\cdot, \cdot)$, and in terms of vanishing of $\mathrm{Tor}_i^{\mathcal{P}}(\cdot, \cdot)$ (flatness for rings) one can define coherence. One result found in Oberst and Röhrl [20] we use for the definition:

Theorem 4.1. The following conditions for $\mathcal{P}$ are equivalent

(a) Each h^P is coherent.

(b) $\mathrm{f.p.}[\mathcal{P}, \mathbf{Ab}] = \mathrm{Coh}[\mathcal{P}, \mathbf{Ab}]$.

(c) The product of flat contravariant functors is flat.

When any of these conditions (among others) is fulfilled, we call $\mathcal{P}$ *left coherent.* If the corresponding conditions for $\mathcal{P}^{op}$ are fulfilled, we say $\mathcal{P}$ is *right coherent*. In case $\mathcal{P}$ is left and right coherent, we will say $\mathcal{P}$ is *coherent.*

In case $\mathcal{P}$ is coherent, we define functors $\alpha : \mathrm{Coh}[\mathcal{P}^{op}, \mathbf{Ab}] \to \mathrm{Coh}[\mathcal{P}, \mathbf{Ab}]$ and $\beta : \mathrm{Coh}[\mathcal{P}, \mathbf{Ab}] \to \mathrm{Coh}[\mathcal{P}^{op}, \mathrm{Ab}]$ which are contravariant and adjoint on the right (i.e., $\mathrm{Hom}(A, \alpha B) \cong \mathrm{Hom}(B, \beta A)$). If $f \in \mathrm{Hom}_{\mathcal{P}}(P, Q)$ is such that $F = \mathrm{coker}\, h_f$ (i.e., F is finitely $\mathcal{P}$–presented) then $\alpha F = \ker h^f$. If $\eta : \mathrm{coker}\, h_f \to \mathrm{coker}\, h_{f'} = F'$, there is induced a natural transformation $\alpha\eta : \alpha F' \to \alpha F$. Thus is defined α, β being similarly defined. One notes that $\alpha h_P = h^P$ and $\beta h^Q = h_Q$ for $P, Q \in \mathcal{P}$, so α, β induce a perfect duality between $h_{\mathcal{P}}$ and $h^{\mathcal{P}}$. The pair α, β has been called a *pseudo–duality* and we will use this terminology (for want of a better term).

In general, given abelian categories $\mathcal{A}$, $\mathcal{B}$ and contravariant, adjoint on the right, functors $\alpha' : \mathcal{A} \to \mathcal{B}$, $\beta' : \mathcal{B} \to \mathcal{A}$, we say that α', β' is a pseudo–duality if there is a small additive coherent $\mathcal{P}$ such that $\mathcal{A} \cong \mathrm{Coh}[\mathcal{P}^{op}, \mathrm{Ab}]$, $\mathcal{B} \cong \mathrm{Coh}[\mathcal{P}, \mathrm{Ab}]$ and $\alpha = \alpha', \beta = \beta'$ (up to isomorphism).

Proposition 4.2. Let $\mathcal{P}$ be coherent. If $h_P \to h_Q \to F \to 0$ is exact in $\mathrm{Coh}[\mathcal{P}^{op}, \mathbf{Ab}]$, let $TF = \mathrm{coker}(\alpha h_Q \to \alpha h_P)$ (i.e.,

$$0 \to \alpha F \to h^Q \to h^P \to TF \to 0$$

is exact). Then there is an exact sequence of functors

$$0 \to \mathrm{Ext}^1(TF, \cdot) \to F \otimes \cdot \to \mathrm{n.t.}[\alpha F, \cdot] \to \mathrm{Ext}^2(TF, \cdot) \to 0.$$

A similar analysis holds for $F \in \mathrm{Coh}[\mathcal{P}, \mathbf{Ab}]$ and β.

$R^j\alpha$ and $R^j\beta$ denote the derived functors of α and β. Then we make the definition: $X \in \mathrm{Coh}[\mathcal{P}^{op}, \mathbf{Ab}]$ is n*–torsion* for an integer $n > 0$, provided $R^j\alpha X' = 0$ for all subobjects X' of X and all integers $0 \leqslant j < n$.

Theorem 4.3. Let $\mathcal{P}$ be a coherent category. Let k be an integer. The following statements are equivalent.

(a) $R^i\alpha X$ is i–torsion for all $X \in \mathrm{Coh}[\mathcal{P}^{op}, \mathbf{Ab}]$ and all i, $1 \leqslant i \leqslant k$.

(b) For all $P \in \mathcal{P}$, if $0 \to h^P \to E_0 \to E_1 \to \cdots$ is a minimal injective resolution of h^P in $[\mathcal{P}, \mathbf{Ab}]$, then flat dim. $E_i \leqslant i$ for $0 \leqslant i \leqslant k - 1$.

(c) $R^i\alpha Y$ is i–torsion for all $Y \in \mathrm{Coh}[\mathcal{P}, \mathbf{Ab}]$ for all i, $1 \leqslant i \leqslant k$.

(d) For all $P \in \mathcal{P}$, if $0 \to h_P \to E_0 \to E_1 \to \cdots$ is a minimal injective resolution of h_P in $[\mathcal{P}^{op}, \mathbf{Ab}]$, then flat dim. $E_i \leqslant i$ for $0 \leqslant i \leqslant k - 1$.

Auslander has shown the equivalence of (a) and (c). Our contribution is the proof that (a) and (b) (and hence by dual arguments) and (c) and (d) are equivalent. When one of these conditions is satisfied, we say $\mathcal{P}$ is k–Gorenstein (and this includes the conditions that $\mathcal{P}$ is small additive and coherent).

A ring R is k–Gorenstein if R is coherent and, as stated before, the ith term in a minimal injective resolution of R (left or right) has flat dimension bounded by i for $0 \leqslant i \leqslant k - 1$. (Every ring is 0–Gorenstein.) R is Gorenstein if R is k–Gorenstein for all integers k.

In the context of our program, to study trivial extensions of categories, or more particularly, $\mathrm{Map}(F\mathcal{A}, \mathcal{B})$, we introduce what seems to be a natural extension of pseudo–dualities on $\mathcal{A}$ and $\mathcal{B}$ to a pseudo–duality on $\mathrm{Map}(F\mathcal{A}, \mathcal{B})$.

Let $\mathcal{P}$, $\mathcal{Q}$ be small additive categories which we suppose to be coherent. Suppose $F : \mathrm{Coh}[\mathcal{P}^{op}, \mathrm{Ab}] \to [\mathcal{Q}^{op}, \mathrm{Ab}]$ is a right exact functor. Define $G : \mathrm{Coh}[\mathcal{Q}, \mathrm{Ab}] \to [\mathcal{P}, \mathrm{Ab}]$ by defining first $(Gh^Q)(P) = \mathrm{n.t.}[h_Q, Fh_P]$ and then $G(\mathrm{coker}(h^Q \to h^S))(P) = \mathrm{coker}(\mathrm{n.t.}[h_Q, Fh_P] \to \mathrm{n.t.}[h_S, Fh_P])$.

$$
\begin{array}{ccc}
 & \mathrm{Coh}[\mathcal{P}^{op}, \mathrm{Ab}] & \\
 F \swarrow & & \\
[\mathcal{Q}^{op}, \mathrm{Ab}] \supseteq \mathrm{Coh}[\mathcal{Q}^{op}, \mathrm{Ab}] & &
\end{array}
\qquad
\begin{array}{cc}
\mathrm{Coh}[\mathcal{P}, \mathrm{Ab}] \subseteq [\mathcal{P}, \mathrm{Ab}] & \\
\nearrow G & \\
\mathrm{Coh}[\mathcal{Q}, \mathrm{Ab}] &
\end{array}
$$

Theorem 4.4. Let $\mathcal{P}$, $\mathcal{Q}$ be coherent categories. If F and G are coherent preserving functors, then

$$\mathrm{Coh\,Map}(F\,\mathrm{Coh}[\mathcal{P}^{op}, \mathrm{Ab}], [\mathcal{Q}^{op}, \mathrm{Ab}])$$

and

$$\mathrm{Coh\,Map}(G\,\mathrm{Coh}[\mathcal{Q}, \mathrm{Ab}], [\mathcal{P}, \mathrm{Ab}])$$

are in pseudo–duality.

Now it is natural to ask whether $\mathrm{Coh\,Map}(G\,\mathrm{Coh}[\mathcal{Q}, \mathrm{Ab}], [\mathcal{P}, \mathrm{Ab}])$ is k–Gorenstein if $\mathrm{Coh}[\mathcal{P}, \mathrm{Ab}]$ and $\mathrm{Coh}[\mathcal{Q}, \mathrm{Ab}]$ are k–Gorenstein.

Example 4.5. Let R be a k–Gorenstein ring. Let $F = (R \oplus R) \otimes \cdot$ Then $\begin{pmatrix} R & 0 \\ R \oplus R & R \end{pmatrix}$ is coherent $(G = \cdot \otimes (R \oplus R))$ but not 1–Gorenstein (i.e., is not any–Gorenstein.)

Thus even when F is a rather nice functor, a Gorenstein property is not preserved. On the other hand, $F = M \otimes_R \cdot$ need not be so nice and R and S need not be Gorenstein but $\begin{pmatrix} R & 0 \\ M & S \end{pmatrix}$ can be 1–Gorenstein. The best we can do at present is to show that Gorenstein properties are reflected.

Proposition 4.6. If F and G are exact and $\mathrm{Coh\,Map}(G\,\mathrm{Coh}[\mathcal{Q}, \mathrm{Ab}], [\mathcal{P}, \mathrm{Ab}])$ is k–Gorenstein, then $\mathrm{Coh}[\mathcal{Q}, \mathrm{Ab}]$ and $\mathrm{Coh}[\mathcal{P}, \mathrm{Ab}]$ are k–Gorenstein.

The original question was: If R is k–Gorenstein, is $T_2(R)$ k–Gorenstein? This we can answer.

Theorem 4.7. Let R be a coherent ring. Then $T_2(R)$ is coherent, and R is k–Gorenstein if and only if $T_2(R)$ is k–Gorenstein.

This is established by investigating the flat dimension of the terms in a minimal injective resolution of $T_2(R)$.

We close this chapter by asking whether necessary and sufficient conditions can be found for the functors F, G which will insure that Gorenstein properties are preserved.

5. Applications to dominant dimension and Gorenstein rings

In this section we consider applications of our categorical machinery to Gorenstein finite dimensional algebras.

After establishing a method of constructing finite dimensional algebras of various dominant dimensions from an arbitrary finite dimensional algebra Λ, we characterize the category of finitely generated (= coherent) reflexive modules over a finite dimensional algebra of dominant dimension $\geqslant 2$ which is right T–stable (see definitions below). Finally, an existence theorem is provided for such algebras.

Several papers appear in the literature which bear on our work in this section. Among these are papers of Kato [8, 9, 10], Müller [17, 18, 19], Morita [14, 15, 16] and Tachikawa [24]. Our point of view will partially be taken from the preceding papers and especially from the papers of Gabriel [6] and Roos [21, 22].

Recall from Section 4 a ring R is n–Gorenstein if, in a minimal (left or right) injective resolution $0 \to R \to E_0 \to E_1 \to \cdots$ of R, flat dim. ${}_R E_i \leqslant i$ for $i < n$. We say that R is of *dominant dimension* $\geqslant n$ (abbreviated dom. dim. $R \geqslant n$) if the stronger property flat dim. ${}_R E_i = 0$ holds for $i < n$. A ring R will be called *right T–stable* if each indecomposable injective in Mod_R is either 1–torsion (see Section 4) or torsion free (that is to say, each finitely generated submodule is torsionless in the sense of Bass [3]). If the "1–torsion" condition in the preceding statement is replaced by "torsion" (i.e., no nonzero homomorphisms into R), we say that R is *right weakly T–stable.* If R is right Noetherian and right T–stable, it is easily observed that the Serre subcategory $\mathcal{T}_1$ of 1–torsion modules in Mod_R is stable in the sense of Roos [21], that is, $\mathcal{T}_1$ is closed with respect to injective envelopes. It is also elementary that every integral domain is T–stable and every right Noetherian, right hereditary ring is right weakly T–stable.

The results which follow rely heavily upon the earlier results of Section 1, 3, and 4. If Λ is a finite dimensional algebra with ground field k, Λ^d denotes the injective cogenerator $\mathrm{Hom}_\Lambda(\Lambda, k)$ considered either as a left or right module.

Proposition 5.1. Let R be a finite dimensional algebra, M a finitely generated injective module in Mod_R and let $A = \mathrm{End}_R M$. Further, let

$$\Omega = \begin{pmatrix} R & 0 \\ M & A \end{pmatrix}.$$

Then

(a) $SA = \begin{pmatrix} 0 & 0 \\ M & A \end{pmatrix}$ is a right projective–injective Ω–module (the functor S is defined in Section 4).

(b) If $M = R^d$, then Ω is 1–Gorenstein [Müller, 18] and is right and left T–stable.

(c) Suppose dom. dim. $R \geqslant m$, R is right weakly T–stable and M is the torsion direct summand of R^d. Then dom. dim. $\Omega \geqslant m + 1$ and Ω is right T–stable. If $m = 0$, then dom. dim. $\Omega \geqslant 2$.

(d) For M as in either (b) or (c),

$$\text{gl. dim. } R \leqslant \text{gl. dim. } \Omega \leqslant 1 + 2\text{gl. dim. } R.$$

Among other things, our next theorem shows that finite dimensional algebras of dominant dimension $\geqslant n$ are not particularly rare. The proof of this result makes use of Proposition 5.1 and an elementary induction step.

Theorem 5.2. Let Λ be a finite dimensional algebra. For every integer $n > 0$, there is a finite dimensional algebra Λ_n and a two–sided ideal I_n of Λ_n satisfying:

(a) $\Lambda \cong \Lambda_n/I_n$ and I_n is a right direct summand of Λ_n (hence right projective).

(b) Λ_n is right T–stable.

(c) dom. dim. $\Lambda_n \geqslant n + 1$.

(d) gl. dim. $\Lambda \leqslant$ gl. dim. $\Lambda_n \leqslant 2^n + (2^n + 1)$gl. dim. Λ.

Now let Λ be a finite dimensional algebra with dom. dim. $\Lambda \geqslant 2$, let $\mathcal{T}_1$ be the Serre subcategory of 1–torsion modules in Mod_Λ and let $\mathfrak{R}$ be the full subcategory of finitely generated (= coherent) reflexive modules in Mod_Λ. It is easy to show (see Morita [14, 15]) that $\mathfrak{R}$ consists precisely of the finitely generated $\mathcal{T}_1$–closed objects of Mod_Λ in the sense of Gabriel [6]. Further, let P be the direct sum of the projective indecomposable modules X in Mod_Λ such that $X/(\text{rad }\Lambda)X \notin \mathcal{T}_1$. Then results of Morita [14, 15] and Gabriel [6] show the following categories are equivalent:

(a) $\text{coh}(\text{Mod}_\Lambda/\mathcal{T}_1)$.

(b) $\text{coh}(\text{Mod}_{\text{End}_\Lambda P})$.

(c) $\mathfrak{R}$.

In particular, $\mathfrak{R}$ is an abelian category. The above equivalence of categories as well as Proposition 5.1 and Theorem 5.2 yield our next result.

Theorem 5.3. Let Λ be a weakly right T–stable finite dimensional algebra. Then Coh Mod_Λ is equivalent to the category of finitely generated reflexive right modules over some right T–stable finite dimensional algebra Γ with dom. dim. $\Gamma \geqslant 2$ and gl. dim. $\Lambda \leqslant$ gl. dim. $\Gamma \leqslant 1 + 2$gl. dim. Λ.

Corollary 5.4. If Λ is an hereditary finite dimensional algebra, then coh(Mod_Λ) can be realized as the category of finitely generated reflexive right modules over some finite dimensional algebra Γ with dom.dim. $\Gamma \geqslant 2$ and gl. dim. $\Gamma \leqslant 3$.

Corollary 5.5. Let Λ be an arbitrary finite dimensional algebra and let

$$\Omega = \begin{pmatrix} \Lambda & 0 \\ \Lambda^d & \Lambda \end{pmatrix}.$$

Then coh(Mod_Ω) can be realized as the category of finitely generated reflexive right modules over some finite dimensional algebra Γ with dom. dim. $\Gamma \geqslant 3$ and gl. dim. $\Lambda \leqslant$ gl. dim. $\Gamma \leqslant 2 + 3$gl. dim. Γ.

Remark. We point out (in the notation of Theorem 5.2) that coh(Mod_{Λ_n}) is equivalent to the category of finitely generated reflexive right modules over Λ_{n+1}.

As a sort of converse of Theorem 5.3, we prove the following result.

Theorem 5.6. Let Λ be a right T–stable finite dimensional algebra with dom. dim. $\Lambda \geqslant 2$. Then the abelian category $\mathfrak{R}$ of finitely generated reflexive right Λ–modules is equivalent to coh(Mod_R) for some weakly right T–stable finite dimensional algebra R.

We refer the reader to Morita [15] for related results concerning the category of reflexive modules over an artin ring of dominant dimension $\geqslant 2$.

6. Representation dimension of finite dimensional algebras

In a course given during the Fall semester (1970-1971) at the University of Illinois, M. Auslander established the following results: Let Λ be a finite dimensional algebra of finite representation type (i.e., a finite number of indecomposable modules up to isomorphism), let M be the direct sum of the distinct indecomposable modules in ${}_\Lambda\text{Mod}$ and let $\Gamma = \text{End}_\Lambda M$. Then dom. dim. $\Gamma \geqslant 2$ and gl. dim. $\Gamma \leqslant 2$ (in fact gl. dim. $\Gamma = 2$, unless Λ is semi-simple). Moreover, ${}_\Lambda\text{Mod}$ is equivalent to $\text{End}_\Gamma E_0\text{Mod}$ where E_0 is the injective envelope of Γ in ${}_\Gamma\text{Mod}$.

Conversely, if Γ is a finite dimensional algebra such that gl. dim. $\Gamma \leqslant 2$ and if P is a finitely generated projective-injective module in ${}_\Gamma\text{Mod}$, then $\text{End}_\Gamma P$ is of finite representation type.

The proof of this result (first part) makes use of the fact $\text{coh}[({}_\Lambda\text{Mod})^{\text{op}}, \text{Ab}]$ is equivalent to $\text{coh}({}_\Gamma\text{Mod})$. The preceding results motivated Auslander to suggest the following method of "measuring how far" a finite dimensional algebra is from being of finite representation type.

Let Λ be a finite dimensional algebra and let $\mathfrak{A}$ be an additive full subcategory of ${}_\Lambda\text{Mod}$ generated by indecomposable modules in ${}_\Lambda\text{Mod}$ such that $\mathfrak{A}$ contains all projective and injective indecomposable modules in ${}_\Lambda\text{Mod}$. Then Auslander shows $\mathfrak{A}$ is coherent in the sense of Section 4 and dom. dim. $\text{coh}[\mathfrak{A}^{\text{op}}, \text{Ab}] \geqslant 2$. (Actually, Morita [15] shows $\mathfrak{A}$ must contain the indecomposable projectives and indecomposable injectives in order for dom. dim. $\text{coh}[\mathfrak{A}^{\text{op}}, \text{Ab}] \geqslant 2$.) Define the left representation dimension of Λ by

$$\text{left rep. dim. } \Lambda = \inf \{\text{gl. dim. coh}[\mathfrak{A}^{\text{op}}, \text{Ab}]\}$$

where $\mathfrak{A}$ is as above. If we let A be the direct sum of each of the distinct indecomposable modules in $\mathfrak{A}$, then

$$\text{gl. dim. End}_\Lambda A = \text{gl. dim. coh}[\mathfrak{A}^{\text{op}}, \text{Ab}].$$

We call A a minimal additive generator of $\mathfrak{A}$. In the terminology of [15], A (with the above properties) is called a generator-cogenerator of ${}_\Lambda\text{Mod}$. Thus

$$\text{left rep. dim. } \Lambda = \inf \{\text{gl. dim. End}_\Lambda A : A \text{ is a coherent, , generator-cogenerator of } {}_\Lambda\text{Mod}\}.$$

Using the duality induced by the ground field it is easily observed that left rep. dim. Λ = right rep. dim. Λ. Thus we speak only of the representation dimension of Λ, denoted rep. dim. Λ. The results that follow exhibit several large classes of finite dimensional algebras having finite representation dimension. We note, from the preceding theory due to Auslander, if rep. dim. $\Lambda \leqslant 2$, then Λ is of finite representation type.

Theorem 6.1. If Λ is right weakly T-stable and if gl. dim. $\Lambda < \infty$, then

$$\text{rep. dim. } \Lambda \leqslant 1 + 2\text{gl. dim. } \Lambda.$$

The construction in 5.1(c) is useful in the proof of Theorem 6.1.

Corollary 6.2. If Λ is an hereditary finite dimensional algebra, then rep. dim. $\Lambda \leqslant 3$. If Λ is not of finite representation type, then rep. dim. $\Lambda = 3$.

Remark. If Λ is a finite dimensional algebra and gl. dim. $\Lambda < \infty$, then (in the notation of Theorem 5.2) rep. dim. $\Lambda_n \leqslant 2^{n+1} + (2^{n+2} + 1)$gl. dim. $\Lambda < \infty$. In particular, every finite dimensional algebra Λ of finite global dimension is the homomorphic image of a fimite dimensional algebra of finite representation dimension.

We also add that Auslander has shown: If Λ is a Q.F. finite dimensional algebra, then rep. dim. $\Lambda \leqslant$ Loewy length of $\Lambda < \infty$.

Clearly, we may extend the notion of representation dimension to an artin algebra Λ over a commutative artin ring R such that Λ is finitely generated as a left and right R–module.

Theorem 6.3. Let Λ be a finitely generated (as a left and right module) artin algebra over a commutative artin ring and let $T_2(\Lambda)$ denote the 2×2 lower triangular matrices over Λ. Then

$$\text{rep. dim. } T_2(\Lambda) \leqslant 2 + \text{rep. dim. } \Lambda.$$

Remark. Let $\mathbf{Z}_{p^n}$ denote the ring of integers module p^n, p a prime in $\mathbf{Z}$. Then Auslander showed rep. dim. $T_2(\mathbf{Z}_{p^2}) = 2$ and J. Janusz further established rep. dim. $T_2(\mathbf{Z}_{p^3}) = 2$. However, the problem remains open for $T_2(\mathbf{Z}_{p^n})$ with $n > 3$. From Theorem 6.3, we do have rep. dim. $T_2(\mathbf{Z}_{p^n}) \leqslant 4$ for all $n \geqslant 1$.

We conclude our paper with the following amusing consequence of our results in Sections 4, 5 and 6. Let Λ be an hereditary artin algebra, let $t\Lambda^d$ be the torsion direct summand of Λ^d and let $S = \text{End}_\Lambda(t\Lambda^d)$. One can show that every indecomposable Λ–module is projective or injective if and only if $\text{Tor}_1^S(S/\text{rad } S, t\Lambda^d)$ is a right Λ–projective module. Examples of such rings are the following matrix rings

$$\Lambda = \begin{pmatrix} k & 0 \\ k & k \end{pmatrix} \quad \text{and} \quad \Gamma = \begin{pmatrix} k & 0 & 0 \\ 0 & k & 0 \\ k & 0 & k \end{pmatrix}$$

where k is a field.

References

1. M. Auslander, On the dimension of modules and algebras III: Global dimension, Nagoya Math. J. **9** (1955), 67-77.
2. M. Auslander, Coherent functors, Proceedings of the Conference on Categorical Algebra, La Jolla , (Springer–Verlag), 1965, 189-231.
3. H. Bass, Finitistic dimension and a homological generalization of semi–primary rings, Trans. Amer. Math. Soc. **95** (1960), 466-488.
4. S. Chase, A generalization of the ring of triangular matrices, Nagoya Math. J. **18** (1961), 13-25.
5. S. Eilenberg, A. Rosenberg, D. Zelinsky, On the dimension of modules and algebras VIII. Dimension of tensor products, Nagoya Math. J. **12** (1957), 71-93.
6. P. Gabriel, Des catégories abéliennes, Bull. Soc. Math. France **90** (1962), 323-448 .
7. M. Harada, On special type of hereditary abelian categories, Osaka J. Math. **4** (1967), 243-255.
8. T. Kato, Rings of dominant dimension $\geqslant 1$, Proc. Japan Acad. **44** (1968), 579-584.
9. T. Kato, Rings of U–dominant dimension $\geqslant 1$, Tohoku Math. J. **21** (1969), 321-327.
10. T. Kato, Dominant modules, J. Algebra **14** (1970), 341-349.
11. B. Mitchell, "Theory of Categories", Academic Press, New York and London, 1965.
12. B. Mitchell, On the dimension of objects and categories I, J. Algebra **9** (1968), 314-340.
13. B. Mitchell, On the dimension of objects and categories II, J. Algebra **9** (1968), 341-368.
14. K. Morita, Duality in Q.F.–3 rings, Math. Z. **108** (1969), 237-252.
15. K. Morita, Localizations in categories of modules I, Math. Z. **114** (1970), 121-144.
16. K. Morita, Localizations in categories of modules II, Crelles J. **242** (1970), 163-169.
17. B. J. Müller, The classification of algebras by dominant dimension, Can. J. Math. **20** (1968), 398-409.
18. B. J. Müller, Dominant dimension of semi–primary rings, Crelles J. **232** (1968), 173-179.
19. B. J. Müller, On Morita duality, Can. J. Math. **21** (1969), 1338-1347.
20. U. Oberst and H. Röhrl, Flat and coherent functors, J. Algebra **14** (1970), 91-105.
21. J. E. Roos, Locally Noetherian categories and generalized strictly linearly compact rings. Applications, Category Theory, Homology Theory and Their Applications II, Lecture Notes in Mathematics **92**, Springer–Verlag, Berlin, 1969.
22. J. E. Roos, Coherence of general matrix rings and non–stable extensions of locally Noetherian categories, mimeographed notes, University of London.
23. L. Small, An example in Noetherian rings, Proc. Nat. Acad. (U.S.A) **54** (1965), 1035-1036.

24. H. Tachickawa, On left QF−3 rings, Pac. J. Math. **32** (1970), 255-268.
25. C. E. Watts, Intrinsic characterization of some additive functors, Proc. Amer. Math. Soc. **11** (1960), 5-8.
F. K. L. Fields, On the global dimension of residue rings, Pac. J. Math. **32** (1970), 345-349.

HIGHER K–FUNCTORS

S. M. Gersten

Rice University

There have been numerous attempts to explain the phenomenon of exact sequences connecting the functors K_0 and K_1 [3]. These theories are now known to be related and the connections between them are beginning to be understood.

We remind the reader of the classical situation, in a formulation due to Milnor [17]. If A is a unital associative ring, then $K_0(A)$ is the Grothendieck group of the category $\mathcal{P}(A)$ of finitely generated projective (left) A–modules. It is a functor via $P \to B \otimes_A P$, if $A \to B$ is a ring homomorphism. The functor $K_1(A)$ is the commutator quotient group of $G\ell(A) = \bigcup_n G\ell_n(A)$. By a result of J. H. C. Whitehead [18], $K_1(A) = G\ell(A)/\mathcal{E}(A)$ where $\mathcal{E}(A)$ is the subgroup generated by elementary transvections $E_{ij}(a) = 1 + ae_{ij}$ $(i \neq j,\ a \epsilon A)$. Milnor has proposed a candidate for $K_2(A)$ which can be shown to be $H_2(\mathcal{E}(A), \mathbf{Z})$.

Theorem [2], [17]. Consider a cartesian diagram of rings

$$\begin{array}{ccc} A & \longrightarrow & B \\ \downarrow & & \downarrow f \\ C & \xrightarrow{g} & D \end{array}$$

If f is surjective, then there is a natural exact sequence

$$\begin{array}{l} K_{i+1}(A) \to K_{i+1}(B) \oplus K_{i+1}(C) \to K_{i+1}(D) \xrightarrow{\partial} K_i(A) \\ \qquad\qquad\qquad\qquad\qquad\qquad \downarrow \\ \qquad\qquad\qquad\qquad K_i(D) \leftarrow K_i(B) \oplus K_i(C) \end{array}$$

for $i = 0$. If both f and g are surjective, then in addition there is an exact sequence for $i = 1$.

All attempts to explain and extend these exact sequences to the left depend on some notion of homotopy of the linear group of a ring (the extension to the right was given by Bass [2]). We describe here three attempts, due to the author and R. G. Swan, to M. Karoubi and O. Villamayor, and to D. Quillen. The K–functors will be denoted respectively by K_i^{G-S}, K_i^{k-v}, and K_i^Q.

If R is a ring (without unit), denote [8] by FR the free ring without unit on the set R. Thus FR is the augmentation ideal of the free associative algebra over **Z** (non–commuting polynomials) on the set R. The functor F sits in a cotriple (F, ϵ, μ), $\epsilon_R : FR \to R$, $\mu_R : FR \to F^2R$, and hence gives rise to a simplicial resolution of the identity

$$F_* : I \xleftarrow{\epsilon} F \leftleftarrows F^2 \leftleftarrows F^3 \cdots .$$

Applying the functor $G\ell$ (generalized to rings without unit) one produces an augmented simplicial group $G\ell F_*R$. We set $K_{i+2}^{G-S}(R) = \tilde{\pi}_i(G\ell F_*R$, $i \geqslant -1$, where $\tilde{\pi}_i$ denotes the augmented homotopy groups [8] ($\tilde{\pi}_i = \pi_i$, $i \geqslant 1$, and $\tilde{\pi}_0$, $\tilde{\pi}_{-1}$ are defined to take into account the augmentation $\epsilon_R : FR \to R$). This simplicial definition was originally given by the author [8], where the functor E', a modified version of $\mathcal{E}$, was used instead of $G\ell$. Swan has since shown [26] that the groups $K_i^{G-S}(R)$ are those defined by a different method in his earlier paper [25]. We can summarize what is known by

Theorem. There is a canonical surjection $K_2(R) \to K_2^{G-S}(R)$, where R is unital. If one knew $K_2(\mathbf{Z}) \cong K_2(\mathbf{Z}\{X\})$ for all sets X, where $\mathbf{Z}\{X\}$ is the free associative algebra on X, then the map $K_2(R) \to K_2^{G-S}(R)$ would be an isomorphism. If Λ is a commutative real Banach algebra, then there is a canonical surjection $K_2^{G-S}(\Lambda) \to \pi_1 S\ell(\Lambda)$.

The second of the theories requires the notions of the path ring. If R is a (non unital) ring, then the path ring ER is the ideal $tR[t]$ in the polynomial ring $R[t]$. Thus one has a short exact sequence of rings

$$ER \to R[t] \xrightarrow{t \to 0} R .$$

By analogy with the Gersten–Swan theory, we imbed E in a cotriple (E, ϵ, μ) where ϵ is "$t \to 1$" and $\mu_R : ER \to E^2R$ is given by $tR[t] \xrightarrow{t \to tu} tuR[t,u]$. This gives rise to a simplicial resolution to the identity

$$E_* : I \xleftarrow{\epsilon} E \leftleftarrows E^2 \leftleftarrows E^3 \cdots .$$

One sets [7] $K_{i+2}^{k-v}(R) = \widetilde{\pi}_i(G\ell E_* R)$, $i \geqslant -1$. These groups can be shown [7] to be exactly those defined by Karoubi and Villamayor [15] by a different procedure. We refer the reader also to [15] for the formulation of the long exact sequences for the Karoubi–Villamayor theory.

There is a natural map $FR \to ER$ that may be described as follows. If $r \in R$, denote by $|r|$ the associated generator of FR. Then $|r| \to rt$ induces the desired ring homomorphism $FR \to ER$. This induces a map of cotriples and hence maps $K_i^{G-S}(R) \to K_i^{k-v}(R)$. In the case of a commutative real Banach algebra Λ, the surjection $K_2^{G-S}(\Lambda) \to \pi_1(S\ell\Lambda)$ factors through $K_2^{k-v}(\Lambda)$ [7]. Thus one has at the K_2 level maps $K_2(R) \to K_2^{G-S}(R) \to K_2^{k-v}(R)$ if R is unital. Furthermore, we have

Theorem [7] [26] [14]. If $K_1(R) \to K_1(R[t])$ is an isomorphism, then $K_2(R) \to K_2^{k-v}(R)$ is surjective. This hypothesis is satisfied, for example, if R is (homologically) regular [4]. In addition [26] [10], if $K_2(R) \to K_2(R[t])$ is an isomorphism, then $K_2(R) \to K_2^{k-v}(R)$ is also an isomorphism. In this case, $K_2(R) \xrightarrow{\cong} K_2^{G-S}(R) \xrightarrow{\cong} K_2^{k-v}(R)$.

The only case known where $K_2(R) \to K_2(R[t])$ is an isomorphism is the case R a field [24] or more generally a division ring [5] (see also Professor P. M. Cohn's second lecture in these proceedings).

The situation at the K_1 level is much easier to describe. For any R, $K_1(R) \xrightarrow{\cong} K_1^{G-S}(R)$. Also, $K_1^{k-v}(R) = G\ell(R)/UP(R)$ where $UP(R) \supset \mathcal{E}(R)$ is the subgroup generated by unipotent matrices.

For many purposes, it is convenient to reformulate the Karoubi–Villamayor theory. One introduces the simplicial ring R_*, where $(R_*)_n = R[t_0,\cdots,t_n]/(t_0+\cdots+t_n-1)$. The face operator ∂_i is given by

$$\partial_i(t_j) = \begin{Bmatrix} t_j, & j < i \\ 0, & j = i \\ t_{j-1}, & j > i \end{Bmatrix}$$

and the degeneracy s_i by

$$s_i(t_j) = \begin{Bmatrix} t_j, & j < i \\ t_i + t_{i+1}, & j = i \\ t_{j+1}, & j > i \end{Bmatrix} .$$

Let $BG\ell R_*$ denote the simplicial classifying space of the simplicial group $G\ell R_*$. Then D. L. Rector has shown [22].

Theorem $\pi_i(BG\ell R_*) \cong K_i^{k-v}(R), \ i \geqslant 1.$

This complex had also been considered by D. W. Anderson [1] who called the homotopy groups $\pi_i(BG\ell R_*)$ the "polynomial K–theory" of R. The history may be even longer, for Bass has told me that Heller considered these groups in the early 1960's.

We now come to the third and possibly most interesting of the theories, due to D. Quillen [20]. If R is a unital ring, then let $BG\ell R$ be the classifying space for the constant simplicial group $G\ell R$. Quillen proves that it is possible to attach 2 and 3 cells to $BG\ell R$ to get $BG\ell(R)^+$ such that if $f: BG\ell R \to BG\ell R^+$ is the inclusion, then $\ker(\pi_1(f)) = \mathcal{E}(R)$ and such that for any local system L of $\pi_1 BG\ell R^+$ modules, f induces an isomorphism

$$H_*(BG\ell R,L) \xrightarrow{\cong} H_*(BG\ell R^+,L).$$

Quillen observes that $\pi_1 BG\ell R^+ = K_1(R)$ and $\pi_2 BG\ell R^+ = K_2(R)$, and defines $K_n^Q R = \pi_n BG\ell R^+$, $n > 0$. It has been shown that $BG\ell R^+$ is a very nice space. It is an H space and even the component of an infinite loop space [1] [11]. Quillen has computed $K_n^Q(R)$ if R is a finite field or algebraic closure of a finite field [20] [21], and has observed some connections with the étale cohomology of R, when R is commutative.

Now the inclusion $G\ell(R) \to G\ell(R_*)$ induces a map $BG\ell(R) \to BG\ell R_*$, which factors through $BG\ell(R)^+$ (since $BG\ell R_*$ is n–simple for all n [10]). A careful analysis of the resulting map $BG\ell(R)^+ \to BG\ell(R_*)$ shows that it is the edge map of a spectral sequence [1] [10].

Theorem. If R is a unital ring, then there is a spectral sequence, whose

$$E^1_{p,q} = K_q^Q(R[t_0,\cdots,t_p]/(t_0+\cdots+t_p-1))$$

and whose $E^\infty_{p,q}$ is the graded group associated to a filtration of $K^{k-v}_{p+q}(R)$.

Analysis of this spectral sequence, together with the earlier quoted results of Silvester and Dennis, give a proof of the assertion made before that $K_2(F) \cong K_2^{G-S}(F) \cong K_2^{k-v}(F)$ if F is a division ring.

Very little is known yet about this theory. One notable result is

Theorem. For any unital ring R,

$$K^Q_{n+1}(R[t,t^{-1}]) \cong K^Q_{n+1}(R) \oplus K^Q_n(R) \oplus (?).$$

The case $n = 0$ is part of the assertion of the "Fundamental Theorem" of K–theory [2] and is due to Bass, Heller, and Swan [4]. The case $n = 1$ has been published by J. Wagoner [28]. The general result is due to the author [11]. The proof entails a detailed description of the infinite loop space $K_0(R) \times BG\ell(R)^+$ (due independently to Gersten [11] and Wagoner [27]) together with an analysis of the multiplicative properties of the Ω–spectrum. There is a corresponding result for the Karoubi–Villamayor groups [14], [9]. The result is of interest, since it is a discrete analog of the periodicity theorem of Bott (see [2], p. 750).

We should mention that there is a quadratic K–theory currently being developed by Karoubi and Villamayor [16] which promises some interesting connections with the linear theories described here. However even in the linear theory there are some directions that still should be pursued. One should be able to extend K–theories to a "generalized cohomology theory of algebraic varieties." Although no one has succeeded yet in doing this, there is one result that suggests this may be possible for the Karoubi–Villamayor theory.

Theorem [6]. If A is a (homologically) regular R–algebra and f and g generate the unit ideal of R, then the diagram

$$\begin{array}{ccc} K_0(A) \times BG\ell A_* & \longrightarrow & K_0(A_f) \times BG\ell(A_f)_* \\ \downarrow & & \downarrow \\ K_0(A_g) \times BG\ell(A_g)_* & \longrightarrow & K_0(A_{fg}) \times BG\ell(A_{fg})_* \end{array}$$

is a fibre product diagram in the sense of homotopy theory.

In particular this gives a Mayer–Vietoris sequence for the localization represented by writing Spec R as the union of Spec R_f and Spec R_g. Also, it yields a rather mysterious exact sequence (which should be compared with a corresponding sequence of Bass and Tate [17, §13]).

$$K_2(A_{fg}) \to K_1(A) \to K_1(A_f) \oplus K_1(A_g) \to K_1(A_{fg}) \to \cdots .$$

Finally, the "topological" K_2 of C. Moore [19], [17] for local fields has not yet been incorporated into an algebraic K–theory of topological rings. Karoubi has a very general procedure for defining K–theories of "Banach rings," but these involve taking completions of polynomial extensions [15]. It seems to me that what is wanted is a Quillen–type definition of the K–theory of local fields that would yield Moore's results for K_2. It would be related to Karoubi's continuous theories by a spectral sequence, analogous to the one exhibited earlier in the discrete case.

References

1. D. W. Anderson, Simplicial K–theory and generalized homology theories I, examples from K–theory, preprint.
2. H. Bass, "Algebraic K–theory," Benjamin, 1969.
3. H. Bass, K–theory and stable algebra, IHÉS Publications Mathematique, 1964.
4. H. Bass, A. Heller, and R. G. Swan, The Whitehead group of a polynomial extension, IHÉS Publications Mathématiques, 1964.
5. K. Dennis, preprint.
6. S. M. Gersten, A Mayer–Vietoris sequence in the K–theory of localizations, preprint.
7. S. M. Gersten, On Mayer–Vietoris functors and algebraic K–theory, J. of Alg. **18** (1971), 51-88.
8. S. M. Gersten, On the functor K_2, I. J. of Alg. **17** (1971), 212-237.
9. S. M. Gersten, On the K–theory of Laurent polynomials, Proc. Amer. Math. Soc., 1971.
10. S. M. Gersten, On the relationship between the K–theory of Quillen and the K–theory of Karoubi and Villamayor, preprint.
11. S. M. Gersten, On the spectrum of algebraic K–theory, to appear in Bull. Amer. Math. Soc.
12. S. M. Gersten and D. L. Rector, A relationship between two simplicial algebraic K–theories, Bull. Amer. Math. Soc. **77** (1971), 397-399.
13. M. Karoubi, Foncteurs dérivés et K–théorie, Séminaire sur la K–theorie, Lecture Notes in Math. Vol. 136, Springer Verlag.
14. M. Karoubi, La Périodicité de Bott en K–théorie générale, Ann. Scient. Ec. Nom. Sup. **270** (1971), 63-95.
15. M. Karoubi, et O. Villamayor, Foncteurs K^n en algèbre et en topologie, C. R. Acad. Sc. Paris **269** (1969), 416-419.
16. M. Karoubi et O. Villamayor, K–théorie hermitienne, C. R. Acad. Sc. Paris **272** (1971), 1237-1240.
17. J. Milnor, Algebraic K–theory, Lecture Notes IAS, Princeton, 1970.
18. J. Milnor, Whitehead torsion, Bull, Amer. Math. Soc. **72** (1966), 358-426.
19. C. C. Moore, Group extensions of p–adic and adelic linear groupes, IHÉS Publ. Math **35** (1969), 5-74.
20. D. Quillen, Proc. Int. Congress of Mathematicians at Nice, 1970.
21. D. Quillen, The K–theory associated to a finite field I, preprint.
22. D. L. Rector, The K–theory of a space with coefficients in a discrete ring, Bull. Amer. Math. Soc. **77** (1971), 571-575.
23. J. R. Silvester, A presentation of $G\ell_n(Z)$ and $G\ell_n(K[x])$, preprint.
24. J. R. Silvester, On $K_2(k[X])$, preprint.

25. R. G. Swan, Non abelian homological algebra and K–theory, Proc. Symp. in Pure Math AMS **17** (1970), 88-123.
26. R. G. Swan, Some relations between higher K–functors, preprint.
27. J. Wagoner, Delooping classifying spaces in algebraic K–theory, preprint.
28. J. Wagoner, On K_2 of the Laurent polynomial ring, Amer. Journ. Math. (1971), 123-138.

PROPERTIES OF THE IDEALISER

A. W. Goldie

University of Leeds

The aim of this lecture is to present methods which are particularly appropriate for the study of domains, prime rings and simple rings in the noetherian case. This is done by considering right ideals, *the* k*–critical ideals* and an associated set of subrings, *the idealisers.* (See the lecture of J. C. Robson and also [6]).

We suppose that S denotes a noetherian ring with unit element. At this stage there is no point in pressing the distinction between right and left noetherian. Always V will be a right ideal of S and R be the *idealiser of* V *in* S:

$$R = II_S(V) = \{ x \in S \mid xV \subseteq V \} .$$

V is an ideal (two–sided) in R and R is the largest subring of S having this property for V. The factor ring R/V is the *eigen–ring* of V. The concept originated with O. Ore and H. Fitting.

Proposition 1.1. R/V is ring isomorphic to $\mathrm{End}_S(S/V)$.

Proof. Let $\theta \in \mathrm{End}_S(S/V)$ so that $\theta[s + V] \to [rs + V]$ $(s \in S)$, where $[r + V] = \theta[1 + V]$. Clearly $r \in R$, since θ is a map. The converse holds, which means that $\theta \leftrightarrow [r + V]$ $(r \in R)$ are in (1–1) correspondence. This is an isomorphism $\mathrm{End}_S(S/V)$ onto R/V.

This well–known result has a useful generalization. Let A,B be right ideals of S and set

$$(B \cdot\!\cdot A) = (B \cdot\!\cdot A)_S = \{ s \in S \mid sA \subseteq B \} .$$

Proposition 1.2. Let B be a right ideal of S and $C = (B \cdot\!\cdot V)$. Then C/B is a right R/V–module and $[c + B] \to \lambda(c)$, $c \in C$, is an R/V–module isomorphism of C/B onto $\mathrm{Hom}_S(S/V, S/B)$. Here $\lambda(c)$: $[s + V] \to [cs + B]$

$(s \in S)$.

Proof. Similar to that of Proposition 1.1, although we do need to bear in mind that the R/V–action is given by

$$\lambda(c)[r + V] = \lambda(cr).$$

In order to define critical ideals we need the concept of Krull dimension intro– duced by P. Gabriel. (See [3].) This is recursively defined for a module M in the following way, and is denoted by K–*dim* M.

(1) K–dim M = 0 if and only if M has a composition series of submodules.

(2) In general consider a chain of submodules $M = M_1 \supset M_2 \supset \cdots$ for which K–dim $(M_i/M_{i+1}) \nleqslant n-1$ for each i. If every such chain terminates then K–dim M $\leqslant$ n.

(3) For a ring S define K–dim S = K–dim S_S.

We require the following results already known in [3].

Proposition 1.3. Let $A \cong B/C$ be modules. Then K–dim B = max $\{$K–dim A, K–dim C$\}$ and K–dim S = sup $\{$K–dim A; A a finitely generated S–module$\}$.

A right S–module M is said to be k–*critical*, where $k = 0,1,2,\cdots$ provided that K–dim M = k and K–dim $M' < k$ for every proper factor module M' of M. We also say that a right ideal I of S is k–critical when the factor module S/I is k–critical. The existence of k–critical modules and one–sided ideals is ensured by the noetherian condition. In particular the following cases should be observed. A right ideal I is 0–critical if and only if I is a maximal right ideal. It is 1–critical if and only if S/I is an S–module of infinite length (has no composition series) but every proper factor module has finite length (has a composition series).

Proposition 1.4. (i) A k–critical module M is a uniform module.
(ii) Each non–zero submodule of M is a k–critical module.

Proof. Let N, N' be non–zero submodules of M such that $N \cap N' = 0$. Since $N \cong (N + N')/N'$, then K–dim $N < k$. Since K–dim $M/N < k$, we obtain K–dim $M < k$, which is a contradiction. Thus M is uniform.

Certainly K–dim N = k, since K–dim $M/N < k$ and K–dim M = k. Let $N \supset N' \neq 0$; then K–dim $N/N' \leqslant$ K–dim $M/N' < k$. Thus N is a k–critical module.

When a ring has infinite Krull–dimension (on the right) then it has k–critical right ideals for $k = 0,1,2,\cdots$. In the case of finite dimension there are *critical series* of maximal length

$$S \supset V_o \supset V_1 \supset \cdots \supset V_k \supsetneq 0$$

where V_i is an i–critical right ideal and the series has stopped, because for all right ideals $J \subset V_k$ we have K–dim $V_k/J \leqslant k$. This implies that K–dim $V_k \leqslant k$, and since K–dim $S/V_k = k$, we can deduce that K–dim $S = k$.

Clearly we have

Proposition 1.5. Let $S \supset V_o \supset V_1 \supset \cdots \supset V_k$ where V_i is i–critical. Then $k \leqslant$ k–dim S and there exists such a series for which $k =$ K–dim S.

Proposition 1.6. Let V be a k–critical right ideal of S and set $V_a = \{ x \in S \mid ax \in V \}$, where $a \in S$, $a \notin V$. Then V_a is a k–critical right ideal of S.

Proof. The map $[s + V_a] \to [as + V]$ is an S–isomorphism of S/V_a onto $(aS + V)/V$. Now $(aS + V)/V$ is k–critical.

Proposition 1.7. When S is a commutative ring the critical ideals are exactly the prime ideals.

Proof. Let V be critical and $ab \in V$, where $a,b \notin V$. As $b \in V_a$ we have $V_a \supsetneq V$. Then K–dim $(S/V_a) < k$, which is a contradiction. So V is a prime ideal of S.

Conversely, let P be a prime ideal with K–dim $P = k$. For any $a \notin P$ consider the sequence.

$$aS + P \supset a^2S + P \supset \cdots \supset a^nS + P \supset \cdots \supset P .$$

Since $P_a = P$ we know that S/P and $aS + P/P$ are S–isomorphic and the natural map induces the following isomorphisms

$$S/aS + P \leftrightarrow aS + P/a^2S + P \leftrightarrow a^2S + P/a^3S + P \leftrightarrow \cdots .$$

If K–dim $S/aS + P = k$, then we have an infinite sequence of factors from S to P, each of Krull dimension k. Then K–dim $S/P > k$; a contradiction. It follows that any factor of S/P has dimension $< k$, so that P is a k–critical ideal.

Proposition 1.8. Let V be a k–critical right ideal of S. Then the factor ring R/V is an integral domain. More generally, End M is an integral domain, whenever M is a k–critical module.

Proof. Let $ab \in V$, where $a \in R$ and $b \in S$. If $a,b \notin V$ then $V_a \supseteq bS + V$ and K–dim $S/V_a \leqslant$ K–dim $S/bS + V < k$, which is a contradiction. The general argument is on the same lines.

The possibility that R/V is an Ore domain leads to another concept for modules. A module M is *compressible* if there is an S–isomorphism of M into every non–zero submodule of M. A compressible module is uniform and finitely generated, since we are supposing that S is a noetherian ring. (Since M can be injected into a cyclic submodule, which is accordingly noetherian, M itself must be noetherian. Then M contains a uniform submodule, it can be injected into that and consequently is uniform itself.)

Proposition 1.9. Let S be a commutative ring and T be an ideal. Then S/T is a compressible S–module if and only if T is a prime ideal.

Proof. Let $a \notin T$, then $S/T_a \cong (aS + T)/T$. Supposing that T is prime, we obtain $T_a = T$ and S/T is isomorphic to the arbitrary cyclic submodule $aS + T/T$ thus S/T is compressible.

Conversely, if $ab \in T$, there exists an isomorphism of S/T into $aS + T/T$ and $[b + T]$ goes to zero, so $b \in T$ and T is a prime ideal.

Proposition 1.10. Let S be a semi–prime ring and U be a uniform right ideal of S. Then U is a compressible S–module.

Proof. Let $V \neq 0$ be a right ideal such that $V \subseteq U$. Certainly $VU \neq 0$, since $V^2 \neq 0$. We know that $r(v) \cap U = 0$ implies that $vU = 0$. (Here $r(v) = \{x \in S \mid vx = 0\}$ and the reader may refer to [4].) Then there must exist $v \in V$ with $r(v) \cap U = 0$. This means that $u \rightarrow vu (u \in U)$ is an S–isomorphism $U \rightarrow vU \subseteq V$.

A discussion of the existence of compressible right ideals occurs in [5].

Proposition 1.11. Let V be a k–critical right ideal of S and S/V be a compressible module. Then R/V is a right Ore domain. For any $s \in S, s \notin V$ there exists $v \in V$ with $sv \in R$ and $sv \notin V$.

Proof. We know already that R/V is a domain. Let $r_1, r_2 \in R$ but

$r_1, r_2 \notin V$. Since S/V can be injected into $((r_1S + V) \cap (r_2S + V))/V$, the image of $[1 + V]$ is $[r + V]$, where $r \in R, r \notin V$. Let $r = r_1s_1 + v_1 = r_2s_2 + v_2$. Then $r_1(s_1V + V) \subset V$. It follows that $s_1V + V = V$, so that $s_1 \in R$. Similarly $s_2 \in R$ and neither s_1 nor s_2 belongs to V, because $r \notin V$. Hence $[r_1 + V]$ and $[r_2 + V]$ have a common right multiplier in R/V, which is the condition for the latter to be a right Ore domain.

The *bound* of a right ideal I in S is the largest (two–sided) ideal of S in I.

Theorem 1.12. Let V be a k–critical right ideal and S/V be a compressible module. Let T be the bound of V and $L = \{ s \in S \mid sV \subseteq T\} = (T \cdot\cdot V)$. The following properties hold:

(i) S/T is a prime ring.

(ii) If $T = L \neq R$, then R/T is a prime ring which has the same quotient ring as S/T.

(iii) If $T \neq L \neq R$ then $L \not\subseteq V$ and $V \not\subseteq L$. The ring R/T has exactly the minimal prime ideals L/T and V/T; also $(L \cap V)^2 \subseteq T$. Moreover T is the bound of L in S and $R = R' = \{ s \in S \mid Ls \subseteq L\}$.

Proof. Let A, B be ideals of S and $AB \subseteq T$. If $A \not\subseteq V$ then $A + V$ contains elements of R not in V, using compressibility. Hence $B \subseteq V$ is clear, as V is a prime of R. Thus either $A \subseteq T$ or $B \subseteq T$ so T is a prime ideal of S.

Next let A,B be ideals of R and $AB \subseteq L$. Then $(SAV)(SBV) \subseteq LV \subseteq T$, so that, either $SAV \subseteq T$ or $SBV \subseteq T$; in other words, either $A \subseteq L$ or $B \subseteq L$. Thus L is a prime of R. Certainly $(L \cap V)^2 \subseteq LV \subseteq T$ and hence L,V are the only minimal primes of T.

When $L \subseteq V$ then $LS \subseteq V$, and yet T = bound V, so that $LS \subseteq T$ and $L = T$. When $V \subseteq L$ then $V^2 \subseteq T$ implies that $(SV)^2 \subseteq T$ and $SV \subseteq T$. Then $V = SV$, V is an ideal in S and $R = S$. This possibility is a triviality, but is covered by the theorem as stated.

In case (iii) L and V are distinct primes of R. Let $sV \subseteq V$ for some $s \in R$, then $LsV \subseteq T$, hence $Ls \subseteq L$. Now let $Ls \subseteq L$, then $LsV \subseteq T$ and as $L \not\subseteq V$ there is an element $r \in L, r \notin V$. Then $r(sV + V) \subseteq V$, $sV \subseteq V$. We have proved that $R = R'$.

Let $T \subset T' \subset L$, where T' is an ideal of S. Then $T'V \subseteq T$ and, T being prime in S, we have, either $T' \subseteq T$ or $V \subseteq T$. The latter case $V = T \subseteq L = R$ has been eliminated from (iii), so T is the bound of L in S.

Somewhat more can be said by way of comment on this theorem. As we shall see later, both cases (ii) and (iii) do occur. For example, (iii) appears when we discuss the case $S = A_2$ and (ii) appears when S is a simple artinian ring and V is a maximal right ideal of S. It is interesting that R is also the idealiser of the left ideal L of S although L need not be a critical left ideal of S.

We can contribute a little more detail in (iii). Taking $T = 0$ for convenience (or going to S/T), we have a noetherian prime ring S with V a non–essential right ideal having L as its left annihilator. Since S/V is a uniform S–module, any right ideal which properly contains V has a regular element of S. Thus V is a maximal right complement in S and hence is a maximal right annihilator in S. Then L is a minimal left annihilator in S and hence is a uniform left ideal in S. It follows that whenever $ts = 0$, where $t \in L$, $t = 0$, and $s \in S$, then $Ls = 0$. Hence L is a minimal left annihilator in the ring R and, consequently, V is a maximal right annihilator in R. See [4] Lemma 2.1 and Theorem 9 for details.

These results naturally raise many problems:

(1) The conditions under which R, R/V, R/T are noetherian rings.

(2) The existence of a quotient ring for R/T.

(3) The relationship between the module properties: k–critical and compressible.

For the case when V is a maximal right ideal, Problem 3 is trivial and the others are easily settled. The result is due to Robson; the proof is different.

Proposition 1.13. Let V be a maximal right ideal of S, then R is right noetherian.

Proof. We can suppose that $R \neq S$, so that $SV = S$. Let A be a right ideal of R and suppose that AS has a minimal set of generators modulo AV say, $a_1, \cdots, a_k$. Thus

$$AS = a_1S + \cdots + a_kS + AV.$$

Let $a \in A$ and $a = a_1s_1 + \cdots + a_ks_k (\mathrm{mod}\ AV)$.

Suppose that $s_1 \notin R$ so that $s_1V \not\subseteq V$. Then $s_1V + V = S$ and

$$a_1 \in a_1(s_1V + V) \subseteq a_2S + \cdots + a_kS + AV$$

which is not allowed by the minimality of the set $a_1, \cdots, a_k$. Hence $s_1 \in R$ and $s_2, \cdots, s_k$ likewise. Hence $A = a_1R + \cdots + a_kR + AV$. However $S = SV$ implies that S is finitely generated over V and hence over R. Also AV is finitely generated over S and hence over R. It follows that A is finitely generated over R.

Robson's theorem covers the case when V is a finite intersection of maximal ideals and this proof adapts to this case also.

Let A be a right ideal of R and $a_1,\cdots,a_k \in A$ be such that

$$AS = a_1S + \cdots + a_kS + AV.$$

Let $J_1 = \{s \in S \mid a_1s \in a_2S + \cdots + a_kS + AV\}$ and let K_1 be a complement of J_1 in S relative to V. Then $J_1 \cap K_1 = V$ and $J_1 + K_1 = S$, where J_1, K_1 are right ideals in S. It follows that $AS = a_1K_1 + a_2S + \cdots + a_kS + AV$. Set $J_2 = \{s \in S \mid a_2s \in a_1K_1 + a_3S + \cdots + a_kS + AV\}$ and let K_2 be a complement of J_2 relative to V. Then $AS = a_1K_1 + a_2K_2 + a_3S + \cdots + a_kS + AV$. The process is repeated and eventually $AS = a_1K_1 + \cdots + a_kK_k + AV$. Take any $a \in A$ and write $a = a_1k_1 + \cdots + a_kk_k \pmod{AV}$. Then $k_1V + V \subseteq J_1 \cap K_1 = V$ and hence $k_1 \in R$. Similarly all $k_i \in R$. It follows that $A = a_1R + \cdots + a_kR + AV$. Now AV is finitely generated over R and we deduce that A is likewise. This proves that R is a right noetherian ring.

The case when V is k–critical for $k > 0$ is much more difficult and counter–examples are to be expected. One possible source of complication arises already when V is 1–critical, because there are two possibilities for the formation of V. It may happen that V is an infinite intersection of maximal right ideals or that it is not. In the latter case let H be the intersection of all maximal right ideals containing V. Then $II(V) \subseteq II(H)$, because $r \in II(V)$ and $V \subset M$, where M is maximal, implies that $V \subseteq M_r = \{ s \in S \mid sr \in M \}$ and M_r is again maximal. Hence $rH \subseteq M$ and, because M is arbitrary of its kind, $rH \subseteq H$. More precisely, we have $II(V) \subseteq II(\cap M_a \mid a \in S)$. In either case we notice that $II(V)$ is a subring of a noetherian ring closely associated with it, because S/H and $S/\cap M_a$ are finite S–modules.

At present the only support for the k–critical case with $k > 0$ is the following example.

Let $S = A_2 = A_1 \otimes A_1$ where $A_1 = F[x,y]$ is the polynomial ring in x,y over a field of characteristic zero, subject to the law $xy - yx = 1$. It is convenient to regard $S = F[x,y, x_1,y_1]$ where $xy - yx = 1 = x_1y_1 - y_1x_1$ and x,y commute with x_1,y_1. We choose $V = xS$. Then

$$R = V + A_1 \text{ and } S = V + A_1[y].$$

where A_1 is the subalgebra $F[x_1,y_1]$. The reason is that every element s of S

can be written uniquely in the form

$$s = x^n\phi_n(y) + \cdots + x\phi_1(y) + \phi_0(y) \; : \; \phi_i(y) \in A_1[y]$$

and $x\phi_i(y) - \phi_i(y)x = \phi_i'(y)$ (which is the usual derivative). Then $sx \in xS$ if and only if $\phi_0(y)x - x\phi_0(y) \in xS$, namely $\phi_0'(y) \in xS$. Thus $\phi_0' = 0$ and $\phi_0(y)$ is a constant in A_1. Since $R/V \cong A_1$ we see that R/V is a noetherian ring.

Any right R–module W in S which contains V has an element $\phi(y)$ and $\phi'(y) = \phi(y)x - x\phi(y) \in W$. Since the derivatives of $\phi(y)$ lie in W and F is a field, the coefficients (in A_1) of $\phi(y)$ belong to W. Hence $W \cap A_1[y]$ is a homogeneous right ideal over A_1. It readily follows that the right ideals W_S of S which contain V are in (1 – 1) correspondence with right ideals of A_1 under $W_S \leftrightarrow (W_S \cap A_1)$, indeed $W_S = (W_S \cap A_1)S$. The same is true of right ideals W_R of R which contain V, thus $W_R = (W_R \cap A_1)R$. It does not hold for R–modules W in general. In particular maximal right S–ideals containing V correspond to maximal right ideals of A_1, and, as a consequence, V is the intersection of the maximal right S–ideals which contain it. Moreover S/V is compressible, and 1–critical. Case (iii) occurs because S is a simple ring.

Let A be a right ideal of R, then AS/AV is finitely generated over S and AV is finitely generated over R (see Robson [6]).

Thus A will be finitely generated over R if we can prove that S/V is a noetherian R–module and this reduces to proving that S/R is a noetherian R–module. Let X be an R–submodule of S with $X \supset R$ and let $z \in X$, $z \notin R$; then $X \supset zV + V \neq V$. As $S/zV + V$ has a composition series as S–module we need only prove that any composition factor of $S/zV + V$ is a noetherian R–module. A simple module is isomorphic to S/M, where M is a maximal right ideal of S. This is a simple R–module if $V \not\subseteq M$, so let $V \subseteq M$. Then S/M has a unique maximal R–submodule $(M \cdot\!\cdot\; V)M \cong \mathrm{Hom}(S/V, S/M)$. Let $B = (M \cdot\!\cdot\; V)$, then we have reduced it to proving that B/M is finitely generated as an R/V–module. However, we can compute that $B = A_1 + M = R + M$, which does it.

References

1. J. E. Björk, Conditions which imply that subrings of semi–primary rings are semi–primary, J. Algebra (to appear).
2. H. Cartan and S. Eilenberg, "Homological Algebra," Princeton , 1956.
3. P. Gabriel and R. Rentschler, Sur la dimension des anneaux et ensembles ordonnés, C. R. Acad. Sci. Paris, **265** (1967), 712–715.
4. A. W. Goldie, The structure of prime rings under ascending chain conditions, Proc. London Math. Soc. **8** (1958), 589–608.
5. A. W. Goldie, Torsion–free modules and rings, J. Algebra, **1** (1964), 268–287.
6. J. C. Robson, Idealizers and hereditary noetherian prime rings, J. Algebra

(to appear).

7. L. Silver, Non–commutative localizations and applications, J. Algebra 7 (1967), 44–76.

STRUCTURE AND CLASSIFICATION OF HEREDITARY NOETHERIAN PRIME RINGS

Arun Vinayak Jategaonkar
Rutgers University and Cornell University

TO LALITA
who made a birthday card for me.

Contents

Introduction

Various classes of hereditary Noetherian prime rings (*HNP–rings*, for short) have been investigated from time to time. A HNP–ring R is a *pseudo–Dedekind ring* if R has only a finite number of idempotent ideals and every non-zero ideal of R contains an invertible ideal. In this paper, we shall be concerned with pseudo–Dedekind rings. It is known [8] that pseudo–Dedekind rings have a rather uncomplicated ideal theory and that they form a very extensive class of HNP–rings. Recently, Robson [32] has constructed a HNP–ring which is not pseudo–Dedekind.

A HNP–ring R is a *Dedekind ring* if R has no proper idempotent ideals. Naturally, a *Dedekind domain* is a domain which is also a Dedekind ring. Notice that our Dedekind domains need *not* be commutative. The simplest examples of Dedekind domains are (not necessarily commutative) PID's, simple hereditary Noetherian domains and the usual commutative Dedekind domains. Dedekind rings are of course pseudo–Dedekind, only better. We refer to [7, 30, 31] for results on Dedekind rings.

The aim of the present paper may be loosely described as follows:

(1) To obtain a structure theory for pseudo–Dedekind rings. More specifically,

to see how pseudo–Dedekind rings can be constructed from Dedekind domains.

(2) To define reasonable equivalence relations on the set of pseudo–Dedekind orders in a fixed simple Artinian ring Q and decide when two such orders belong to the same orbit under the given relation. We are particularly interested in the equivalence relation defined by conjugacy in Q. In general, this is a very tight equivalence relation. As approximations to conjugacy, we treat 'similarity' which is a special kind of category equivalence and a new equivalence relation called 'formal conjugacy' which is half–way between similarity and conjugacy but defined only over a special type of pseudo–Dedekind orders called almost Dedekind orders. See §1 for definitions.

Such questions were previously attempted only in the classical situation which we proceed to define. Let Ω be a commutative Dedekind domain with quotient field K, Σ be a finite dimensional central division K–algebra and $Q = M_n(\Sigma)$. A Ω–subalgebra R of Q is called an *Ω–order in* Q if R is finitely generated as an Ω–module and $RK = Q$. If such a ring R is hereditary then it is pseudo–Dedekind (cf. [8]). These rings have been intensively investigated. The main references to the literature on these rings are Auslander and Goldman [2], Brumer's thesis [4, 5], a series of papers by Harada out of which we pick [12, 13, 14], Murtha's thesis [28] and a paper by Drozd–Kiricenko–Roiter [6]. These results are summarized in §13 of Reiner's survey [29] (see also §7 of [29] and Chapter IX of [33]). The approach here is via localization and completion. These results are particularly pleasing when Ω is a complete DVR. When Ω is just a DVR, the classification is fine but the structure is expressed in terms over which there is very little algebraic hold. (We remedy this in §6.) When one comes to an arbitrary commutative Dedekind domain Ω, these results say very little; they essentially describe what happens to R if one localizes at a prime in Ω. Such information is difficult to utilize in classification; indeed, we have not been able to locate any results on classification except when Ω is a DVR [29] or Ω is a PID and $\Sigma = K$ [28].

The structure of semi–perfect HNP–rings was obtained by Michler [26]; these rings are like HNP–rings in the classical situation with Ω a complete DVR.

In the present paper, we obtain a structure theorem for arbitrary pseudo–Dedekind rings (Theorem 6.4). It includes all the previously known structure theoretic results for various classes of pseudo–Dedekind rings and provides new information even in the classical case. We also classify pseudo–Dedekind orders under similarity (Theorem 7.6) and almost Dedekind orders under formal conjugacy (Theorem 8.6). These results do not require any restrictions. However, we find it necessary to impose an additional restriction for classification under conjugacy; roughly, a certain Dedekind domain is assumed to be a PID. In the classical situation described in the above paragraph, this restriction means that a maximal Ω–order in Σ should be a PID. Within this framework, our results are almost complete (Theorem 9.14). This also includes the known results in the classical situation which are indicated above. We also classify finitely generated projectives within the same framework.

Our approach may have some merit by itself. Recall that, in the classical situation, the approach is via localization and completion. Now, in an unpublished note, the present author has shown that the localization developed in Kuzmanovich [23] for Dedekind rings can be extended to pseudo–Dedekind rings. Thus, if one wishes, one could try the localization and completion technique for pseudo–Dedekind rings too. However, in that case, one cannot hope to do much better than in the classical case. The main trouble is the well–known one: it is easy to localize but difficult to put things back. So we avoid localization completely. Roughly, this is done as follows: in the classical situation, the aim in localizing a hereditary Ω–order R at a prime **p** in Ω is to keep only one prime (viz. **p**) in Ω at hand. We develop a class of HNP–rings, called almost Dedekind rings, which are so transparently constructed from Dedekind domains that one can keep only one prime at hand by 'omitting' the rest; e.g., $R = \begin{pmatrix} Z & Z \\ 6Z & Z \end{pmatrix}$ is an almost Dedekind **Z**–order in $M_2(Q)$; if we want to keep 2**Z** at hand, we omit 3**Z** from R and obtain $\begin{pmatrix} Z & Z \\ 2Z & Z \end{pmatrix}$. Of course all pseudo–Dedekind rings are not almost Dedekind. This is where our structure theorem comes into play: Every pseudo–Dedekind ring is (in a special way) Morita equivalent to an almost Dedekind ring. Our strategy is now clear. We develop various things for almost Dedekind rings by a primewise examination if necessary and transfer them to pseudo–Dedekind rings via a category equivalence. The generality of our results thus stems from the fact that it is easier to get over a category equivalence than to recover from the effects of localization.

We now describe the layout of the paper – Section 1 contains a brief review of the Morita theorems and contains the definitions of 'similarity', 'formal conjugacy', and 'almost Dedekind rings'. Sections 2–5 are devoted to establishing the structure of almost Dedekind rings (Theorem 5.4). We take a Dedekind domain D with quotient skew field F and for every triangular frame **t** in D, we define a ring $\Delta_n(D, \mathbf{t})$ which is an order in $M_n(F)$. Section 2 shows that $\Delta_n(D, \mathbf{t})$ is almost Dedekind; these rings are our models for almost Dedekind rings. The notion of omission of primes is introduced in §3 and utilized to show that a bit more complicated rings than $\Delta_n(D, \mathbf{t})$ are in fact conjugate to rings like $\Delta_n(D, \mathbf{t})$. Section 4 contains a preliminary version of the structure theorem. Finally, in §5, we introduce the notion of the subbasic ring of a pseudo–Dedekind ring and use it to get the structure of almost Dedekind rings. Section 6 contains our main result on the structure of pseudo–Dedekind rings.

The remaining three sections are devoted to classification under similarity, formal conjugacy and conjugacy, respectively. The main new concept is that of **p**–invariants of a pseudo–Dedekind ring; this is defined in §7. In §9, we give two examples to illustrate some of the difficulties in extending the main results of this paper.

As usual, **Z** is the ring of integers; we let I_n stand for the set $\{1,2,\cdots,n\}$. We keep a commutative ring Ω fixed throughout the rest of the paper. All rings, modules, homomorphisms and category equivalences respect Ω. We need this only to show that the known results in the classical case can be obtained from our results by an appropriate choice of Ω; as a rule, we do not mention Ω. We impose symmetric conditions on rings; e.g., A Noetherian ring means a left and right Noetherian ring*

1. Preliminaries

We begin with a review of equivalences of module categories. Our standing reference for this topic (which we call the 'Morita theorems') is Bass [3, Chapter II]. A familiarity with this aspect of category theory is presupposed. Our main object is to point out certain simplifications that are available while dealing with categories of modules over orders in semi–simple rings. Some familiarity with orders in general and HNP–rings in particular is assumed. See [7, 8, 10, 11, 20, 25, 30, 36].

The following lemma is essentially in Levy [25].

Lemma 1.1. Let $R \subseteq S$ be orders in a semi–simple ring Q. Let M_S and N_R be modules and $f : M \to N$ be an R–homomorphism. If N_R is torsion–free then $f(M)$ is naturally a right S–module and $f : M \to f(M)$ is an S–homomorphism.

Proof. Since N_R is torsion–free, there exists K_Q such that $NQ = K$ (cf. [25]). Given $x \in M$ and $s \in S$, express s as $s = rc^{-1}$ with $r, c \in R$ and observe that in K we have

$$[f(xs) - f(x)s]\,c = 0.$$

So, $f(xs) = f(x)s$ in $f(M)$. ∎

Let R be an order in a semi–simple ring Q and M_R be an essential submodule of Q_R. As usual, we let $\mathbb{C}_\ell(M : R) = \{ q \in Q \mid qM \subseteq R \}$ and $\mathbf{o}_\ell(M) = \{ q \in Q \mid qM \subseteq M \}$. We observe that $\mathbb{C}_\ell(M : R) \overset{\text{canon}}{\cong} \operatorname{Hom}(M_R, R_R)$. For, given an element $q \in \mathbb{C}_\ell(M : R)$, we have $f_q \in \operatorname{Hom}(M_R, R_R)$ defined by $f_q(x) = qx$, $x \in M$. Evidently, the map $\psi : \mathbb{C}_\ell(M : R) \to \operatorname{Hom}(M_R, R_R)$ defined by $q \to f_q$ is a homomorphism of left R–modules. Since M_R is essential in Q_R, $MQ = Q$, so that M contains a unit of Q. This makes ψ an injection. To see that it is a surjection, observe that every $f : M_R \to R_R$ can be uniquely extended to $\hat{f} : Q_R \to Q_R$ and by Lemma 1.1, $\hat{f}$ is a Q–homomorphism. This gives us $q \in Q$

*The main results of the first six sections were presented at ring theory conferences at Plattsburg and Oberwolfach in 1970; the present version was presented at the Park City Conference. I wish to thank Professors Robert Gordon, F. Kasch, J. A. Riley and Alex Rosenberg for giving me these opportunities. Hearty thanks are due to Professor S. U. Chase with whom I had many profitable discussions.

such that $\hat{f}(q') = qq'$, $\forall q' \in Q$; clearly, $q \in \mathcal{C}_\ell(M : R)$ and $\psi(q) = f$.

We shall frequently use arguments similar to the above one without supplying details.

It is now clear that $\mathbf{o}_\ell(M) \overset{\text{canon}}{\cong} \operatorname{End} M_R$ and that we have the following commutative diagrams with exact top rows:

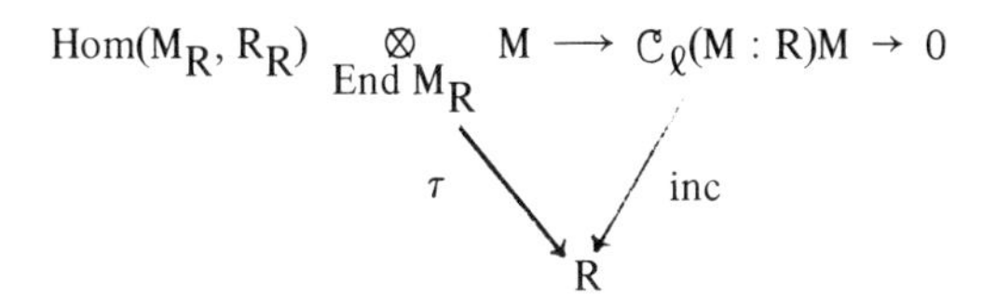

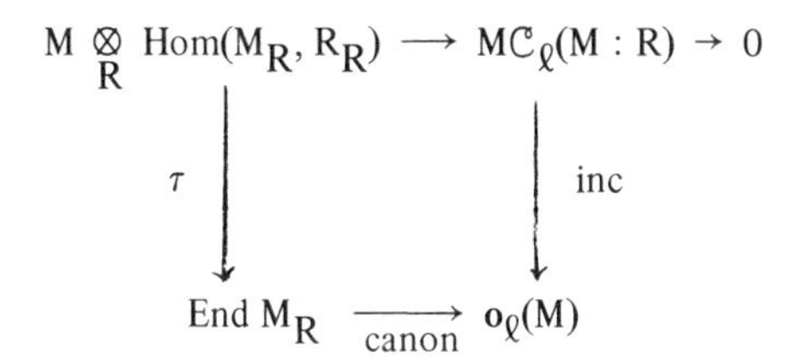

In these diagrams, τ is the trace map and μ is the map defined by $\mu(x \otimes f)(y) = xf(y)$ for $x,y \in M$ and $f \in \operatorname{Hom}(M_R, R_R)$. It is then immediate that M_R is a generator in mod–R if and only if $\mathcal{C}_\ell(M : R)M = R$, and that M_R is a finitely generated projective if and only if $M\mathcal{C}_\ell(M : R) = \mathbf{o}_\ell(M)$.

We shall frequently encounter bimodules over two different orders in the same simple Artinian ring and we shall want to prove that they are invertible (see definition below). Although we shall not need the generality gained by considering orders in semi–simple rings, our next result may be of some interest in the general setting.

Let R, S be orders in a semi–simple ring Q and let ${}_SM_R$ be an (S, R)–bisubmodule of Q. Assume that ${}_SM$ is faithful. Let $X = \{q \in Q \mid qM \subseteq M\}$. Since $S \subseteq X \subseteq Q$, X must be an order in Q and an essential extension of ${}_SS$. We have the canonical map $\psi : X \to \operatorname{End} M_R$ and it is easy to see that ψ is a surjection. If $\ker \psi \neq (0)$, then $\ker \psi \cap S \neq (0)$ and ${}_SM$ becomes unfaithful. So, ψ is an isomorphism. It follows [16, 20, 36] that $\dim M_R = \dim X = \dim Q$. Therefore M_R contains a unit in Q. Now we see that M is faithful and essential in Q on both sides. This observation is needed in the following result.

Proposition 1.2. Let R, S be orders in a semi–simple ring Q and let M be an (S, R)–bisubmodule of Q. Then the following conditions on M are equivalent:

(1) ${}_SM_R$ is *invertible*; i.e., there exist a (R, S)–bimodule L such that

$L \underset{S}{\otimes} M \cong R$ and $M \underset{R}{\otimes} L \cong S$ as bimodules.

(2) M_R is a progenerator in mod–R and $S = \mathbf{o}_\ell(M)$.

(3) $\mathcal{C}_\ell(M : R)M = R$ and $M\mathcal{C}_\ell(M : R) = S$.

(4) There exists a (R, S)–bisubmodule N of Q such that $NM = R$ and $MN = S$.

(5) M_R is a generator in mod–R and ${}_SM$ is a generator in S–mod.

Proof. Notice that each of these five conditions implies that M_R is faithful; so, as observed above, M is faithful and essential in Q from both sides. Now the Morita theorems [3] and our discussion so far in this section suffice to prove the following implications: (1) ⇒ (2) ⇒ (3) ⇒ (4) ⇒ (1) ⇒ (5).

We now show that (5) ⇒ (2). Put $\mathbf{o}_\ell(M) = E$. Since ${}_SM$ is a generator in S–mod, there exist $x_1, \cdots, x_n \in M$ and $f_1, \cdots, f_n \in \mathrm{Hom}({}_SM, {}_SS)$ such that $1 = \Sigma_{i=1}^n (x_i)f_i$. Since $S \subseteq E$ and M is naturally a left E–module, Lemma 1.1 shows that $f_1, \cdots, f_n$ are indeed E–homomorphisms of ${}_EM$ into E; so, $1 = \Sigma_{i=1}^n (x_i)f_i$ now shows that ${}_EM$ is a generator in E–mod. Since M_R is a generator in mod–R, it is a balanced R–module and a finitely generated projective left E–module. So, ${}_EM$ is a progenerator in E–mod and $R \overset{\text{canon}}{\cong} \mathrm{End}_E M$. Consequently, M_R is a progenerator in mod–R. Similarly, ${}_SM$ is a progenerator in S–mod; so it is balanced which yields $E = S$. ∎

The equivalence of condition (5) with others in the above proposition seems to be new.

We now define the equivalence relations on orders with which we shall be concerned throughout this paper. Let R, S be orders in a simple Artinian ring Q. R is classically *equivalent* with S in Q if there exist units u, u', v, v' in Q such that $uSv \subseteq R$ and $u'Rv' \subseteq S$. Classical equivalence is clearly an equivalence relation on the set of all orders in Q. Its orbits are quite big and algebraically stable. For example, in the classical situation(described in the introduction), the set of all Ω–orders in Q is a single orbit under classical equivalence. The other equivalence relations we shall study are all subordinate to classical equivalence in Q.

Let R, S be orders in a simple Artinian ring Q. We say R is *similar* to S in Q, denoted as $R \underset{Q}{\overset{\text{sim}}{\sim}} S$ or $R \overset{\text{sim}}{\approx} S$, if there exists an invertible (S, R)–bisubmodule of Q. In view of Proposition 1.2, it is immediate that similarity is an equivalence relation on the set of orders in Q. We note that similarity is subordinate to classical equivalence. For, if $R \underset{Q}{\overset{\text{sim}}{\sim}} S$ and ${}_SM_R$ is an invertible (S, R)–bisubmodule of Q then, as noted before Proposition 1.2, M_R is essential in Q_R. So, M contains a regular element v of Q. Since M_R is finitely generated, using the left common multiple property in R, we get a unit u in Q such that $uM \subseteq R$.

Thus $uSv \subseteq R$. Similarly, $u'Rv' \subseteq S$ for some units u', v' in Q.

We shall have many occasions to use the uniform *dimension* of a module. We refer to Goldie [11] for relevant details. Note that every Noetherian module has finite dimension.

Let R be a HNP ring and let the dimension of R_R be n. It is easy to see that R has n orthogonal idempotents $e_1, \cdots, e_n$ and that each e_iRe_i is a hereditary Noetherian domain (cf. Lemma 1.4 of [7] and [30, 34]). If R has n orthogonal idempotents $e_1, \cdots, e_n$ such that e_iRe_i is a Dedekind domain, then R is called an *almost Dedekind ring.* It is easily seen that a Dedekind ring is almost Dedekind [30]. We shall see in §2 that an almost Dedekind ring need not be Dedekind. In §6, we shall see that almost Dedekind rings are pseudo–Dedekind; an example in §9 shows that the converse is not true.

We now define formal conjugacy. This relation is defined on a special type of orders which include all almost Dedekind orders but not all pseudo–Dedekind orders. The definition proceeds in three steps.

Let Q be a fixed simple Artinian ring of uniform dimension n. Let R, S be orders in Q. R is said to be *diagonally formally conjugate* with S in Q if there exists a set of orthogonal idempotents $e_1, \cdots, e_n$ in $R \cap S$ and there exists an invertible (S, R)–bisubmodule H of Q subject to the following conditions:

(1) e_iRe_i and e_iSe_i are Dedekind domains for all $i \in I_n$.

(2) $H = \sum_{i=1}^{n} e_iHe_iR = \sum_{i=1}^{n} Se_iHe_i$.

Let R, S be orders in Q. R is *immediately formally conjugate* with S in Q if there exist units u, v in Q such that $u^{-1}Ru$ is diagonally formally conjugate with $v^{-1}Sv$ in Q.

R is said to be *formally conjugate* with S in Q, denoted as $R \overset{f}{\underset{Q}{\sim}} S$ or $R \overset{f}{\sim} S$, if there exists a sequence

$$R = R_0, R_1, \cdots, R_k = S$$

of orders in Q such that R_i is immediately formally conjugate with R_{i+1} in Q for $0 \leqslant i \leqslant k-1$.

Proposition 1.3. Let Q be a simple Artinian ring of uniform dimension n. Then 'formal conjugacy' is an equivalence relation on the set of those orders R in Q which contain a set $e_1, \cdots, e_n$ of orthogonal idempotents such that each e_iRe_i is a Dedekind domain.

Proof. It suffices to show that diagonal formal conjugacy is a reflexive relation. Let R, S, H and $e_1, \cdots, e_n$ be as in the definition of diagonal conjugacy. Evidently, e_iHe_i is a (e_iSe_i, e_iRe_i)–bisubmodule of e_iQe_i. Since H has a.c.c. on submodules from either side, it follows that e_iHe_i has a.c.c. on submodules from either side. Consequently, it is a progenerator on either side. By Proposition 1.2, we have a (e_iRe_i, e_iSe_i)–bisubmodule $e_iK_ie_i$ of e_iQe_i such that $e_iHe_iK_ie_i = e_iSe_i$ and $e_iK_ie_iHe_i = e_iRe_i$. Put $K = \Sigma\, e_iK_ie_iS$. It is straightforward to check that $KH = R$, $HK = S$ and $K = \Sigma_{i=1}^n e_iKe_iS = \Sigma_{i=1}^n Re_iKe_i$. ∎

It is clear that formally conjugate orders are similar. Later, we shall see that similar almost Dedekind orders need not be formally conjugate.

For later reference, we note the following triviality.

Proposition 1.4. Let R, S be orders in a simple Artinian ring Q. If R is almost Dedekind and $R \underset{Q}{\overset{f}{\sim}} S$, then S is almost Dedekind.

Proof. R and S are Morita equivalent; so, S is hereditary. ∎

Finally, for the sake of completeness, we define conjugacy. Let R, S be orders in a simple Artinian ring Q. R is *conjugate* with S in Q if there exists a unit u in Q such that $uRu^{-1} = S$. Evidently, conjugate orders are similar and formally conjugate if the latter is defined; the converse is not true.

2. A Model for Almost Dedekind Rings

Throughout this section, D is a fixed Dedekind domain with quotient skew field F and $Q = M_n(F)$ for some fixed positive integer n.

Definition. An indexed set $\mathbf{a} = \{\mathbf{a}_{ij} : i, j \in I_n\}$ of fractional D–ideals is called a *fractional frame* of order n in D if $\mathbf{a}_{ii} = D$, $\mathbf{a}_{ik}\mathbf{a}_{kj} \subseteq \mathbf{a}_{ij}$ and $\mathbf{a}_{ij}\mathbf{a}_{ji}$ is a non–zero semi–prime ideal of D for all $i, j, k \in I_n$. A fractional frame $\{\mathbf{a}_{ij} : i, j \in I_n\}$ is an *integral frame* in D if each $\mathbf{a}_{ij}$ is an integral ideal of D. It is a *triangular frame* if $\mathbf{a}_{ij} = D$ whenever $i \leqslant j$.

We shall usually denote a triangular frame as $\mathbf{t} = \{\mathbf{t}_{ij} : i, j \in I_n\}$.

Definition. Let $\mathbf{a} = \{\mathbf{a}_{ij} : i, j \in I_n\}$ be a fractional frame in D. Let $(D, \mathbf{a}) = \{(x_{ij}) \in M_n(F) : x_{ij} \in \mathbf{a}_{ij};\ i, j \in I_n\}$. Then $(D, \mathbf{a})$ is a ring, called the *associated ring* of the fractional frame $\mathbf{a}$.

Recall that if $\{X_{ij} : i, j \in I_n\}$ is an indexed set of non–empty subsets of a ring R then (X_{ij}) denotes the set of all $n \times n$ matrices (x_{ij}) with $x_{ij} \in X_{ij}$. We shall retain this *convention* throughout this paper. Thus, $(D, \mathbf{a}) = (\mathbf{a}_{ij})$.

It is immediate that $(D, \mathbf{a})$ is a Noetherian prime ring and, using [20], an order in $M_n(F)$.

Notation. If $\mathbf{a}$ is an integral frame in D, we let $\|(D,\mathbf{a})\| = $ l.c.m. $\mathbf{a}$.

Let $\mathbf{t} = \{\mathbf{t}_{ij} : i,j \in I_n\}$ be a triangular frame in D. Our favored notation for $(D,\mathbf{t})$ is $\Delta_n(D,\mathbf{t})$ or $\Delta(D,\mathbf{t})$. Observe that the ideals $\mathbf{t}_{ij}$ increase as we move upwards along a column of $\Delta_n(D,\mathbf{t})$ or along a row from left to right. This will be repeatedly used without further mention.

The following sequence of lemmas is designed to show that $\Delta_n(D,\mathbf{t})$ is almost Dedekind.

Lemma 2.1. Let Λ be a simple Artinian ring and let $\{\lambda_1,\cdots,\lambda_n\}$ be an indexed set of elements of Λ such that

$$\lambda_1\Lambda \subseteq \lambda_2\Lambda \subseteq \cdots \subseteq \lambda_n\Lambda = \Lambda.$$

Then there exists an indexed set $\{\mu_1,\cdots,\mu_n\}$ of elements of Λ such that $\Sigma_{i=1}^n \lambda_i\mu_i = 1$ and $\mu_i\lambda_j = 0$ whenever $i > j$.

Proof. Choose integers $1 = k_1 < k_2 < \cdots < k_s \leqslant n$ such that

$$\lambda_{k_i}\Lambda \subsetneqq \lambda_{k_{i+1}}\Lambda \quad \text{for } 1 \leqslant i \leqslant s-1, \text{ and}$$

$$\lambda_{k_i}\Lambda = \lambda_t\Lambda \quad \text{for } k_i \leqslant t \leqslant k_{i+1}-1.$$

Choose orthogonal idempotents $\epsilon_1,\cdots,\epsilon_s$ in Λ such that, putting $e_i = \Sigma_{t=1}^i \epsilon_t$, we have $e_i\Lambda = \lambda_{k_i}\Lambda$; this can be done since Λ is a simple Artinian ring (cf. [1, p. 33]). Clearly, $\epsilon_i e_j = 0$ whenever $i > j$. Choose $\alpha_{k_i} \in \Lambda$ such that $e_i = \lambda_{k_i}\alpha_{k_i}$. Put $\mu_{k_i} = \alpha_{k_i}\epsilon_i$ and $\mu_t = 0$ if $t \neq k_1,\cdots,k_s$. Then $\Sigma_{i=1}^n \lambda_i\mu_i = \Sigma_{i=1}^s \lambda_{k_i}\alpha_{k_i}\epsilon_i = \Sigma_{i=1}^s e_i\epsilon_i = \Sigma_{i=1}^s \epsilon_i = 1$. If $i \neq k_1,\cdots,k_s$ then $0 = \mu_i = \mu_i\lambda_j$ for all j. Let $i = k_\ell > j$; choose m with $\lambda_{k_m}\Lambda = \lambda_j\Lambda$ so that $m < \ell$. Since $\lambda_{k_m} \in e_m\Lambda$, therefore $\lambda_{k_m} = e_m\lambda_{k_m}$; so, $\mu_i\lambda_{k_m} = \alpha_{k_\ell}\epsilon_\ell e_m\lambda_{k_m} = 0$. Thus, $\mu_i\lambda_{k_m}\Lambda = (0)$ which yields $\mu_i\lambda_j = 0$. ∎

Lemma 2.2. Let Λ be a semi-simple ring, $\{A_{ij} : i,j \in I_n\}$ be an indexed set of ideals of Λ and $\{\lambda_i : i \in I_n\}$ be an indexed set of elements of Λ subject to the following conditions:

$$A_{ik}A_{kj} \subseteq A_{ij}; \quad A_{ij} = \Lambda \text{ if } i \leqslant j;$$

$$\lambda_i A_{ij} \subseteq \lambda_j\Lambda; \quad \sum_{i=1}^n \lambda_i\Lambda = \Lambda.$$

Then there exists an indexed set $\{\mu_i : i \in I_n\}$ of elements of Λ such that

$\Sigma_{i=1}^{n} \lambda_i \mu_i = 1$ and $\mu_i \lambda_j \in A_{ij}$ for all $i, j \in I_n$.

Proof. We may assume without loss of generality that Λ is a simple Artinian ring; so, whenever $i \leqslant j$, $A_{ij} = \Lambda$ which yields $\lambda_i \Lambda \subseteq \lambda_j \Lambda$. Further, $\Lambda = \Sigma_{i=1}^{n} \lambda_i \Lambda = \lambda_n \Lambda$. The lemma is now immediate from Lemma 2.1. ∎

For the rest of this section, we take a fixed triangular frame $\mathbf{t} = \{ \mathbf{t}_{ij} : i, j \in I_n \}$ in D and denote $\Delta_n(D, \mathbf{t})$ as T. We let $\{ e_{ij} : i, j \in I_n \}$ be the usual matrix units in $M_n(D)$. The uniform right ideal $e_{11}T$ of T will be called the first row of T. If I is a right ideal of T contained in $e_{11}T$, then we have a chain $K_1 \subseteq \cdots \subseteq K_n$ of right ideals of D such that $K_i \mathbf{t}_{ij} \subseteq K_j$ and $I = (X_{ij})$ with $X_{1j} = K_j$, $X_{ij} = 0$ otherwise. **Notation.** We shall denote the right ideal I as $I = (K_1, \cdots, K_n)$.

Lemma 2.3. Let $I = (K_1, \cdots, K_n)$ be a right ideal of T with $K_n = D$. Then I is a cyclic uniform projective right T-module.

Proof. We have $\mathbf{t}_{n1} = K_n \mathbf{t}_{n1} \subseteq K_1 \subseteq \cdots \subseteq K_n$. Since D is a Dedekind domain, there exist $z_i \in K_i$, $1 \leqslant i \leqslant n-1$, such that $K_i = \mathbf{t}_{n1} + z_i D$ (cf. [7, Cor. 3.6]). Put $z_n = 1$ and

$$z = \begin{pmatrix} z_1 & z_2 & \cdots & z_n \\ 0 & 0 & \cdots & 0 \\ \vdots & \vdots & & \vdots \\ 0 & 0 & \cdots & 0 \end{pmatrix}.$$

It is immediate that $I = zT$.

If $\mathbf{t}_{n1} = D$, then $I = e_{11}T$ and $T = M_n(D)$ which is hereditary; thus the lemma holds. Assume that $\mathbf{t}_{n1} \neq D$. Since $\mathbf{t}_{n1}$ is a non-zero semi-prime ideal of the Dedekind domain D, it follows [30, Theorem 3.5] that $\Lambda = D/\mathbf{t}_{n1}$ is a semi-simple ring. Put $A_{ij} = \mathbf{t}_{ij}/\mathbf{t}_{n1}$ and $\lambda_i = z_i + \mathbf{t}_{n1}$; then Lemma 2.2 gives an indexed set $x_1, \cdots, x_n$ of elements of D such that $x_i z_j \in \mathbf{t}_{ij}$ and $1 = \Sigma_{i=1}^{n} z_i x_i + a$ for some $a \in \mathbf{t}_{n1}$. Since $z_n = 1$, we may replace x_n by $x_n + a$ and get $x_i z_j \in \mathbf{t}_{ij}$ for all i, j and $1 = \Sigma_{i=1}^{n} z_i x_i$. Let

$$x = \begin{pmatrix} x_1 & 0 & \cdots & 0 \\ x_2 & 0 & \cdots & 0 \\ \vdots & \vdots & & \vdots \\ x_n & 0 & \cdots & 0 \end{pmatrix}.$$

Evidently, $xzT \subseteq T$ and $zxz = z$. By the dual basis theorem, I_T is projective. ∎

Lemma 2.4. Let $I_T = (L_1, \cdots, L_n)$ be a right ideal of T. Then I_T is a uniform right projective T–module.

Proof. Since $e_{11}T$ is a uniform T–module and $I \subseteq e_{11}T$, I_T is uniform.

Let $L = L_n$, $E = \mathbf{o}_\ell(L)$ and $L^* = \mathcal{C}_\ell(L : D)$. Since L is a progenerator in mod–D and $L^* \overset{\text{canon}}{\cong} \text{Hom}_D(L, D)$, it follows that $L^*L = D$ and $LL^* = E$. Also, E is a Dedekind order in F and, by the Morita theorems, there exists an indexed set $\mathbf{a} = \{ \mathbf{a}_{ij} : i, j \in I_n \}$ of integral ideals of E such that $\mathbf{a}_{ij}L = L\mathbf{t}_{ij}$, and $L^*\mathbf{a}_{ij} = \mathbf{t}_{ij}L^*$. Further, we have an indexed set $\{ K_i : i \in I_n \}$ of right ideals of E such that $K_iL = L_i$ for all i. Evidently, $K_i\mathbf{a}_{ij}L = K_iL\mathbf{t}_{ij} = L_i\mathbf{t}_{ij} \subseteq L_j = K_jL$ so that $K_i\mathbf{a}_{ij} \subseteq K_j$. Further, $K_n = E$. It is now straightforward to check that $\mathbf{a}$ is a triangular frame in E and that $K = (K_1, \cdots, K_n)$ is a uniform right ideal of $R = \Delta_n(E, \mathbf{a})$. By Lemma 2.3, K_R is projective.

Let ${}_RH_T = (L\mathbf{t}_{ij})$ and ${}_TG_R = (\mathbf{t}_{ij}L^*)$ as submodules of $M_n(F)$. It is immediate that $HG = R$ and $GH = T$. Thus, ${}_RH_T$ is an invertible (R, T)–bimodule. (Indeed, R and T are diagonally formally conjugate although we do not need this extra information here). Consider the sequence

$$K \otimes_R H \xrightarrow{\text{inc} \otimes \text{id}} R \otimes_R H \overset{\text{canon}}{\cong} H_T.$$

Since ${}_RH$ is projective, inc ⊗ id is a monomorphism. Thus $K \otimes_R H \cong KH = I$ as right T–modules. Since K_R is projective and H_T is projective, so is I_T. ∎

The following proposition is one–half of our structure theorem for almost Dedekind rings.

Proposition 2.5. Let $\mathbf{t} = \{\mathbf{t}_{ij} : i, j \in I_n \}$ be a triangular frame in a Dedekind domain D. Then $\Delta_n(D, \mathbf{t})$ is an almost Dedekind and a pseudo–Dedekind order in $M_n(F)$ where F is the quotient skew field of D.

Proof. We continue to denote $\Delta_n(D, \mathbf{t})$ as T. Routine arguments concerning generalized matrix rings show that T is a Noetherian prime ring and it follows from [21] that T is an order in $Q = M_n(F)$. So, T is of uniform dimension n. Clearly, $\{e_{ii} : i \in I_n\}$ is a set of orthogonal idempotents in T such that each $e_{ii}Te_{ii} \cong D$, a Dedekind domain. Since T is a prime Noetherian ring, all uniform right ideals of T are mutually subisomorphic [7]; Lemma 2.4 shows that they are all projective. Since $T = \oplus_{i=1}^n e_{ii}T$ where each $e_{ii}T$ is uniform, Kaplansky's Theorem [24, page 85] shows that T is right hereditary. To see that T is left hereditary, either adopt the last four lemmas for columns of T or use the left–right symmetry of the global dimension of a Noetherian ring. We have thus shown that T is an almost Dedekind order in Q.

It is clear that any non–zero ideal I of T has the form $I = (\theta_{ij})$ where θ_{ij} are non–zero ideals of D. Let $\theta = \cap_{i,j=1}^n \theta_{ij}$, $H = (\theta \mathbf{t}_{ij})$ and $G = (\theta^{-1}\mathbf{t}_{ij})$. Since fractional ideals of D commute [30, Theorem 2.1], it follows that $GH = HG = T$. Thus H is an invertible ideal of T and is contained in I. Now, $M_n(D)$ is a Dedekind order in $M_n(F)$, $T \subseteq M_n(D)$ and $M_n(D)$ is a finitely generated torsion–free T–module from either side. Therefore, $M_n(D)_T$ is T–projective (cf. [25]). By Proposition 1.8 of [8], the trace ideal of $M_n(D)_T$ is a minimal idempotent ideal of T so, by Theorem 4.8 of [8], T is a pseudo–Dedekind ring. ∎

We postpone the discussion of ideals and modules of $\Delta_n(D, \mathbf{t})$ until §7 and §9, respectively.

3. Omission of Primes

Throughout this section, D is a fixed Dedekind domain with quotient skew field F and $Q = M_n(F)$ for some fixed positive integer n. Our object is to relate the associated rings of integral frames of order n in D with those of triangular frames. This is done by a useful device which we proceed to define.

Let $\{g_{ij} : i, j \in I_n\}$ be an indexed set of (not necessarily non–zero) ideals of D and let $G = (g_{ij}) \subseteq M_n(D)$. Let $\mathbf{p}$ be a given *prime* in D, i.e., $\mathbf{p}$ is a non–zero prime ideal in D. We define an indexed set $\{\mathbf{h}_{ij} : i, j \in I_n\}$ of D as follows: if $g_{ij} \neq (0)$, then $\mathbf{h}_{ij} = \mathbf{p}^{m_{ij}}$ where m_{ij} is the largest integer such that $\mathbf{p}^{m_{ij}} \mid g_{ij}$. If $g_{ij} = (0)$, then $\mathbf{h}_{ij} = (0)$. As usual, $\mathbf{p}^0 = D$. **Notation**: We put $G^{(\mathbf{p})} = (\mathbf{h}_{ij})$. We have tacitly used the commutativity and uniqueness of prime factorization of ideals in D (cf. [30]).

We note that if R is the associated ring of an integral or triangular frame in D then so is $R^{(\mathbf{p})}$. Evidently, $R^{(\mathbf{p})} \neq M_n(D)$ if and only if $\mathbf{p} \mid \|R\|$, in which case $\|R^{(\mathbf{p})}\| = \mathbf{p}$. Also,

$$R = \bigcap_{\mathbf{p}} R^{(\mathbf{p})} = \bigcap_{\mathbf{p} \mid \|R\|} R^{(\mathbf{p})}.$$

The following lemma is crucial in relating the associated rings of integral frames with those of triangular frames and in the classification of almost Dedekind orders under similarity done in §7.

Lemma 3.1. Let $T = \Delta_n(D, t)$ and let M_T be a right T–module of the form $M = (g_{ij})$ where $\{ g_{ij} : i, j \in I_n \}$ is an indexed set of non–zero ideals of D with $g_{in} = D$ for all i. Let $||M|| = \text{l.c.m.} \{ g_{ij} : i, j \in I_n \}$. For each prime $\mathbf{p}$ in D with $\mathbf{p} \mid ||M||$, assume that the rows of $M^{(\mathbf{p})}$ are ordered in a monotonically decreasing manner under inclusion, and let $\pi_\mathbf{p}$ be a given permutation matrix in $M_n(D)$. Let

$$N = \bigcap_{\mathbf{p} \mid ||M||} \pi_\mathbf{p} M^{(\mathbf{p})} .$$

Then N is naturally a right T–module and $N_T \cong M_T$.

Proof. It is easily seen that each $M^{(\mathbf{p})}$ and so N is naturally a right T–module. Since $||T|| \subseteq ||M|| = ||N||$, therefore $||M||$ is a non–zero semi–prime ideal of D. So, $||M|| = \mathbf{p}_1 \cdots \mathbf{p}_\kappa$ where $\mathbf{p}_\lambda$ are distinct primes in D. If $\kappa = 0$ then $M = N = M_n(D)$ and the lemma trivially holds. We assume that $\kappa > 0$ and choose a fixed indexing for the prime divisors of $||M||$. Let $\mathbf{p}_\mu$ be the first prime divisor of $||M||$ for which $\pi_{\mathbf{p}_\mu} M^{(\mathbf{p}_\mu)} \neq M^{(\mathbf{p}_\mu)}$. To simplify notation, put $\pi_{\mathbf{p}_\lambda} = \pi_\lambda$ and $\mathbf{p}_\mu = \mathbf{p}$; we shall denote the ith row of N as N_i and the ith row of $\pi_\lambda M^{(\mathbf{p}_\lambda)}$ as $(\pi_\lambda M^{(\mathbf{p}_\lambda)})_i$ and treat them as right T–submodules of the first row of T. It is evident that $N \cong \oplus_{i=1}^n N_i$, etc.

Since the rows of $M^{(\mathbf{p}_\lambda)}$ are assumed to be ordered in a monotonically decreasing order under inclusion and $\pi_\lambda M^{(\mathbf{p}_\lambda)} = M^{(\mathbf{p}_\lambda)}$ for $\lambda < \mu$, we have obtained the following formulas: if $\lambda < \mu$ and $1 \leqslant i < j \leqslant n$, then

$$(N_i \cap N_j)^{(\mathbf{p}_\lambda)} = N_j^{(\mathbf{p}_\lambda)} = j^{th} \text{ row of } M^{(\mathbf{p}_\lambda)},$$

$$(N_i + N_j)^{(\mathbf{p}_\lambda)} = N_i^{(\mathbf{p}_\lambda)} = i^{th} \text{ row of } M^{(\mathbf{p}_\lambda)}.$$

Similarly,

$$(N_i \cap N_j)^{(\mathbf{p})} = (\pi_\mu M^{(\mathbf{p})})_i \cap (\pi_\mu M^{(\mathbf{p})})_j,$$

$$(N_i + N_j)^{(\mathbf{p})} = (\pi_\mu M^{(\mathbf{p})})_i + (\pi_\mu M^{(\mathbf{p})})_j.$$

Since π_μ is a permutation matrix, there is an inclusion relation between the ith and j^{th} rows of $\pi_\mu M^{(\mathbf{p})}$. If $(\pi_\mu M^{(\mathbf{p})})_i \supseteq (\pi_\mu M^{(\mathbf{p})})_j$ then

$$(N_i \cap N_j)^{(p)} = (\pi_\mu M^{(p)})_j; \quad (N_i + N_j)^{(p)} = (\pi_\mu M^{(p)})_i.$$

On the other hand, if $(\pi_\mu M^{(p)})_i \subseteq (\pi_\mu M^{(p)})_j$ then

$$(N_i \cap N_j)^{(p)} = (\pi_\mu M^{(p)})_i; \quad (N_i + N_j)^{(p)} = (\pi_\mu M^{(p)})_j .$$

Let L be the right T–submodule of $M_n(D)$ whose i^{th} row is $N_i + N_j$, j^{th} row is $N_i \cap N_j$ and the kth row is N_k for $k \neq i, j$. We have the usual split–exact sequence

$$0 \to N_i \cap N_j \to N_i \oplus N_j \to N_i + N_j \to 0.$$

It follows that $N \cong L$ as right T–modules. The formulas in the preceding paragraph show that

$$L = (\bigcap_{\lambda<\mu} M^{(p_\lambda)}) \cap \pi_\mu^* M^{(p)} \cap (\bigcap_{\lambda>\mu} \pi_\lambda^* M^{(p_\lambda)})$$

where π_λ^*, π_μ^* are certain permutation matrices in $M_n(D)$. Further, for the fixed i,j we have been working with, we have $(L_i)^{(p)} \supseteq (L_j)^{(p)}$ whereas, for $k \neq i, j$, $(L_k)^{(p)} = (N_k)^{(p)}$. In a finite number of such steps, we can obtain permutation matrices π' in $M_n(D)$ such that putting

$$K = (\bigcap_{\lambda<\mu} M^{(p_\lambda)}) \cap \pi'_\mu M^{(p)} \cap (\bigcap_{\lambda>\mu} \pi'_\mu M^{(p_\lambda)}),$$

we have $K \cong N$ and the rows of $\pi'_\mu M^{(p)}$ are arranged in a monotonically decreasing order. So, $\pi'_\mu M^{(p)} = M^{(p)}$. Since $M = \cap_{\lambda=1}^{\kappa} M^{(p_\lambda)}$, an induction on μ is now available to conclude that $N \cong M$ as right T–modules. ∎

Corollary 3.2. Let $T = \Delta_n(D, t)$. For each prime $p \mid \|T\|$, let π_p be a permutation matrix in $M_n(D)$ and let $N = \cap_{p \mid \|T\|} \pi_p T^{(p)}$. Then $N_T \cong T_T$.

Proof. The rows of $T^{(p)}$ are linearly ordered under inclusion so Lemma 3.1 applies. ∎

Lemma 3.3. Let R be the associated ring of an integral frame $\mathbf{a} = \{ a_{ij} : i, j \in I_n \}$ in D such that $\|R\| = p$, a prime in D. Then there exists a permutation matrix π in $M_n(D)$ and a triangular frame $\mathbf{t}$ in D such that $\Delta_n(D, \mathbf{t}) = \pi R \pi^{-1}$ and $\|\Delta_n(D, \mathbf{t})\| = p$.

Proof. If $\mathbf{a}$ is already triangular, we take $\mathbf{a} = \mathbf{t}$ and $\pi = 1$. Assume that $\mathbf{a}$ is not triangular. Lexicographically order the set $U = \{ (i,j) \mid 1 \leqslant i \leqslant j \leqslant n \}$. Let $(i,j) \in U$ be the first element in U for which $\mathbf{a}_{ij} \neq D$. Since $\|R\| = \mathbf{p}$, $\mathbf{a}_{ij} = \mathbf{p}$. Also, $\mathbf{a}_{\ell m} = D$ whenever $(\ell, m) \in U$ and $\ell < i$ or $\ell = i \leqslant m \leqslant j-1$. Let $i \leqslant k \leqslant j-1$. Then $\mathbf{a}_{ik} = D$ and $\mathbf{p} = \mathbf{a}_{ij} \supseteq \mathbf{a}_{ik}\mathbf{a}_{kj} = \mathbf{a}_{kj} \supseteq \|R\| = \mathbf{p}$ yields $\mathbf{a}_{kj} = \mathbf{p}$. Since $\mathbf{a}_{jk}\mathbf{a}_{kj}$ is a semi–prime ideal of D and $\|R\| = \mathbf{p} = \mathbf{a}_{kj}$, we get $\mathbf{a}_{jk} = D$. Thus, R has the following form:

$$\begin{pmatrix} & \vdots & \vdots & & \vdots & \vdots & \\ \cdots & D & D & \cdots & D & \mathbf{p} & \cdots \\ & \vdots & \vdots & & \vdots & \vdots & \\ \cdots & D & D & \cdots & D & D & \cdots \\ & \vdots & \vdots & & \vdots & \vdots & \end{pmatrix} \begin{matrix} \leftarrow i^{th} \\ \\ \leftarrow j^{th} \end{matrix}$$

$$\begin{matrix} \uparrow & & \uparrow \\ i^{th} & & j^{th} \end{matrix}$$

Let π_1 be the permutation matrix in $M_n(D)$ obtained by exchanging the ith and j^{th} rows of the identity matrix in $M_n(D)$. Evidently, $\pi_1 R \pi_1^{-1} = (D, \mathbf{b})$ where $\mathbf{b} = \{ \mathbf{b}_{ij} : i,j \in I_n \}$ is an integral frame in D and $\|(D, \mathbf{b})\| = \mathbf{p}$. Using the above paragraph, we see that $\mathbf{b}_{\ell m} = D$ whenever $\ell < i$ or $\ell = i \leqslant m \leqslant j$. An induction is now available to conclude the proof. ■

We note that the above simple–minded procedure does not triangulate $\begin{pmatrix} \mathbf{Z} & 3\mathbf{Z} \\ 2\mathbf{Z} & \mathbf{Z} \end{pmatrix}$.

Lemma 3.4. Let R be the associated ring of an integral frame $\mathbf{a} = \{ \mathbf{a}_{ij} : i,j \in I_n \}$ in D. For each prime $\mathbf{p} \mid \|R\|$, let $\pi_\mathbf{p}$ be a permutation matrix in $M_n(D)$ and let $M = \cap_{\mathbf{p} \mid \|R\|} R^{(\mathbf{p})} \pi_\mathbf{p}$. Then M is a finitely generated left R–submodule of $M_n(D)$ and $\mathbf{o}_\ell(M) = R$.

Proof. Since each $R^{(\mathbf{p})} \pi_\mathbf{p}$ is a left R–submodule of $M_n(D)$, so is M. Since ${}_R M_n(D)$ is finitely generated and R is Noetherian, ${}_R M$ is finitely generated. Clearly, ${}_R M$ is an essential submodule of ${}_R M_n(D)$; so R acts faithfully on M and $R \subseteq \mathbf{o}_\ell(M)$.

Let $\{ e_{ij} : i,j \in I_n \}$ be the usual system of matrix units in $M_n(D)$. Since $e_{ii} \in R$ for all $i \in I_n$, it follows that given an element $q \in Q$, $q \in \mathbf{o}_\ell(M)$ if $e_{ii} q e_{jj} \in \mathbf{o}_\ell(M)$ for all $i,j \in I_n$. Thus $\mathbf{o}_\ell(M) = (\mathbf{b}_{ij})$ where $\mathbf{b}_{ij}$ are certain non–zero additive subgroups of F. We also have an indexed set $\{ M_{ij} : i,j \in I_n \}$ of non–zero ideals of D such that $M = (M_{ij})$. Clearly, $\mathbf{b}_{ij} M_{jk} \subseteq M_{ik}$. Since $M_{ik} \subseteq D$, we have $\mathbf{b}_{ij}(\Sigma_{k=1}^n M_{jk}) \subseteq D$. Evidently, $\Sigma_{k=1}^n M_{jk}$ is a non–zero ideal of D; if it is a proper ideal, we can choose a prime $\mathbf{p} \mid M_{jk}$ for all $k \in I_n$; this would imply $\mathbf{p} \mid \mathbf{a}_{jk}$ for all k, which is absurd since $\mathbf{a}_{jj} = D$. Consequently, $\mathbf{a}_{ij} \subseteq \mathbf{b}_{ij} \subseteq D$. It is then immediate that $\{ \mathbf{b}_{ij} : i, j \in I_n \}$ is an integral frame in D.

Suppose $\mathbf{a}_{ij} \neq \mathbf{b}_{ij}$ for some $i, j \in I_n$. Then there exists a prime $\mathbf{p}$ in D with $\mathbf{p} \mid \mathbf{a}_{ij}$ but $\mathbf{p} \nmid \mathbf{b}_{ij}$. Suppose $\pi_{\mathbf{p}}$ sends the jth column of $R^{(\mathbf{p})}$ to the kth column of $R^{(\mathbf{p})}\pi_{\mathbf{p}}$. Then $\mathbf{p} \mid M_{ik}$; so $\mathbf{b}_{ij}M_{jk} \subseteq M_{ik} \subseteq \mathbf{p}$, which yields $M_{jk} \subseteq \mathbf{p}$. However, this is impossible since the $(j,k)^{th}$ entry of $R^{(\mathbf{p})}\pi_{\mathbf{p}}$ is D. It follows that $R = o_\ell(M)$. ∎

Proposition 3.5. Let R be the associated ring of an integral frame of order n in a Dedekind domain D. Then there exists a unit u in $M_n(D)$ and a triangular frame $\mathbf{t}$ of order n in D such that $uRu^{-1} = \Delta_n(D, \mathbf{t})$ and $\|R\| = \|\Delta_n(D, \mathbf{t})\|$. Consequently, R is an almost Dedekind and pseudo–Dedekind order in $M_n(F)$, F being the quotient skew field of D.

Proof. Given a prime $\mathbf{p}$ in D with $\mathbf{p} \mid \|R\|$, we use Lemma 3.3 to obtain a permutation matrix $\pi_{\mathbf{p}}$ in $M_n(D)$ and a triangular frame $\mathbf{t}_{\mathbf{p}}$ in D such that $\pi_{\mathbf{p}}^{-1} R^{(\mathbf{p})}\pi_{\mathbf{p}} = \Delta_n(D, \mathbf{t}_{\mathbf{p}})$ and $\|\Delta_n(D, \mathbf{t}_{\mathbf{p}})\| = \mathbf{p}$. Let $T = \cap_{\mathbf{p} \mid \|R\|} \Delta_n(D, \mathbf{t}_{\mathbf{p}})$. Then evidently, $T = \Delta_n(D, \mathbf{t})$ for a triangular frame $\mathbf{t}$ in D and $\|T\| = \|R\|$. Further, $T^{(\mathbf{p})} = \pi_{\mathbf{p}}^{-1} R^{(\mathbf{p})}\pi_{\mathbf{p}}$ for each $\mathbf{p} \mid \|R\|$. So, $\cap_{\mathbf{p} \mid \|T\|} \pi_{\mathbf{p}} T^{(\mathbf{p})} = \cap_{\mathbf{p} \mid \|R\|} R^{(\mathbf{p})}\pi_{\mathbf{p}} = M$, say. Using injectivity and torsion–freeness of $M_n(F)_T$ and using Corollary 3.2, we have $M = uT$ for some unit u in $M_n(F)$. Since $TM_n(D) = MM_n(D) = M_n(D)$, it follows that u is a unit in $M_n(D)$ Clearly, $o_\ell(uT) = uTu^{-1}$. By Lemma 3.4, $o_\ell(uT) = R$. The rest follows from Proposition 2.5. ∎

4. The Canonical Form

In this section, we shall obtain a preliminary version of a canonical form for almost Dedekind orders under formal conjugacy.

Lemma 4.1. Let F be a skew field and let $\{ \mathbf{a}_{ij} : i, j \in I_n \}$ be an indexed set of non–zero additive subgroups of F satisfying the following conditions:

(i) $\mathbf{a}_{ik}\mathbf{a}_{kj} \subseteq \mathbf{a}_{ij}$;

(ii) $\mathbf{a}_{ii}$ is a Dedekind order in F;

(iii) $\mathbf{a}_{ij}\mathbf{a}_{ji}$ is a semi–prime ideal of $\mathbf{a}_{ii}$ for all $i, j, k \in I_n$.

Let $Q = M_n(F)$ and $R = (\mathbf{a}_{ij}) \subseteq M_n(F)$. Then R is an almost Dedekind order in Q and there exists a triangular frame $\mathbf{t}$ in $D = \mathbf{a}_{11}$ such that $R \underset{Q}{\overset{f}{\sim}} \Delta_n(D, \mathbf{t})$.

Proof. For fixed $i, j \in I_n$, $\mathbf{a}_{ij}\mathbf{a}_{ji} \subseteq \mathbf{a}_{ii}$ yields $\mathbf{a}_{ij}\alpha \subseteq \mathbf{a}_{ii}$ for all $\alpha \in \mathbf{a}_{ji}$.

Since $a_{ji} \neq (0)$, a_{ij} is isomorphic with a left ideal of a_{ii} and so a progenerator in a_{ii} – mod. Similarly, a_{ij} is a progenerator in mod – a_{jj}. By Proposition 1.2, a_{ij} is an invertible (a_{ii}, a_{jj})-bisubmodule of F.

Since $a_{1i}a_{ij} \subseteq a_{1j}$, by the Morita theorems, there exists a non–zero two–sided ideal b_{ij} of $D = a_{11}$ such that $a_{1i}a_{ij} = b_{ij}a_{1j}$ for all $i, j \in I_n$. We shall show that $b = \{ b_{ij} : i, j \in I_n \}$ is an integral frame in D. Observe that $b_{ii}a_{1i} = a_{1i}a_{ii} = a_{1i} = a_{11}a_{1i} = b_{1i}a_{1i}$; the lattice isomorphism in the Morita theorems shows that $b_{ii} = b_{1i} = D$. Also, $b_{ik}b_{kj}a_{1j} = b_{ik}a_{1k}a_{kj} = a_{1i}a_{ik}a_{kj} \subseteq a_{1i}a_{ij} = b_{ij}a_{1j}$; so, the lattice isomorphism used above also gives $b_{ik}b_{kj} \subseteq b_{ij}$. Since $a_{ij}a_{ji}$ is a semi–prime ideal of the Dedekind domain a_{ii}, it is a product of distinct maximal ideals of a_{ii}; so, $b_{ij}b_{ji}$ is a product of distinct maximal ideals of D and thus a semi–prime ideal of D (cf. [30]).

Put $S = (b_{ij}) \subseteq M_n(D)$. Let $\{ e_{ij} : i, j \in I_n \}$ be the usual system of matrix units in $M_n(D)$. Both R and S are Noetherian orders in $M_n(F)$ (cf. [21]), $\{ e_{ii} : i \in I_n \}$ is a set of orthogonal idempotents in $R \cap S$ and $e_{ii}Re_{ii}$, $e_{ii}Se_{ii}$ are Dedekind domains. Let $H = (a_{1i}a_{ij}) \subseteq Q$. Then H is evidently an (S, R)–bisubmodule of Q with $H = \Sigma_{i=1}^{n} Se_{ii}He_{ii} = \Sigma_{i=1}^{n} e_{ii}He_{ii}R$. Further, in a self–evident notation, if we put $G = R(\text{diag}\, a_{1i}^{-1})$ then G is an (R, S)–bisubmodule of Q. A straightforward computation shows that $GH = R$ and $HG = S$. Hence $R \overset{f}{\sim} S$. Using Proposition 3.5, we get a triangular frame t in D with $R \underset{Q}{\overset{f}{\sim}} \Delta_n(D, t)$. Since $\Delta_n(D, t)$ is almost Dedekind, so is R. ■

Corollary 4.2 In the above lemma, if D is a PID then $R \underset{Q}{\overset{conj}{\sim}} \Delta_n(D, t)$.

Proof. We have $a_{1i} = Dx_i$ for some $x_i \in F$. Let $u = \text{diag}\, x_i$. Then $u^{-1}Su = R$ and, by Proposition 3.5, $S \overset{conj}{\sim} \Delta_n(D, t)$. ■

Remarks :

(1) Lemma 4.1 brings out the type of structural distortions introduced by a diagonal formal conjugacy (which is the only new ingredient of formal conjugacy).

(2) Corollary 4.2 leads us to expect that whenever we have a domain connected with our ring that should have been just a Dedekind domain but turns out to be a PID, then we should expect conjugacy rather than formal conjugacy (cf. Theorem 5.4).

The following lemma essentially shows that every almost Dedekind order satisfies the hypothesis of Lemma 4.1 up to an isomorphism. (See proof of Proposition 4.4).

Lemma 4.3 Let F be a skew field and $\{ a_{ij} : i, j \in I_n \}$ be an indexed set of non–zero additive subgroups of F such that $a_{ik}a_{kj} \subseteq a_{ij}$ and a_{ii} is a Dedekind order in F for all $i, j, k \in I_n$. Let $R = (a_{ij}) \subseteq M_n(F)$. Then R is a Noetherian order in $M_n(F)$. If R is hereditary then $\mathbf{a}_{ij}\mathbf{a}_{ji}$ is a semi–prime ideal of $\mathbf{a}_{ii}$ for all $i, j \in I_n$.

Proof. Since $\mathbf{a}_{ij}\mathbf{a}_{ji} \subseteq \mathbf{a}_{ii}$, it follows that each $\mathbf{a}_{ij}$ is a finitely generated left $\mathbf{a}_{ii}$–module and a finitely generated right $\mathbf{a}_{jj}$–module. Thus R is a Noetherian order in $M_n(F)$.

Assume that R is hereditary. Let $\{ e_{ij} : i, j \in I_n \}$ be the usual system of matrix units in $M_n(F)$. Given $i < j$, the ring $(e_{ii} + e_{jj})R(e_{ii} + e_{jj})$ is naturally isomorphic with the ring

$$\begin{pmatrix} \mathbf{a}_{ii} & \mathbf{a}_{ij} \\ \mathbf{a}_{ji} & \mathbf{a}_{jj} \end{pmatrix}$$

which is a HNP–order in $M_2(F)$ (cf. [30, 34]). We may thus assume without loss of generality that $n = 2$. In this case, we proceed to show that $\mathbf{a}_{12}\mathbf{a}_{21}$ is a semi–prime ideal of $\mathbf{a}_{11}$. That $\mathbf{a}_{21}\mathbf{a}_{12}$ is a semi–prime ideal of $\mathbf{a}_{22}$ will follow in a similar manner.

Evidently, $\mathbf{a}_{12}\mathbf{a}_{21}$ is a non–zero ideal of $\mathbf{a}_{11}$. If $\mathbf{a}_{12}\mathbf{a}_{21} = \mathbf{a}_{11}$, it is already a semi–prime ideal of $\mathbf{a}_{11}$. Assume that $\mathbf{a}_{12}\mathbf{a}_{21} \neq \mathbf{a}_{11}$ and let $\mathbf{b}$ be an ideal of $\mathbf{a}_{11}$ such that $\mathbf{b}^2 \subseteq \mathbf{a}_{12}\mathbf{a}_{21} \subseteq \mathbf{b}$. To finish the proof, we have to show that $\mathbf{b} = \mathbf{a}_{12}\mathbf{a}_{21}$. Put

$$S = \{\eta \in F \mid \eta\mathbf{b} \subseteq \mathbf{a}_{11};\ \ \eta\mathbf{a}_{12} \subseteq \mathbf{a}_{12}\},$$

$$T = \{\theta \in F \mid \theta\mathbf{b} \subseteq \mathbf{a}_{21};\ \ \theta\mathbf{a}_{12} \subseteq \mathbf{a}_{22}\},$$

$$I = \begin{pmatrix} \mathbf{b} & \mathbf{a}_{12} \\ 0 & 0 \end{pmatrix}$$

It is easily seen that I is a right ideal of R and that $\{ q \in Q \mid qI \subseteq R \} = \begin{pmatrix} S & F \\ T & F \end{pmatrix}$. Now using injectivity of Q_R and Lemma 1.1, we have an epimorphism

$$\begin{pmatrix} S & F \\ T & F \end{pmatrix} \xrightarrow{\text{canon}} \mathrm{Hom}(I_R, R_R).$$

By the dual basis theorem, we have $x_1, \cdots, x_k \in I$ and $q_1, \cdots, q_k \in \begin{pmatrix} S & F \\ T & F \end{pmatrix}$ such that $x = \Sigma_{i=1}^{k} x_i q_i x$ for all $x \in I$. Consequently,

$$\begin{pmatrix} \mathbf{b} & \mathbf{a}_{12} \\ 0 & 0 \end{pmatrix} = \begin{pmatrix} \mathbf{b} & \mathbf{a}_{12} \\ 0 & 0 \end{pmatrix}\begin{pmatrix} S & F \\ T & F \end{pmatrix}\begin{pmatrix} \mathbf{b} & \mathbf{a}_{12} \\ 0 & 0 \end{pmatrix}$$

which yields $b = bSb + a_{12}Tb$. Since $Tb \subseteq a_{21}$, it follows that $a_{12}Tb \subseteq a_{12}a_{21}$. Further, $a_{11} \subseteq S \subseteq o_\ell(a_{12}a_{21}) = a_{11}$ since a_{11} is Dedekind (cf. [30]). Thus $S = a_{11}$ and so $a_{12}a_{21} \subseteq b \subseteq b^2 + a_{12}a_{21} = a_{12}a_{21}$, i.e., $a_{12}a_{21} = b$. ∎

We now give a preliminary version of the canonical form for almost Dedekind orders.

Proposition 4.4. Let F be a skew field and R be an order in $Q = M_n(F)$. Then R is an almost Dedekind ring if and only if there exists a Dedekind order D in F and a triangular frame **t** of order n in D such that $R \underset{Q}{\overset{f}{\sim}} \Delta_n(D, \mathbf{t})$.

Proof. Let R be an almost Dedekind ring. Since the uniform dimension of R is n, there exist n orthogonal idempotents $f_1,\cdots,f_n$ in R such that each f_iRf_i is a Dedekind domain. Let $\{ e_{ij} : i, j \in I_n \}$ be the usual system of matrix units in $M_n(F)$. There exists a unit u in Q such that $u^{-1}f_iu = e_{ii}$ for all $i \in I_n$ (cf. [19, page 59]). Let $u^{-1}Ru = S$. Then S is an almost Dedekind order in Q and $e_{11},\cdots,e_{nn}$ are orthogonal idempotents in S such that each $e_{ii}Se_{ii}$ is a Dedekind domain. The two–sided Peirce decomposition of S with respect to $\{e_{ii} : i \in I_n\}$ yields an indexed set $\{ a_{ij} : i, j \in I_n \}$ of non–zero additive sub–groups of F such that $a_{ik}a_{kj} \subseteq a_{ij}$, a_{ii} is a Dedekind order in F and $S = (a_{ij})$. By Lemma 4.3, $a_{ij}a_{ji}$ is a semi–prime ideal of a_{ii} for all $i, j \in I_n$. By Lemma 4.1, $S \underset{Q}{\overset{f}{\sim}} \Delta_n(D, \mathbf{t})$ where $D = a_{11}$ and **t** is a triangular frame in D. Thus $R \underset{Q}{\overset{f}{\sim}} \Delta_n(D, \mathbf{t})$. The converse follows from Propositions 1.4 and 2.5. ∎

We now give a characterization of almost Dedekind rings.

Proposition 4.5. Let R be a Noetherian prime ring of uniform dimension n containing a set $\{ e_1,\cdots,e_n \}$ of orthogonal idempotents such that each e_iRe_i is a Dedekind domain. Then R is an almost Dedekind ring if and only if $e_iRe_jRe_i$ is a semi–prime ideal of e_iRe_i for all $i, j \in I_n$.

Proof. Let Q be the simple Artinian quotient ring of R. Choose a complete system of matrix units $\{ e_{ij} : i, j \in I_n \}$ in Q with $e_i = e_{ii}$ for all $i \in I_n$. Let F be the centralizer of $\{ e_{ij} : i, j \in I_n \}$ in Q and identify Q with $M_n(F)$ by the usual isomorphism [19, page 52]. Then $R = (a_{ij}) \subseteq M_n(F)$. Now apply Lemmas 4.1 and 4.3. ∎

5. The Canonical Form (continued)

We shall presently show that the Dedekind order D in F given in Proposition

4.4 is uniquely determined up to a similarity in F. In fact, a much stronger uniqueness is available. To state it, we introduce the concept of a subbasic ring. A preliminary lemma is needed.

Lemma 5.1. Let F be a skew field and R be a Dedekind order in $M_n(F) = Q$. Then there exists a Dedekind order D in F such that $R \underset{Q}{\overset{f}{\sim}} M_n(D)$. D is uniquely determined by R up to a similarity in F.

Proof. By Proposition 4.4, we have a Dedekind order D in F and a triangular frame **t** of order n in D such that $R \overset{f}{\sim} \Delta_n(D, \mathbf{t})$. So, $\Delta_n(D, \mathbf{t})$ is a Dedekind order in Q. Clearly, $M_n(D) \supseteq \Delta_n(D, \mathbf{t})$ and $M_n(D)$ is a finitely generated $\Delta_n(D, \mathbf{t})$–module from either side. So, $M_n(D)$ and $\Delta_n(D, \mathbf{t})$ are classically equivalent in Q. It follows [30] that $\Delta_n(D, \mathbf{t}) \overset{f}{=} M_n(D) \overset{f}{\sim} R$.

If E is an order in F such that $M_n(E) \overset{f}{\sim} R$ then $M_n(E) \overset{f}{\sim} M_n(D)$. So, E is Dedekind. Also, we have an invertible $(M_n(E), M_n(D))$–subbimodule H of $M_n(F)$. It is immediate that $H = M_n(K)$ where K is a (E, D)–subbimodule of F which is finitely generated from either side. Using Proposition 1.2, it follows that K is invertible. Hence $E \overset{sim}{\sim} D$. ∎

Let us note a corollary which brings out the naturality of formal conjugacy.

Corollary 5.2. If R and S are Dedekind orders in a simple Artinian ring Q then $R \underset{Q}{\overset{sim}{\sim}} S$ if and only if $R \underset{Q}{\overset{f}{\sim}} S$.

Proof. The 'if' part is trivial. We prove the 'only if' part. Let $Q = M_n(F)$ for a skew field F. By Lemma 5.1, we may as well assume that $R = M_n(D)$ and $S = M_n(E)$ where D, E are Dedekind orders in F. The last part of the proof of Lemma 5.1 shows that $M_n(D) \underset{Q}{\overset{sim}{\sim}} M_n(E)$ gives $D \underset{F}{\overset{sim}{\sim}} E$, so $M_n(D) \underset{Q}{\overset{f}{\sim}} M_n(E)$. ∎

We now introduce the concept of a 'subbasic ring' which will play an important role in the rest of the paper.

Let Q be a simple Artinian ring. Let us assume that Q is given a concrete representation as $Q = M_n(F)$ with F a skew field. Let $\mathcal{E}$ be an orbit of orders in Q under the classical equivalence. Assume that $\mathcal{E}$ contains at least one Dedekind order in Q. Then all Dedekind orders belonging to $\mathcal{E}$ are similar in Q [30]; as shown above, they are all formally conjugate. Further, they determine a unique orbit $\mathfrak{D}$ of Dedekind orders in F under similarity in F. $\mathfrak{D}$ is called the subbasic class of $\mathcal{E}$. Any $D \in \mathfrak{D}$ is called a *subbasic ring* of $\mathcal{E}$ or a subbasic ring of any order $R \in \mathcal{E}$.

Trivially, if one subbasic ring of $\mathcal{E}$ is a PID then all rings in the subbasic class $\mathfrak{D}$ are conjugate PID's.

We have made some choices in defining the subbasic ring D of an equivalence class $\mathcal{E}$ and D depends upon these choices. However, this is unimportant to us because we shall use only the ideal theory of D, which is shared by all rings in $\mathcal{D}$, (cf. [30]), and we can pass from one choice of F to another by inner automorphisms of Q (cf. [19, page 59]).

We need one more observation.

Lemma 5.3. Let F be a skew field, $Q = M_n(F)$ and $\mathcal{E}$ be an orbit of orders in Q under classical equivalence. If $\mathcal{E}$ contains either an almost Dedekind ring or a pseudo–Dedekind ring then $\mathcal{E}$ contains a Dedekind ring and so has a subbasic ring D which is a Dedekind order in F.

Proof. If $R \in \mathcal{E}$ and R is pseudo–Dedekind, then it is known [8] that $\mathcal{E}$ contains a Dedekind order. If R is almost Dedekind then, by Proposition 4.4, $R \underset{Q}{\overset{f}{\sim}} \Delta_n(D, \mathbf{t})$ where D is a Dedekind order in F. Clearly, $M_n(D) \in \mathcal{E}$. ■

We can now state the existence of the canonical form the way we want it.

Theorem 5.4. Let F be a skew field and $Q = M_n(F)$. Let $\mathcal{E}$ be an orbit of orders in Q under classical equivalence. Assume that $\mathcal{E}$ has a subbasic ring D which is Dedekind order in F. Then a given order $R \in \mathcal{E}$ is almost Dedekind if and only if $R \underset{Q}{\overset{f}{\sim}} \Delta_n(D, \mathbf{t})$ for some triangular frame $\mathbf{t}$ in D. If D is a PID then R is almost Dedekind if and only if $R \underset{Q}{\overset{\text{conj}}{\sim}} \Delta_n(D, \mathbf{t})$ for some triangular frame $\mathbf{t}$ in D.

Proof. Let $\mathcal{D}$ be the subbasic class of $\mathcal{E}$. By Proposition 4.4, $R \overset{f}{\sim} \Delta_n(D', \mathbf{t}')$ for some $D' \in \mathcal{D}$ and, as observed before Lemma 5.3, $D' \underset{F}{\overset{\text{sim}}{\sim}} D$. If K is an invertible (D', D)–bisubmodule of F then $\Delta_n(D', \mathbf{t}')(\text{diag } K) = H$ defines a diagonal formal conjugacy $\Delta_n(D', \mathbf{t}') \underset{Q}{\overset{f}{\sim}} \Delta_n(D, \mathbf{t})$ for some triangular frame $\mathbf{t}$ in D. The rest is clear. ■

Observe that Proposition 4.4 is contained in the above theorem: given an almost Dedekind order R in Q, we take $\mathcal{E}$ to be the orbit of R under classical equivalence and apply Lemma 5.3 to get a subbasic class for $\mathcal{E}$.

We postpone the examination of the uniqueness of our canonical form; it requires more machinery. We proceed to show that the following result due to Michler [26] follows immediately from Theorem 5.4.

Proposition 5.5. Let F be a skew field and R be an order in $M_n(F) = Q$. Then R is a semi–perfect HNP–ring if and only if there exists a local PID D which is an order in F and a triangular frame $\mathbf{t}$ of order n in D such that $R \underset{Q}{\overset{\text{conj}}{\sim}} \Delta_n(D, \mathbf{t})$.

Proof. Let R be a semi–perfect HNP order in Q. Since the uniform dimension of R is n, R contains n orthogonal idempotents, say $e_1,\cdots,e_n$. Since each e_i is a uniform idempotent, e_iRe_i is a Noetherian domain and since R is semi–perfect, each e_iRe_i is local (cf. [27]). By Kaplansky's theorem, every one–sided ideal of e_iRe_i is free. It follows that each e_iRe_i is a PID and thus R is almost Dedekind. Now apply Theorem 5.4 to get $R \underset{Q}{\overset{\text{conj}}{\approx}} \Delta_n(D, t)$ with D a local PID. The converse is immediate from the characterization of semi–perfect rings (cf. [27] and Theorem 5.4). ■

Let us note that when D is a local PID, $\Delta_n(D, t)$ is very transparent since D has only one prime, viz. J(D).

6. Structure of Pseudo–Dedekind Rings

The main result of this section establishes a very simple connection between almost Dedekind and pseudo–Dedekind rings via category equivalence.

The following lemma is needed to show that 'pseudo–Dedekind' is a Morita invariant property.

Lemma 6.1. Let R, S be arbitrary rings and ${}_SM_R$ be an invertible (S,R)–bimodule. Then for a given ideal I of R, there exists a unique ideal $I^\#$ of S such that $I^\#M = MI$. The correspondence $I \leftrightarrow I^\#$ is a multiplication preserving lattice isomorphism of the ideal lattice of R with the ideal lattice of S. I is maximal, idempotent or invertible if and only if $I^\#$ is respectively so.

Proof. The existence of the required lattice isomorphism is a part of the Morita theorems. It is immediate that I is maximal or idempotent if and only if $I^\#$ is so. Suppose I is invertible and $I^\#M = MI$. Then $I^\# \otimes_S M \overset{\text{canon}}{\cong} I^\#M = MI \overset{\text{canon}}{\cong} M \otimes_R I$. Let M^{-1} be an (R, S)–bimodule such that $M \otimes_R M^{-1} \cong S$ and $M^{-1} \otimes_S M \cong R$ as bimodules. Consider $M \otimes_R I \otimes_R M^{-1}$. It is clearly an invertible (S, S)–bimodule and $M \otimes_R I \otimes_R M^{-1} \cong MI \otimes M^{-1} \cong I^\#M \otimes M^{-1} \cong I^\#$ as an (S, S)–bimodule, so $I^\#$ is invertible.The converse is similarly proved by observing that $IM^{-1} = M^{-1}I^\#$ and that M^{-1} is an invertible (R, S)–bimodule. ■

Proposition 6.2. 'Pseudo–Dedekind' is a Morita invariant property.

Proof. As is well–known, 'hereditary', 'Noetherian' and 'prime' are Morita invariant properties. It is immediate from Lemma 6.1 that 'every non–zero ideal contains an invertible ideal' and 'there exist only a finite number of idempotent ideals' are Morita invariant properties. ■

Corollary 6.3. Every almost Dedekind ring is pseudo–Dedekind.

Proof. This is immediate from Propositions 2.5, 6.2, and Theorem 5.4. ∎

We now prove the main theorem of this section.

Theorem 6.4. Let F be a skew field, $Q = M_n(F)$, and R be an order in Q. Then the following conditions on R are equivalent:

(1) R is a pseudo–Dedekind ring.

(2) There exists a Dedekind order D in F, a triangular frame $\mathbf{t}$ of order m $(m \in \mathbf{Z}^+)$ in D and a finitely generated essential right $\Delta_m(D, \mathbf{t})$–submodule M of $M_{n\times m}(F)$ such that $R = \mathbf{o}_\ell(M)$.

(3) There exists a Dedekind order D in F which is a subbasic ring of R. Further, if E is any Dedekind order in D which is a subbasic ring of R, then there exists a triangular frame $\mathbf{t}'$ of order m $(m \in \mathbf{Z}^+)$ in E and a progenerator M' in mod–$\Delta_m(D, \mathbf{t}')$ such that M' is an essential right $\Delta_m(D, \mathbf{t}')$–submodule of $M_{n\times m}(F)$ and $\mathbf{o}_\ell(M') = R$.

Proof. (1) ⇒ (3): Let $Y_1, \cdots, Y_k$ be all the distinct idempotent ideals of R which are minimal among idempotent ideals of R. Let $X_i = Y_i Y_{i-1} \cdots Y_1$, $S_i = \mathbf{o}_\ell(Y_i)$, $1 \leqslant i \leqslant k$ and let $S = \mathbf{o}_r(X_1)$. It is known [8] that S and S_i are Dedekind orders in Q, they contain R and are finitely generated R–modules from either side, and that $R = \cap_{i=1}^k S_i$. Evidently, each X_i is an integral R–ideal, so it contains a unit of Q. Thus X_i is an essential left ideal of S_i and an essential right ideal of S. By Proposition 1.2, X_i is an invertible (S_i, S)–bisubmodule of Q. Consequently, $\mathbf{o}_\ell(X_i) = S_i$, $\mathbf{o}_r(X_i) = S$ for $1 \leqslant i \leqslant k$. Choose an invertible ideal K of R such that $K \subseteq \cap_{i=1}^k X_i$. Using the inclusion reversing bijection between subrings between R and K^{-1} and the idempotent ideals of R (cf. [8]), we have $S = \mathbf{o}_\ell(Y_s)$ for some s. Let $A = Y_s X_k$. Then A is an integral S–ideal and we have a chain $A \subseteq X_k \subseteq X_{k-1} \subseteq \cdots \subseteq X_1$ of right ideals of S. Let $\Sigma = M_k(Q) =$.(F) where $m = kn$. Let $X = (X_k, X_{k-1}, \cdots, X_1)_{1\times k}$ and

$$B = \begin{pmatrix} S & S & \cdots & S & S \\ A & S & \cdots & S & S \\ \vdots & \vdots & & \vdots & \vdots \\ A & A & \cdots & S & S \\ A & A & \cdots & A & S \end{pmatrix}_{k\times k}.$$

It is immediate that B is a Noetherian order in Σ and that X is a finitely generated right B–module. Let $M_{n\times m}(F)$ be naturally considered as a (Q,Σ)–bimodule. Counting uniform dimensions, we see that X is an essential submodule of $M_{n\times m}(F)$ from both sides. It follows that $\mathbf{o}_\ell(X) = \cap_{i=1}^{k} \mathbf{o}_\ell(X_i) = R$. Let $W = \mathbf{o}_r(X)$ and $W_{ij} = \{ q \in Q \mid X_{k-i+1}q \subseteq X_{k-j+1} \}$. Since $B \subseteq W \subseteq \Sigma$, W must be an order in Σ. As indicated after Lemma 1.1, $W \overset{\text{canon}}{\cong} \text{End}\,{}_RX$. Since R is a HNP and ${}_RX$ is a finitely generated torsion–free left R–module, W must be a HNP. It is easily seen that $W = (W_{ij}) \subseteq M_k(Q)$. Since $W_{ii} = \mathbf{o}_r(X_{k-i+1}) = S$ which is a Dedekind order in Q, it follows that W is an almost Dedekind order in Σ.

Consider X_W. Since $R = \mathbf{o}_\ell(X) \overset{\text{canon}}{\cong} \text{End}\,X_W$ and $W \overset{\text{canon}}{\cong} \text{End}\,{}_RX$, X_W is a balanced right W–module which is finitely generated projective over its endomorphism ring. Hence [9] X_W is a generator in mod–W. Since X_B is finitely generated, so is X_W; further X_W is torsion–free over the HNP ring W. Hence X is a progenerator in mod–W. By Theorem 5.4, there exists a Dedekind order D in F, a triangular frame $\mathbf{t}$ of order m in D and an invertible $(W, \Delta_m(D,\mathbf{t}))$–bisubmodule N of Σ. Let $M = XN$. Then M is a progenerator in mod–$\Delta_m(D,\mathbf{t})$ and $R = \mathbf{o}_\ell(M)$. Consider

$$M^{\#} = \begin{bmatrix} M \\ M \\ \vdots \\ M \end{bmatrix} \Bigg\} \text{ k terms}$$

naturally as a right $\Delta_m(D,\mathbf{t})$–submodule of $M_{m\times m}(F)$. Evidently, $M^{\#}$ is a progenerator in mod–$\Delta_m(D,\mathbf{t})$ and an essential right $\Delta_m(D,\mathbf{t})$–submodule of $M_{m\times m}(F)$ with $\mathbf{o}_\ell(M^{\#}) = M_k(R) \subseteq \Sigma$. Hence $M^{\#}$ is an invertible $(M_k(R), \Delta_m(D,\mathbf{t}))$–bisubmodule of Σ. It is now clear that D is the subbasic ring of $M_k(R)$ and consequently of R.

If E is a given Dedekind order in F which is also a subbasic ring of R then $D \overset{\text{sim}}{\sim} E$. Let L be an invertible (D,E)–subbimodule of F, $\mathbf{t} = \{ \mathbf{t}_{ij} : i,j \in I_m \}$, $H = (\mathbf{t}_{ij}L)$ and $\mathbf{t}' = \{ L^{-1}\mathbf{t}_{ij}L : i,j \in I_m \}$. Evidently, H is an invertible $(\Delta_m(D,\mathbf{t}), \Delta_m(E,\mathbf{t}'))$–bisubmodule of Σ. Also, $MH = M'$ is a progenerator in mod $\Delta_m(E,\mathbf{t}')$ which is essential in $M_{n\times m}(F)$ and $\mathbf{o}_\ell(M') = R$.

(3) $\Rightarrow$ (2): Trivial.

(2) $\Rightarrow$ (1): Let D be a Dedekind order in F, $\Sigma = M_m(F)$, $T = \Delta_m(D,\mathbf{t})$, M_T a finitely generated essential right T–submodule of $M_{n\times m}(F)$, $R = \mathbf{o}_\ell(M)$ and $S = \mathbf{o}_r(M)$. It is easily seen that $R \overset{\text{nat}}{\cong} \text{End}\,M_T \overset{\text{nat}}{\cong} \text{End}\,M_S$ and $S \overset{\text{nat}}{\cong} \text{End}\,{}_RM$. Since $T \subseteq S \subseteq \Sigma$, S must be a HNP–order in Σ (cf. [23]). Also, M_S is a finitely generated torsion–free right S–module, so is S–projective.

Thus, $R \cong \operatorname{End} M_S$ is a HNP–order in Q (cf. [30, 34]).

Let I be the S–trace ideal of M_S. Since M_S is non–zero finitely generated projective, $MI = M$ and $I^2 = I \neq (0)$. Since $S \subseteq \mathbf{o}_r(I) \subseteq \Sigma$ and $I\mathbf{o}_r(I) = I$, we have $M\mathbf{o}_r(I) = MI\mathbf{o}_r(I) = MI = M$. It follows that $\mathbf{o}_r(I) \subseteq \mathbf{o}_r(M) = S$; so, $S = \mathbf{o}_r(I)$. Since S is HNP, it follows that I is a progenerator in mod–S. Thus, $I^*I = S$ where $I^* = \{\sigma \in \Sigma \mid \sigma I \subseteq S\}$. Now, using $I^2 = I$, we have $I = SI = (I^*I)I = I^*I = S$. Hence M_S is a progenerator in mod–S and $\mathbf{o}_\ell(M) = R$.

In view of Proposition 6.2 and Corollary 6.3, it now suffices to show that S is an almost Dedekind ring. Since M_S is finitely generated projective and S is a direct sum of uniform right ideals, it follows [24] that M_S is a direct sum of uniform submodules. Let K_S be a uniform direct summand of M_S. Since the injective hulls of K as a right S–module and as a right T–module agree, K_T is also uniform. Since M_T is finitely generated, so is K_T. Thus $K_T \cong (K_1, \cdots, K_m)$ where $0 \neq K_1 \subseteq K_2 \subseteq \cdots \subseteq K_m$ are right ideals of D. Let $\{e_{ij} : i, j \in I_m\}$ be the usual matrix units in $M_m(F)$. Since $T \subseteq S$, it follows that $e_{ii} \in S$ for all i. Let $S_{ii} = \{x \in F \mid xe_{ii} \in S\}$. Clearly, $D \subseteq S_{ii}$. Also, for every $x \in S_{ii}$, $K_i x \subseteq K_i$ so that $S_{ii} \subseteq \mathbf{o}_r(K_i) = D$. Hence $S_{ii} = D$ for all i. It follows that S is an almost Dedekind ring. As indicated, this completes the proof. ∎

In §9, we give an example of a pseudo–Dedekind ring which is not almost Dedekind. In view of the above theorem, it shows that 'almost Dedekind' is *not* a Morita invariant property.

We have to recall some definitions before we can state our next result. An order R in a simple Artinian ring Q is said to be *bounded* if every essential one–sided ideal contains a non–zero two–sided ideal. When $R \neq Q$, R is *quasi–local, semi–local* or *local* if $R/J(R)$ is respectively a semi–simple ring, a simple Artinian ring or a skew field.

Observe that if R is a HNP order in Q, then every proper factor ring of R is Artinian [7]. Consequently, a non–zero ideal is primitive if and only if it is maximal. Hence, if $J(R) \neq (0)$, then R is already quasi–local and has only a finite number of maximal ideals.

The following proposition shows that the behavior of a pseudo–Dedekind order is strongly influenced by its subbasic ring.

Proposition 6.6. Let F be a skew field and $Q = M_n(F)$. Let $\mathcal{E}$ be an equivalence class of orders in Q under classical equivalence and let $\mathfrak{S}$ be the set of all pseudo–Dedekind rings in $\mathcal{E}$. Let D be a Dedekind order in F which is a subbasic ring of $\mathfrak{S}$. Then D is bounded, primitive, semi–local, or semi–perfect if and only if at least one ring in $\mathfrak{S}$ is, respectively so, if and only if every ring in $\mathfrak{S}$ is respectively so.

Proof. Let $\mathbf{t}$ be a triangular frame of order m in D, $m \in \mathbf{Z}^+$, and $T =$

$\Delta_m(D, t)$. We show first that D is bounded, primitive, semi–local, or semi–perfect if and only if T is so.

Let $\{ e_{ij} : i, j \in I_m \}$ be the usual system of matrix units in $M_m(F)$. Then $D \cong e_{11}Te_{11}$. It is now immediate that D is bounded, etc., if T is (cf. [19] and [20]).

Suppose D is bounded. Let X be an essential right ideal of T. It is easily seen that $\{ \text{diag } d \mid d \in D^* \}$ is an Ore set in T and the quotient ring of T with respect to this set is $M_m(F)$. Since X is essential in T, it contains a regular element c of T. Now c^{-1} can be expressed as $c^{-1} = t(\text{diag } d)^{-1}$ for some $t \in T$ and non–zero $d \in D$. Hence $\text{diag } d \in X$. Let K be a non–zero two–sided ideal of D contained in $d \cdot \|T\|$. Then $M_m(K)$ is a non–zero two–sided ideal of T such that $M_m(K) \subseteq \Sigma\, e_{ii}dT = (\text{diag } d)T \subseteq X$. Similarly, every essential left ideal contains a non–zero two–sided ideal of T. Hence T is bounded.

Suppose D is right primitive. Let L be a maximal right ideal of D which does not contain any non–zero two–sided ideal of D. Let $V = (L, \cdots, L)_{1 \times m} \subseteq e_{11}T$ and $M = V + \Sigma_{i=2}^m e_{ii}T$. Let N be a proper right ideal of T containing M. Since $\Sigma_{i=2}^m e_{ii}T \subseteq N$, it follows that $V \subseteq e_{11}N \subseteq N$. Thus, $e_{11}N = (L_1, \cdots, L_m)$ where $L \subseteq L_1 \subseteq L_2 \subseteq \cdots \subseteq L_m$ is a chain of right ideals of D. Since N is proper and L is maximal, we have $L = L_1 \supseteq L_n\|T\|$. If $L_n = D$, we have $L \supseteq \|T\|$, a contradiction. It follows that $L_n = L$, so $M = N$. Thus M is a maximal right ideal of T. If W is a non–zero two–sided ideal of T contained in M, then $e_{11}We_{11}$ is a non–zero ideal of $e_{11}Te_{11}$ ($\cong D$) contained in $e_{11}Me_{11}$ ($\cong L$), a contradiction. Hence T is right primitive. Similarly, if D is left primitive then so is T. (Note: D is right primitive if and only if D is left primitive (cf. [7]).)

If D is semi–local, then $J(D) \neq (0)$. So, $0 \neq e_{11}J(T)e_{11} \subseteq J(T)$. As already observed above, T is semi–local.

Suppose D is semi–perfect. Then D must be local since it is a domain. Then $\{ e_{ii} : i \in I_n \}$ gives an orthogonal decomposition of the identity of T into local idempotents. Hence (cf. [27]) T is semi–perfect.

It is easily seen that the four properties stated in the proposition are Morita invariant. The proposition now follows from Theorem 6.4 and what has been proved above. ∎

7. Invariants for Similarity

Let F be a skew field, $\mathcal{E}$ be an equivalence class of orders in $M_n(F) = Q$ under classical equivalence, $\mathfrak{S}$ be the set of all pseudo–Dedekind orders in $\mathcal{E}$ and D be a Dedekind order in F which is a subbasic ring of $\mathcal{E}$. Let $R \in \mathfrak{S}$. In this section, we establish a natural bijection between primes in D and ideals in R which are maximal among invertible ideals of R. It is known that there is a natural bijection between the later type of ideals in R and the cycles in R. Thus, for every prime $\mathbf{p}$ in D, we get a unique cycle of R, called its $\mathbf{p}$–cycle. The length and invariants of this cycle are called the $\mathbf{p}$–length and $\mathbf{p}$–invariants of R. These concepts are fundamental for the classification problems treated in the rest of this paper.

In this section we classify $\mathfrak{g}$ under similarity; the appropriate invariants turn out to be $\mathbf{p}$–lengths.

First, we state some known results regarding ideals in a pseudo–Dedekind ring.

Let R be a pseudo–Dedekind order in a simple Artinian ring Q. A *cycle* in R is either a maximal ideal of R which is invertible or a sequence $\mathbf{m}_1, \cdots, \mathbf{m}_\ell$ of distinct maximal ideals of R which are idempotent and satisfy the relations $\mathbf{o}_r(\mathbf{m}_i) = \mathbf{o}_\ell(\mathbf{m}_{i+1})$, the indexing of the ideals $\mathbf{m}_i$ being modulo ℓ. The *length* of the cycle $\mathbf{m}_1, \cdots, \mathbf{m}_\ell$ is defined to be ℓ. We shall not distinguish between two cycles which are cyclic permutations of each other. Let n_i be the uniform dimension of the simple Artinian ring $R/\mathbf{m}_i, i \in I_\ell$. The sequence $(n_1, \cdots, n_\ell)$ and all cyclic permutations of it are called *invariants* of the cycle $\mathbf{m}_1, \cdots, \mathbf{m}_\ell$.

The following theorem is due to Eisenbud and Robson [8].

Theorem 7.1. Let R be a pseudo–Dedekind ring. Then the following assertions hold:

(1) Let $\mathbf{m}$ be a maximal ideal of R. Then $\mathbf{m}$ is either invertible or idem–potent. If $\mathbf{m}$ is invertible then $\cap_{n=1}^{\infty} \mathbf{m}^n = (0)$. Moreover, an ideal which is maximal among idempotent ideals of R is necessarily maximal.

(2) Every maximal idempotent ideal belongs to a unique cycle of R. If $\mathbf{m}_1, \cdots, \mathbf{m}_\ell$ is a cycle of maximal idempotent ideals of R, then $\cap_{i=1}^{\ell} \mathbf{m}_i$ is maximal among invertible ideals of R and $\mathbf{m}_1, \cdots, \mathbf{m}_\ell$ are the only maximal ideals of R which contain $\cap_{i=1}^{\ell} \mathbf{m}_i$. If X is an invertible ideal of R and $X \subseteq \mathbf{m}_i$ for some i, then $X \subseteq \cap_{i=1}^{\ell} \mathbf{m}_i$.

(3) Invertible fractional ideals of R form an abelian group under multi–plication.

Proof. Let $\mathbf{m}_1, \cdots, \mathbf{m}_\ell$ be a cycle of maximal idempotent ideals in R. Let $I = \cap_{j=1}^{\ell} \mathbf{m}_j$ and $\bar{R} = R/I$. Then $\bar{R}$ is an Artinian ring and $\bar{\mathbf{m}}_j = \mathbf{m}_j/I$, $1 \leqslant j \leqslant \ell$, are maximal ideals of $\bar{R}$ such that $\cap_{j=1}^{\ell} \bar{\mathbf{m}}_j = (0)$. Thus $\bar{R}$ is semi–simple and the $\bar{\mathbf{m}}_j$ are *all* the maximal ideals of $\bar{R}$. Hence $\mathbf{m}_1, \cdots, \mathbf{m}_\ell$ are the only maximal ideals of R which contain I. The rest is proved in [8]. ∎

It is important to note that the above theorem establishes a bijection between cycles and ideals which are maximal among invertibles.

We proceed to study the ideal theory of almost Dedekind rings with special attention to cycles.

Until further notice, D is a fixed Dedekind domain with quotient skew field F and $Q = M_n(F)$ for some fixed positive integer n.

A *partition* ρ of n is a sequence $\rho = (\rho_1, \cdots, \rho_\ell)$ of positive integers such that $\Sigma_{i=1}^{\ell} \rho_i = n$; unless explicitly stated otherwise, we shall assume that $\ell > 1$.

Notation. Let $\mathbf{t} = \{ \mathbf{t}_{ij} : i, j \in I_n \}$ be a triangular frame in D with $\|\Delta_n(D, \mathbf{t})\| = \mathbf{p}$, a prime in D. Then there exists a uniquely determined partition $\rho = (\rho_1, \cdots, \rho_\ell)$ of n such that the following condition holds: Put $\tau_s = \Sigma_{i=s}^{\ell} \rho_i$ for $1 \leqslant s \leqslant \ell$ and $\tau_{\ell+1} = 0$. Then

$$\mathbf{t}_{ij} = \begin{cases} D & \text{if } \tau_{s+1} < i \leqslant \tau_s \text{ and } \tau_{s+1} < j \leqslant n \text{ for } s = 1,2,\cdots,\ell. \\ \mathbf{p} & \text{otherwise.} \end{cases}$$

In this situation, we shall find it convenient to denote $\Delta_n(D, \mathbf{t})$ as $\Delta_n(D, \mathbf{p}, \rho)$ or $\Delta_n(D, \mathbf{p}, \rho_1, \cdots, \rho_\ell)$. The shape of $\Delta_n(D, \mathbf{t}, \rho)$ is indicated in Figure 1.

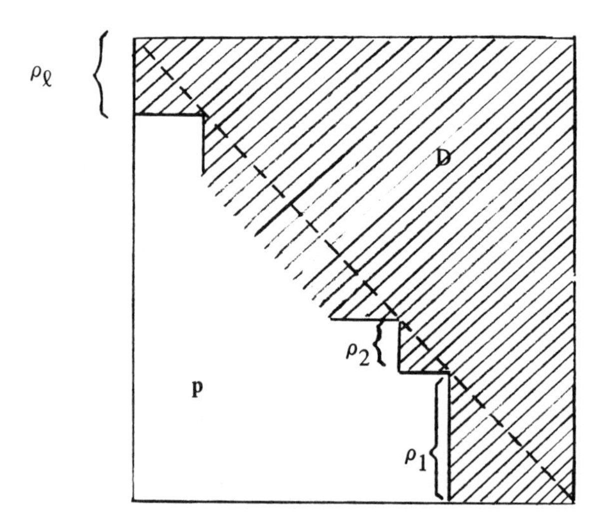

Figure 1.

Let $T = \Delta_n(D, \mathbf{t}, \rho)$ where $\rho = (\rho_1, \cdots, \rho_\ell)$. We shall frequently treat elements of T as $\ell \times \ell$ blocked matrices in an obvious manner. The same blocking arrangement will apply to elements of $M_n(F)$ if they are referred to as blocked matrices while dealing with T. Note that although we start indexing the numbers ρ_i from below (cf. Figure 1), our indexing of blocks is as usual, e.g., the $(1,1)^{th}$ block of T in Figure 1 is the left hand top corner of size $\rho_\ell \times \rho_\ell$. We let $\mathbf{m}_s(T)$, $1 \leqslant s \leqslant \ell$, stand for all those elements of T in which the $(\ell-s+1, \ell-s+1)^{th}$ block is filled with elements of $\mathbf{p}$. For $1 \leqslant s < \ell$, we let $\Omega_s(T)$ stand for all those matrices in $M_n(F)$ in which the $(\ell-s+1, \ell-s)^{th}$ block is filled with elements of D while all other blocks are as in elements of T. The set $\Omega_\ell(T)$ consists of all those blocked matrices in $M_n(F)$ in which the $(1,\ell)^{th}$ block is filled with elements of $\mathbf{p}^{-1}$ while all other blocks are as in elements of T.

The following theorem establishes a close connection between primes in D and cycles in $\Delta_n(D, \mathbf{t})$.

Theorem 7.2. Let $T = \Delta_n(D, \mathbf{t})$ and $\|T\| = \mathbf{p}_1 \cdots \mathbf{p}_\kappa$ be the prime factorization of $\|T\|$ so that $\mathbf{p}_\lambda$ are distinct primes in D. Let $\rho_\lambda = (\rho_{1\lambda}, \cdots, \rho_{\ell_\lambda \lambda})$ be the partition of n such that $T^{(\mathbf{p}_\lambda)} = \Delta_n(D, \mathbf{p}_\lambda, \rho_\lambda)$ for $1 \leqslant \lambda \leqslant \kappa$. For $1 \leqslant s \leqslant \ell_\lambda$, $1 \leqslant \lambda \leqslant \kappa$, we put $\mathbf{m}_s(T, \mathbf{p}_\lambda) = T \cap \mathbf{m}_s(T^{(\mathbf{p}_\lambda)})$. Then the following assertions hold:

(1) The maximal ideals of R which are invertible are precisely those of the form $(\text{diag } \mathbf{p})T$ where $\mathbf{p}$ is a prime in D with $\mathbf{p} \nmid \|T\|$. The maximal idempotent ideals of T are precisely those of the form $\mathbf{m}_s(T, \mathbf{p}_\lambda)$ for $1 \leqslant s \leqslant \ell_\lambda$, $1 \leqslant \lambda \leqslant \kappa$.

(2) T has precisely κ distinct cycles of maximal idempotent ideals. Each such cycle has the form $\{ \mathbf{m}_s(T, \mathbf{p}_\lambda) : 1 \leqslant s \leqslant \ell_\lambda \}$ for some uniquely determined λ, $1 \leqslant \lambda \leqslant \kappa$. There are precisely κ distinct invertible ideals of T which are maximal among invertibles but not themselves maximal; these are $\mathscr{I}_{\mathbf{p}_\lambda}(T) = \cap_{s=1}^{\ell_\lambda} \mathbf{m}_s(T, \mathbf{p}_\lambda)$, $1 \leqslant \lambda \leqslant \kappa$. Further, $\{ \mathscr{I}_{\mathbf{p}_\lambda}(T) \}^{\ell_\lambda} = (\text{diag } \mathbf{p}_\lambda)T$.

(3) For $1 \leqslant \lambda \leqslant \kappa$, $1 \leqslant s \leqslant \ell_\lambda$, we have

$$\mathscr{I}_{\mathbf{p}_\lambda}(T)\mathbf{m}_s(T, \mathbf{p}_\lambda) = \mathbf{m}_{s+1}(T, \mathbf{p}_\lambda)\mathscr{I}_{\mathbf{p}_\lambda}(T)$$

where the indexing of the ideals $\mathbf{m}_s(T, \mathbf{p}_\lambda)$ is modulo ℓ_λ. If $\mu \neq \lambda$, $1 \leqslant \mu \leqslant \kappa$, then $\mathscr{I}_{\mathbf{p}_\mu}(T)\mathbf{m}_s(T, \mathbf{p}_\lambda) = \mathbf{m}_s(T, \mathbf{p}_\lambda)\mathscr{I}_{\mathbf{p}_\mu}(T)$. For every prime $\mathbf{p}$ in D, we have $[(\text{diag } \mathbf{p})T]\,\mathbf{m}_s(T, \mathbf{p}_\mu) = \mathbf{m}_s(T, \mathbf{p}_\mu)[(\text{diag } \mathbf{p})T]$.

Proof. Let $\{ e_{ij} : i, j \in I_n \}$ be the usual system of matrix units in $M_n(D)$. Evidently, $e_{ij} \in T$ whenever $i \leqslant j$. Let $\mathbf{m}$ be a maximal ideal of T. Then $\mathbf{m}$ has the form $\mathbf{m} = (X_{ij}) \subseteq M_n(D)$ where each X_{ij} is a non-zero ideal of D, $X_{ij} \subseteq \mathbf{t}_{ij}$ and the ideals X_{ij} increase as we move upwards along a column or from left to right along a row of T. Moreover, maximality of $\mathbf{m}$ implies that $T/\mathbf{m}$ is a simple Artinian ring. The form of $\mathbf{m}$ shows that $T/\mathbf{m}$ can be represented as a generalized matrix ring $T/\mathbf{m} \overset{nat}{\cong} (t_{ij}/X_{ij})$. Consequently, t_{ii}/X_{ii} is either a simple Artinian ring or the zero ring; i.e., $X_{ii} = D$ or a maximal ideal of D for all $i \in I_n$.

For $i < j$, we put

$$S_{ij} = \begin{pmatrix} D & D \\ t_{ij} & D \end{pmatrix}, \quad \mathbf{n}_{ij} = \begin{pmatrix} X_{ii} & X_{ij} \\ X_{ji} & X_{jj} \end{pmatrix}.$$

Evidently, $S_{ij} \overset{nat}{\cong} (e_{ii} + e_{jj})T(e_{ii} + e_{jj})$ and $\mathbf{n}_{ij} \overset{nat}{\to} (e_{ii} + e_{jj})\mathbf{m}(e_{ii} + e_{jj})$. Thus S_{ij} is a HNP order in $M_2(F)$ (cf. [34]), and $\mathbf{n}_{ij} = S_{ij}$ or $S_{ij}/\mathbf{n}_{ij}$ is a simple Artinian ring. In the later case X_{ii} and X_{jj} are prime ideals of D with at least one of them proper, and if X_{ii} and X_{jj} are both proper then $X_{ii} = X_{jj} = X_{ij}$. Since $\mathbf{m} \neq T$, we have at least one $i \in I_n$ such that $X_{ii} \neq D$. It is now easy to see that $\mathbf{m}$ determines a unique prime $\mathbf{p}$ in D such that $X_{ii} = \mathbf{p}$ or D for all $i \in I_n$, $X_{ii} = \mathbf{p}$ for at least one $i \in I_n$, and whenever $X_{ii} = X_{jj} = \mathbf{p}$ with $i < j$ then $X_{ij} = \mathbf{p}$.

Case 1. Suppose $X_{ii} = \mathbf{p}$ for all $i \in I_n$. If possible, let $\mathbf{p} = \mathbf{p}_\lambda$ for some λ, $1 \leqslant \lambda \leqslant \kappa$. Then

$$\mathbf{n}_{1n} = \begin{pmatrix} \mathbf{p} & \mathbf{p} \\ X_{n1} & \mathbf{p} \end{pmatrix} \subsetneqq \begin{pmatrix} D & D \\ t_{n1} & \mathbf{p} \end{pmatrix} \subsetneqq S_{1n}$$

and using $\mathbf{p}_\lambda \supseteq t_{n1}$, it is easily seen that $\begin{pmatrix} D & D \\ t_{n1} & \mathbf{p} \end{pmatrix}$ is an ideal of S_{1n}, contrary to the maximality of $\mathbf{n}_{1n}$. Hence $\mathbf{p} \nmid \|T\|$. Let $1 \leqslant i < j \leqslant n$. Then, as noted above, $X_{ij} = \mathbf{p}$ and $t_{ji} \supseteq X_{ji} \supseteq X_{jj}t_{ji} = \mathbf{p}t_{ji}$. If $X_{ji} = t_{ji}$, then

$$S_{ij}/\mathbf{n}_{ij} \overset{nat}{\cong} \begin{pmatrix} D/\mathbf{p} & D/\mathbf{p} \\ 0 & D/\mathbf{p} \end{pmatrix}$$

which is not simple. Hence $X_{ji} = \mathbf{p}t_{ji}$ so $\mathbf{m} = (\text{diag } \mathbf{p})T$, which is clearly an invertible ideal of T.

Case 2. Suppose $X_{ii} = D$ for some $i \in I_n$. We have $X_{jj} = \mathbf{p}$ for some $j \in I_n$. If $i < j$ then $\|T\| \subseteq t_{ji} = t_{ji}X_{ii} \subseteq X_{ji} \subseteq X_{jj} = \mathbf{p}$, and if $i > j$ then $\|T\| \subseteq t_{ij} = X_{ii}t_{ij} \subseteq X_{ij} \subseteq X_{jj} = \mathbf{p}$. Thus, in any case, we have $\mathbf{p} \mid \|T\|$; so, $\mathbf{p} =$

$\mathbf{p}_\lambda$ for some λ, $1 \leqslant \lambda \leqslant \kappa$. Let $T^{(\mathbf{p}_\lambda)} = \Delta_n(D, \mathbf{p}_\lambda, \rho_\lambda)$ where $\rho_\lambda = (\rho_{1\lambda}, \cdots, \rho_{\ell_\lambda \lambda})$, and let $\tau_s = \Sigma_{i=s}^{\ell_\lambda} \rho_{i\lambda}$ for $1 \leqslant s \leqslant \ell_\lambda$.

Let $X_{ii} = X_{jj} = \mathbf{p}_\lambda$ with $1 \leqslant i < j \leqslant n$. If possible, let $i \leqslant \tau_s < j$. So, $\mathbf{t}_{ji} \subseteq \mathbf{p}_\lambda$. Clearly, $\mathbf{t}_{ji} \supseteq X_{ji} \supseteq X_{jj}\mathbf{t}_{ji} = \mathbf{p}_\lambda \mathbf{t}_{ji}$ and the maximality of $\mathbf{n}_{ij} = \begin{pmatrix} \mathbf{p}_\lambda & \mathbf{p}_\lambda \\ X_{ji} & \mathbf{p}_\lambda \end{pmatrix}$ in S_{ij} shows that $X_{ji} \neq \mathbf{t}_{ji}$. Thus $X_{ji} = \mathbf{p}_\lambda \mathbf{t}_{ji}$. But then $\mathbf{n}_{ij}$ is properly contained in the maximal ideal $\begin{pmatrix} D & D \\ \mathbf{t}_{ji} & \mathbf{p}_\lambda \end{pmatrix}$ of S_{ij}, a contradiction.

If possible, let $X_{ii} = \mathbf{p}_\lambda, X_{jj} = D$ and $\tau_{s+1} < i, j \leqslant \tau_s$ for some s. If $i < j$ then $\mathbf{t}_{ji} \supseteq X_{ji} \supseteq X_{jj}\mathbf{t}_{ji} = \mathbf{t}_{ji}$ so that $\mathbf{t}_{ji} = X_{ji} \subseteq X_{ii} = \mathbf{p}_\lambda$, contrary to the definition of the partition ρ_λ. If $i > j$ then $\mathbf{t}_{ij} \supseteq X_{ij} \supseteq \mathbf{t}_{ij}X_{jj} = \mathbf{t}_{ij}$ so that $\mathbf{t}_{ij} = X_{ij} \subseteq X_{ii} \subseteq \mathbf{p}_\lambda$ which again contradicts the choice of ρ_λ. To sum up, we have a unique integer s, $1 \leqslant s \leqslant \ell_\lambda$, such that

$$X_{ii} = \begin{cases} \mathbf{p}_\lambda & \text{if } \tau_{s+1} < i \leqslant \tau_s \\ D & \text{otherwise.} \end{cases}$$

We now claim that $\mathbf{m} = \mathbf{m}_s(T, \mathbf{p}_\lambda)$. Put $\upsilon = \tau_{s+1}$ and for $1 \leqslant \alpha, \beta \leqslant \rho_{s\lambda}$, put $\mathbf{a}_{\alpha\beta} = \mathbf{t}_{\alpha+\upsilon, \beta+\upsilon}$, $Y_{\alpha\beta} = X_{\alpha+\upsilon, \beta+\upsilon}$. Evidently, $\mathbf{a} = \{\mathbf{a}_{\alpha\beta} : 1 \leqslant \alpha, \beta \leqslant \rho_{s\lambda}\}$ is a triangular frame in D. Let $\Gamma = \Delta_{\rho_{s\lambda}}(D, \mathbf{a})$. Then $\mathbf{p}_\lambda \nmid \|\Gamma\|$. Let $Y = (Y_{\alpha\beta})$ $\subseteq \Gamma$. Then Y is an ideal of Γ and there is an obvious isomorphism $T/\mathbf{m} \cong \Gamma/Y$. So, Y is a maximal ideal of Γ. Since $Y_{\alpha\alpha} = \mathbf{p}_\lambda$ for $1 \leqslant \alpha \leqslant \rho_{s\lambda}$, Case 1 shows that $Y = (\text{diag } \mathbf{p}_\lambda)\Gamma$. Hence $\mathbf{m} = \mathbf{m}_s(T, \mathbf{p}_\lambda)$, as claimed.

It is easy to see that the ideals $\mathbf{m}_s(T, \mathbf{p}_\lambda)$ and $(\text{diag } \mathbf{p})T$ with $\mathbf{p} \nmid \|T\|$ are indeed maximal ideals of T.

We now show that $\{\mathbf{m}_s(T,\mathbf{p}_\lambda) : 1 \leqslant s \leqslant \ell_\lambda\}$ is a cycle of maximal idempotent ideals of T. Put $\mathbf{m}_{s\lambda} = \mathbf{m}_s(T, \mathbf{p}_\lambda)$. For $1 \leqslant s < \ell_\lambda$, let $\Sigma_s = \Omega_s(T^{(\mathbf{p}_\lambda)}) \cap [\cap_{1 \leqslant \mu \leqslant \kappa, \mu \neq \lambda} T^{(\mathbf{p}_\mu)}]$. Let Σ_{ℓ_λ} be the set of those matrices in $M_n(F)$ whose upper right hand corner of size $\rho_{\ell_\lambda \lambda} \times \rho_{1\lambda}$ is filled with elements of $\mathbf{p}_\lambda^{-1}$ and all other entries are as in elements of T. A straightforward computation shows that $\mathbf{m}_{s\lambda}\Sigma_s \subseteq \mathbf{m}_{s\lambda}$ and $\Sigma_{s-1}\mathbf{m}_{s\lambda} \subseteq \mathbf{m}_{s\lambda}$, the indexing being modulo ℓ_λ. Since $\Sigma_s \supsetneq T$, $\mathbf{m}_{s\lambda}$ cannot be invertible; but it is maximal so it must be an idempotent maximal ideal of T. This proves part (1).

Proposition 2.5 shows that T has enough invertible ideals. Choose an invertible $I \subseteq \cap_{s=1}^{\ell_\lambda} \mathbf{m}_{s\lambda}$. We already have $T \subsetneq \Sigma_s \subseteq \mathbf{o}_\ell(\mathbf{m}_{s+1,\lambda})$. If $\Sigma_s \neq \mathbf{o}_\ell(\mathbf{m}_{s+1,\lambda})$, then the inclusion reversing bijection between idempotent ideals of T over I and the subrings between T and I^{-1} (cf. [8]) shows that $\mathbf{m}_{s+1,\lambda}$ is not maximal in Σ_s, a contradiction. So, $\Sigma_s = \mathbf{o}_\ell(\mathbf{m}_{s+1,\lambda})$. Similarly, $\Sigma_s = \mathbf{o}_r(\mathbf{m}_{s\lambda})$. Hence $\{\mathbf{m}_{s\lambda} : 1 \leqslant s \leqslant \ell_\lambda\}$ is a cycle of maximal idempotent ideals of T. It is immediate that these are the only cycles of maximal idempotent ideals of T and that $\mathscr{I}_{p_\lambda}(T)$, $1 \leqslant \lambda \leqslant \kappa$, are the only invertible ideals of T which are maximal among invertibles but not themselves maximal (cf. [8]). A straightforward computation shows that $[\mathscr{I}_{p_\lambda}(T)]^{\ell_\lambda} = (\operatorname{diag} \mathbf{p})T$. This proves (2).

Let $1 \leqslant \lambda \leqslant \kappa$ and $1 \leqslant s \leqslant \ell_\lambda$. Since $\mathscr{I}_{p_\lambda}(T) \subseteq \mathbf{m}_{s\lambda} \subseteq T$, therefore $\mathscr{I}_{p_\lambda}(T) \subseteq \mathscr{I}_{p_\lambda}(T)\mathbf{m}_{s\lambda}[\mathscr{I}_{p_\lambda}(T)]^{-1} \subseteq T$. Since the middle term is a proper ideal of T, using parts (1) and (2), we can choose an integer f, $1 \leqslant f \leqslant \ell_\lambda$, such that $\mathscr{I}_{p_\lambda}(T)\mathbf{m}_{s\lambda}[\mathscr{I}_{p_\lambda}(T)]^{-1} \subseteq \mathbf{m}_{f\lambda}$. If this inclusion is proper, then $\mathbf{m}_{s\lambda} \subsetneqq [\mathscr{I}_{p_\lambda}(T)]^{-1}\mathbf{m}_{f\lambda}\mathscr{I}_{p_\lambda}(T) \subsetneqq T$, contrary to the maximality of $\mathbf{m}_{s\lambda}$. Hence $\mathscr{I}_{p_\lambda}(T)\mathbf{m}_{s\lambda} = \mathbf{m}_{f\lambda}\mathscr{I}_{p_\lambda}(T)$. Clearly, f is uniquely determined by s. Let $T^{(p_\lambda)} = \Delta_n(D, p_\lambda, \rho_\lambda)$ with $\rho_\lambda = (\rho_{\lambda 1}, \cdots, \rho_{\ell_\lambda \lambda})$ and $\tau_k = \Sigma_{i=k}^{\ell_\lambda} \rho_{i\lambda}$. Assume for a moment that the indexing of rows and columns of T is cyclic modulo n. A comparison of $(\tau_s, \tau_{s+1})^{th}$ entries of $\mathscr{I}_{p_\lambda}(T)\mathbf{m}_{s\lambda}$ and $\mathbf{m}_{f\lambda}\mathscr{I}_{p_\lambda}(T)$ shows that $f \equiv s+1 (\operatorname{mod} \ell_\lambda)$. So, $\mathscr{I}_{p_\lambda}(T)\mathbf{m}_{s\lambda} = \mathbf{m}_{s+1,\lambda}\mathscr{I}_{p_\lambda}(T)$.

Assume that $\lambda \neq \mu$, $1 \leqslant \mu \leqslant \kappa$. As above, given $1 \leqslant s \leqslant \ell_\lambda$, there is a unique f, $1 \leqslant f \leqslant \ell_\lambda$, such that $\mathscr{I}_{p_\mu}(T)\mathbf{m}_{s\lambda} = \mathbf{m}_{f\lambda}\mathscr{I}_{p_\mu}(T)$. Since $\mathscr{I}_{p_\lambda}(T)$ and $\mathscr{I}_{p_\mu}(T)$ commute, using the above paragraph, we have

$$[\mathscr{I}_{p_\lambda}(T)]^{\ell_\lambda - s}\mathscr{I}_{p_\mu}(T)\mathbf{m}_{s\lambda}[\mathscr{I}_{p_\lambda}(T)]^{s-\ell_\lambda} = \mathscr{I}_{p_\mu}(T)\mathbf{m}_{\ell_\lambda \lambda}$$

$$= \mathbf{m}_{f+\ell_\lambda - s,\lambda}\,\mathscr{I}_{p_\mu}(T).$$

A straightforward computation shows that the $(1,1)^{th}$ entries of the terms in the last equation match only if $f = s$. The remaining assertion is trivial. ■

We retain the *notation* $\mathscr{I}_p(T)$ introduced in the statement of Theorem 7.2. If $\mathbf{p}$ is a prime in D and $\mathbf{p} \nmid \|T\|$, we shall put $\mathscr{I}_p(T) = T(\operatorname{diag} \mathbf{p})$.

We now prove a technical looking lemma which plays a crucial role in defining the 'p' part of the p–invariants.

Lemma 7.3. Let $T = \Delta_n(D, \mathbf{t})$, $S = \Delta_n(D, \mathbf{a})$ and let H be an invertible (S, T)–bisubmodule of Q. Then $\mathscr{I}_p(S)H = H\mathscr{I}_p(T)$ for all primes $\mathbf{p}$ in D. Also, $\|S\| = \|T\|$.

Proof. Let $\{ e_{ij} : i, j \in I_n \}$ be the usual system of matrix units in $M_n(F)$. Since ${}_SH_T$ is finitely generated from both sides and $e_{ij} \in T \cap S$ whenever $i \leqslant j$, it follows that $H = (\mathbf{h}_{ij}) \subseteq M_n(F)$ for certain fractional ideals $\mathbf{h}_{ij}$ of D. Let $\mathbf{p}$ be a prime divisor of $\|T\|$. By Theorem 7.2, $\mathscr{I}_p(T)$ is an ideal of T which is maximal among invertible ideals of T. By using Lemma 6.1 and Theorem 7.2 it is seen that $H\mathscr{I}_p(T) = \mathscr{I}_{p'}(S)H$ for some uniquely determined prime divisor p' of $\|S\|$. Using Lemma 6.1, it is clear that there is a bijection between the set of maximal ideals of T over $\mathscr{I}_p(T)$ and the set of maximal ideals of S over $\mathscr{I}_{p'}(S)$. By Theorem 7.2, these sets are finite; if their cardinality is ℓ, then $H(\text{diag } \mathbf{p}) = H\{\mathscr{I}_p(T)\}^\ell = \{\mathscr{I}_{p'}(S)\}^\ell H = (\text{diag } \mathbf{p}')H$. Since fractional ideals of D commute under multiplication, it follows that $\mathbf{p} = \mathbf{p}'$; Also, $\|S\| \subseteq \|T\|$. By a similar argument, we get $\|S\| \supseteq \|T\|$.

If $\mathbf{p} \nmid \|T\|$, then $\mathscr{I}_p(T) = T(\text{diag } \mathbf{p})$ and the lemma is clear. ∎

We now define $\mathbf{p}$–invariants of a pseudo–Dedekind ring.

Let F be a skew field and $Q = M_n(F)$. Let R be a pseudo–Dedekind order in Q, $\mathcal{E}$ be the equivalence class of R under classical equivalence in Q and D be a Dedekind order in F which is also a subbasic ring of $\mathcal{E}$. Let X be an ideal of R which is maximal among invertible ideals of R. We proceed to show that X uniquely determines a prime $\mathbf{p}$ in D.

By Theorem 6.4, there exists a triangular frame $\mathbf{t}$ of order m in D, $m \in \mathbf{Z}^+$, and a progenerator M in mod–$\Delta_m(D, \mathbf{t})$ which is essential in $M_{n\times m}(F)$ with $\mathbf{o}_\ell(M) = R$. Lemma 6.1 yields an ideal $\mathscr{I}$ of $\Delta_m(D, \mathbf{t})$ which is maximal among invertible ideals of $\Delta_m(D, \mathbf{t})$ and has the property $XM = M\mathscr{I}$. By Theorem 7.2, there exists a unique prime $\mathbf{p}$ of D such that $\mathscr{I} = \mathscr{I}_p(\Delta_m(D, \mathbf{t}))$.

We now show that $\mathbf{p}$ is independent of the choices made in obtaining it from X. Thus, let $\mathbf{a}$ be a triangular frame of order s in D and let K be a progenerator in mod–$\Delta_s(D, \mathbf{a})$ which is essential in $M_{n\times s}(F)$ with $\mathbf{o}_\ell(K) = R$. As above, we have a unique prime ideal $\mathbf{p}'$ of D such that $XK = K\mathscr{I}_{p'}(\Delta_s(D, \mathbf{a}))$. We shall show that $\mathbf{p} = \mathbf{p}'$.

Let $M^\# = (M, \cdots, M)_{1\times s}$ and $K^\# = (K, \cdots, K)_{1\times m}$. Clearly, $M^\#$ (resp. $K^\#$) is a progenerator in mod–$M_s(\Delta_m(D, \mathbf{t}))$ (resp. mod–$M_m(\Delta_s(D, \mathbf{a}))$), which is an essential submodule of $M_{n\times ms}(F)$ and has the property $\mathbf{o}_\ell(M^\#) = R$ (resp. $\mathbf{o}_\ell(K^\#) = R$). There are obvious triangular frames $\tilde{\mathbf{t}}$ and $\tilde{\mathbf{a}}$ of order ms in D and obvious permutation matrices π_1 and π_2 in $M_{ms}(F)$ such that $\pi_1^{-1}M_s(\Delta_m(D, \mathbf{t}))\pi_1 = \Delta_{ms}(D, \tilde{\mathbf{t}}) = T$, say, and $\pi_2^{-1}M_m(\Delta_s(D, \mathbf{a}))\pi_2 = \Delta_{ms}(D, \tilde{\mathbf{a}}) = A$, say. Let $\tilde{M} = M^\#\pi_1$ and $\tilde{K} = K^\#\pi_2$. Then $\tilde{M}$ is an invertible (R, T)–bimodule and $\tilde{K}$ is an invertible (R, A)–bimodule. It is easily seen that $\pi_1^{-1}M_s(\mathscr{I}_p(\Delta_m(D, \mathbf{t})))\pi_1 = \mathscr{I}_{\tilde{p}}(T)$ and $\pi_2^{-1}M_m(\mathscr{I}_{p'}(\Delta_s(D, \mathbf{a})))\pi_2 = \mathscr{I}_{p'}(A)$. Thus, $X\tilde{M} = \tilde{M}\mathscr{I}_p(T)$ and $X\tilde{K} = \tilde{K}\mathscr{I}_{p'}(A)$. Let $\mathscr{L} = \{\alpha \in M_{ms\times n}(F) \mid \alpha\tilde{K} \subseteq A\}$. Since $\tilde{K}$ is an invertible (R, A)–bimodule and $\mathscr{L} \overset{\text{canon}}{\cong} \operatorname{Hom}_A(\tilde{K}, A)$, it follows that $\mathscr{L}\tilde{K} = A$ and $\tilde{K}\mathscr{L} = R$. Hence $\mathscr{L}\tilde{M}$ is an invertible (A, T)–bisubmodule of $M_{ms}(F)$. Now $X\tilde{K} = \tilde{K}\mathscr{I}_{p'}(A)$ yields $\mathscr{L}X = \mathscr{I}_{p'}(A)\mathscr{L}$; so, $\mathscr{L}\tilde{M}\mathscr{I}_p(T) = \mathscr{L}X\tilde{M} = \mathscr{I}_{p'}(A)\mathscr{L}\tilde{M}$. We now invoke Lemma 7.3 to obtain $\mathbf{p} = \mathbf{p}'$.

Let $\{\mathbf{m}_i : 1 \leqslant i \leqslant \ell\}$ be a cycle in R. Then Theorem 7.1 shows that $\cap_{i=1}^{\ell} \mathbf{m}_i$ is maximal among invertible ideals of R and uniquely determines the given cycle $\{ \mathbf{m}_i : 1 \leqslant i \leqslant \ell\}$. Let $\mathbf{p}$ be the prime ideal of D determined by $\cap_{i=1}^{\ell} \mathbf{m}_i$ as indicated above. With some justification, we can now call the cycle $\{ \mathbf{m}_i : 1 \leqslant i \leqslant \ell \}$ of R the $\mathbf{p}$*–cycle* of R, its length the $\mathbf{p}$*–length* of R, denoted as $\ell_{\mathbf{p}}(R)$, and its invariants the $\mathbf{p}$*–invariants* of R.

Before we give an illustration, we need a definition. The *capacity* of a prime $\mathbf{p}$ in a Dedekind domain is the uniform dimension of the simple Artinian ring $D/\mathbf{p}$ and is denoted as $c(\mathbf{p})$.

Let us look at $R = \Delta_n(D, \mathbf{p}, \rho_1, \cdots, \rho_\ell)$. For any prime $\mathbf{g}$ of D other than $\mathbf{p}$, the $\mathbf{g}$–cycle of R consists of $\{ R(\text{diag } \mathbf{g})\}$, so $\ell_{\mathbf{g}}(R) = 1$. Since $R/R(\text{diag } \mathbf{g}) \cong M_n(D/\mathbf{g})$, the $\mathbf{g}$–invariants of R are $\{ nc(\mathbf{p}) \}$. The $\mathbf{p}$–cycle of R is $\mathbf{m}_1(R), \cdots, \mathbf{m}_\ell(R)$, as defined earlier in this section. Thus $\ell_{\mathbf{p}}(R) = \ell$ and the $\mathbf{p}$–invariants of R are $(\rho_1 c(\mathbf{p}), \cdots, \rho_\ell c(\mathbf{p}))$. All this follows immediately from Theorem 7.2. This sequence of invariants explains why we count ρ's from below (cf. Figure 1.) rather than in the natural way.

We comment on the use of $\mathbf{p}$ in $\mathbf{p}$–lengths and $\mathbf{p}$–invariants. We shall fix our attention on the set $\mathfrak{S}$ of all pseudo–Dedekind orders in a certain equivalence class $\mathcal{E}$ of orders in a simple Artinian ring Q. Given $R, S \in \mathfrak{S}$, we would like to say that $R \overset{\text{sim}}{\sim} S$ (or $R \overset{\text{conj}}{\sim} S$) if and only if some numbers connected with R and S are the same. In case of similarity, these numbers are lengths of the cycles of R and cycles of S (cf. Theorem 7.6). But R and S may have cycles of differing lengths. In that case, which lengths do we try to match? The answer is $\mathbf{p}$–lengths for various primes $\mathbf{p}$ in D. It should now be clear that the concrete representation of Q as $M_n(F)$ and the choice of subbasic order D for $\mathcal{E}$ play an entirely subsidiary role so far as classification is concerned.

We proceed to classify equivalent pseudo–Dedekind orders under similarity.

Lemma 7.4. Let D be a Dedekind order in a skew field F, $Q = M_n(F)$ and $T = \Delta_n(D, t)$. Let $A = \cap_{\lambda=1}^{\kappa} \Delta_n(D, \mathbf{p}_\lambda, \overbrace{1, \cdots, 1}^{\ell_\lambda - 1}, n-\ell_\lambda+1)$, where $\|T\| = \mathbf{p}_1 \cdots \mathbf{p}_\kappa$ and $\ell_\lambda = \ell(\mathbf{p}_\lambda)$. Then $A \underset{Q}{\overset{\text{sim}}{\sim}} T$.

Proof. Using Theorem 7.2, we have $T^{(\mathbf{p}_\lambda)} = \Delta_n(D, \mathbf{p}_\lambda, \rho_\lambda)$, where ρ_{λ} is a partition of n of the form $\rho_\lambda = (\rho_{1\lambda}, \cdots, \rho_{\ell_\lambda \lambda})$, $\ell_\lambda = \ell(\mathbf{p}_\lambda)$. Let $\tau_{s\lambda} = \Sigma_{i=s}^{\ell_\lambda} \rho_{i\lambda}$ for $1 \leqslant s \leqslant \rho_{\ell_\lambda \lambda}$. Define $\epsilon_{\alpha\beta}^{(\lambda)} = 0$ whenever $1 \leqslant \alpha \leqslant \rho_{\ell_\lambda \lambda}$ and $1 \leqslant \beta \leqslant n$ or whenever $\tau_{s+1} < \alpha \leqslant \tau_s$ and $n-s < \beta \leqslant n$ for $s = 1, \cdots, \ell_\lambda - 1$; let $\epsilon_{\alpha\beta}^{(\lambda)} = 1$ otherwise. Let $M_\lambda = (\mathbf{p}_\lambda^{\epsilon_{\alpha\beta}^{(\lambda)}}) \subseteq M_n(D)$. Thus the last $\rho_{1\lambda}$ rows of M_λ are the same as the nth row of $A^{(\mathbf{p}_\lambda)}$, the preceding $\rho_{2\lambda}$ rows of M_λ are the same as $(n-1)^{\text{th}}$ row of $A^{(\mathbf{p}_\lambda)}$ and so on. Let $M = \cap_{\lambda=1}^{\kappa} \mathbf{M}_\lambda$. Clearly, M is a (T, A)–bisubmodule of Q. In the notation explained in §3, $M_\lambda = M^{(\mathbf{p}_\lambda)}$ for $1 \leqslant \lambda \leqslant \kappa$. It is now clear that Lemma 3.1 is applicable to M_A.

We proceed to show that M_A is a progenerator in mod–A. Consider the ith row of A as a right A–submodule of the first row of A. Since $A = \cap_{\lambda=1}^{\kappa} A^{(p_\lambda)}$, we have i^{th} row of $A = \cap_{\lambda=1}^{\kappa}$ (i^{th} row of $A^{(p_\lambda)}$). Now, the ith row of $A^{(p_\lambda)}$ occurs as the $\theta_{ij}^{(\lambda)th}$ row of M_λ for some integer $\theta_{ij}^{(\lambda)}$. Let $\pi_\lambda^{(i)}$ be the permutation matrix in $M_n(D)$ obtained by exchanging the first and $\theta_{ij}^{(\lambda)th}$ rows of the identity matrix of $M_n(D)$. Then the first row of $N_i = \cap_{\lambda=1}^{\kappa} \pi_\lambda^{(i)} M^{(p_\lambda)}$ is the ith row of A. By Lemma 3.1, $N_i \cong M$ as right A–modules, $1 \leqslant i \leqslant n$. Since A_A is a direct summand of $\oplus_{i=1}^{n} N_i \cong M^{(n)}$ as a right A–module, it follows that M_A is a generator in mod–A. Similarly, ${}_TM$ is a generator in T–mod. Proposition 1.2 shows that M is an invertible (T, A)–bisubmodule of Q. Hence $A \underset{Q}{\overset{\text{sim}}{\sim}} T$. ∎

Lemma 7.5. Let F be a skew field, $Q = M_n(F)$ and $\mathfrak{A}$ be the set of all almost Dedekind orders in a certain equivalence class $\mathcal{E}$ of orders in Q. Let D be a Dedekind order in F which is a subbasic ring of $\mathcal{E}$. Let $R, S \in \mathfrak{A}$. Then $R \underset{Q}{\overset{\text{sim}}{\sim}} S$ if and only if $\ell_p(R) = \ell_p(S)$ for all primes p in D.

Proof. Suppose $R \overset{\text{sim}}{\sim} S$. By Theorem 5.4, we have $R \overset{f}{\sim} \Delta_n(D, t)$ and $S \overset{f}{\sim} \Delta_n(D, a)$ for some triangular frames t and a in D. Let M be an invertible $(R, \Delta_n(D, t))$–bisubmodule of Q and let X be an ideal of R with $XM = M\mathcal{J}_p(\Delta_n(D, t))$. It is easily seen that there exists a bijection between the maximal ideals of T over X and maximal ideals of $\Delta_n(D, t)$ over $\mathcal{J}_p(\Delta_n(D, t))$. Hence $\ell_p(R) = \ell_p(\Delta_n(D, t))$ for all p. Similarly, $\ell_p(S) = \ell_p(\Delta_n(D, a))$ for all p. Using Lemma 7.3, we get $\ell_p(R) = \ell_p(S)$ for all p.

Suppose $\ell_p(R) = \ell_p(S)$ for all p in D. In view of Theorem 5.4 and the above half of the lemma, we may assume without loss of generality that $R = \Delta_n(D, t)$ and $S = \Delta_n(D, a)$ for some triangular frames t and a in D. By Lemma 7.4, both R and S are similar to

$$\bigcap_{p_\lambda \mid \|R\|} \Delta_n(D, p, \overbrace{1, \cdots, 1}^{\ell_p(R)-1 \text{ terms}}, n-\ell_p(R)+1).$$

Hence $R \overset{\text{sim}}{\sim} S$. ∎

The following theorem classifies pseudo–Dedekind orders under similarity.

Theorem 7.6. Let F be a skew field and $Q = M_n(F)$. Let $\mathfrak{S}$ be the set of all pseudo–Dedekind orders in an equivalence class $\mathcal{E}$ of orders in Q and D be a Dedekind order in F which is a subbasic ring of $\mathcal{E}$. Let $R, S \in \mathfrak{S}$. Then the following conditions are equivalent:

(1) $R \overset{\text{sim}}{\sim} S$.

(2) $\ell_p(R) = \ell_p(S)$ for all primes p in D.

(3) For some positive integer m, there exists a triangular frame t of order m in D and essential right $\Delta_m(D, t)$–submodules M, N of $M_{n\times m}(F)$ such that M, N are progenerators in mod–$\Delta_m(D, t)$, $o_\ell(M) = R$ and $o_\ell(N) = S$.

Proof. (1) ⇒ (3): Let ${}_SH_R$ be an invertible (S, R)–bisubmodule of Q. By Theorem 6.4, we can choose a triangular frame t in D of order m, say, and an essential right $\Delta_m(D, t)$–submodule M of $M_{n\times m}(F)$ such that M is a progenerator in mod–$\Delta_m(D, t)$ and $o_\ell(M) = R$. Then $N = HM$ is a progenerator in mod–$\Delta_m(D, t)$ and $o_\ell(N) = S$. Since the right dimensions of N and $M_{n\times m}(F)$ agree, N is an essential submodule of $M_{n\times m}(F)$.

(3) ⇒ (1): Let M, N be as stated and let $M^* = \{ f \in M_{m\times n}(F) \mid fM \subseteq \Delta_m(D, t) \}$. By the Morita theorems, M^* is an invertible $(\Delta_m(D, t), R)$–bimodule. So, NM^* is an invertible (S, R)–bisubmodule of Q.

(3) ⇒ (2): The p–cycle of $\Delta_m(D, t)$ is transferred to the p–cycle of R (resp. S) by M (resp. N). So, $\ell_p(R) = \ell_p(\Delta_m(D, t)) = \ell_p(S)$ for all primes p in D.

(2) ⇒ (3): As seen while defining p–invariants, we can choose an integer m and triangular frames t, a of order m in D with the following properties: There exists an essential $\Delta_m(D, t)$–submodule M of $M_{n\times m}(F)$ and an essential $\Delta_m(D, a)$–submodule N of $M_{n\times m}(F)$ such that M, N are progenerators on the right side, $o_\ell(M) = R$ and $o_\ell(N) = S$. It follows that $\ell_p(\Delta_m(D, t)) = \ell_p(\Delta_m(D, a))$ for all primes p in D, so, by Lemma 7.5, $\Delta_m(D, t) \overset{\text{sim}}{\sim} \Delta_m(D, a)$. If H is an invertible $(\Delta_m(D, t), \Delta_m(D, a))$–bisubmodule of $M_m(F)$ then NH is an essential $\Delta_m(D, t)$–submodule of $M_{n\times m}(F)$ which is a progenerator in mod–$\Delta_m(D, t)$ with $o_\ell(NH) = S$. ∎

8. Invariants for Formal Conjugacy

We shall use p–invariants to classify equivalent almost Dedekind orders under formal conjugacy. Our results also provide a uniqueness theorem regarding the canonical form obtained in §5.

Until further notice, D is a Dedekind domain with quotient skew field F and $Q = M_n(F)$.

Lemma 8.1. Let K be a finitely generated right D–submodule of F and $E = o_\ell(K)$. Then E is a Dedekind order in F and K is an invertible (E, D)–bisubmodule. Given a prime p in D, there exists a unique prime p' in E such that $p'K = Kp$. In this situation, we have $c(p) = c(p')$.

Proof. It is well known [16, 20, 36] that E is an order in F. Using the left common multiple property of D, it may be seen that K_D is isomorphic with a right ideal of D; so, K_D is a progenerator in mod–D. Since $E \cong \operatorname{End} K_D$, E is Dedekind and K is an invertible (E, D)–bimodule. By Lemma 6.1, for each prime $\mathbf{p}$ in D, we have a unique prime $\mathbf{p}'$ in E such that $\mathbf{p}'K = K\mathbf{p}$.

It remains to show that $c(\mathbf{p}) = c(\mathbf{p}')$ (which is what we really need later on). Since K_D is isomorphic with a right ideal of D, using Theorem 3.3 of [7], it follows that $K/K\mathbf{p}$ is a cyclic right $D/\mathbf{p}$–module. Since $D/\mathbf{p}$ is a simple Artinian ring of dimension $c(\mathbf{p})$, we have $\dim(K/K\mathbf{p}) \leqslant c(\mathbf{p})$ so that $\dim \operatorname{End}(K/K\mathbf{p}) \leqslant c(\mathbf{p})$. Now we observe that $K/K\mathbf{p}$ is indeed an $(E/\mathbf{p}', D/\mathbf{p})$–bimodule. Since $E/\mathbf{p}'$ is simple, there is a monomorphism $E/\mathbf{p}' \to \operatorname{End}_{D/\mathbf{p}}(K/K\mathbf{p})$. Since the dimension of a semi–simple ring equals the number of primitive orthogonal idempotents in a decomposition of its identity, it follows that $c(\mathbf{p}') = \dim(E/\mathbf{p}') \leqslant c(\mathbf{p})$. Exchanging the roles of D and E, we obtain $c(\mathbf{p}) \leqslant c(\mathbf{p}')$. ■

Lemma 8.2. Let $\mathbf{a} = \{\mathbf{a}_{ij} : i, j \in I_n\}$ be an integral frame in D with associated ring $A = (\mathbf{a}_{ij}) \subseteq M_n(D)$. Let $K_1, \cdots, K_n$ be non–zero finitely generated right D–submodules of F and $R = (\operatorname{diag} K_i)A(\operatorname{diag} K_i^{-1}) = (K_i\mathbf{a}_{ij}K_j^{-1})$. Let $\mathbf{p}$ be a prime ideal of D such that $\mathbf{p} \nmid \|A\|$. Let $X = (\operatorname{diag} \mathbf{p})A$ and $Y = (\operatorname{diag} K_i)X(\operatorname{diag} K_i^{-1})$. Then X is an invertible maximal ideal of A and $\dim(A/X) = \dim(R/Y)$.

Proof. By Proposition 3.5, $A \overset{\text{conj}}{\sim} \Delta_n(D, \mathbf{t}) = T$ and $\|A\| = \|T\|$. Let ${}_AM_T$ be an invertible (A, T)–bisubmodule of Q. It is easily seen that $M = (\theta_{ij}) \subseteq M_n(F)$ where θ_{ij} are fractional ideals of D. Since the fractional ideals of D commute under multiplication, we have $XM = M[T(\operatorname{diag} \mathbf{p})]$. Since $\mathbf{p} \nmid \|A\| = \|T\|$, Theorem 7.2 shows that $T(\operatorname{diag} \mathbf{p}) = \mathscr{P}_\mathbf{p}(T)$ is a maximal and invertible ideal of T. By Lemma 6.1, X is a maximal and invertible ideal of A.

Clearly, $A/X = (\mathbf{a}_{ij}/\mathbf{a}_{ij}\mathbf{p})$ is a simple Artinian ring; so, its dimension is the sum of the dimensions of the rings $\mathbf{a}_{ii}/\mathbf{a}_{ii}\mathbf{p} \cong D/\mathbf{p}$ which occur in the generalized matrix ring representation $A/X \cong (\mathbf{a}_{ij}/\mathbf{a}_{ij}\mathbf{p})$. Thus $\dim A/X = nc(\mathbf{p})$.

Let $H = (K_i\mathbf{a}_{ij}) \subseteq M_n(F)$. It is easily seen that H is an invertible (R, A)–bisubmodule of Q and $YH = HX$. By Lemma 6.1, Y is maximal and invertible in R. Also, $R/Y \cong (K_i\mathbf{a}_{ij}K_j^{-1}/K_i\mathbf{p}\mathbf{a}_{ij}K_j^{-1})$. Let $E_i = \mathbf{o}_\ell(K_i)$ and let $\mathbf{p}_i$ be the prime ideal of E_i such that $\mathbf{p}_iK_i = K_i\mathbf{p}$. By Lemma 8.1, $c(\mathbf{p}_i) = c(\mathbf{p})$. Also, $K_i\mathbf{a}_{ii}K_i^{-1}/K_i\mathbf{p}\mathbf{a}_{ii}K_i^{-1} \cong E_i/\mathbf{p}_i$. Hence $\dim R/Y = \Sigma_{i=1}^n c(\mathbf{p}_i) = nc(\mathbf{p})$. ■

Lemma 8.3. Let $\{\mathbf{a}_{ij} : i, j \in I_n\}$ be an integral frame in D with associated ring A. Suppose $\mathbf{p}$ is a prime ideal in D such that $\mathbf{p} \mid \|A\|$. Further, suppose that $A^{(\mathbf{p})}$ has the form $A^{(\mathbf{p})} = \Delta_n(D, \mathbf{p}, \rho_1, \cdots, \rho_\ell)$. Let $\mathbf{m}_\lambda = \mathbf{m}_\lambda(A^{(\mathbf{p})}) \cap A$ for $1 \leqslant \lambda \leqslant \ell$. Let $K_1, \cdots, K_n$ be non–zero finitely generated right D–sub–

modules of F, $R = (\text{diag } K_i)A(\text{diag } K_i^{-1})$ and $H = (\text{diag } K_i)A$. Then H is an invertible (R, A)–bisubmodule of Q. Put $\mathbf{n}_\lambda = H\mathbf{m}_\lambda H^{-1}$. Then there exists a permutation φ on I_n such that $\mathbf{m}_{\varphi(1)}, \cdots, \mathbf{m}_{\varphi(\ell)}$ and $\mathbf{n}_{\varphi(1)}, \cdots, \mathbf{n}_{\varphi(\ell)}$ are the $\mathbf{p}$–cycles of A and R, respectively, $\mathbf{n}_\lambda H = H\mathbf{m}_\lambda$ and $\dim R/\mathbf{n}_\lambda = \dim A/\mathbf{m}_\lambda$ for $1 \leqslant \lambda \leqslant \ell$. Consequently, the $\mathbf{p}$–invariants of R and A are the same.

Proof. Using the generalized matrix representation of A, the shape of $\mathbf{m}_\lambda(A^{(\mathbf{p})})$ and Lemma 8.2, it is easily seen that $\mathbf{m}_\lambda$ is indeed a maximal ideal of A. As in Proposition 3.5, $A \overset{\text{conj}}{\sim} \Delta_n(D, \mathbf{t}) = T$ with $\|A\| = \|T\|$ and $A^{(\mathbf{p})} = T^{(\mathbf{p})}$. Let M be an invertible (A, T)–bisubmodule of Q so that $M = (\theta_{ij}) \subseteq M_n(F)$ for some invertible fractional ideals θ_{ij} of D. If $\mathbf{m}_\lambda$ is an invertible ideal of A then Theorem 7.2 and Lemma 6.1 show that $\mathbf{m}_\lambda M = M(\text{diag } \mathbf{g})$ for some prime ideal $\mathbf{g}$ of D with $\mathbf{g} \nmid \|T\| = \|A\|$; this is clearly impossible. Since $\mathbf{m}_\lambda$ is maximal but not invertible, it must be idempotent. Also, $(\text{diag } \mathbf{p})A \subseteq \cap_{\lambda=1}^{\ell} \mathbf{m}_\lambda$, $(\text{diag } \mathbf{p})T = \cap_{\lambda=1}^{\ell} \mathbf{m}_\lambda(T, \mathbf{p})$ and $[(\text{diag } \mathbf{p})A]M = M[T(\text{diag } \mathbf{p})]$. By Theorem 7.2, $(\text{diag } \mathbf{p})T = [\mathcal{J}_{\mathbf{p}}(T)]^{\ell}$. Let X be the ideal of A such that $XM = M\mathcal{J}_{\mathbf{p}}(T)$. Then X is maximal among invertible ideals of A but not itself maximal and $X^{\ell} = (\text{diag } \mathbf{p})A \subseteq \mathbf{m}_\lambda$ for all $1 \leqslant \lambda \leqslant \ell$. Thus, $X \subseteq \cap_{\lambda=1}^{\ell} \mathbf{m}_\lambda$. Since there are only ℓ maximal ideals over $\mathcal{J}_{\mathbf{p}}(T)$ in T, there are only ℓ maximal ideals over X in A. By Theorem 7.2, we have a permutation φ on I_n such that $\mathbf{m}_{\varphi(1)}, \cdots, \mathbf{m}_{\varphi(\ell)}$ is a cycle in A; this is clearly the $\mathbf{p}$–cycle of A.

H is obviously an invertible (R, A)–bisubmodule of Q and $\mathbf{n}_\lambda H = H\mathbf{m}_\lambda$. So, each $\mathbf{n}_\lambda$ is a maximal idempotent ideal of R. Clearly $H^{-1} = \{ q \in Q \mid qH \subseteq A \}$, $H^{-1}H = A$ and $HH^{-1} = R$. Since $H\mathbf{o}_\ell(\mathbf{m}_\lambda)H^{-1}$ is a subring of Q and contains A, it is an order in Q. Further, $H\mathbf{o}_\ell(\mathbf{m}_\lambda)H^{-1}\mathbf{n}_\lambda = H\mathbf{o}_\ell(\mathbf{m}_\lambda)\mathbf{m}_\lambda H^{-1} = \mathbf{n}_\lambda$ so that $H\mathbf{o}_\ell(\mathbf{m}_\lambda)H^{-1} \subseteq \mathbf{o}_\ell(\mathbf{n}_\lambda)$. Similarly, using H^{-1} instead of H, we get $H^{-1}\mathbf{o}_\ell(\mathbf{n}_\lambda)H \subseteq \mathbf{o}_\ell(\mathbf{m}_\lambda)$. Consequently, $H^{-1}\mathbf{o}_\ell(\mathbf{m}_\lambda)H = \mathbf{o}_\ell(\mathbf{n}_\lambda)$. A similar relation holds among right orders. It follows that $\mathbf{n}_{\varphi(1)}, \cdots, \mathbf{n}_{\varphi(\ell)}$ is the $\mathbf{p}$–cycle of R.

It remains to show that $\dim A/\mathbf{m}_\lambda = \dim R/\mathbf{n}_\lambda$ for $1 \leqslant \lambda \leqslant \ell$. Let $\{ e_{ij} : i, j \in I_n \}$ be the usual system of matrix units in $M_n(F)$. Let $\tau_\lambda = \Sigma_{i=\lambda}^{\ell} \rho_i$, and let $f = \Sigma_{i=\tau_{\lambda+1}}^{\tau_\lambda} e_{ii}$. Then $fAf \overset{\text{nat}}{\cong} A^{\#}$ where $A^{\#}$ is the associated ring of an integral frame of order ρ_λ and $\mathbf{p} \nmid \|A^{\#}\|$. Further, there is an obvious isomorphism $A/\mathbf{m}_\lambda \cong A^{\#}/A^{\#}(\text{diag } \mathbf{p})$. Since $R = (K_i a_{ij} K_j^{-1})$, it follows that $fRf \overset{\text{nat}}{\cong} (\text{diag } L_i)A^{\#}(\text{diag } L_i^{-1})$ where $L_i = K_{i+\tau_{\lambda+1}}$, $1 \leqslant i \leqslant \rho_\lambda$. This isomorphism sends $f\mathbf{n}_\lambda f$ to $(\text{diag } L_i)A^{\#}(\text{diag } \mathbf{p})(\text{diag } L_i^{-1})$. The shape of $\mathbf{m}_\lambda$ enables us to get an obvious isomorphism $R/\mathbf{n}_\lambda \cong fRf/f\mathbf{n}_\lambda f$. Lemma 8.2 now completes the proof. ■

Lemma 8.4. Let $\{ \mathbf{a}_{ij} : i, j \in I_n \}$ be an integral frame in D with associated ring A. Let $K_1, \cdots, K_n$ be non–zero finitely generated right D–submodules of F and let $R = (\text{diag } K_i)A(\text{diag } K_i^{-1})$. Then the $\mathbf{p}$–invariants of A and R agree for all primes $\mathbf{p}$ in D.

Proof. Let p be a prime in D. If $p \nmid \|A\|$ then Lemma 8.2 shows that the p–invariants of A and R agree.

Suppose $p \mid \|A\|$. By Lemma 3.3, there exists a permutation matrix π in $M_n(D)$ such that $\pi A^{(p)}\pi^{-1} = \Delta_n(D, p, \rho)$. Let $B = \pi A \pi^{-1}$. Then B is the associated ring of an integral frame of order n in D, $\|A\| = \|B\|$ and $B^{(p)} = \pi A^{(p)}\pi^{-1} = \Delta_n(D, p, \rho)$. Further, if φ is the permutation on I_n such that the ith row of $M_n(F)$ goes to the $\varphi(i)^{th}$ row under π^{-1}, then

$$R = (\text{diag } K_{\varphi(i)})B(\text{diag } K_{\varphi(i)}^{-1}).$$

The p–invariants of A and B clearly agree and Lemma 8.3 shows that the p–invariants of B and R agree. ■

Lemma 8.5. Let $T = \Delta_n(D, t)$, $R = \Delta_n(D, a)$, and $\|T\| = \|R\| = p_1 \cdots p_\kappa$. Suppose μ is an integer, $1 \leqslant \mu \leqslant \kappa$, such that $R^{(p_\lambda)} = T^{(p_\lambda)}$ for $\lambda \neq \mu$. Further, let $R^{(p_\mu)} = \Delta_n(D, p_\mu, \rho)$ and $T^{(p_\mu)} = \Delta_n(D, p_\mu, \psi)$ where $\rho = (\rho_1, \rho_2, \cdots, \rho_{\ell-1}, \rho_\ell)$ and $\psi = (\rho_\ell, \rho_1, \rho_2, \cdots, \rho_{\ell-1})$. Then $T \overset{f}{\sim} R$. If D is a PID then $T \overset{\text{conj}}{\sim} R$.

Proof. Let π be the permutation matrix obtained by cyclically permuting the rows of the identity matrix of $M_n(D)$ upwards until the top ρ_ℓ rows become the bottom ρ_ℓ rows. Let $A = \cap_{\lambda \neq \mu} R^{(p_\lambda)}$. Put $H = (\pi R^{(p_\mu)}) \cap A$ and $S = (\pi R^{(p_\mu)}\pi^{-1}) \cap A$. Then S is obviously the associated ring of an integral frame of order n in D. Corollary 3.2 and Lemma 3.4 show that $S \overset{\text{conj}}{\sim} R$. Let $\epsilon_i = 0$ for $1 \leqslant i \leqslant \Sigma_{j=1}^{\ell-1} \rho_j$ and $\epsilon_i = 1$ for $\Sigma_{j=1}^{\ell-1} \rho_j < i \leqslant n$. Consider $B = (\text{diag } p_\mu^{\epsilon_i})S(\text{diag } p_\mu^{-\epsilon_i})$. Clearly, $R \overset{f}{\sim} B$. A straightforward computation shows that $B^{(p_\lambda)} = R^{(p_\lambda)}$ for $\lambda \neq \mu$ and $B^{(p_\mu)} = T^{(p_\mu)}$. Thus $B = T$. If D is a PID, clearly, $R \overset{\text{conj}}{\sim} B = T$. ■

We now prove the main result of this section.

Theorem 8.6. Let F be a skew field, $Q = M_n(F)$ and $\mathfrak{A}$ be the set of all almost Dedekind orders in an equivalence class $\mathcal{E}$ of orders in Q. Let D be a Dedekind order in F which is a subbasic ring of $\mathcal{E}$. Given $R, S \in \mathfrak{A}$, $R \overset{f}{\sim} S$ if and only if the p–invariants of R and S agree for all primes p in D. If D is PID, then $R \overset{f}{\sim} S$ if and only if $R \overset{\text{conj}}{\sim} S$.

Proof. Suppose R, S are diagonally formally conjugate orders in Q. We proceed to show that the p–invariants of R and S agree for all p. Since the uniform dimensions of R and S are n, we have n orthogonal idempotents $f_1, \cdots, f_n$ in $R \cap S$ and an invertible (S, R)–bisubmodule H of Q such that $H = \Sigma_{i=1}^n f_i H f_i R = \Sigma_{i=1}^n S f_i H f_i$. Choose a complete system of matrix units $\{f_{ij} : i, j \in I_n\}$ in Q with $f_i = f_{ii}$ for all $i \in I_n$. Let $\{e_{ij} : i, j \in I_n\}$ be the usual system of matrix units in $M_n(F) = Q$ and choose a unit u in Q such that $u^{-1}f_{ij}u = e_{ij}$ for all $i,j \in I_n$ (cf. [19, page 59]). Then $u^{-1}Ru = (\mathbf{b}_{ij})$ where $\{\mathbf{b}_{ij} : i,j \in I_n\}$ is an indexed set of non–zero additive subgroups of F. Using Lemmas 4.1 and 4.3, we obtain a Dedekind order E in F, $D \underset{F}{\overset{\text{sim}}{\sim}} E$, an integral frame $\{\mathbf{a}^{\#}_{ij} : i,j \in I_n\}$ in E with associated ring $A^{\#}$ and non–zero finitely generated right E–submodules $K_1, \cdots, K_n$ of F such that $u^{-1}Ru = (\text{diag } K_i)A^{\#}(\text{diag } K_i^{-1})$. Let K be an invertible (E,D)–bisubmodule of F and $\mathbf{a}_{ij} = K^{-1}\mathbf{a}^{\#}_{ij}K$. Then $\{\mathbf{a}_{ij} : i,j \in I_n\}$ is an integral frame in D with associated ring A and $u^{-1}Ru = (\text{diag } K_iK)A(\text{diag } K^{-1}K_i^{-1})$. It follows from Lemma 8.4 that the $\mathbf{p}$–invariants of R and A agree for all primes $\mathbf{p}$ in D.

Since $u^{-1}Su$ and $u^{-1}Ru$ are diagonally formally conjugate via $u^{-1}Hu$, we have non–zero finitely generated right E–submodules $L_1, \cdots, L_n$ of F such that $u^{-1}Su = (\text{diag } L_i)u^{-1}Ru(\text{diag } L_i^{-1}) = (\text{diag } L_iK_iK)A(\text{diag } K^{-1}K_i^{-1}L_i^{-1})$. By Lemma 8.4, the $\mathbf{p}$–invariants of A and S agree. Hence the $\mathbf{p}$–invariants of S and R agree.

It is now immediate that the $\mathbf{p}$–invariants of R and S agree if $R \sim S$.

Now assume that $R, S \in \mathfrak{G}$ and that the $\mathbf{p}$–invariants of R and S agree for all $\mathbf{p}$. We shall show that $R \overset{f}{\sim} S$. In view of Theorem 5.4 and what has been proved above, we may assume that $R = \Delta_n(D, \mathbf{a})$ and $S = \Delta_n(D, \mathbf{b})$. Clearly $\mathbf{p} \mid \|R\|$ if and only if $\ell_{\mathbf{p}}(R) > 1$ and $\mathbf{p} \mid \|S\|$ if and only if $\ell_{\mathbf{p}}(S) > 1$. Thus, $\|R\| = \|S\| = \mathbf{p}_1 \cdots \mathbf{p}_\kappa$, say. Let $R^{(\mathbf{p}_\lambda)} = \Delta_n(D, \mathbf{p}_\lambda, \rho_\lambda)$ and $S^{(\mathbf{p}_\lambda)} = \Delta_n(D, \mathbf{p}_\lambda, \psi_\lambda)$ where ρ_λ and ψ_λ are partitions of n. We have already seen how the $\mathbf{p}_\lambda$–invariants of $R^{(\mathbf{p}_\lambda)}$ and $S^{(\mathbf{p}_\lambda)}$ are determined in terms of ρ_λ and ψ_λ. It follows that ρ_λ and ψ_λ are cyclic permutations of each other.

We now repeatedly apply Lemma 8.5 to R taking $\mu = 1$ and get a sequence $R = R_0, R_1, \cdots, R_{f_1}$ of associated rings of triangular frames in D such that the successive orders in this sequence are formally conjugate, $R^{(\mathbf{p}_\lambda)} = R_\upsilon^{(\mathbf{p}_\lambda)}$ for $2 \leqslant \lambda \leqslant \kappa$ and $0 \leqslant \upsilon \leqslant f_1$, and $R_{f_1}^{(\mathbf{p}_1)} = S^{(\mathbf{p}_1)}$. Repeat the argument for $\mu = 2$, starting with R_{f_1}. In κ steps like this, we get a sequence of orders $R = R_0, R_1, \cdots R_f$ such that the successive orders are formally conjugate and R_f is an associated ring of a triangular frame in D such that $R_f^{(\mathbf{p})} = S^{(\mathbf{p})}$ for all $\mathbf{p}$. Hence $R \overset{f}{\sim} R_f = S$. The remaining assertion is easily proved using Lemma 8.5. ∎

Corollary 8.7. Let D be a Dedekind domain with quotient skew field F, $Q = M_n(F)$, $R = \Delta_n(D, \mathbf{t})$ and $S = \Delta_n(D, \mathbf{a})$. Then the following conditions are equivalent:

(1) $R \overset{f}{\sim} S$.

(2) $R^{(\mathbf{p})} \overset{f}{\sim} S^{(\mathbf{p})}$ for all primes $\mathbf{p}$ in D.

(3) The $\mathbf{p}$–invariants of R and S agree for all primes $\mathbf{p}$ in D.

Proof. It suffices to observe that for any fixed prime $\mathfrak{p}$ in D, the $\mathfrak{p}$-invariants of R and $R^{(\mathfrak{p})}$ are the same. ∎

The above corollary provides the information about uniqueness of the canonical form obtained in §5.

The following proposition along with Theorems 5.4 and 8.6 contains some structure theorems obtained by Brumer [5], Harada [13] and Murtha [28] for the classical situation. See also Jacobinski [17].

Proposition 8.8. Let F be a skew field, $\mathcal{S}$ be the set of all pseudo-Dedekind rings in an equivalence class $\mathcal{E}$ of orders in $M_n(F) = Q$ and let D be a Dedekind order in F which is also a subbasic ring of $\mathcal{S}$ Then the following conditions are equivalent:

(1) Every $R \in \mathcal{S}$ is an almost Dedekind ring.

(2) Every bounded maximal right ideal of D is two-sided.

Proof. (2) ⇒ (1). Let $R \in \mathcal{S}$. By Theorem 6.4, we have a ring $T = \Delta_m(D, \mathbf{t})$ and a progenerator M in mod–T such that M is an essential right T–submodule of $M_{n\times m}(F)$ and $\mathbf{o}_\ell(M) = R$. Since $\dim M_T = n$, we have $M = U_1 \oplus \cdots \oplus U_n$ where each U_i is uniform right T–projective. Le e_i be the projection of M onto U_i along $\oplus_{j\neq i} U_j$. Evidently, $\{ e_i : i \in I_n \}$ is a set of orthogonal idempotents in R and $e_iRe_i \cong \text{End } U_i$. It thus suffices to show that the endomorphism ring of each uniform projective right T–module U is a Dedekind domain. We may take U as a T–submodule of $M_{1\times m}(D)$ without loss of generality. So, $U = (A_1, \cdots, A_m)$ where $A_m\|T\| \subseteq A_1 \subseteq \cdots \subseteq A_m$. Let $E = \mathbf{o}_\ell(A_m)$, $B_i = A_iA_m^{-1}$ and $S = (A_m t_{ij} A_m^{-1})$ where $\mathbf{t} = \{t_{ij} : i, j \in I_n\}$. Then $V = (B_1, \cdots, B_m)$ is a uniform right S–submodule of $M_{1\times m}(E)$, $\mathbf{o}_\ell(U) = \mathbf{o}_\ell(V)$ and $B_m = E$. Since $D \underset{F}{\overset{sim}{\sim}} E$, E also has property (2) from which it is easily seen that any right ideal of E with a non–zero semi–prime bound is two–sided. Since the semi–prime non–zero ideal $A_m\|T\|A_m^{-1}$ of E is contained in each B_i, it follows that $\mathbf{o}_\ell(V) = E$. Hence, as indicated, R is an almost Dedekind ring.

(1) ⇒ (2). Let X be a bounded maximal right ideal of D which is not two–sided. The bound of X must clearly be a prime ideal of D, say bound X = $\mathfrak{p}$. Let $T = \begin{pmatrix} D & D \\ \mathfrak{p} & D \end{pmatrix}$ and $N_T = (D, D)^{(n-1)} \oplus (X, D)$ as a natural right T–submodule of the $(M_n(F), T)$–bimodule $M_{n\times 2}(F)$. Let $R = \mathbf{o}_\ell(N)$. Using Theorem 6.4, it follows that $R \in \mathcal{S}$.

We claim that R is not almost Dedekind. First notice that

$$\mathbf{Nm}_1(T) = N \begin{pmatrix} D & D \\ \mathbf{p} & \mathbf{p} \end{pmatrix} = (D, D)^{(n-1)} \oplus (X, X)$$

and

$$\mathbf{Nm}_2(T) = N \begin{pmatrix} \mathbf{p} & D \\ \mathbf{p} & D \end{pmatrix} = (\mathbf{p}, D)^n.$$

Second, observe that $T \begin{pmatrix} X & D \\ 0 & 0 \end{pmatrix} = T$; so, $(X, D)_T$ is a progenerator in mod–T. Consequently, N_T is also a progenerator in mod–T. Now, using the Morita theorems, it is immediate that the **p**–invariants of R are $\{\dim(N/\mathbf{Nm}_1(T))_T, \dim(N/\mathbf{Nm}_2(T))_T\}$. Using our first observation, the **p**–invariants of R are $\{d, (n-1)c(\mathbf{p}) + d\}$ where $d = \dim D/X$ and $c(\mathbf{p}) = \dim(D/\mathbf{p})$.

Now, if possible, let R be an almost Dedekind ring. By Theorem 5.4, $R \underset{Q}{\overset{f}{\sim}} \Delta_n(D, \mathbf{a})$. Since the only non–trivial cycle of T is the **p**–cycle, it follows that the only non–trivial cycle of $\Delta_n(D, \mathbf{a})$ is its **p**–cycle. By Theorem 7.2, $\|\Delta_n(D, \mathbf{a})\| = \mathbf{p}$. But then the **p**–invariants of R and $\Delta_n(D, \mathbf{a})$ do not agree since all the terms of the **p**–invariants of $\Delta_n(D, \mathbf{a})$ are non–zero multiples of $c(\mathbf{p})$; this contradicts Theorem 8.6 and establishes our claim. ∎

Remark. Taking $n = 2$ in the above example we see that 'almost Dedekind' is *not* invariant under similarity. The following example shows that similar almost Dedekind orders need not be formally conjugate:

$$R = \begin{pmatrix} D & D & D & D \\ D & D & D & D \\ D & D & D & D \\ \mathbf{p} & \mathbf{p} & \mathbf{p} & D \end{pmatrix}, \quad S = \begin{pmatrix} D & D & D & D \\ D & D & D & D \\ \mathbf{p} & \mathbf{p} & D & D \\ \mathbf{p} & \mathbf{p} & D & D \end{pmatrix},$$

where D is a Dedekind domain and **p** is a prime in D.

9. Invariants and Conjugacy

Let $\mathfrak{S}$ be the set of all pseudo–Dedekind orders in an equivalence class $\mathcal{E}$ of orders in $M_n(F)$, F a skew field. We assume that $\mathcal{E}$ has a PID D as a subbasic ring and attempt to classify rings in $\mathfrak{S}$ under conjugacy. If D is semi–local, we do get such a classification. However, in general we have to settle for a little less and we provide some examples to show that the anomalies permitted in our classification do occur.

The first step is to classify finitely generated projectives over $\Delta_n(D, t)$, D a PID. This is of some interest in itself.

Let us recall a definition. If A, B are non-zero right ideals of a domain D, then A is a *total divisor* of B if $DB \subseteq A$. The following theorem gives a very convenient form for finitely generated projectives over $\Delta_n(D, t)$.

Theorem 9.1. Let D be a Dedekind domain, $T = \Delta_n(D, t)$ and M be a projective right T–module of uniform dimension $k > 0$. Then there exists an indexed set $\{A_{ij} : (i, j) \in I_k \times I_n\}$ of non-zero right ideals of D satisfying the following conditions: Set $A = (A_{ij}) \subseteq M_{k\times n}(D)$. Then

(1) A is a right T–submodule of $M_{k\times n}(D)$ and $A \cong M$.

(2) Each A_{ij} is a total divisor of $A_{i+1,j}$.

(3) $A_{in} = D$ for $1 \leqslant i \leqslant k-1$. If D is a PID then $A_{in} = D$ for $1 \leqslant i \leqslant k$.

Proof. Evidently, T is a direct sum of uniform projectives (viz. its rows) and all these projectives are isomorphic to some right T–submodules of $M_{1\times n}(D)$, the first row of T. Thus [24, page 85] we can choose submodules U_i of $M_{1\times n}(D)$ such that $M \cong \oplus_{i=1}^{k} U_i$. Each U_i is of the form $U_i = (U_{i1}, \cdots, U_{in})$ where $U_{i1} \subseteq \cdots \subseteq U_{in}$ is a chain of right ideals of D. If D is a PID then $U_{in} = dD$ and $U_{ij} = dV_{ij}$ for some right ideals V_{ij} of D; thus, changing U_{in} up to isomorphism if necessary, we may assume that $U_{in} = D$.

Let **X** be the set of all those right T–submodules V of $M_{1\times n}(D)$ which are isomorphic with some direct summand of M; if D is a PID, we further require that V be of the form $V = (V_1, \cdots, V_n)$ with $V_n = D$. We may assume without loss that $k \geqslant 2$.

Since $M_{1\times n}(D)$ is a Noetherian right T–module, **X** has a maximal member, say $A_1 = (A_{11}, \cdots, A_{1n})$. So, $M \cong A_1 \oplus B_2 \oplus \cdots \oplus B_k$ for some $B_2, \cdots, B_k \in$ **X**. Let $B_i = (B_{i1}, \cdots, B_{in})$. Since D is a Dedekind domain and $B_{2n} \neq (0)$, we can choose an element α in the quotient skew field of D such that $\alpha B_{2n} + A_{1n} = D$ (cf. [7,Cor. 3.8]). Since $B_{21} \subseteq B_{22} \subseteq \cdots \subseteq B_{2n}$ and $\alpha B_{2n} \subseteq D$, we have a T–isomorphism $B_2 \to \alpha B_2 \subseteq M_{1\times n}(D)$ defined by $(b_{21}, \cdots, b_{2n}) \to (\alpha b_{21}, \cdots, \alpha b_{2n})$ where $b_{2j} \in B_{2j}$. Thus $\alpha B_2 \in$ **X**. The usual exact sequence

$$0 \to A_1 \cap \alpha B_2 \to A_1 \oplus \alpha B_2 \to A_1 + \alpha B_2 \to 0$$

immediately yields $A_1 \oplus B_2 \cong (A + \alpha B_2) \oplus (A_1 \cap \alpha B_2)$. Evidently, $A_1 \subseteq A_1 + \alpha B_2 \in X$ so $A_1 = A_1 + \alpha B_2$. Consequently, $A_{1n} = D$.

If possible, let $DB_{ij} \not\subseteq A_{1j}$ for some i, j. Choose $d \in D$ such that $dB_{ij} \not\subseteq A_{1j}$. Then, as above, $A_1 \oplus B_i \cong (A_1 + B_i) \oplus (A_1 \cap B_i)$ where $A_1 \subsetneqq (A_1 +$

$B_i) \in \mathbf{X}$, contrary to our choice of A_1.

Our proof so far shows that the theorem is true for $k = 2$. A straightforward induction on k using the properties of A_1 established above suffices to complete the proof. ∎

Until further notice, we shall work in the following *context*: D is a PID with quotient skew field F, $\mathbf{t} = (t_{ij})$ is a triangular frame of order n in D, $T = \Delta_n(D, \mathbf{t})$, $\|T\| = \mathbf{p}_1 \cdots \mathbf{p}_\kappa$ the prime factorization of $\|T\|$ [so, $\mathbf{p}_\lambda$ $(1 \leqslant \lambda \leqslant \kappa)$ are necessarily distinct primes in D.] We let $T^{(\mathbf{p}_\lambda)} = \Delta_n(D, \mathbf{p}_\lambda, \rho_\lambda)$ where $\rho_\lambda = (\rho_{1\lambda}, \cdots, \rho_{\ell_\lambda \lambda})$ is a partition of n. Let $\tau_{i\lambda} = \Sigma_{j=i}^{\ell_\lambda} \rho_{j\lambda}$ for $1 \leqslant i \leqslant \ell_\lambda$ and $\tau_{\ell_\lambda+1,\lambda} = 0$. Recall that $T \cap \mathbf{m}_i(T^{(\mathbf{p}_\lambda)}) = \mathbf{m}_i(T, \mathbf{p}_\lambda)$ for $1 \leqslant i \leqslant \ell_\lambda$.

Suppose M is a finite dimensional projective right T–module.

The right T–module $M/M\mathbf{m}_i(T, \mathbf{p}_\lambda)$ is a finitely generated semi–simple right T–module since it is really a module over the simple Artinian ring $T/\mathbf{m}_i(T, \mathbf{p}_\lambda)$. Let $f_{i\lambda}$ be the dimension of the module $M/M\mathbf{m}_i(T, \mathbf{p}_\lambda)$. Then the sequence $\{ f_{1\lambda}, \cdots, f_{\ell_\lambda \lambda} \}$ of non–negative integers is called the *sequence of* $\mathbf{p}_\lambda$*–invariants* of M. This sequence is said to be *trivial* if all its terms, except possibly the first, are zero. *Caution:* While considering sequences of invariants of a finite dimensional projective, we do *not* identify sequences that differ by a cyclic permutation.

Trivially, isomorphic finite dimensional projectives have the same sequences of $\mathbf{p}_\lambda$–invariants for $1 \leqslant \lambda \leqslant \kappa$; a partial converse of this triviality is our main result on classification of projectives.

We need one more definition. A projective right T–module M of dimension k is in *normal form* if $M = (A_{ij}) \subseteq M_{k \times n}(D)$ where $\{ A_{ij} : (i, j) \in I_k \times I_n \}$ is an indexed set of non–zero right ideals of D such that A_{ij} is a total divisor of $A_{i+1, j}$ and $A_{in} = D$ for all i, j involved. Theorem 9.1 shows that every finite dimensional projective is isomorphic with a projective module in normal form. An example at the end of this section shows that the normal form need not be uniquely determined up to isomorphism. Observe that if $M = (A_{ij})$ is in normal form, then $\|T\| \subseteq \cap_{i,j} A_{ij}$. The largest ideal of D contained in $\cap_{i,j} A_{ij} = A_{k1}$ is denoted as $\|M\|$. Clearly, $\|T\| \subseteq \|M\|$.

We begin by looking at a special kind of finite dimensional projectives in normal form.

Lemma 9.2. Let $X = (X_{ij})$ be a projective right T–module of dimension k and in normal form. Further, let $\|X\| = \mathbf{p}_\alpha$ for some α, $1 \leqslant \alpha \leqslant \kappa$. Then

(1) X is naturally a right $T^{(\mathbf{p}_\lambda)}$–module.

(2) For $\lambda \neq \alpha$, $1 \leqslant \lambda \leqslant \kappa$, the sequence of $\mathbf{p}_\lambda$–invariants of X is $\{kc(\mathbf{p}_\lambda), \overbrace{0, \cdots, 0}^{(\ell_\lambda - 1) \text{ terms}}\}$.

(3) Let $\{f_1, \cdots, f_{\ell_\alpha}\}$ be the sequence of $\mathbf{p}_\alpha$–invariants of X. Let W_υ denote the υth column of X naturally considered as a right D–submodule of $M_{k\times 1}(D)$. Let W_0 denote the right D–submodule of $M_{k\times 1}(D)$ all whose entries are $\mathbf{p}_\alpha$. Then $W_\upsilon = W_{\tau_{i\alpha}}$ if $1 + \tau_{i+1,\alpha} \leqslant \upsilon \leqslant \tau_{i\alpha}$ and $W_{\tau_{i\alpha}}/W_{\tau_{i+1,\alpha}}$ is a semi–simple right D–module of dimension f_i, $1 \leqslant i \leqslant \ell_\alpha$. Thus, $\Sigma_{i=1}^{\ell_\alpha} f_i = kc(\mathbf{p}_\alpha)$.

(4) Let U_i be the ith row of X treated naturally as a right T–submodule of $M_{1\times n}(D)$. Then the sequence of $\mathbf{p}_\alpha$–invariants of U_i is uniquely determined by the sequence of $\mathbf{p}_\alpha$–invariants of X. Further, the dimension of D/X_{ij} is uniquely determined by (i, j) and the sequence of $\mathbf{p}_\alpha$–invariants of X.

Proof. By reindexing the prime divisors of $\|T\|$ if necessary, we may assume that $\alpha = 1$.

(1): Since $\mathbf{p}_1 \subseteq X_{ij}$ for all i, j, it is clear that X is naturally a right module over the ring $\Delta_n(D, \mathbf{p}_1, \overbrace{1, \cdots, 1}^{\ell_1 \text{ terms}})$. Since $T^{(\mathbf{p}_1)} = T + \Delta_n(D, \mathbf{p}_1, \overbrace{1, \cdots, 1}^{\ell_1 \text{ terms}})$, X is naturally a right $T^{(\mathbf{p}_1)}$–module.

(2): Let $2 \leqslant \lambda \leqslant \kappa$ and $1 \leqslant \mu \leqslant \ell_\lambda - 1$. We claim that $X\mathbf{m}_\mu(T, \mathbf{p}_\lambda) = X$. Clearly, we can express $\mathbf{m}_\mu(T, \mathbf{p}_\lambda)$ as $(\theta_{ij})_{n\times n}$ where the θ_{ij} are certain non–zero ideals of D. The $(i, j)^{th}$ entry of $X\mathbf{m}_\mu(T, \mathbf{p}_\lambda)$ is $\Sigma_{h=1}^{n} X_{ih}\theta_{hj}$. Trivially, $\Sigma_{h=1}^{n} X_{ih}\theta_{hj} \subseteq X_{ij}$. If $1 \leqslant j \leqslant \tau_{\mu+1,\lambda}$ or if $\tau_{\mu\lambda} < j \leqslant n$, then $\theta_{jj} = D$; in these cases, trivially $\Sigma_{h=1}^{n} X_{ih}\theta_{hj} = X_{ij}$. In the remaining case, $\theta_{jj} = \mathbf{p}_\lambda$ and $\theta_{1j} = D$. (Here we use the assumption $\mu \neq \ell_\lambda$.) Since $\mathbf{p}_1 + \mathbf{p}_\lambda = D$, we get $X_{ij} = X_{ij}\mathbf{p}_1 + X_{ij}\mathbf{p}_\lambda \subseteq \mathbf{p}_1 + X_{ij}\theta_{jj} \subseteq X_{i1}\theta_{1j} + X_{ij}\theta_{jj} \subseteq \Sigma_{h=1}^{n} X_{ih}\theta_{hj}$. This proves our claim.

We can express $\mathbf{m}_{\ell_\lambda}(T, \mathbf{p}_\lambda)$ as $(\psi_{ij})_{n\times n}$ where the ψ_{ij} are certain non–zero ideals of D. Then $X\mathbf{m}_{\ell_\lambda}(T, \mathbf{p}_\lambda) = (Y_{ij})_{k\times n}$ where $Y_{ij} = \Sigma_{h=1}^{n} X_{ih}\psi_{hj} \subseteq X_{ij}$ for all i, j. If $\ell_\lambda < j$, then $\psi_{jj} = D$; consequently, $Y_{ij} = X_{ij}$ for $i \in I_k$. We claim that $Y_{ij} = X_{ij}\mathbf{p}_\lambda$ for $1 \leqslant j \leqslant \ell_\lambda$. Notice that the X_{ih} increase with h and $\psi_{hj} = \mathbf{p}_\lambda$ if $1 \leqslant h \leqslant j \leqslant \ell_\lambda$. So, $\Sigma_{h=1}^{j} X_{ih}\psi_{hj} = X_{ij}\mathbf{p}_\lambda$. For $j + 1 \leqslant h \leqslant \ell_\lambda$, we have $\psi_{hj} = \mathbf{t}_{hj}\mathbf{p}_\lambda$ and $X_{ih}\mathbf{t}_{hj} \subseteq X_{ij}$. So, $\Sigma_{h=j+1}^{\ell_\lambda} X_{ih}\psi_{hj} \subseteq X_{ij}\mathbf{p}_\lambda$. For $h > \ell_\lambda$, we have $\psi_{hj} = \mathbf{t}_{hj}$. We proceed to show that $\Sigma_{h=\ell_\lambda+1}^{n} X_{ih}\mathbf{t}_{hj} \subseteq X_{ij}\mathbf{p}_\lambda$; once this is proved, our claim will readily follow. If $\mathbf{t}_{hj} \subseteq \mathbf{p}_1\mathbf{p}_\lambda$, then trivially $X_{ih}\mathbf{t}_{hj} \subseteq X_{ij}\mathbf{p}_\lambda$. Suppose $\mathbf{t}_{hj} \not\subseteq \mathbf{p}_1\mathbf{p}_\lambda$. Since $j \leqslant \ell_\lambda < h$, it follows that $\mathbf{t}_{hj} \subseteq \mathbf{p}_\lambda$. So, $\mathbf{t}_{hj} \not\subseteq \mathbf{p}_1$. In this case, (j, j) and (h, h) belong to the same diagonal block of $T^{(\mathbf{p}_1)}$. Since X is naturally a $T^{(\mathbf{p}_1)}$–module, we have $X_{ij} = X_{ih}$. So, $X_{ih}\mathbf{t}_{hj} = X_{ij}\mathbf{t}_{hj} \subseteq X_{ij}\mathbf{p}_\lambda$. This establishes our claim.

It is now immediate that the sequence of $\mathbf{p}_\lambda$–invariants of M is $\{kc(\mathbf{p}_\lambda), \overbrace{0, \cdots, 0}^{\ell_\lambda - 1 \text{ terms}}\}$.

(3): Let us simplify notation by writing τ_i for τ_{i1}. Since X is naturally a $T^{(p_1)}$–module, it is immediate that $W_{\tau_i} = W_\upsilon$ if $1 + \tau_{i+1} \leqslant \upsilon \leqslant \tau_i$.

Let $1 \leqslant \mu \leqslant \ell_1$. It is clear that $Xm_\mu(T, p_1) = (Y_{ij}) \subseteq (X_{ij})$ where Y_{ij} are certain non–zero right ideals of D. We claim that $Xm_\mu(T, p_1)$ is the T–submodule of X in which the υth column is $W_{\tau_{\mu+1}}$ for $\tau_{\mu+1} + 1 \leqslant \upsilon \leqslant \tau_\mu$ and all other columns are as in X.

We have to consider two cases. First, assume that $1 \leqslant \mu \leqslant \ell_1 - 1$. If (j, j) is not in the μth diagonal block of $T^{(p_1)}$, then the $(j, j)^{th}$ entry of $m_\mu(T^{(p_1)}, p_1)$ is D, which yields $Y_{ij} = X_{ij}$. If (j, j) does not belong to the μth diagonal block of $T^{(p_1)}$, then

$$Y_{ij} = \sum_{h=1}^{\tau_{\mu+1}} X_{ih} + \sum_{h=\tau_{\mu+1}+1}^{\tau_\mu} X_{ih} t_{hj} p_1$$

$$+ \sum_{h=\tau_\mu+1}^{n} X_{ih} t_{hj} \quad .$$

Since X_{ih} increases with h, $\sum_{h=1}^{\tau_{\mu+1}} X_{ih} = X_{i, \tau_{\mu+1}}$. Since $p_1 \subseteq X_{i\tau_{\mu+1}}$ and $t_{hj} \subseteq p_1$ for $h > \tau_\mu$, it follows that $Y_{ij} = X_{i\tau_{\mu+1}}$. This proves our claim for $1 \leqslant \mu \leqslant \ell_1 - 1$ The remaining case, viz. $\mu = \ell_1$, is similarly handled. The assertion (3) of the lemma is immediate from our claim.

(4): Clearly, $U_i = (X_{1i}, X_{2i}, \cdots, X_{ni}) \subseteq M_{1 \times n}(D)$ and $\|U_i\| \supseteq p_1$. Also, from part (3) of the lemma, we have $X_{i\upsilon} = X_{i\tau_j}$ if $\tau_{j+1} + 1 \leqslant \upsilon \leqslant \tau_j$. Let us put $X_{i\tau_{\ell_1+1}} = p_1$. Let $g_j = \dim X_{i\tau_j}/X_{i\tau_{j+1}}$ and $h_j = \dim D/X_{i\tau_{j+1}}$ for $1 \leqslant j \leqslant \ell_1$. If $\|U_i\| = p_1$ then part (3) shows that the sequence of p_1–invariants of U_i is $\{ g_1, \cdots, g_{\ell_1} \}$. If $\|U_i\| = D$ then $U_i = M_{1 \times n}(D)$ and again the sequence of p_1–invariants of U_i is $\{ g_1, \cdots, g_{\ell_1} \}$. Now observe that $h_j = \Sigma_{s=1}^{j} g_s$. Thus to prove (4), it suffices to show that $\dim D/X_{i\tau_j}$ is uniquely determined by (i, j) and the sequence of p_1–invariants of X.

Choose integers $q \geqslant 0$ and $0 \leqslant r < c(p_1)$ such that $\Sigma_{\upsilon=1}^{j} f_\upsilon = qc(p_1) + r$. Note that the entries of $W_{\tau_{j+1}}$ form a descending chain of total divisors and all of them contain the maximal ideal p_1 of D. Thus at most one entry of $W_{\tau_{j+1}}$ can be distinct from D and p_1. Since $\dim(M_{1 \times n}(D)/W_{\tau_{j+1}}) = qc(p_1) + r$, it follows that the last q entries of $W_{\tau_{j+1}}$ are p_1 and the first $k-q-1$ entries are D; the remaining $(k-q)^{th}$ entry is D if $r = 0$; otherwise, it is a right ideal, say V, of D such that $V \supseteq p_1$ and $\dim D/V = r$. The lemma is now clear. ∎

The next lemma relates arbitrary projectives in normal form with the type of projectives treated in the above lemma.

Lemma 9.3. Let $M = (A_{ij})$ be a projective right T–module of uniform dimension k and in normal form. Suppose that $\|M\| = p_1 \cdots p_\alpha$ where $1 \leqslant \alpha \leqslant \kappa$. Let $X_{ij}^{(\lambda)} = A_{ij} + p_\lambda$ for $(i, j) \in I_k \times I_n$ and $1 \leqslant \lambda \leqslant \alpha$. Let $X_\lambda = (X_{ij}^{(\lambda)}) \subseteq M_{k\times n}(D)$. Then each X_λ is a projective right T–module in standard form with uniform dimension k and $M = \cap_{\lambda=1}^{\alpha} X_\lambda$; further for each λ, $1 \leqslant \lambda \leqslant \alpha$, $\|X_\lambda\| = p_\lambda$, and the sequences of p_λ–invariants of X_λ and M agree.

Proof. Since $(p_\lambda)_{k\times n}$ is a right T–submodule of $M_{k\times n}(D)$ and $X_\lambda = M + (p_\lambda)_{k\times n}$, it follows that X_λ is a right T–module. Since X_λ is finitely generated and torsion–free, it is projective; since X_λ is essential in $M_{k\times n}(D)$ which has uniform dimension k, X_λ has uniform dimension k. Since $A_{i+1,j} \subseteq DA_{i+1,j} \subseteq A_{ij}$, we have $X_{i+1,j}^{(\lambda)} \subseteq DA_{i+1,j} + p_\lambda \subseteq X_{ij}^{(\lambda)}$. Thus $X_\lambda = (X_{ij}^{(\lambda)})$ is in normal form. Evidently $\|X_\lambda\| \supseteq p_\lambda$ and $M \subseteq \cap_{\lambda=1}^{\alpha} X_\lambda$.

Since $\|M\|$ is a proper semi–prime ideal of the PID D, the ring $\bar{D} = D/\|M\|$ is a semi–simple ring. It is easily seen that $\bar{D} = \oplus_{\lambda=1}^{\alpha}(\|M\|p_\lambda^{-1})/\|M\|$ is the block decomposition of $\bar{D}$. Let $\sigma : D \to \bar{D}$ be the canonical map. Then $\sigma(X_{ij}^{(\lambda)}) = \sigma(p_\lambda) + \sigma(A_{ij})$. Since $X_{ij}^{(\lambda)} \supseteq \|M\|$, we have

$$\sigma\left(\bigcap_{\lambda=1}^{\alpha} X_{ij}^{(\lambda)}\right) = \bigcap_{\lambda=1}^{\alpha} \sigma(X_{ij}^{(\lambda)}).$$

Also, using the block decomposition of $\bar{D}$, we get

$$\sigma(A_{ij}) = \bigcap_{\lambda=1}^{\alpha} \{\sigma(p_\lambda) + \sigma(A_{ij})\}.$$

Hence $A_{ij} = \cap_{\lambda=1}^{\alpha} X_{ij}^{(\lambda)}$, and so $M = \cap_{\lambda=1}^{\alpha} X_\lambda$. Now if $\|X_\lambda\| = D$, then $\sigma(p_\lambda) + \sigma(A_{k1}) = \sigma X_{k1}^{(\lambda)} = \bar{D}$; so, $\sigma(A_{k1}) \supseteq \sigma(\|M\|p_\lambda^{-1})$, i.e., $A_{k1} \supseteq \|M\|p_\lambda^{-1}$, and so $\|M\| \supseteq \|M\|p_\lambda^{-1}$, which is absurd. Thus $\|X_\lambda\| = p_\lambda$.

We shall now show that, for a given λ, the sequences of p_λ–invariants of X_λ and M agree. This is trivial if $\alpha = 1$. Assume that $\alpha > 1$. Then

$$X_\lambda + \bigcap_{\substack{\mu=1 \\ \mu\neq\lambda}}^{\alpha} X_\mu = M_{k\times n}(D).$$

Using this observation and the usual exact sequence we get

$$M \oplus [M_{k\times n}(D)]^{\alpha-1} \cong X_\lambda \oplus \left[\bigcap_{\substack{\mu=1 \\ \mu\neq\lambda}}^{\alpha} X_\mu\right] \oplus [M_{k\times n}(D)]^{\alpha-2}.$$

Continuing in this manner, we obtain

$$M \oplus [M_{k \times n}(D)]^{\alpha-1} \cong \bigoplus_{\lambda=1}^{\alpha} X_\lambda.$$

It is easily seen that the sequence of p_λ–invariants of $M_{k \times n}(D)$ is $\{ kc(p_\lambda), 0, \cdots, 0 \}$. It is equally easy to see that the sequence of p_λ–invariants of a direct sum is the termwise addition of the sequences of p_λ–invariants of the factors. The lemma is now immediate from Lemma 9.2. ∎

We now point out two immediate consequences of Lemmas 9.2 and 9.3. To put the first one in perspective, we recall our earlier remark that the normal form is not necessarily unique.

Corollary 9.4. Let $M = (A_{ij})$ be a finite dimensional projective right T–module in normal form. Then $\|M\|$ is the product of those primes p_λ for which the sequence of p_λ–invariants of M is non–trivial.

Proof. If $p_\lambda \mid \|M\|$ then using Lemmas 9.2 and 9.3, it is immediate that the sequence of p_λ–invariants of M is non–trivial. On the other hand, if $p_\lambda \nmid \|M\|$, then using $p_\lambda + \|M\| = D$, it is straightforward to see that the sequence of p_λ–invariants of M is trivial. The proof is completed by noting that $\|M\|$ is a semi–prime ideal of D. ∎

The following lemma is one of the main steps in the proof of our classification theorem on finite dimensional projectives.

Lemma 9.5. With the notation of Lemma 9.3, if $k > n\alpha$ then all the entries of at least one row of $M = (A_{ij})$ are two–sided ideals of D.

Proof. Let us call a row of M a bad row if it contains at least one entry which is not a two–sided ideal of D. By Lemma 9.3, $M = \cap_{\lambda=1}^{\alpha} X_\lambda$. Clearly, the ith row of M is not bad if the ith row of X_λ is not bad for all λ, $1 \leqslant \lambda \leqslant \alpha$. Since $\|X_\lambda\| = p_\lambda$ and since X_λ is in normal form, therefore each column of X_λ contains at most one entry which is not a two–sided ideal of D. Thus X_λ has at most n bad rows. Consequently, M has at most $n\alpha$ bad rows. ∎

In the next few lemmas, we compare projectives in normal form with the same sequences of p_λ–invariants.

Lemma 9.6. For each $\lambda \in I_\kappa$, let A_λ, B_λ be right ideals of D such that $A_\lambda \cap B_\lambda \supseteq p_\lambda$ and $D/A_\lambda \cong D/B_\lambda$. Let $A = \cap_{\lambda=1}^{\kappa} A_\lambda$ and $B = \cap_{\lambda=1}^{\kappa} B_\lambda$.

Then $D/A \cong D/B$.

Proof. Since $\|T\| = p_1 \cdots p_\kappa$ is a semi–prime ideal of D, the ring $\bar{D} = D/\|T\|$ is a semi–simple ring. Let $\sigma : D \to \bar{D}$ be the canonical map. Let $\sigma(A) = U_1 \oplus \cdots \oplus U_\kappa$, $\sigma(B) = V_1 \oplus \cdots \oplus V_\kappa$ where U_λ and V_λ are those homo–geneous components of $\sigma(A)$ and $\sigma(B)$ respectively which are annihilated by p_λ. Then $\dim U_\lambda = \dim\{(U_\lambda + \sigma(p_\lambda))/\sigma(p_\lambda)\} = \dim(A_\lambda/p_\lambda) = \dim(B_\lambda/p_\lambda) = \dim V_\lambda$. Hence $\sigma(A) \cong \sigma(B)$; so $D/A \cong D/B$. ∎

Lemma 9.7. Let $A_1 \subseteq A_2 \subseteq D$ and $B_1 \subseteq B_2 \subseteq D$ be chains of right ideals of D and let u be a given element of D such that the formula $f(d+A_1) = ud + B_1$, $d \in D$, defines a D–isomorphism $f : D/A_1 \to D/B_1$ with $f(A_2/A_1) = B_2/B_1$. Then $uA_2 + B_1 = B_2$ and $uA_2 \cap B_1 = uA_1$.

Proof. Trivially, $uA_2 + B_1 = B_2$ and $uA_1 \subseteq uA_2 \cap B_1$. Let $x \in uA_2 \cap B_1$. Then $x = uy$, $y \in A_2$, so that $f(y + A_1) = x + B_1 = 0 + B_1$. Since f is an isomorphism, this yields $y \in A_1$; so $x \in uA_1$. ∎

Lemma 9.8. Let $M = (A_{ij})$ and $N = (B_{ij})$ be finite dimensional projective right T–modules in normal form. Suppose that for each $\lambda \in I_\kappa$, the sequences of p_λ–invariants of M and N agree. Then

(1) $\|M\| = \|N\|$ and $\dim M = \dim N = k$, say.

(2) $A_{ij}/A_{k1} \cong B_{ij}/B_{k1}$ for all i, j.

(3) Given $i \in I_k$ and $\lambda \in I_\kappa$, the sequences of p_λ–invariants of the ith rows of M and N agree.

Proof. (1): By Corollary 9.4, $\|M\| = \|N\|$. Reindexing the prime divisors of $\|T\|$ if necessary, we may assume that $\|M\| = p_1 \cdots p_\alpha$, $0 \leqslant \alpha \leqslant \kappa$. If $\alpha = 0$ then $M = M_{k_1 \times n}(D)$ and $N = M_{k_2 \times n}(D)$; obviously, their p_1–sequences begin with $k_1 c(p_1)$ and $k_2 c(p_1)$ respectively; thus $k_1 = k_2$. If $\alpha > 0$, using Lemmas 9.2 and 9.3, it is immediate that $\dim M = \dim N$.

For the rest of the proof, we may assume without loss that $\alpha > 0$.

(2): Let $X_\lambda = (X_{ij}^{(\lambda)}) = M + (p_\lambda)_{k \times n}$, $Y = (Y_{ij}^{(\lambda)}) = N + (p_\lambda)_{k \times n}$ where $1 \leqslant \lambda \leqslant \alpha$. By Lemma 9.3, the sequences of p_λ–invariants of X_λ and Y_λ are the same. By Lemma 9.2, $\dim D/X_{ij}^{(\lambda)} = \dim D/Y_{ij}^{(\lambda)}$ for all i, j involved. Also, $A_{ij} = \cap_{\lambda=1}^{\alpha} X_{ij}^{(\lambda)}$, $B_{ij} = \cap_{\lambda=1}^{\alpha} Y_{ij}^{(\lambda)}$ and $X_{ij}^{(\lambda)} \cap Y_{ij}^{(\lambda)} \supseteq p_\lambda$. Set $X_{ij}^{(\lambda)} = Y_{ij}^{(\lambda)} = D$ for $\alpha + 1 \leqslant \lambda \leqslant \kappa$. By Lemma 9.6, we have $D/A_{ij} \cong D/B_{ij}$ for all i, j. Clearly, $A_{k1} = \cap_{i,j} A_{ij}$ and $B_{k1} = \cap_{i,j} B_{ij}$. So, $A_{ij}/A_{k1} \cong B_{ij}/B_{k1}$ for all i, j.

(3): Let U_i and $U_i^{(\lambda)}$ be the ith rows of M and X_λ respectively. Then

$\|U_i^{(\lambda)}\| \supseteq p_\lambda$ and $U_i = \cap_{\lambda=1}^{\alpha} U_i^{(\lambda)}$. Lemmas 9.2 and 9.3 show that the sequence of p_λ–invariants of $U_i^{(\lambda)}$ is uniquely determined by that of M. Clearly, $\|U_i\| \supseteq \|M\|$. Choose notation such that $\|U_i\| = p_1 \cdots p_\beta$, $\beta \leqslant \alpha$. It is clear that $\|U_i^{(\lambda)}\| = D$ if $\lambda > \beta$. Thus $U_i = \cap_{\lambda=1}^{\beta} U_i^{(\lambda)}$. Since $U_i^{(\lambda)} = U_i + (p_\lambda)_{1 \times n}$, Lemma 9.3 now shows that the sequences of p_λ–invariants of U_i are uniquely determined by i and the sequences of p_λ–invariants of M. ∎

Lemma 9.9. Let

$$M = \begin{pmatrix} A_{11} & A_{12} \cdots A_{1n} \\ A_{21} & A_{22} \cdots A_{2n} \end{pmatrix}, \quad N = \begin{pmatrix} B_{11} & B_{12} \cdots B_{1n} \\ B_{21} & B_{22} \cdots B_{2n} \end{pmatrix}$$

be projective right T–modules in normal form. Assume that all the entries of at least one row of M or N are two–sided ideals of D. Further, assume that the sequences of p_λ–invariants of M and N agree for all $\lambda \in I_\kappa$. Then $M \cong N$.

Proof. Suppose that all the entries of the first row of M are two–sided ideals of D. By Lemma 9.8, the p_λ–invariants of the first rows of M and N agree for $1 \leqslant \lambda \leqslant \kappa$. It follows that these rows are the same (this equality crucially uses the two–sidedness of entries of the first row of M). Let this common first row of M and N be $(K_1, \cdots, K_n)$. By Corollary 9.4, we have $\|M\| = \|N\| = K$, say. Then D/K is a cyclic semi–simple T–module. By Lemma 9.8, $A_{ij}/A_{21} \cong B_{ij}/B_{21}$. Using the two–sidedness of the entries of the first rows of M and N, we can find an automorphism φ of the T–module D/K such that $\varphi(K_j/K) = K_j/K$ and $\varphi(A_{2j}/K) = B_{2j}/K$ for all $j \in I_n$. Thus, we have an element $u \in D$ such that $\varphi(d + K) = ud + K$ for all $d \in D$. This isomorphism induces an isomorphism φ_j: $D/A_{2j} \to D/B_{2j}$ defined by $\varphi_j(d + A_{2j}) = ud + B_{2j}$ and $\varphi_j(K_j/A_{2j}) = K_j/B_{2j}$. By Lemma 9.7, we have $uK_j + B_{2j} = K_j$ and $uK_j \cap B_{2j} = uA_{2j}$ for all $j \in I_n$. Now using the usual exact sequence, we get

$$N \cong \begin{pmatrix} uK_1 & uK_2 \cdots uK_n \\ B_{21} & B_{22} \cdots B_{2n} \end{pmatrix} \cong \begin{pmatrix} K_1 & K_2 & \cdots K_n \\ uA_{21} & uA_{22} \cdots uA_{2n} \end{pmatrix} \cong M.$$

A similar argument works when all entries of the second row of M are two–sided ideals. ∎

The following theorem contains our main result on classification of finitely generated projectives.

Theorem 9.10. Let D be a PID, $T = \Delta_n(D, t)$ and $\|T\| = p_1 \cdots p_\kappa$ be a prime factorization of $\|T\|$. Let M, N be finite dimensional projective right T-modules. Consider the following statements:

(1) $M \cong N$.

(2) $M^{(n\kappa+1)} \cong N^{(n\kappa+1)}$.

(3) For each $\lambda \in I_\kappa$, the sequences of p_λ-invariants of M and N agree.

Then (1) $\Rightarrow$ (2) $\Leftrightarrow$ (3). If $\dim M > n\kappa$ or if $J(D) \neq (0)$ then (1) $\Leftrightarrow$ (3).

Proof. Let M and N satisfy (3) and let $k = \dim M > n\kappa$. By Theorem 9.1, we may assume that M and N are in normal form. By Lemma 9.5, we can find $i \in I_k$ such that all the entries in the ith row of M are two-sided ideals of D. By Lemma 9.8 (3), the sequences of p_λ-invariants of jth rows of M and N agree for each $j \in I_k$ and $\lambda \in I_\kappa$. Now start with M and repeatedly use Lemma 9.9 to replace successively the rows of M by the corresponding rows of N; this yields a sequence of isomorphic modules which begins with M and ends with N; so, $M \cong N$.

The implications (1) $\Rightarrow$ (2) $\Leftrightarrow$ (3) are now immediate.

Now assume that $J(D) \neq (0)$ and that (3) holds. Since sequences of p_λ-invariants of the jth rows of M and N agree for every $j \in I_k$ and $\lambda \in I_\kappa$ we may as well assume that M and N are uniform projectives in normal form. Since $J(D)$ is the unique ideal minimal among all semi-prime ideals of D, we get $J(D) \subseteq \|M\| = \|N\|$. Using Lemma 9.8, we have a unit $\bar{u}$ in $D/J(D)$ such that $\bar{u}(A_i/J(D)) = B_i/J(D)$, where A_i and B_i are the ith entries of M and N respectively. Lift $\bar{u}$ to a unit u. Then $uM = N$. ∎

We remark that, in view of Theorem 6.4, all the information obtained above can be transferred to finite dimensional projectives over an arbitrary pseudo-Dedekind ring R and can be utilized to obtain a description of $K_0(R)$.

An example at the end of this section shows that, in general, (3) is insufficient to imply (1) in Theorem 9.10.

Unless indicated otherwise, we shall continue to work in the situation indicated after Theorem 9.1. We now want to determine how finite dimensional projective right T-modules are related if, for each $\lambda \in I_\kappa$, the sequence of p_λ-invariants of M is a cyclic permutation of the sequence of p_λ-invariants of N.

Lemma 9.11. Let M be a finite dimensional projective right T-module. For each $\mu \in I_\kappa$, let $\{ f_{1\mu}, f_{2\mu}, \cdots, f_{\ell_\mu \mu} \}$ be the sequence of p_μ-invariants of M.

(1) Given a $\lambda \in I_\kappa$, let $N = M\mathscr{I}_{p_\lambda}(T)$. Then the sequences of p_μ-

invariants of M and N agree for all $\mu \neq \lambda$. The sequence of p_λ–invariants of N is $\{ f_{2\lambda}, f_{3\lambda}, \cdots, f_{\ell_\lambda \lambda}, f_{1\lambda} \}$.

(2) For any prime p in D and any $\lambda \in I_\kappa$, the sequences of p_λ–invariants of M and $M[(\text{diag } p)T]$ agree.

Proof. (1): Let $1 \leqslant j \leqslant \ell_\lambda$. Since $M/Mm_j(T, p_\lambda)$ is a semi–simple T–module of dimension $f_{j\lambda}$, we have a non–refinable chain $Mm_j(T, p_\lambda) = K_0 \subset K_1 \subset \cdots \subset K_{f_{i\lambda}} = M$ of right T–submodules of M. Since $\mathscr{I}_{p_\lambda}(T)$ is an invertible ideal of T, it is easily seen that $Mm_j(T, p_\lambda)\mathscr{I}_{p_\lambda}(T) = K_0\mathscr{I}_{p_\lambda}(T) \subset K_1\mathscr{I}_{p_\lambda}(T) \subset \cdots \subset K_{f_{i\lambda}}\mathscr{I}_{p_\lambda}(T) = M\mathscr{I}_{p_\lambda}(T)$ is a non–refinable chain of T–submodules of $M\mathscr{I}_{p_\lambda}(T)$. By Theorem 7.2 (3), we have $m_j(T, p_\lambda)\mathscr{I}_{p_\lambda}(T) = \mathscr{I}_{p_\lambda}(T)m_{j-1}(T, p_\lambda)$, the indexing being modulo ℓ_λ. Hence the sequence of p_λ–invariants of N is as claimed.

The remaining part of the lemma is similarly proved. ■

Proposition 9.12. Let D be a PID, $T = \Delta_n(D, t)$ and $\|T\| = p_1 \cdots p_\kappa$ be a prime factorization of $\|T\|$. Let M, N be finite dimensional projective right T–modules. Consider the following statements:

(1) There exists an invertible fractional ideal $\mathscr{I}$ of T such that $M \otimes_T \mathscr{I} \cong N$.

(2) There exists an invertible fractional ideal $\mathscr{I}$ of T such that $M^{(n\kappa+1)} \otimes_T \mathscr{I} \cong N^{(n\kappa+1)}$.

(3) For each $\lambda \in I_\kappa$, the sequence of p_λ–invariants of N is a cyclic permutation of the sequence of p_λ–invariants of M.

Then (1) ⇒ (2) ⇔ (3). If $\dim M > n\kappa$ or if $J(D) \neq (0)$, then (1) ⇔ (3).

Proof. Suppose $k = \dim M > n\kappa$ and that condition (1) holds. In view of Theorem 9.1, we may as well assume that M, N are in standard form. Clearly, $M \otimes_T \mathscr{I} \cong M\mathscr{I}$ for any invertible ideal $\mathscr{I}$ of T. Further, by using Theorem 7.1, we can find an invertible ideal K of D such that $\mathscr{I} = [(\text{diag } K)T]\mathscr{I}_0$ for some invertible ideal $\mathscr{I}_0$. It is now easy to see that a use of Theorems 7.1, 7.2, 9.10 and Lemma 9.11 yields condition (3). The remaining part of the implications (1) ⇒ (2) ⇒ (3) is now trivial.

Suppose $\dim M > n\kappa$ or $J(D) \neq (0)$ and that (3) holds. Using Lemma 9.11, we can choose ϵ_λ, $0 \leqslant \epsilon_\lambda \leqslant \ell_\lambda - 1$, such that N and $M\Pi_{\lambda=1}^{\kappa}[\mathscr{I}_{p_\lambda}(T)]^{\epsilon_\lambda}$ have identical sequences of p_λ–invariants for each $\lambda \in I_\kappa$. Theorem 9.10 shows that $N \cong M\Pi_{\lambda=1}^{\kappa}[\mathscr{I}_{p_\lambda}(T)]^{\epsilon_\lambda}$. It is now easy to complete the proof. ■

The next lemma brings out the connection between Proposition 9.12 and classification of pseudo–Dedekind orders under conjugacy. A related result appears in Knus [22].

Lemma 9.13. Let Q, Λ be semi–simple rings and X be an invertible (Q, Λ)–bimodule. Let T be an arbitrary order in Λ, M and N be essential right T–submodules of X which are also progenerators in mod–T and let $R = \mathbf{o}_\ell(M)$, $S = \mathbf{o}_\ell(N)$. Then R, S are orders in Q. Further, $R \overset{\text{conj}}{\sim} S$ if and only if there exists an invertible fractional T–ideal $\mathcal{J}$ such that $M \otimes_T \mathcal{J} \cong N$ as right T–modules.

Proof. It is well known [16, 20, 36] that R, S are orders in Q.

Let $\mathcal{J}$ be an invertible ideal of T such that $M \otimes_T \mathcal{J} \cong N$. Since $M \otimes_T \mathcal{J} \cong M\mathcal{J}$ as right T–modules, we have a T–isomorphism $f : M\mathcal{J} \to N$. Since X_T is the T–injective hull of $M\mathcal{J}$ and N, f extends uniquely to a Λ–isomorphism $\hat{f} : X \to X$ (cf. Lemma 1.1). Thus, $\hat{f}$ is induced by a unit u of Q such that $uM\mathcal{J} = N$. Clearly, $\mathbf{o}_\ell(M) = \mathbf{o}_\ell(M\mathcal{J})$. Consequently, $u^{-1}Su = R$.

Conversely, suppose v is a unit in Q such that $v^{-1}Sv = R$. Let $K = v^{-1}N$ so that $\mathbf{o}_\ell(K) = R$. Let $X^* = \operatorname{Hom}_\Lambda(X, \Lambda)$ and $M^* = \{ g \in X^* \mid gM \subseteq T \}$. By the Morita theorems, M^* is an invertible (T, R)–bimodule; so, $M^* \otimes_R K \cong M^*K \subseteq T$ and $\mathcal{J} = M^*K$ is an invertible fractional ideal of T. Clearly, $M\mathcal{J} = MM^*K = RK \cong N$ as right T–modules. ■

The following theorem contains our results concerning classification of pseudo–Dedekind orders under conjugacy.

Theorem 9.14. Let D be a PID with quotient skew field F and $Q = M_n(F)$. Let R, S be equivalent pseudo–Dedekind orders in Q with D as a subbasic ring. Let $\mathbf{p}_1, \cdots, \mathbf{p}_\kappa$ be all the distinct primes in D which are associated with the cycles of maximal idempotent ideals of R or S. Let $m = \max\{ \ell_{\mathbf{p}_\lambda}(R), \ell_{\mathbf{p}_\lambda}(S) : \lambda \in I_\kappa \}$. Consider the following statements:

(1) R and S are conjugate in Q.

(2) $M_f(R)$ and $M_f(S)$ are conjugate in $M_f(Q)$ whenever $nf > m\kappa$.

(3) The $\mathbf{p}$–invariants of R and S agree for all primes $\mathbf{p}$ in D.

Then (1) $\Rightarrow$ (2) $\Leftrightarrow$ (3). If $n > m\kappa$ or if $J(D) \neq (0)$, then (1) $\Leftrightarrow$ (3).

Proof. It is easy to see that (1) $\Rightarrow$ (2) $\Rightarrow$ (3).

Assume that R and S have the same $\mathbf{p}$–invariants for all primes $\mathbf{p}$ in D. Then, evidently, $\ell_\mathbf{p}(R) = \ell_\mathbf{p}(S)$ for all $\mathbf{p}$. By Theorem 7.6, we have a triangular frame $\mathbf{t}$ of order h, say, with associated ring $T = \Delta_h(D, \mathbf{t})$ and progenerators M, N in mod–T such that M, N are essential right T–submodules of $M_{n\times h}(F)$, $\mathbf{o}_\ell(M) = R$ and $\mathbf{o}_\ell(N) = S$. We claim that we can choose $h = m$. We may clearly change T up to similarity so long as we replace it with an associated ring of a triangular frame of order h in D. In view of Lemma 7.4, we may assume that

$$T = \bigcap_{\lambda=1}^{\kappa} \Delta_h(D, \mathbf{p}_\lambda, \overbrace{1, \cdots, 1}^{(\ell_\lambda - 1)\text{ terms}}, h - \ell_\lambda + 1)$$

where $\ell_\lambda = \ell_{\mathbf{p}_\lambda}(T)$. We now cut off the first $h-m$ rows and columns of T. More precisely, we can choose a triangular frame $\mathbf{a}$ of order m in D and an invertible $(T, \Delta_m(D, \mathbf{a}))$–bisubmodule K of $M_{h\times m}(F)$. Then MK, NK are progenerators in mod–$\Delta_m(D, \mathbf{a})$, they are essential submodules of $M_{n\times m}(F)$ and $\mathbf{o}_\ell(MK) = R$, $\mathbf{o}_\ell(NK) = S$. Hence, we could have taken $T = \Delta_m(D, \mathbf{a})$ to begin with. This proves our claim.

We now assume that $h = m$. Our discussion after Lemma 7.3 shows that, for each $\lambda \in I_\kappa$, the $\mathbf{p}_\lambda$–invariants of R (resp. S) is the set of all cyclic permutations of the sequence of $\mathbf{p}_\lambda$–invariants of M (resp. N). Thus, the sequence of $\mathbf{p}_\lambda$–invariants of M is a cyclic permutation of the sequence of $\mathbf{p}_\lambda$–invariants of N. We now apply Proposition 9.12. Thus, if $\dim R = \dim M = n > m\kappa$ or if $J(D) \neq (0)$ then we have an invertible fractional ideal $\mathcal{J}$ of T such that $M\mathcal{J} \cong N$; by Lemma 9.13, $R \overset{\text{conj}}{\sim} S$. In the general case, $\dim M_f(R) = \dim M_f(S) = fn$. Thus, if $fn > m\kappa$ then $M_f(R) \overset{\text{conj}}{\sim} M_f(S)$ in $M_f(Q)$. This completes the proof. ∎

We now discuss two examples which point out some of the difficulties in extending the main theorems of this section.

Example 1. We let H stand for the ring of Hurwitz integral quaternions. Recall that H is a $\mathbf{Z}$–order in the skew field $\mathbf{K}$ of rational quaternions; the elements of H are of the form $m_0\xi + m_1 i + m_2 j + m_3 k$ where $1, i, j, k$ are the usual quaternions, $\xi = (1 + i + j + k)/2$ and all the m's are rational integers. It is well known that H is a PID (in fact, a non–commutative Euclidean domain). If p is an odd rational prime, then $H/pH \cong M_2(Z_p)$ (cf. [35]); so, p is a reducible element of H and any factorization of p contains precisely two irreducible elements. Indeed, if x is an irreducible factor of p then $p = x\bar{x}$, where $\bar{x}$ is the conjugate of x in $\mathbf{K}$; also, xH and Hx are respectively maximal right and left ideals of H with bound pH (cf. [18]). Apart from the prime ideals of the form pH, p an odd rational prime, the only other (non–zero) prime ideal of H is $\mathbf{p}_0 = (1 + i)H$; clearly, $\mathbf{p}_0^2 = 2H$.

We choose a fixed rational prime $p > 48$. Let $\mathbf{p} = pH$ and consider the ring

$$T = \begin{pmatrix} H & H \\ p & H \end{pmatrix}.$$

By Proposition 2.5, T is an almost Dedekind $\mathbf{Z}$–order in $M_2(K)$. Let x be an irreducible factor of p; so, $p = x\bar{x}$. Let $M = \begin{pmatrix} xH & H \\ 0 & 0 \end{pmatrix}$. Evidently, M is a uniform right T–projective module. Also, since $x, \bar{x}$ are irreducible elements and $pH \subseteq xH$, $pH \subseteq \bar{x}H$, we have $H = HxH = H\bar{x}H$. Now,

$$\begin{pmatrix} H & 0 \\ H\bar{x} & 0 \end{pmatrix} \begin{pmatrix} xH & H \\ 0 & 0 \end{pmatrix} = \begin{pmatrix} HxH & H \\ H\bar{x}xH & H\bar{x}H \end{pmatrix} = T$$

shows that M is a progenerator in mod–T.

Since $HxH = H$, we can choose $d \in H$ such that $dxH \not\subseteq xH$; so $dxH + xH = H$. Since $dxH \cap xH \subseteq dH$, we have $f \in H$ with $dxH \cap xH = dfH$. Since $dx\bar{x} = dp = pd = x\bar{x}d \in dxH \cap xH$, we get $pH \subseteq fH$. If $fH = H$, then $d \in dxH$, i.e., $H = xH$, contrary to our choice of x. If $pH \subsetneq fH$, then f is an irreducible element of H. Now, $dxH \supseteq dfH$ yields $xH \supseteq fH$; so, $xH = fH$. Consequently, $xH \cap dxH = dfH = dxH$ which means $xH \supseteq dxH$, contrary to our choice of d. Hence $dxH \cap xH = pH$. Now, the usual split exact sequence argument shows that

$$M_T^{(2)} \cong \begin{pmatrix} xH & H \\ xH & H \end{pmatrix} \cong \begin{pmatrix} xH & H \\ dxH & dH \end{pmatrix}$$

$$\cong \begin{pmatrix} dxH + xH & H + dH \\ dxH \cap xH & H \cap dH \end{pmatrix} \cong T_T.$$

Let y be any irreducible factor of p in H ($x = y$ is permitted). Let $N = \begin{pmatrix} yH & H \\ 0 & 0 \end{pmatrix}$. Since x and y are similar elements of H (cf. [18, page 33]), there exists an element $h \in H$ such that $hxH = hH \cap yH$ and $H = hH + yH$. Then, as right T–modules, we have

$$\begin{pmatrix} H & H \\ yH & H \end{pmatrix} \cong \begin{pmatrix} hH & hH \\ yH & H \end{pmatrix} \cong \begin{pmatrix} hH + yH & hH + H \\ hH \cap yH & hH \cap H \end{pmatrix}$$

$$\cong \begin{pmatrix} H & H \\ hxH & hH \end{pmatrix} \cong \begin{pmatrix} H & H \\ xH & H \end{pmatrix}.$$

It is now immediate that $M^{(n)} \cong N^{(n)}$ for all $n > 1$. (After we appropriately choose x and y, this will show that the normal form of Theorem 9.1 is not necessarily unique).

Treat $M_{1\times 2}(\mathbf{K})$ naturally as a $(\mathbf{K},T)$–bimodule. Then M, N are essential right T–submodules of $M_{1\times 2}(\mathbf{K})$ and progenerators in mod–T. Let $R = \mathbf{o}_\ell(M)$, $S = \mathbf{o}_\ell(N)$. By Theorem 6.4, R and S are pseudo–Dedekind $\mathbf{Z}$–orders in $\mathbf{K}$. They are equivalent to H and are properly contained in H (e.g., $R = H \cap xHx^{-1}$). So, R and S are not Dedekind; being domains, they can not be almost Dedekind either. Since $M^{(n)} \cong N^{(n)}$, Lemma 9.13 shows that $M_n(R)$ and $M_n(S)$ are conjugate in $M_n(\mathbf{K})$ for all $n > 1$ and that $R \overset{\text{conj}}{\underset{\mathbf{K}}{\approx}} S$ if and only if $M \cong N\mathscr{I}$ for some invertible fractional ideal $\mathscr{I}$ of T.

We proceed to show how x, y can be chosen such that $M \not\cong N\mathscr{I}$ for any invertible fractional ideal $\mathscr{I}$ of T. Since $\mathfrak{p}_0 = (1+i)H = H(1+i)$, for a given $y \in H$, we have $y_0 \in H$ such that $y(1+i) = (1+i)y_0$. For the moment, assume that x, y are chosen to satisfy the following condition:

(*) x cannot be expressed as uyv or uy_0v for any choice of units u,v in H.

Let q be an odd rational prime, $q \neq p$. Let $\mathfrak{g} = qH$ so $\mathscr{I}_\mathfrak{g}(T) = \begin{pmatrix} qH & qH \\ pqH & qH \end{pmatrix}$ and $N\mathscr{I}_\mathfrak{g}(T) = \begin{pmatrix} qyH & qH \\ 0 & 0 \end{pmatrix} \cong N$ as right T–modules. Similar arguments show that $N \cong N[\mathscr{I}_\mathfrak{g}(T)]^{-1} \cong N[\mathscr{I}_{\mathfrak{p}_0}(T)]^{2k} \cong N[\mathscr{I}_\mathfrak{p}(T)]^{2k}$ for all $k \in \mathbf{Z}$. Thus, if $M \cong N\mathscr{I}$ for some invertible fractional ideal $\mathscr{I}$ of T, then using Theorem 7.1 and the above observation, we may assume that $\mathscr{I} = [\mathscr{I}_{\mathfrak{p}_0}(T)]^{\epsilon}[\mathscr{I}_\mathfrak{p}(T)]^{\delta}$, $0 \leqslant \epsilon,\delta \leqslant 1$.

Suppose $\epsilon = \delta = 1$. We then have an element $\alpha \in \mathbf{K}$ such that

$$\begin{pmatrix} xH & H \\ 0 & 0 \end{pmatrix} = \begin{pmatrix} \alpha & 0 \\ 0 & 0 \end{pmatrix}\begin{pmatrix} yH & H \\ 0 & 0 \end{pmatrix}\begin{pmatrix} pH & H \\ pH & pH \end{pmatrix}\begin{pmatrix} 1+i & 0 \\ 0 & 1+i \end{pmatrix}$$

$$= \begin{pmatrix} \alpha p(1+i)H & \alpha y(1+i)H \\ 0 & 0 \end{pmatrix}.$$

Thus $\alpha y(1+i)$ is a unit in H which makes y a unit in H, contrary to our choice; $\epsilon = 0$ and $\delta = 1$ lead to a similar contradiction. If $\epsilon = 1$ and $\delta = 0$, then we have $\begin{pmatrix} xH & H \\ 0 & 0 \end{pmatrix} = \begin{pmatrix} \alpha y(1+i)H & \alpha(1+i)H \\ 0 & 0 \end{pmatrix}$. Then $u = \alpha(1+i)$ is a unit in H and $x = \alpha y(1+i)v = uy_0 v$ for some unit v in H; this contradicts (*); $\epsilon = \delta = 0$ leads to a similar contradiction.

It remains to show that we can choose irreducible factors x, y of p satisfying (*). Let y be any irreducible divisor of p. The condition (*) certainly holds if xH/pH is not isomorphic with uyH/pH or uy_0H/pH for any unit u in H. Since $H/pH \cong M_2(\mathbf{Z}_p)$ and since H has only 24 units, at most 48 minimal right ideals of $M_2(\mathbf{Z}_p)$ are excluded as permissible images of xH. Since $\mathbf{Z}_p^{(2)}$ has $p+1 > 48$ one dimensional sub–spaces and these are in bijection with the minimal right ideals of $M_2(\mathbf{Z}_p)$, we have an element x as we want.

We conclude that R and S are not conjugate in $\mathbf{K}$ although $M_n(R)$ and $M_n(S)$ are conjugate in $M_n(\mathbf{K})$ for all $n > 1$. Using the Skolem–Noether theorem, it can be seen that R and S are non–isomorphic as rings.

We also conclude that $M \ncong N\mathcal{J}$ for any invertible ideal $\mathcal{J}$ of T although $M^{(n)} \cong N^{(n)}$ for all $n > 1$.

Hereditary $\mathbf{Z}$–orders in $\mathbf{K}$ have been considered by Harada [15]; his results may be used to show that $R \ncong S$ in the above example.

The following example is essentially in Knus [22].

Example 2. Let D be a commutative Dedekind domain with quotient field F and finite class number $n > 1$. Let X be a non–principal ideal of D and s be the least positive integer such that X^s is principal. Let $M = D^{(n)}$, $N = D^{(n-1)} \oplus X$, $Q = M_n(F)$. Treat M, N as D–submodules of the (Q, D)–bimodules $M_{n\times 1}(F)$. Let $R = \mathbf{o}_\ell(M)$, $S = \mathbf{o}_\ell(N)$. By Lemma 9.13, $M_k(R)$ and $M_k(S)$ are conjugate in $M_k(Q)$ if and only if $M^{(k)} \cong N^{(k)}\mathcal{J}$ for some fractional ideal $\mathcal{J}$ of D. Now, $M^{(k)} = D^{(nk)}$ and $N^{(k)}\mathcal{J} \cong D^{nk-1} \oplus X^k\mathcal{J}^{nk} \cong D^{nk-1} \oplus X^k$. Thus, $M^{(k)} \cong N^{(k)}\mathcal{J}$ if and only if $X^k \cong D$ if and only if $s \mid k$.

This shows that if we allow D to be an arbitrary Dedekind domain in Theorem 9.14, then $\mathbf{p}$–invariants are insufficient to imply ‘eventual conjugacy’. Of course the remedy in this example is to take the class of X as an additional invariant. We do not know whether any such trick will extend the theorem to a significantly large class of Dedekind domains.

The same example shows that formally conjugate almost Dedekind rings need not be conjugate.

References

1. E. Artin, C. J. Nesbitt and R. M. Thrall, “Rings with Minimum Condition”, University of Michigan Press, Ann Arbor, 1944.
2. M. Auslander and O. Goldman, Maximal orders, Trans. Amer. Math. Soc. **97**

(1960), 1-24.

3. H. Bass, "Algebraic K–Theory", Math. Lecture Notes Series, Benjamin, New York, 1968.
4. A. Brumer, The structure of hereditary orders, thesis (unpublished) Princeton University, 1963.
5. A. Brumer, Structure of hereditary orders, Bull. Amer. Math. Soc. **69** (1963), 721-724; Addendum, *ibid.* **70** (1964), 185.
6. Ju. A. Drozd, V. V. Kiricenko and A. V. Roiter, On hereditary and Bass orders, Math. U.S.S.R. Izv. **1** (1967), 1357-1376.
7. D. Eisenbud and J. C. Robson, Modules over Dedekind prime rings, J. Algebra **16** (1970), 67-85.
8. D. Eisenbud and J. C. Robson, Hereditary Noetherian prime rings, J. Algebra **16** (1970), 86-104.
9. C. Faith, A general Wedderburn theorem, Bull. Amer. Math. Soc. **73** (1967), 65-67.
10. A. W. Goldie, "Lectures on Non–commutative Noetherian Rings", Canadian Math. Congress, York University, Toronto, 1967.
11. A. W. Goldie, Some aspects of ring theory, Bull. London Math. Soc. **1** (1969), 129-154.
12. M. Harada, Hereditary orders, Trans. Amer. Math. Soc. **107** (1963), 273-290.
13. M. Harada, Structure of hereditary orders over local rings, J. Math. Osaka City Univ. **14** (1963), 1-22.
14. M. Harada, Multiplicative ideal theory in hereditary orders, J. Math. Osaka City Univ. **14** (1963), 83-106.
15. M. Harada, Hereditary orders in generalized quaternions D_T, J. Math. Osaka City Univ. **14** (1963), 71-81.
16. R. Hart, Endomorphisms of modules over semi–prime rings, J. Algebra **4** (1966), 46-51.
17. H. Jacobinski, Two remarks about hereditary orders (to appear).
18. N. Jacobson, "The Theory of Rings", Amer. Math. Soc. Surveys VI, New York, 1943.
19. N. Jacobson, "Structure of Rings", Amer. Math. Soc. Colloq. Publ., **37**, 1964.
20. A. V. Jategaonkar, Endomorphism rings of torsionless modules, Trans. Amer. Math. Soc. (to appear).
21. R. E. Johnson, Prime matrix rings, Proc. Amer. Math. Soc. **16** (1965), 1099-1105.
22. M. A. Knus, Algèbres d'Azumaya et modules projectifs, Commentarii Math. Helv. **45** (1970), 372-383.
23. J. Kuzmanovich, Localizations of Dedekind prime rings (to appear).
24. J. Lambek, "Lectures on Rings and Modules", Blaisdell, London, 1966.
25. L. Levy, Torsion–free and divisible modules over non–integral domains, Canad. J. Math. **15** (1963), 132-151.

26. G. Michler, Structure of semi–perfect hereditary Noetherian rings, J. Algebra **13** (1969), 327-344.
27. B. J. Mueller, On semi–perfect rings, Illinois J. Math. **14** (1970), 464-467.
28. J. A. Murtha, Hereditary orders over principal ideal domains, thesis (unpublished), University of Wisconsin, 1964.
29. I. Reiner, A survey of integral representation theory, Bull. Amer. Math. Soc. **76** (1970), 159-227.
30. J. C. Robson, Noncommutative Dedekind rings, J. Algebra **9** (1968), 247-265.
31. J. C. Robson, A note on Dedekind prime rings (to appear).
32. J. C. Robson, Idealizers and hereditary Noetherian prime rings (to appear).
33. K. W. Roggenkamp,"Lattices Over Orders, II", Lecture Notes in Math. **142** , Springer–Verlag, Berlin–Heidelberg, New York, 1970.
34. F. L. Sandomierski, A note on the global dimension of subrings, Proc. Amer. Math. Soc. **23** (1969), 478-480.
35. B. L. van der Waerden, "Modern Algebra, II", Ungar, New York, 1950.
36. J. M. Zelmanowitz, Endomorphism rings of torsionless modules, J. Algebra **5** (1967), 325-341.

ON THE REPRESENTATION OF MODULES

BY SHEAVES OF MODULES OF QUOTIENTS

J. Lambek

McGill University

Elsewhere we have shown how a module M_R over an associative ring R with unity may be represented by a sheaf of certain factor modules of M_R. We now obtain a similar theorem, where the stalks are modules of quotients of M_R with respect to prime ideals of the center of R. This result, in the commutative case, is due to Grothendieck, and our argument differs from his only in minor points. However it is intended to be accessible to ring theorists who are afraid to delve into a tome on algebraic geometry.

Given an associative ring R with unity, we let $\Pi(R)$ be the set of prime ideals of the center $C(R)$ of R, made into a compact topological space by the usual Stone–Zariski topology: Open sets of $\Pi(R)$ have the form $\Gamma(A) = \{ P \in \Pi(R) \mid A \not\subseteq P\}$, where A is any subset of $C(R)$. Without loss in generality A may be taken to be an ideal of $C(R)$. Regrettably there is no obvious way in which Π may be made into a functor from the category of rings to the category of topological spaces.

If Δ is a multiplicative subset of R, it is well known that one obtains a torsion theory by declaring that a module B_R is torsion if and only if

$$\forall_{b \in B} \exists_{\delta \in \Delta} \; b\delta = 0.$$

In particular, for a right ideal D of R, R_R/D is torsion if and only if

$$\forall_{r \in R} \; r^{-1}D \cap \Delta \neq \phi.$$

When Δ is in the center C of R, this simplifies to

$$D \cap \Delta \neq \phi.$$

In particular, when $\Delta = C - P$, P being a prime ideal of the center, this amounts to

$$\exists_{c \in C-P} \; c \in D.$$

From now on we consider the torsion theory determined by $C - P$.

For any module M_R we have the torsion submodule $T_P(M_R)$ and the module of quotients $Q_P(M_R)$, the latter being equipped with a canonical homomorphism $h_P(M) : M \to Q_P(M_R)$ with kernel $T_P(M_R)$. In particular, $T_P(R_R) = 0_P$ is an ideal, the so-called P–component of 0, $Q_P(R_R)$ is a ring, the so-called ring of quotients of R with respect to the multiplicative set $C - P$, and $h_P(R_R)$ is a ring homomorphism.

Following Grothendieck, we wish to study the sheaf of rings

$$\Sigma(R) = \mathop{\dot{\bigcup}}_{P \in \Pi} Q_P(R)$$

and the sheaf of modules

$$\Sigma(M_R) = \mathop{\dot{\bigcup}}_{P \in \Pi} Q_P(M_R) .$$

We note that for each $m \in M$ there is a section $\hat{m} : \Pi \to \Sigma$ defined by $\hat{m}(P) = (P, h_P(m))$, where we write h_P for $h_P(M) : M \to Q_P(M)$. It is easily seen that the correspondence $m \to \hat{m}$ is one–to–one; we shall see that it is also onto.

Lemma. The section $\hat{m}$ vanishes on the basic open set $\Gamma(c)$, $c \in C(R)$, if and only if there exists a natural number n such that $mc^n = 0$.

Proof. To say that $\hat{m}(P) = 0$ for all $P \in \Gamma(c)$ means that

$$\forall_{P \in \Pi}(c \notin P \Rightarrow m^{-1}\ 0 \cap C \not\subseteq P),$$

that is to say, c belongs to the radical of $m^{-1}\,0 \cap C$ in C, that is, $c^n \in m^{-1}\,0 \cap C$ for some natural number n.

Theorem. An associative ring R is isomorphic to the ring of sections of the sheaf of rings $Q_P(R)$, where P ranges over the prime ideals of the center of R. A module M_R is isomorphic to the module of sections of the sheaf of modules $Q_P(M)$.

Proof. We present the crucial part of the argument as follows. Assume $f : \Pi \to \Sigma(M)$ is a section of the sheaf of modules. Then, for each $P \in \Pi$, $f(P) \in Q_P(M)$, hence

$$f(P)h_P(c_P) = h_P(m_P),$$

where $c_P \in C{-}P$, $m_P \in M$, and $h_P : M \to Q_P(M)$ is the canonical homomorphism.

The above equation may be written

$$(\widehat{fc_P})(P) = \hat{m}_P(P),$$

hence $\widehat{fc_P}$ and $\hat{m}_P$ also agree on a basic open neighborhood of P, say on $\Gamma(d_P)$, where $d_P \in C$. A finite number of these open sets will cover the compact space Π, say the neighborhoods of $P_1,\cdots,P_k$. We shall replace the subscript P_i by i, thus

$$(\widehat{fc_i} - \hat{m}_i)\,\Gamma(c_i d_i) = 0,$$

for $i = 1,\cdots,k$. Take any $j = 1,\cdots,k$, then $\widehat{fc_i}\hat{c}_j$ agrees with $\hat{m}_i\hat{c}_j$ on $\Gamma(c_i d_i)$ and with $\hat{m}_j\hat{c}_i$ on $\Gamma(c_j d_j)$, hence

$$(\hat{m}_i\hat{c}_j - \hat{m}_j\hat{c}_i)\,\Gamma(c_i d_i c_j d_j) = 0.$$

Now, by the lemma, $\hat{m}(\Gamma(c)) = 0$ if and only if there exists a natural number n such that $mc^n = 0$. In the present case, we obtain a natural number $n(i,j)$ such that

$$(m_i c_j - m_j c_i)(c_i d_i c_j d_j)^{n(i,j)} = 0.$$

Let $n = \max\{\, n(i,j) \mid i,j = 1,\cdots,k \,\}$. Then

$$m_i c_i^n d_i^n c_j^{n+1} d_j^n = m_j c_i^{n+1} d_i^n c_j^n d_j^n.$$

Since the $\Gamma(c_i d_i) = \Gamma(c_i^{n+1} d_i^{n+1})$ cover Π, there exist elements $t_i \in R$ such that

$$1 = \sum_{i=1}^{k} c_i^{n+1} d_i^{n+1} t_i.$$

Put

$$m = \sum_{i=1}^{k} m_i c_i^n d_i^n t_i,$$

then we have

$$\begin{aligned} mc_j^{n+1} d_j^n &= \sum_{i=1}^{k} m_j c_i^{n+1} d_i^n c_j^n d_j^n t_i \\ &= m_j c_j^n d_j^n. \end{aligned}$$

Thus $\hat{m}\hat{c}_j$ agrees with $\hat{m}_j$ on $\Gamma(c_j d_j)$. But $\widehat{fc_j}$ also agrees with $\hat{m}_j$ on $\Gamma(c_j d_j)$, hence $(\widehat{fc_j})(P) = (\hat{m}\hat{c}_j)(P)$ for all P such that $c_j d_j \notin P$. Thus $f(P)c_j = \hat{m}(P)c_j$, hence $f(P) = \hat{m}(P)$, because $Q_P(M)$ is torsionfree and $c_j \in C-P$. Since the $\Gamma(c_j d_j)$ cover Π, we conclude that $f = \hat{m}$.

References

1. A. Grothendieck and J. Dieudonné, Eléments de géométrie algébrique I, I.HE.S. Publications math. **4**, (1960).
2. J. Lambek, On the representation of modules by sheaves of factor modules, Canadian Mathematical Bulletin, (to appear).

SOME REMARKS ON RINGS WITH SOLVABLE UNITS

Charles Lanski

University of Southern California

The condition that a ring have a solvable group of units does not in general seem to yield much further information about the ring. Indeed, many diverse examples can be given in which the group of units is abelian. However, if one assumes that there is a "reasonable" richness of units, then additional information can be obtained. With such assumptions there seems to be a relation between solvability and commutativity related conditions on the ring. For example, in the case of an Artinian ring R with a solvable group, the result of Eldridge [1,Theorem 1] on the structure of R modulo its radical shows that R must satisfy a polynomial identity, in fact, some power of the standard identity of degree four. The results in [4] show that for a semi-prime ring, any solvable normal subgroup of units must commute elementwise with all idempotents.

Our first goal is to show that solvability implies the local nilpotence of nil subrings. R will always denote a ring with identity and U(R) its group of units. For any ring $S, \mathcal{L}(S)$ will denote the locally nilpotent radical of S. We recall that if $\mathcal{L}(S) = 0$ then S contains no nonzero one sided ideals which are locally nilpotent.

Lemma 1. Let N be a nonzero nil ring with $\mathcal{L}(N) = 0$. If $k > 1$ is an integer then there exist $a_1, a_2, \cdots, a_k \in N$ with $a_j a_i = 0$ for $j \geqslant i$ and $a_1 a_2 \cdots a_k \neq 0$.

Proof. N cannot be of bounded index since otherwise $\mathcal{L}(N) \neq 0$ by Levitzki's Theorem [2, p. 1]. Hence there is an $x \in N$ with $x^n = 0$, $x^{n-1} \neq 0$, and $n > 4k$. If $2 \leqslant t < n/2$, then $x^t N x^{n-t} \neq 0$, for otherwise $(Nx^{n-t})^2 = 0$ and since $\mathcal{L}(N) = 0$ we would have $Nx^{n-t} = 0$. But this implies $x^{n-1} = 0$, a contradiction. Consider now the nonzero sets $A_0 = \{ x^{n-2k} \}$ and $A_i = x^{2(k-i)} N x^{n-2(k-i)}$ for $1 \leqslant i \leqslant k-1$. Clearly $A_0 A_0 = 0$, and for $i \geqslant 1$, $A_i A_0 = x^{2(k-i)} N x^{n-2(k-i)} x^{n-2k} = 0$ since $n-2(k-i) + n-2k = 2n - 4k + 2i > 2n - 4k > n$. Also, if $j \geqslant i$, then $A_j A_i = x^{2(k-j)} N x^{n-2(k-j)} x^{2(k-i)} N x^{n-2(k-i)} = 0$ because $n - 2(k-j) + 2(k-i) = n + 2(j-i) \geqslant n$. If $A_0 A_1 \cdots A_{k-1} = 0$ then we would have $(x^{n-2}N)^k = 0$ and so $x^{n-2}N = 0$, since $\mathcal{L}(N) = 0$. But again this implies that $x^{n-1} = 0$. Hence $A_0A_1 \cdots A_{k-1} \neq 0$. Let $a_1 = x^{n-2k}$ and pick $a_i \in A_{i-1}$ so that $a_1a_2 \cdots a_k \neq 0$. These a_i are the required elements.

If N is a nil ring then by N° we will mean the quasi regular group of N; that is, N considered as a group under the operation $a \circ b = a + b + ab$. We note that if N is a subring of a ring with identity, then N° is isomorphic to the group of elements of the form $1 + a$ where $a \in N$.

Theorem 2. Let N be a nil ring. If N° is a solvable group, then $N = \mathcal{L}(N)$.

Proof. If I is an ideal of N then the natural homomorphism of N onto N/I induces a homomorphism of N° onto $(N/I)^\circ$ with kernel I°. Hence $(N/I)^\circ$ is a homomorphic image of N°, so is solvable. In particular $\bar{N}^\circ = (N/\mathcal{L}(N))^\circ$ is solvable, and furthermore, $\mathcal{L}(\bar{N}) = 0$. Hence we can assume to begin with that $\mathcal{L}(N) = 0$.

Now if N is not zero, let k be the length of the derived series for N°. That is, $N^\circ \supset (N^\circ)' \supset \cdots \supset (N^\circ)^{(k-1)} \supset 0$. Note that if $a^2 = 0$ then $a \circ (-a) = (-a) \circ a = 0$, and so $a^2 = b^2 = ba = 0$ implies that $a \circ b \circ (-a) \circ (-b) = ab$. Since $\mathcal{L}(N) = 0$, by Lemma 1 there are $a_1, a_2, \cdots, a_{2^k}$ with $a_ja_i = 0$ if $j \geqslant i$ and the product $a_1a_2 \cdots a_{2^k} \neq 0$. Now $a_ia_{i+1} = a_i \circ a_{i+1} \circ (-a_i) \circ (a_{i+1}) \in (N^\circ)'$. Similarly $a_ia_{i+1}a_{i+2}a_{i+3} = (a_ia_{i+1}) \circ (a_{i+2}a_{i+3}) \circ (-a_ia_{i+1}) \circ (-a_{i+2}a_{i+3}) \in (N^\circ)^{(2)}$. Continuing we obtain $a_1a_2 \cdots a_{2^k} \in (N^\circ)^{(k)} = 0$, contradicting the choice of $a_1, \cdots, a_{2^k}$. Hence $N = 0$ and the theorem is proved.

Denote by C_1 the class of nilpotent rings, by C_2 the class of nil rings with solvable quasi-regular group, and by C_3 the class of locally nilpotent rings.

Corollary 3. $C_1 \subset C_2 \subset C_3$ and all inclusions are proper.

Proof. If N is a nil ring then $(N^\circ)^k \subset N^{2^k}$. Hence if $N^s = 0$ and k is minimal with $s \leqslant 2^k$, then N° has derived series of length at most k, so is solvable. That $C_2 \subset C_3$ is the statement of Theorem 2.

Consider the ring N generated by $\{ x_i \}$ with $x_jx_i = 0$ if $j \geqslant i$. This well known example of Sadiada is a locally nilpotent ring, but since $x_1x_2 \cdots x_k \neq 0$ for all k, the proof of Theorem 2 shows that N° cannot be solvable. If K is any commutative nil ring which is not nilpotent, then $(K^\circ)' = 0$, so $K \in C_2$ but $K \notin C_1$.

Corollary 4. If R is a ring with $U(R)$ solvable, then every nil subring of R is locally nilpotent.

Proof. If K is a nil subring of R then the group of units of the form $1 + k$ where $k \in K$ is solvable since it is a subgroup of $U(R)$. As noted above, this group is isomorphic to K°. By Theorem 2, K is locally nilpotent.

We now note some easy consequences of the result mentioned above for semi-prime rings. For convenience we state a special useful case of this result [4, Theorem 9].

Theorem. Let R be semi-prime and 6-torsion-free. If $U(R)$ is solvable, then every idempotent of R centralizes $U(R)$.

Lemma 5. Let R be semi-prime and 6-torsion-free. If $U(R)$ is solvable, then all idempotents of R are central.

Proof. This is immediate from the Theorem since if e is a noncentral idempotent it must commute with the units $1 + es(1-e)$ and $1 + (1-e)se$ for $s \in R$. But then e must be central.

It will also be convenient to refer to the following obvious fact.

Lemma 6. If N is a nil ideal of R, then the natural homomorphism of R onto R/N induces a homomorphism of $U(R)$ onto $U(R/N)$.

Lemma 7. If R is n-torsion-free and $U(R)$ is solvable, then $R/\mathcal{L}(R)$ is n-torsion-free.

Proof. By Lemma 6, $U(R/\mathcal{L}(R))$ is solvable, so any nil ideal in $R/\mathcal{L}(R)$ is locally nilpotent by Corollary 4. But $\mathcal{L}(R/\mathcal{L}(R)) = 0$, so any nil ideal of $R/\mathcal{L}(R)$ is zero. Let $\overline{T} = \{ x \in R/\mathcal{L}(R) \mid nx = 0 \}$. $\overline{T}$ is an ideal of $R/\mathcal{L}(R)$ and corresponds to an ideal T of R. If $y \in T$ then $ny \in \mathcal{L}(R)$ so $(ny)^k = 0$. Thus $n^k y^k = 0$ and since R is n-torsion-free, $y^k = 0$. Thus T is a nil ideal of R, which implies that $\overline{T} = 0$.

Putting these observations together yields

Lemma 8. Let R be 6-torsion-free with $U(R)$ solvable. If E is the additive subgroup generated by the idempotents of R, then $[E, R]$ is locally nilpotent, where $[E, R]$ is the additive subgroup of R generated by all $er - re$ with $e^2 = e$ and $r \in R$.

Proof. Let $R = R/\mathcal{L}(R)$. Then $\overline{R}$ is semi-prime and 6-torsion-free by Lemma 7 and $U(\overline{R})$ is solvable by Lemma 6. Thus Lemma 5 implies that the idempotents of $\bar{R}$ are central. But then we have $[E, R] \subset \mathcal{L}(R)$, which is locally nilpotent.

Lemma 8 says that $U(R)$ cannot be solvable if R has any idempotents which are not central in $R/\mathcal{L}(R)$. This will be true, for example, if R contains an idempotent e with $eR(1-e)Re = eRe$. We call such an idempotent a *semi-connecting idempotent* of R. If $U(R)$ were solvable in the presence of such an idempotent then by Lemma 8, $[E,R]$ would be locally nilpotent. Now $eR(1-e) = [e,R(1-e)] \subset \mathcal{L}(R)$. Since e is semi-connecting we would have $eR(1-e)\cdot(1-e)Re = eR(1-e)Re = eRe$ contained in $\mathcal{L}(R)$. But $e \in eRe$ and is not nilpotent. Hence $U(R)$ could not be solvable. We state this as

Theorem 9. If R is 6–torsion–free with semi–connecting idempotent e, then $U(R)$ is not solvable.

Corollary 10. Let R be a 6–torsion–free ring and R_n the $n \times n$ matrix ring over R. If $n \geqslant 2$ then $U(R_n)$ is not solvable.

Proof. Since any of the standard matrix unit idempotents is semi–connecting, the result follows from Theorem 9.

If R is an algebra generated by its idempotents, then prime ideals of R are maximal if $U(R)$ is solvable. In fact we have the following.

Theorem 11. Let R be an algebra over a field F where char $F \neq 2,3$. If R is generated as an algebra by its idempotents and if $U(R)$ is solvable, then for any prime ideal P of R, $R/P \cong F$.

Proof. Let N be the prime radical of R. Then $\bar{R} = R/N$ is semi–prime, 6–torsion–free, and generated by its idempotents. Further, by Lemma 6, $U(\bar{R})$ is solvable and so all idempotents of $\bar{R}$ are central by Lemma 5. Thus $[R, R] \subset N$. If P is any prime ideal of R, then R/P is commutative, so is a domain. Since R/P is generated by its idempotents as an algebra we must have $R/P \cong F$.

We recall that a ring R is called regular if given $x \in R$ there is a $y \in R$ with $xyx = x$. Clearly this condition gives R a richness of units.

Theorem 12. Let R be regular and 6–torsion–free. If $U(R)$ is solvable then R is commutative.

Proof. Since R is semi–prime all idempotents of R are central by Lemma 5. If $x \in R$ is nilpotent and if $xyx = x$, then we have $0 = x^k = x^k y^k =$

$x^{k-1}(xy)y^{k-1} = x^{k-1}y^{k-1}(xy) = (xy)^k = xy$. Thus $xyx = x = 0$.

Let P be any prime ideal of R. If e is any idempotent of R then since $eR(1-e) = 0$ we must have either e or $1-e$ contained in P. Let $\bar{R} = R/P$. Clearly $\bar{R}$ is regular. If $\bar{x} \in \bar{R}$ is nilpotent and x maps to $\bar{x}$, then $x^k \in P$, $x^{k-1} \notin P$ for some $k \geqslant 1$. If $x \notin P$ and $xyx = x$, then $e = yx \notin P$. But $yx^k = ex^{k-1} = x^{k-1} = x^{k-1} \in P$, a contradiction. Hence $\bar{R}$ contains no nonzero nilpotent elements, and so all of its idempotents are central. But $\bar{R}$ is regular and prime so is a division ring.

If $\bar{x} \in \bar{R}$, x maps to $\bar{x}$, and $x \notin P$, then $1-e \in P$ where $e = yx$ and $xyx = x$. Suppose that $t \in R$ and $(x + (1-e))t = 0$. Then $0 = (1-e)xt + (1-e)t = (1-e)t$, since $x = ex$. But then $0 = xt = yxt = et$ and $t = 0$. Since $x + (1-e)$ is not a left zero divisor in R it must be a unit. We can thus conclude that $U(\bar{R})$ is a homomorphic image of U(R), and so, is solvable. But $\bar{R}$ is a division ring so must be commutative [3]. Since R is semi–prime it is a subdirect sum of its prime images, so is commutative.

Theorem 13. Let R be an algebraic algebra over F, with char $F \neq 2,3$. If U(R) is solvable and $\mathscr{L}(R) = 0$, then R is commutative.

Proof. Since $\mathscr{L}(R) = 0$, R is semi–prime. By Lemma 5 all idempotents of R are central. If $x \in R$, $x \neq 0$, and x is nil, then xR cannot be nil, for U(R) solvable says that $xR \subset \mathscr{L}(R)$ by Corollary 5. Since R is algebraic there is a nonzero idempotent $e = xy$ which is central. Since x is nil, $0 = x^k = x^k y^k = e^k$, a contradiction. Thus R has no nonzero nil elements. Since R is algebraic, the minimal polynomial for any $s \in R$ must have the form $x^n + \cdots + a_1 x + a_0$, where not both a_1 and a_0 are zero. Hence R is regular and so commutative by Theorem 12.

Theorem 14. Let R be an algebraic algebra over F, with char $F \neq 2,3$. If U(R) is solvable then R is locally finite.

Proof. $R/\mathscr{L}(R)$ is algebraic over F and $U(R/\mathscr{L}(R))$ is solvable by Lemma 6. Theorem 13 implies that $R/\mathscr{L}(R)$ is commutative. Since $\mathscr{L}(R)$ is locally nilpotent, it is locally finite. Thus R is locally finite.

Corollary 15. Let R be an algebra over F, with char $F \neq 2,3$. If U(R) is solvable then any algebraic subring of R is locally finite.

Proof. If A is an algebraic subring of R, so is the subalgebra A_1 generated by A and 1. Since $U(A_1)$ is a subgroup of U(R), it is solvable, so by Theorem 14, A_1 is locally finite. Clearly, this implies that A is locally finite.

References

1. K. E. Eldridge, On ring structures determined by groups, Proc. Amer. Math. Soc. **23** (1969), 472-477.
2. I. N. Herstein, "Topics in Ring Theory," University of Chicago Press, Chicago, 1969.
3. Loo–Keng Hua, On the multiplicative group of a field, Acad. Sinica Science Reports **3** (1950), 1-6.
4. C. Lanski, Subgroups and conjugates in semi–prime rings, (to appear in Math. Ann.).

QUASI–SIMPLE MODULES AND WEAK TRANSITIVITY

A. C. Mewborn

University of North Carolina at Chapel Hill

1. Introduction

Definition 1.1. Let ${}_RV_D$ be an R,D–bimodule, where D is a division ring, R is a ring and ${}_RV$ is a uniform module. We say that R acts *weakly transitively* in the vector space V_D provided there is a left order S in D and an R,S–submodule U of V such that $UD = V$ and such that if $\{v_1, v_2, \cdots, v_n\}$ is a D–linearly independent subset of U and $\{y_1, y_2, \cdots, y_n\} \subseteq U$ then there exists $r \in R$ and $0 \neq s \in S$ such that $rv_i = y_i s$, $1 \leqslant i \leqslant n$.

In [6] it was shown that if R is a prime ring with a maximal annihilator left ideal and a uniform left ideal U, then R acts weakly transitively in V_D, where V is the quasi–injective hull of U and $D = \mathrm{Hom}_R(V, V)$. Moreover, a ring R has a faithful weakly transitive representation in a vector space V_D such that ${}_RV$ is uniform and R has elements represented by non–zero linear transformations of finite rank if and only if R is prime with maximal annihilator left ideal, V is the quasi–injective hull of a uniform left ideal, and $D = \mathrm{Hom}_R(V, V)$.

In general, if ${}_RV$ is uniform then R acts weakly transitively in V_D if and only if ${}_RV$ is the quasi–injective hull of a *quasi–simple* module (see Definition 2.1) and $D = \mathrm{Hom}_R(V, V)$.

In this paper we show that a module ${}_RV$ is the quasi–injective hull of a quasi–simple module if and only if (1) ${}_RV$ is a minimal quasi–injective module, and (2) $\mathrm{Hom}_R(V, V)$ is a division ring. For a (left) Noetherian ring, Condition (2) is redundant.

If I is a left ideal of a ring then R/I is quasi–simple if and only if I is *almost maximal* as defined in [8] (See §3). Storrer [11] has shown that there is a correspondence between the prime kernel functors of Goldman [4] and the uniform quasi–injective modules ${}_RV$ which are maximal with respect to the property $\mathrm{Hom}_R(V, V)$ is a division ring. If R is Noetherian this leads to an identification of the minimal quasi–injective modules with a subset of the Goldman primes. Whether every Goldman prime corresponds to a minimal quasi–injective reduces to the question whether every uniform module contains a quasi–simple module. This question is open for Noetherian rings. Each almost maximal left ideal of a Noetherian ring is a critical prime left ideal as defined by Goldie [3].

In the last section we prove a general result for prime rings with maximal annihilator left ideal which leads to simple proofs of theorems of Faith and Utumi and Goldie.

2. Quasi-simple modules

Throughout this section R will denote a ring with unity; all modules will be unital and will be left modules unless otherwise specified. If M is a module then $\hat{M}$ will denote the injective hull of M and $\tilde{M}$ the quasi-injective hull of M. The following definition is due to Kwangil Koh [10].

Definition 2.1. A non-zero module M is *quasi-simple* provided (a) each non-zero submodule of M contains an isomorphic copy of M, and (b) $\mathrm{Hom}_R(\tilde{M},\tilde{M})$ is a division ring.

Every simple module is quasi-simple, and every commutative integral domain is quasi-simple as module over itself. Condition (b) of 2.1 can be replaced by: (b′) M is uniform and M contains no homomorphic image of a submodule of M.

If M is quasi-simple then $S = \mathrm{Hom}_R(M, M)$ is a left Öre domain and S is a left order in $D = \mathrm{Hom}_R(\tilde{M}, \tilde{M})$.

If M and N are quasi-simple modules, we say that M *is related to* N, and we write $M \sim N$, if M is isomorphic to a submodule of N. The relation "$\sim$" is an equivalence relation on the class of quasi-simple modules.

Proposition 2.2. Let M be quasi-simple. Then $\tilde{M}$ is the unique minimal, non-zero, quasi-injective submodule of $\hat{M}$, and every non-zero cyclic submodule of $\tilde{M}$ is quasi-simple. $\tilde{M}$ is contained in every non-zero quasi-injective submodule of $\hat{M}$.

Proof. Let $(0) \neq N \subseteq M$. There is a monomorphism of M into N which extends to an automorphism α of $\tilde{M}$. Clearly then $\tilde{N} = \tilde{M}$. If M_1 is a quasi-simple submodule of $\hat{M}$, then $\tilde{M}_1 = \widetilde{M_1 \cap M} = \tilde{M}$. If T is any non-zero quasi-injective submodule of $\hat{M}$, then $\tilde{M} = \widetilde{M \cap T} \subseteq T$.

Now suppose $0 \neq x \in \tilde{M}$. Since $\tilde{M} = MD$, there exist $m \in M$ and $\alpha \in D$ such that $m\alpha = x$. Thus Rm is isomorphic to Rx, and so Rx is quasi-simple. The Proposition is proved.

Corollary 2.3. If M and N are quasi-simple, then $M \sim N$ if and only if $\tilde{M}$ is isomorphic to $\tilde{N}$.

Proposition 2.4. Let $\{M_\lambda\}_{\lambda\in\Lambda}$ be an indexed collection of quasi-simple modules and let $M = \Sigma_{\lambda\in\Lambda} \oplus M_\lambda$. If N is quasi-simple and if there is a monomorphism $\phi : N \to M$, then there exist $(0) \neq N_1 \subseteq N$ and a finite subset $\{\lambda_1,\dots,\lambda_n\}$ of Λ such that $N \sim M_{\lambda_i}$, $1 \leqslant i \leqslant n$, and $\phi(N_1) \subseteq \Sigma_{i=1}^n \oplus M_{\lambda_i}$.

Proof. Let $0 \neq x \in N$, $x\phi = \Sigma_{i=1}^r m_i$, $m_i \in M_{\lambda_i}$. For $1 \leqslant i \leqslant r$, let $K_i = \ker(\phi\pi_i) \cap Rx$, where π_i is the canonical projection of M onto M_{λ_i}. Then $K_i = (0)$, some i, since $\cap_{i=1}^r K_i = (0)$ and N is uniform. Assume $\{\lambda_1,\dots\lambda_r\}$ is ordered so that $K_i = (0)$, $1 \leqslant i \leqslant n$, and $K_i \neq (0)$, $n + 1 \leqslant i \leqslant r$. Let $N_1 = \cap_{i=n+1}^r K_i$, if $n \neq r$, and $N_1 = Rx$, if $n = r$. Then $\phi\pi_i \mid N_1$ is a monomorphism, $1 \leqslant i \leqslant n$, which implies $N \sim M_{\lambda_i}$, and $\phi(N_1) \subseteq \Sigma_{i=1}^n \oplus M_{\lambda_i}$. This completes the proof.

If M is a module and $m \in M$, then $m^\ell = \{r \in R \mid rm = 0\}$ is the *annihilator* of m. Now suppose M is uniform and injective, and let $E = \mathrm{Hom}_R(M, M)$. Then E is local and $\mathrm{Rad}\ E = \{\alpha \in E \mid \ker(\alpha) \neq (0)\}$. If $(0) \neq N \subseteq M$, then $\tilde{N} = NE$. If $m_1, m_2 \in M$, there exists $\alpha \in E$ such that $m_1\alpha = m_2$ if and only if $m_1^\ell \subseteq m_2^\ell$; and $\alpha \in \mathrm{Rad}\ E$ if and only if $m_1^\ell \subset m_2^\ell$. If N is a non-zero quasi-injective submodule of M, then $\mathrm{Hom}_R(N, N) = E/K$. where $K = \{\alpha \in E \mid N\alpha = (0)\}$.

If M is a module we set $a^*(M) = \{m^\ell \mid m \neq 0$ and $m^\ell \not\subset m_1^\ell$, for all $0 \neq m_1 \in M\}$.

Proposition 2.5. Assume M is uniform and injective. Let $N = \{0\} \cup \{m \in M \mid m^\ell \in a^*(M)\}$. Then $N = \cap\{\ker \alpha \mid \alpha \in \mathrm{Rad}\ E\}$ and N is quasi-injective. $\mathrm{Hom}_R(N, N) = E/\mathrm{Rad}\ E$, if $N \neq (0)$.

Proof. If $0 \neq m \in N$ and $\alpha \in \mathrm{Rad}\ E$, then $\ker(\alpha) \cap Rm \neq (0)$. Hence $m^\ell \subset (m\alpha)^\ell$, which implies $m\alpha = 0$. If $0 \neq m \notin N$, then there exists $0 \neq m_1 \in M$ such that $m^\ell \subset m_1^\ell$. Thus there exists $\alpha \in \mathrm{Rad}\ E$ such that $m\alpha = m_1$. This proves $N = \cap\{\ker \alpha \mid \alpha \in \mathrm{Rad}\ E\}$. It is clear that $NE = N$, so that N is quasi-injective. The last statement follows because $\mathrm{Rad}\ E = \{\alpha \in E \mid N\alpha = (0)\}$, if $N \neq (0)$.

Proposition 2.6. A quasi-injective module M is the quasi-injective hull of a quasi-simple module if and only if M is minimal quasi-injective and $D = \mathrm{Hom}_R(M, M)$ is a division ring.

Proof. Suppose M is minimal quasi-injective and D is a division ring. Let $0 \neq x \in M$. Then $M = \widetilde{Rx}$. Let $(0) \neq N \subseteq Rx$. Then $\tilde{N} = M$. Hence

$ND = M$. Thus there exists $\alpha \in D$ and $y \in N$ such that $y\alpha = x$. α induces an isomorphism of Ry onto Rx. So N contains an isomorphic copy of Rx. Hence Rx is quasi–simple. The converse follows from 2.2.

If R is (left) Noetherian and M is uniform and injective then $a^*(M)$ is not empty. Hence M contains a quasi–injective submodule N such that $\mathrm{Hom}_R(N, N)$ is a division ring. This result has also been obtained by A. Hudry [5]. Suppose M contains a non–zero submodule N_1 satisfying Condition (a) of 2.1. Then $N_1 \subseteq N$, so that $\widetilde{N}_1 \subseteq N$. Thus it is easy to see that $\mathrm{Hom}_R(\widetilde{N}_1, \widetilde{N}_1)$ is a division ring. This shows that if R is a Noetherian ring, Condition (b) of Definition 2.1 is a consequence of Condition (a), for a uniform module.

It follows from the above remarks that a quasi–injective module N is the quasi–injective hull of a quasi–simple module if and only if N is minimal quasi–injective, if R is left Noetherian. It is an open question whether this is still true when R is not Noetherian. Another open question is whether, for R Noetherian, every uniform injective module contains a minimal quasi–injective submodule, i.e., whether it contains a quasi–simple submodule.

The following lemma is well–known.

Lemma 2.7. Let M be quasi–injective, and let $D = \mathrm{Hom}_R(M, M)$. If $\{m, m_1, m_2, \cdots, m_n\} \subseteq M$, then $m = \Sigma_{i=1}^n m_i\alpha_i$, $\alpha_i \in D$, if and only if $m^\ell \supseteq \cap_{i=1}^n m_i^\ell$.

Proof. The proof is by induction on n. We have already observed the result for $n = 1$. Hence assume $n > 1$ and the Lemma is true for $n - 1$. If $J = \cap_{i=1}^{n-1} m_i^\ell \subseteq m^\ell$, the result is true by inductive hypothesis. Hence, suppose $J \not\subseteq m^\ell$. Define a map $\bar{\alpha} : Jm_n \to Jm$, by $\bar{\alpha} : am_n \to am$, $a \in J$. Then $\bar{\alpha}$ is well defined because $J \cap m_n^\ell \subseteq m^\ell$. $\bar{\alpha}$ has an extension α in D. Furthermore $J \subseteq (m - m_n\alpha)^\ell$. Hence $m - m_n\alpha = \Sigma_{i=1}^{n-1} m_i\alpha_i$, $\alpha_i \in D$. The result follows.

Proposition 2.8. If R is left and right Noetherian, then every uniform left ideal of R has a quasi–simple submodule.

Proof. Let U be a uniform left ideal, $M = \hat{U}$, $E = \mathrm{Hom}_R(M, M)$, and $N = \cap\{\ker(\alpha) \mid \alpha \in \mathrm{Rad}\, E\}$. Then $N \neq (0)$. Let N_1 be a non–zero submodule of $N \cap U$ with maximal annihilator. We may assume $N_1 = Rx$, $0 \neq x \in U$. We will show that N_1 is quasi–simple. Since R is right Noetherian, it satisfies the minimum condition on annihilator left ideals. Let $(0) \neq K \subseteq N_1$. There is a finite subset $\{x_1, x_2, \cdots, x_n\}$ of K such that $\cap_{i=1}^n x_i^\ell = K^\ell = N_1^\ell$. Hence $\cap_{i=1}^n x_i^\ell \subseteq x^\ell$. By the Lemma, $x \in KE$. Hence $\widetilde{K} = \widetilde{Rx}$. Thus $\widetilde{Rx}$ is minimal quasi–injective; $\mathrm{Hom}_R(\widetilde{Rx}, \widetilde{Rx})$ is a division ring because $\widetilde{Rx} \subseteq N$. Therefore Rx is quasi–simple.

3. The weak radical of a ring

In [8] a left ideal I of R was called *almost maximal* provided

(1) R/I is uniform,

(2) if J_1 and J_2 are left ideals properly containing I, then $N(I) \cap J_1 \cap J_2 \supset I$, and

(3) if $J \supset I$ and $a \in R$ such that $Ia \subseteq J$, then $[J : a] \supset I$.

Here $N(I) = \{ r \in R \mid Ir \subseteq I \}$ is the normalizer of I. If I is a left ideal, then I is almost maximal if and only if R/I is quasi–simple. Furthermore, if M is quasi–simple and $0 \neq m \in M$, then m^{ℓ} is almost maximal. The weak radical $W(R)$ of R is defined as $\cap \{ I \mid I$ is an almost maximal left ideal$\}$. Equivalently, $W(R) = \cap \{ M^{\ell} \mid M$ is quasi–simple$\}$. First, $W(R)$ is an ideal of R which contains every nil one–sided ideal. Also, $W(R) \subseteq \text{Rad } R$, where $\text{Rad } R$ denotes the Jacobson radical, and the containment is sometimes proper. If R is commutative or Noetherian then $W(R)$ is the prime radical. (For details of the above, see [8], [9] and [10].)

The following characterization of $W(R)$ follows easily from the results of §2.

Proposition 3.1. $W(R) = \cap \{ M^{\ell} \mid M$ is a minimal quasi–injective module such that $\text{Hom}_R(M, M)$ is a division ring$\}$.

The following proposition gives a connection between almost maximal left ideals and quasi–simple modules in a Morita context.

Proposition 3.2. Let $(R, {}_RV_S, {}_SW_R, S)$ be a Morita context and let $(\, , \,) : V \otimes_S W \to R$ be one of the bilinear maps associated with the context. Let I be an almost maximal left ideal of R and let $W_0 = \{ w \in W \mid (V, w) \subseteq I \}$. Then W/W_0 is a quasi–simple left S–module, if $W_0 \neq W$.

Proof. (1) W/W_0 *is uniform*: Let $w_1, w_2 \in W$, $w_1, w_2 \notin W_0$. Then $(V, w_i) \not\subseteq I$, $i = 1,2$. Hence $((V, w_1) + I) \cap ((V, w_2) + I) \supsetneqq I$. Let $(v_1, w_1) + a_1 = (v_2, w_2) + a_2 \notin I$. $R((v_1, w_1) + I)$ contains an isomorphic copy of R/I as left R–module. $W_0 \neq W$ implies $(V, W) \not\subseteq I$, which implies $(V, W)R \not\subseteq I$. Hence $(V, W)(R/I) \neq (0)$. Therefore $(V, W)R[(v_1, w_1) + I] \neq (0)$. So $(V, W)(v_1, w_1) \not\subseteq I$. Let $v \in V$, $w \in W$ such that $(v, w)(v_1, w_1) \notin I$. Then $w(v_1, w_1) + wa_1 = w(v_2, w_2) + wa_2$; or $[w, v_1]w_1 + wa_1 = [w, v_2]w_2 + wa_2 \notin W_0$. (Here $[\, , \,] : W \otimes_R V \to S$ is the other bilinear map associated with the context.) But $[w, v_1]w_1 + wa_1 \in S(w_1 + W_0) \cap S(w_2 + W_0)$; hence W/W_0 is uniform.

(2) *Each non–zero submodule of* W/W_0 *contains an isomorphic copy of* W/W_0: Let $w \in W$, $w \notin W_0$. Then $(V, w) \not\subseteq I$, so there exists $v \in V$ such that $(v, w) \in N(I)$ but $(v, w) \notin I$. Let $\phi : W \to W(v, w) = [W, v]w$ be defined by $\phi : w' \to w'(v, w)$, all $w' \in W$. Then $w' \in W_0 \Leftrightarrow (V, w') \subseteq I \Leftrightarrow (V, w')(v, w) \subseteq I \Leftrightarrow (V, w'(v, w)) \subseteq I \Leftrightarrow w'(v, w) \in W_0$. Hence ϕ induces a monomorphism of W/W_0 into $W(v, w)/W(v, w) \cap W_0 \subseteq S(w + W_0)$.

(3) W/W_0 *contains no proper homomorphic image of an S–submodule of* W/W_0: Let $w \in W$, $w \notin W_0$, and $\phi : Sw + W_0/W_0 \to W/W_0$ a nonzero S–map. We wish to show that $\ker(\phi) = (0)$. Assume $\phi(w + W_0) = w' + W_0$. Define $\psi : (V, w) + I/I \to (V, w') + I/I$ by $\psi : (v, w) + I \to (v, w') + I$, $v \in V$. Clearly ψ is an R–homomorphism if it is well defined. If ψ is not well defined, there exists $v \in V$ such that $(v, w) \in I$, $(v, w') \notin I$. But $(v, w') \notin I \Rightarrow R(v, w') \not\subseteq I \Rightarrow$ there exists $r \in R$ such that $(rv, w') \in N(I)$, $(rv, w') \notin I$, $(rv, w) \in I$. Now $(V, w) \not\subseteq I$; hence $I \not\supseteq (V, w)(rv, w') = (V, w(rv, w'))= (V,[w,rv]w')$, which implies $[w,rv]w' \notin W_0$. But $I \supseteq (V,w)(rv,w) = (V,[w, rv]w)$, which implies $[w, rv]w \in W_0$. This is impossible, since $\phi([w, rv]w + W_0) = [w, rv]\phi(w + W_0) = [w, rv]w' + W_0 \neq 0$. This completes the proof.

Corollary 3.3. Assume the Morita context (R, V, W, S) has the property $(V, sW) = (0)$ if and only if $s = 0$. Then if R/I is a faithful left R–module, W/W_0 is a faithful left S–module.

Proof. If $sW \subseteq W_0$, then (V, sW) is an ideal of R contained in I. Hence $(V, sW) = (0)$, which implies $s = 0$.

4. Prime rings with maximal annihilators

Throughout this section R will denote a prime ring with a maximal annihilator left ideal N. The following lemma is well known.

Lemma 4.1. The (left) singular ideal Z of R is zero.

Proof. Let $N = x^\ell$. R prime implies $xZ \neq (0)$, if $Z \neq (0)$. If $z \in Z$ such that $xz \neq 0$, then $z^\ell \cap Rx \neq (0)$ implies $(xz)^\ell \supset x^\ell$, a contradiction of the maximality of x^ℓ. Hence $Z = (0)$.

The above Lemma implies that N is not a large left ideal of R, and we let V be a complement of N. We set $W = N^r$, $S = WV$. R prime implies $W^r = (0)$. Also, the maximality of N implies that $vw \neq 0$ if $0 \neq v \in V$, $0 \neq w \in W$, which in turn implies that S is a ring (perhaps without unity) with no zero divisors.

Proposition 4.2. Assume $\{v_1, v_2, \cdots, v_n\} \subseteq V$ is such that $J_i = \cap_{j \neq i} (v_j)^\ell \not\subseteq (v_i)^\ell$, $1 \leqslant i \leqslant n$. Let $0 \neq w_i \in WJ_i$, $w_i \notin v_i^\ell$. Then $\Sigma_{i=1}^n v_i S$ and $\Sigma_{i=1}^n V w_i$ are direct sums.

Proof. If $\Sigma_{i=1}^n v_i s_i = 0$, then $0 = \Sigma_{i=1}^n w_j v_i s_i = w_j v_j s_j$, which implies $s_j = 0$ since $0 \neq w_j v_j \in S$, $1 \leqslant j \leqslant n$. If $\Sigma_{i=1}^n v_i' w_i = 0$, then $0 = \Sigma_{i=1}^n v_i' w_i v_j = v_j' w_j v_j$, which implies $v_j' = 0$ since $w_j v_j \neq 0$, $1 \leqslant j \leqslant n$. The proof is complete.

Now assume $\{v_i\}_{i=1}^n$ and $\{w_i\}_{i=1}^n$ are as in 4.2. Let $R_1 = \Sigma_{i,j} v_i W w_j$ and define $\phi : R_1 \to M_n(S)$ by $\phi : \Sigma_{i,j} v_i w_{ij} w_j \to (w_{ij} w_j v_j)$. Let $S^{(i)} = W w_i v_i$, $1 \leqslant i \leqslant n$. Then $S^{(i)}$ is a left ideal of S.

Proposition 4.3. ϕ is a ring homomorphism and $\phi(R_1)$ contains all matrices $(s_{ij}) \in M_n(S)$ such that $s_{ij} \in S^{(j)}$, $1 \leqslant i,j \leqslant n$. If $\cap_{i=1}^n (v_i)^\ell = (0)$, then ϕ is monic.

Proof. If $\Sigma_{i,j} v_i w_{ij} w_j = 0$, then $0 = \Sigma_{i,j} v_i w_{ij} w_j w_k = \Sigma_{i=1}^n v_i w_{ik} w_k v_k$, $1 \leqslant k \leqslant n$. Then, by 4.2, $w_{ik} w_k v_k = 0$, $1 \leqslant i,k \leqslant n$. Since ϕ is clearly additive, ϕ is well-defined. Next we show ϕ is multiplicative. $\phi(\Sigma_{i,j} v_i w_{ij} w_j)(\Sigma_{k,l} v_k w_{kl}' w_l) = \phi(\Sigma_{i,j,l} v_i w_{ij} w_j v_j w_{jl}' w_l) = (\Sigma_{j=1}^n (w_{ij} w_j v_j)(w_{jl}' w_l v_l)) = \phi(\Sigma_{i,j} v_i w_{ij} w_j)\phi(\Sigma_{k,l} v_k w_{kl}' w_l)$. The statement about $\phi(R_1)$ is obvious. Now assume $\cap_{i=1}^n (v_i)^\ell = (0)$, and $\phi(\Sigma_{i,j} v_i w_{ij} w_j) = 0$. Then $w_{ij} w_j v_j = 0$, which implies $w_{ij} w_j \in \cap_{k=1}^n (v_k)^\ell$, $1 \leqslant i,j \leqslant n$. Hence $w_{ij} w_j = 0$, $1 \leqslant i,j \leqslant n$. Then $\Sigma_{i,j} v_i w_{ij} w_j = 0$, and ϕ is a monomorphism. This completes the proof.

Proposition 4.4. If V is uniform then S is (left) uniform. If V is simple, then S is a division ring.

Proof. Assume V is uniform. Let s_1, s_2 be non-zero elements of S. Since $S \subseteq V$, $Vs_1 \cap Vs_2 \neq (0)$. Let $v_1 s_1 = v_2 s_2 \neq 0$. There exists $w \in W$ such that $wv_1 s_1 = wv_2 s_2 \neq 0$, which implies $Ss_1 \cap Ss_2 \neq (0)$.

Now assume V is simple. Let $0 \neq s \in S$. Then $Vs \neq (0)$, hence $Vs = V$. Then $Ss = WVs = WV = S$, and S is a division ring. The proof is complete.

When V is uniform, the quasi-injective hull $\widetilde{V}$ is a uniform quasi-injective module with zero singular submodule, and $D = \mathrm{Hom}_R(\widetilde{V}, \widetilde{V})$ is a division ring. S is easily seen to be a left order in D. Thus if $0 \neq \alpha \in D$, there exists $v \in V$ such that $0 \neq v\alpha \in V$. If $w \in W$ is such that $wv\alpha \neq 0$, then $wv \in S$ and $0 \neq (wv)\alpha \in S$. A subset $\{v_1, v_2, \cdots, v_n\} \subseteq \widetilde{V}$ is linearly independent over D if and only if $\cap_{j \neq i} (v_j)^\ell \not\subseteq (v_i)^\ell$ $1 \leqslant i \leqslant n$. If R is (left) finite dimensional (R does not contain an infinite direct sum of non–zero left ideals), then V does not contain an infinite linearly independent set, by 4.2. Since $VD = \tilde{V}$, $\tilde{V}$ is a finite dimensional vector space over D in this case.

Let $\{v_1, v_2, \cdots, v_n\}$ be a basis for $\tilde{V}$ contained in V. Then $\cap_{i=1}^n (v_i)^\ell = (0)$ and $\cap_{j \neq i} (v_j)^\ell \not\subseteq (v_i)^\ell$, $1 \leqslant i \leqslant n$. It is easy to see that the map $\phi : R_1 \to M_n(S)$ in 4.3 is just the matrix representation map with respect to the basis $\{v_1, v_2, \cdots, v_n\}$. Thus R is isomorphic to a subring R' of $M_n(D)$ such that $\phi(R_1) \subseteq R' \subseteq M_n(D)$. Since S is left uniform, in 4.3 we get $\bar{S} = \cap_{j=1}^n S^{(j)}$ is a non–zero left ideal of S, and hence a left order in D. Thus $M_n(\bar{S}) \subseteq R' \subseteq M_n(D)$, which is a result of Faith and Utumi [1]. This in turn gives immediately the result of Goldie [2] that R is isomorphic to a left order in $M_n(D)$.

We can obtain from 4.3 an extension of the result of Faith and Utumi to rings which are not finite dimensional.

Proposition 4.5. Let R be a prime ring with maximal annihilator left ideal and uniform left ideal. Then there is a division ring D such that for each positive integer n there is a left order S_n in D, a subring R_n of R, and a map $\phi_n : R_n \to M_n(D)$ such that $M_n(S_n) \subseteq \phi(R_n) \subseteq M_n(D)$.

Proposition 4.5 can also be deduced from [6: Th. 2] using the Faith–Utumi Theorem.

References

1. C. Faith and Y. Utumi, On Notherian Prime Rings, Trans. Amer. Math. Soc. **114** (1965), 53-60.
2. A. W. Goldie, Semiprime rings with maximum condition, Proc. Lond. Math. Soc. **10** (1960), 201-220.
3. A. W. Goldie, Properties of the idealizer, this Proceedings.
4. O. Goldman, Rings and modules of quotients, J. of Algebra **13** (1969), 10-47.
5. A. Hudry, Sur les anneaux localement homogenes, C. R. Acad. Sci. Paris **271** (1970), 1214-1217.
6. Kwangil Koh and A. C. Mewborn, Prime rings with maximal annihilator and maximal complement right ideals, Proc. Amer. Math. Soc. **16** (1965), 1073-1076.
7. Kwangil Koh and A. C. Mewborn, A class of prime rings, Can. Math. Bull. **9** (1966), 63-72.
8. Kwangil Koh and A. C. Mewborn, The weak radical of a ring, Proc. Amer. Math. Soc. **18** (1967), 554-559.
9. Kwangil Koh, On almost maximal right ideals, Proc. Amer. Math. Soc. **25** (1970), 266-272.

10. Kwangil Koh and Jiang Luh, On a finite dimensional quasi-simple module, Proc. Amer. Math. Soc. **25** (1970), 801-807.
11. Hans H. Storrer, On Goldman's Primary Decomposition, Tulane Ring Year Lecture Notes (to appear).

PRIME RIGHT IDEALS AND RIGHT NOETHERIAN RINGS

Gerhard O. Michler

McGill University and University of Tübingen

1. Introduction

A well-known theorem due to I. S. Cohen [1] asserts that a commutative ring R is noetherian if and only if every prime ideal of R is finitely generated. Using the following definition of a prime right ideal we show in this note that Cohen's theorem holds for *any* ring (Theorem 6).

Definition (due to J. Lambek). The right ideal A of the ring R is *prime,* if $sRt \nsubseteq A$ whenever s and t do not belong to A.

As an application of our result we give an easy proof for the fact that the power series ring $R[[X, \alpha]]$ of the right noetherian ring R with the surjective endomorphism α is right noetherian (Corollary 8).

The author is indebted to J. Lambek for many interesting discussions.

2. Proof of the Theorem

In this paper, all rings have an identity element and are associative; ring homomorphisms and modules are unitary. The right (or two-sided) ideal A of the ring R is finitely generated as a right R-module.

Lemma 1. If A is a finitely generated right ideal and B is a finitely generated two-sided ideal, then AB is finitely generated.

Proof. If $A = \Sigma_{i=1}^{n} a_iR$ and $B = \Sigma_{j=1}^{m} b_jR$, then the right ideal AB is generated by $\{x_iy_j : i = 1,2,\cdots n$ and $j = 1,2,\cdots m\}$.

Lemma 2. If $P_2,P_2,\cdots,P_k$ are finitely generated two-sided ideals of the ring R such that R/P_i is right noetherian for $i = 1,2,\cdots,k$, then every right A of R satisfying

$$P_1P_2\cdots P_k \leqslant A \leqslant \bigcap_{i=1}^{k} P_i$$

is finitely generated.

Proof. By Lemma 1 the descending chain

$$P_1 \geqslant P_1P_2 \geqslant \cdots \geqslant P_1P_2\cdots P_{k-1} \geqslant P_1P_2\cdots P_{k-1}P_k$$

consists of finitely generated right ideals of R. Since

$$P_1P_2\cdots P_i/P_1P_2\cdots P_iP_{i+1}$$

is a finitely generated R/P_{i+1}-module over the right noetherian ring R/P_{i+1} for every $i = 1,2,\cdots,k\text{-}1$, it follows that every submodule of the right R-module $P_1/P_1P_2\cdots P_k$ is finitely generated. Thus $A/P_1P_2\cdots P_k$ is finitely generated. Therefore the right ideal A of R is finitely generated.

Lemma 3. If every two-sided ideal of R and every prime right ideal of R is a finitely generated right ideal of R, then R is right noetherian.

Proof. Since every two-sided ideal of R is finitely generated, R satisfies the ascending chain condition on two-sided ideals. Thus, if R is not right noetherian, then there is an ideal M of R which is maximal among the ideals X of R such that R/X is not right noetherian. We may assume that $M = 0$, and that the set $\mathfrak{T}$ of right ideals Y of R which are not finitely generated is not empty. By Zorn's Lemma $\mathfrak{T}$ has a maximal element G. Since G is not finitely generated, G is not a prime ideal. Thus there are $a \notin G$ and $b \notin G$ such that $aRb \leqslant G$. Clearly $A = aR + G$ is finitely generated. Since $U = RbR$ is a finitely generated right ideal of R, AU is a finitely generated right ideal of R contained in G. If $R = RbR$, then

$$a \in aR = aRbR \leqslant G,$$

a contradiction! Thus R/U is right noetherian, and G/AU is a finitely generated R-module. Hence G is a finitely generated right ideal of R. This contradiction shows that R is right noetherian.

Lemma 4. If every two-sided prime ideal of R is a finitely generated right of R, then every two-sided ideal A of R contains a product of finitely many prime ideals P_i of R each containing S.

Proof. Suppose Lemma 3 is false. Then the set $\mathfrak{R}$ of all two-sided ideals A of R not containing a product of finitely many prime ideals $P_i \geqslant A$ of R is not empty. If $X_1 < X_2 < \cdots$ is an infinite properly ascending chain of elements $X_i \in \mathfrak{R}$, then $X = \cup_{i=1}^{\infty} X_i$ also belongs to $\mathfrak{R}$, because otherwise there would be finitely many prime ideals $Q_j \geqslant X$, $j = 1,2,\cdots,k$, such that $Q_1Q_2\cdots Q_k \leqslant X$. Since $Q_1Q_2\cdots Q_k$ is a finitely generated right ideal of R by Lemma 1, it follows that

$$T = Q_1Q_2\cdots Q_k \leqslant X_s$$

for some integer s. Hence $X_s \in \mathfrak{R}$, a contradiction! Therefore Zorn's Lemma applies, and $\mathfrak{R}$ has a maximal element G. Clearly G is not a prime ideal. Hence there are two-sided ideals $A_i > G$, $i = 1,2$, such that $A_1A_2 \leqslant G$. Since $A_i \notin \mathfrak{R}$, there are prime ideals P_{ij} of R such that

$$P_{i1}P_{i2}\cdots P_{ik_i} \leqslant A_i \leqslant \bigcap_{j=1}^{k_i} P_{ij},\ i = 1,2.$$

Thus

$$P_{11}P_{12}\cdots P_{1k_1}P_{21}P_{22}\cdots P_{2k_2} \leqslant A_1A_2 \leqslant G < (\bigcap_{j=1}^{k_1} P_{ij}) \cap (\bigcap_{j=1}^{k_2} P_2),$$

$G \in$. This contradiction proves Lemma 4.

Lemma 5. If every two-sided prime ideal of the ring R is a finitely generated right ideal of R, then R satisfies the ascending chain condition on two-sided prime ideals.

Proof. Otherwise there exists an infinite strictly ascending chain of two-sided prime ideals P_i of R. Let $X = \cup_{i=1}^{\infty} P_i$. By Lemma 4 there are finitely many prime ideals Q_j, $j = 1,2,\cdots,k$, of R such that

$$T = Q_1Q_2\cdots Q_k \leqslant X \leqslant \bigcap_{j=1}^{k} Q_j.$$

T is a finitely generated right ideal of R by Lemma 1. Thus

$$T = Q_1Q_2\cdots Q_k \leqslant P_s$$

for some integer s. Since P_s is a prime ideal of R, it follows that

$$Q_j \leqslant P_s < X \leqslant Q_j$$

for some $j \in \{1,2,\cdots,k.\}$ This contradiction proves Lemma 5.

Theorem 6. The ring R is right noetherian if and only if every prime right ideal of R is finitely generated.

Proof. If every prime right ideal of R is finitely generated, then by Lemma 3 it suffices to show that every two-sided ideal of R is a finitely generated right ideal. This follows at once from Lemma 4 and Lemma 2, if R/P is right noetherian for every two-sided prime ideal P of R.

Let $\mathcal{P}$ be the set of all prime ideals P of R such that R/P is not right noetherian. Since every prime right ideal is a finitely generated right ideal of R, R satisfies the ascending chain condition on two-sided prime ideals by Lemma 5. Thus, if $\mathcal{P}$ is not empty, $\mathcal{P}$ contains a maximal element M. We may assume that $M = 0$. Since R is not right noetherian there exists a two-sided ideal $A \neq 0$ of R which by Lemma 3 is not a finitely generated right ideal of R. By Lemma 4 there are finitely many prime ideals $P_i > A$, $i = 1,2,\cdots,k$, of R such that $P_1P_2\cdots P_k \leqslant A$. Since $P_i \neq 0$ for all i, the rings R/P_i are right noetherian by the choice of M. Hence Lemma 2 applies and A is a finitely generated right ideal of R, a contradiction! Thus $\mathcal{P}$ is empty, and R is right noetherian. The converse is obvious.

As in the commutative case Theorem 6 has the following corollary.

Corollary 7. If R is right noetherian, then R[[X]] is right noetherian.

Actually, a more general result holds. Let α be a surjective endomorphism of the ring R, and let X be an indeterminate over R. As usual $R[[X, \alpha]]$ denotes the ring of all power series

$$f(X) = \sum_{i=0}^{\infty} r_i X^i, \quad r_i \in R,$$

with coefficientwise addition and the following (distributively extended) multiplication:

$$Xr = \alpha(r)X \quad \text{for all } r \in R.$$

Corollary 8. If R is right noetherian, then $S = R[[X, \alpha]]$ is right noetherian.

Proof (cf. Kaplansky [2], p. 48). By Theorem 6 it suffices to show that every prime right ideal P of S is finitely generated. Let $\phi: S \to R$ be the natural ring epimorphism obtained by mapping X onto zero. Let $\phi(P) = \Sigma_{i=1}^{n} a_i R$, where

$A_i \in \phi(P)$. Then there are power series $f_i \in P$ such that a_i is the constant term of f_i. If $X \in P$, then

$$P = \sum_{i=1}^{n} f_i S + XS.$$

Thus we may assume that $X \notin P$. If $g \in P$ and g_0 is the constant term of g, then $g_0 = \Sigma_{i=1}^{n} a_i r_{i0}$ for some $r_{i0} \in R$. Thus $g - \Sigma_{i=1}^{n} f_i r_{i0} = g_1 X$ for some $g_1 \in S$. Furthermore, $g_1 X \in P$. Therefore $g_1 X^n \in P$ and $g_1 Xr = g_1 \alpha(r)X \in P$ for all $r \in R$. Since α is surjective, it follows that $g_1 RX \leqslant P$. Thus $g_1 \in P$, because $X \notin P$ and P is a prime right ideal of S. In the same way we obtain $g_1 = \Sigma_{i=1}^{n} f_i r_{i1} + g_2 X$ with $g_2 \in P$. Thus there are $h_i \in R[[X, \alpha]]$, where $h_i = \Sigma_{j=1}^{\infty} r_{ij} X^j$, such that $g = \Sigma_{i=1}^{n} f_i h_i$. Hence P is a finitely generated right ideal of S.

References

1. I. S. Cohen, Rings with restricted minimum condition, *Duke Math. J.* 17 (1950), 27-42.
2. I. Kaplansky, "Commutative rings", Allyn and Bacon, Boston, 1970.

QUOTIENT RINGS

Kiiti Morita

Tokyo Kyoiku Daigaku

In 1962, P. Gabriel [7] developed a general theory on localizations in abelian categories and by applying it to the case of categories of modules he obtained a method of constructing quotient rings.

On the other hand, in [16] we have shown that quotient rings which are constructed by Gabriel's method are nothing but the double centralizers of "finitely cogenerating" injective modules.

In this paper we shall give an exposition of the results on localizations in categories of modules, quotient rings, and flat epimorphisms in the category of rings, which have been obtained in our recent papers [16], [17] and forthcoming ones [19], [20], and supplement them with a number of new results; among the latter a theorem determining the structure of a full reflective subcategory of ${}_A\mathfrak{M}$ which is itself a Grothendieck category is included.

In the final section we shall discuss the double centralizer property of modules in connection with the above results.

Throughout this paper it will be assumed that every ring has an identity, every module, every ring homomorphism and every subring are unitary, every functor is covariant and additive, and that every endomorphism of a module operates on the opposite side to the scalars. The endomorphism ring (resp. the opposite ring of the endomorphism ring) of a right (resp. left) A–module X will be denoted by $\mathrm{End}(X_A)$ (resp. $\mathrm{End}({}_AX)$), and $E({}_AX)$ is the injective envelope of a left A–module X. We assume always that A and B are rings, and ${}_A\mathfrak{M}$ means the category of all left A–modules.

1. Localizations in categories of modules

Definition 1.1. A right A–module U_A is called a module of type FP if conditions (a) and (b) below are satisfied:

(a) ${}_BU_C \cong {}_BU_A \otimes {}_AC_C$,

(b) U_C is finitely generated and projective,
where $B = \mathrm{End}(U_A)$, $C = \mathrm{End}({}_BU)$. U_A is called a module of type F_t if condition (a) is satisfied.

Then we have

Theorem 1.2 ([16]). For a B–A–bimodule ${}_BU_A$ the following conditions are equivalent, where we set $C = \mathrm{End}({}_BU)$.

I. U_A is a right A–module of type FP and $B = \mathrm{End}(U_A)$.

II. ${}_BU$ is a generator in ${}_B\mathfrak{M}$ and ${}_BU_A \otimes {}_AC_C \cong {}_BU_C$.

III. The natural homomorphism

$$\phi(Y): {}_BU_A \otimes \mathrm{Hom}_B({}_BU_A, Y) \to Y \quad \text{for} \quad Y \in {}_B\mathfrak{M}$$

defined by $\phi(Y)(u \otimes f) = f(u)$ for $u \in U$, $f \in \mathrm{Hom}_B({}_BU_A, Y)$, is an isomorphism.

Corollary 1.3. Every finitely generated, projective right A–module is of type FP.

Corresponding to modules of type FP we have the notion of modules of type FI.

Definition 1.4. A left A–module X is called finitely cogenerating, if there exist a finite number of elements $f_i \in \mathrm{Hom}_A({}_AA, X)$, $i = 1,\cdots,n$, such that

$$\cap \{\mathrm{Ker}\ g \mid g \in \mathrm{Hom}_A({}_AA, X)\} = \cap\{\mathrm{Ker}\ f_i \mid i = 1,\cdots,n\}.$$

Definition 1.5. A left A–module V is called a module of type FI, if conditions (a) and (b) below are satisfied:

(a) ${}_CV$ is injective and finitely cogenerating,

(b) ${}_CV_B \cong \mathrm{Hom}_A({}_AC_C, {}_AV_B)$,

where $B = \mathrm{End}({}_AV)$, $C = \mathrm{End}(V_B)$. ${}_AV$ is called a module of type F_h if condition (b) is satisfied.

Here we have

Lemma 1.6. For an A–B–bimodule ${}_AV_B$ the following statements are equivalent, where $C = \mathrm{End}(V_B)$.

I. ${}_CV_B \cong \mathrm{Hom}_A({}_AC_C, {}_AV_B)$.

II. The canonical map $\zeta: {}_C V_B \to \mathrm{Hom}_A({}_A C_C, {}_A V_B)$ defined by $\zeta(v)(c) = cv$ for $v \in V$, $c \in C$, is an isomorphism.

If II holds and ${}_C V$ is a finitely cogenerating injective left C–module such that $B = \mathrm{End}({}_C V)$, then V is a left A–module of type FI with C as its double centralizer.

Proof. The equivalence I ⇔ II is proved in [16]. If II holds, by [16, Lemma 3.1] we have $\mathrm{End}({}_A V) = \mathrm{End}({}_C V)$.

Corresponding to Corollary 1.3 we have

Theorem 1.7. Every finitely cogenerating, injective left A–module is of type FI.

For the case of A being a left Artinian ring, every left A–module is finitely cogenerating, as was shown in [15]. This fact for the case of faithful modules led Beachy [2] and Kato [10] to introduce cofaithful modules and finitely–faithful modules respectively; these are nothing but finitely cogenerating faithful modules.

Let V be a left A–module.

Definition 1.8. The V–dominant dimension of a left A–module X is said to be greater than or equal to n, V–dom. dim X $\geqslant$ n, where n is a non–negative integer, if there exists an exact sequence

$$0 \to X \to X_1 \to \cdots \to X_n$$

such that each X_i is a direct product of copies of V. In case V–dom. dim X $\geqslant$ n for any positive integer n, the V–dominant dimension of X is said to be ∞.

For the special case of V being injective, this notion was introduced by Tachikawa [26].

Let us consider the following full subcategories of ${}_A\mathcal{M}$:

$$\mathcal{T}(V) = \{X \in {}_A\mathcal{M} \mid \mathrm{Hom}_A(X, V) = 0\}.$$

$$\mathcal{D}(V) = \{X \in {}_A\mathcal{M} \mid V\text{–dom. dim } X \geqslant 2\}.$$

$$\mathcal{D}'(V) = \{X \in {}_A\mathcal{M} \mid V\text{–dom. dim } X \geqslant 1\}.$$

Here our concern lies in $\mathcal{D}(V)$. An abelian category with a generator and with exact direct limits is called a Grothendieck category.

Lemma 1.9 ([16]). In case V is a module of type FI, the subcategory $\mathcal{Q}(V)$ is a Grothendieck category.

As for localizations in the category of modules we have the following theorem.

Theorem 1.10 ([16], [17]). For a ring A and a Grothendieck category $\mathcal{B}$ the following statements are equivalent.

I. There exist two functors $S: {}_A\mathcal{M} \to \mathcal{B}$, $T: \mathcal{B} \to {}_A\mathcal{M}$ such that

(a) S is a left adjoint of T,

(b) ST is naturally equivalent to the identity functor on $\mathcal{B}$ (or T is full and faithful).

II. There exists a left A–module V of type FI such that $\mathcal{B} \cong \mathcal{Q}(V)$.

In case $\mathcal{B} = {}_B\mathcal{M}$ for a ring B, each of I and II is equivalent to III below:

III. There exists a right A–module U of type FP such that $B = \mathrm{End}(U_A)$.

Moreover, in case II and III hold, the modules V in II and U in III can be chosen so that the double centralizers of U and V may be identified. In case I and II hold, S is an exact functor if and only if ${}_AV$ can be chosen to be injective, and in case II and III hold, ${}_AV$ is injective if and only if U_A is flat.

As an application of Theorem 1.10 we see that in case A is a finite dimensional algebra over a field K and V is a finitely generated left A–module of type FI, $\mathcal{Q}(V) \cong {}_B\mathcal{M}$ for some ring B; because the K–dual of V is a right A–module of type FP.

Now we shall prove our main theorem in this section.

Theorem 1.11. Let $\mathcal{B}$ be a full subcategory of ${}_A\mathcal{M}$ which is a Grothendieck category closed under isomorphic images. Then the following statements are equivalent.

I. $\mathcal{B}$ is a reflective subcategory of ${}_A\mathcal{M}$ in the sense of [6]. In other words, there exists a functor $S: {}_A\mathcal{M} \to \mathcal{B}$ such that S is a left adjoint of the inclusion functor $T: \mathcal{B} \to {}_A\mathcal{M}$.

II. There exists a left A–module V of type FI such that $\mathcal{B} = \mathcal{Q}(V)$.

In case I and II hold, S is exact if and only if V can be chosen to be injective.

Proof. The implication II $\Rightarrow$ I will be shown in §3 below, although it has been proved already in [17, Theorem 1.2]. To prove the implication I $\Rightarrow$ II, assume I holds. Let us put

$$U = S({}_A A),\ C = \mathrm{End}({}_A U).$$

Then U is an A–C–bimodule, and since U is an A–module there is a ring homomorphism from A to C. On the other hand, if we put

$$T'(Y) = \mathrm{Hom}_A({}_A U_C, Y) \quad \text{for} \quad Y \in \mathfrak{B},$$

then T' is a functor from $\mathfrak{B}$ to ${}_C\mathfrak{M}$ and

$$T(Y) = Y \cong {}_A C_C \otimes T'(Y), \quad \text{for} \quad Y \in \mathfrak{B},$$

since we have natural isomorphisms:

$$\mathrm{Hom}_A(S({}_A A), Y) \cong \mathrm{Hom}_A({}_A A, Y) \cong Y \quad \text{for} \quad Y \in \mathfrak{B} .$$

Thus we have ${}_A U_C \cong {}_A C_C$.

Next, let us define two functors $P: {}_A\mathfrak{M} \to {}_C\mathfrak{M}$, $Q: {}_C\mathfrak{M} \to {}_A\mathfrak{M}$ by

$$P(X) = \mathrm{Hom}_A({}_A C_C, X),\ X \in {}_A\mathfrak{M},$$

$$Q(Z) = {}_A C_C \otimes Z, \qquad Z \in {}_C\mathfrak{M}.$$

Then $P \cong T'$, $T \cong QT'$.

Let W be an injective cogenerator in $\mathfrak{B}$ such that ${}_A U$ is a subobject of W. Since U is a generator in $\mathfrak{B}$ the injective envelope of the direct sum of all quotients of U in $\mathfrak{B}$ can be taken as W. Let us put

$${}_C V = T'(W).$$

Then ${}_C V_D \cong \mathrm{Hom}({}_A C_C, {}_A W_D)$, where $D = \mathrm{End}({}_A W)$. Since $T \cong QT'$, we have ${}_A W_D \cong {}_A V_D$ and hence

$${}_C V_D \cong \mathrm{Hom}_A({}_A C_C, {}_A V_D).$$

According to a theorem of Gabriel and Popescu there is a functor $S': {}_C\mathfrak{M} \to \mathfrak{B}$

such that

(a)′ S' is a left adjoint of T',

(b)′ S' is exact,

(c)′ the associated natural transformation from $S'T'$ to the identity functor on $\mathfrak{B}$ is an equivalence.

Since W is injective in $\mathfrak{B}$, it follows from (b)′ that ${}_CV$ is injective. Since U is a subobject of W, ${}_CV$ is finitely cogenerating. Since W is a cogenerator in $\mathfrak{B}$, we have $\operatorname{Ker} S' = \mathcal{T}({}_CV)$. Hence by [16, Theorem 5.4] we get $T'(\mathfrak{B}) \subset \mathfrak{D}({}_CV)$ and every module in $\mathfrak{D}({}_CV)$ has an isomorphic copy in $T'(\mathfrak{B})$. Since ${}_CC \in T'(\mathfrak{B}) \subset \mathfrak{D}({}_CV)$, the double centralizer of ${}_CV$ coincides with C (cf. Corollary 9.2 below). Thus ${}_AV$ is a left A–module of type FI and hence $Q(\mathfrak{D}({}_CV)) = \mathfrak{D}({}_AV)$ by [16, Theorem 3.3]. Since $T \cong QT'$, a left A–module X is in $\mathfrak{D}({}_AV)$ if and only if X is isomorphic to some module in $\mathfrak{B}$. Therefore II holds.

As an application of Theorem 1.11 we can prove the following.

Corollary 1.12 (Lambek [13, Proposition 4.5]). Let $\mathfrak{B}$ be a full reflective sub–category of ${}_A\mathfrak{M}$ which is a Grothendieck category and contains ${}_AA$. Then there exists a finitely cogenerating, injective left A–module V such that a left A–module X is in $\mathfrak{D}(V)$ if and only if X is isomorphic to some module in $\mathfrak{B}$.

Proof. From the above proof of Theorem 1.11 we see that there is a left A–module V of type FI such that $X \in \mathfrak{D}(V)$ if and only if $X \cong Y$ for some Y in $\mathfrak{B}$. By assumption ${}_AA \in \mathfrak{D}(V)$ and hence V is faithful. By Corollary 9. 2 below V has the double centralizer property. This shows that V is injective.

Remark. The assumption that $\mathfrak{B}$ is a Grothendieck category is not redundant in Theorem 1.11. Indeed, $\mathfrak{D}'(V)$ with a left A–module V of type FI is a full reflective subcategory of ${}_A\mathfrak{M}$ and it is a Grothendieck category if and only if V is an injective cogenerator in ${}_{A/I}\mathfrak{M}$ for some two–sided ideal I of A; this is seen from Theorem 5.4 below.

2. A characterization of the subcategory $\mathfrak{D}(V)$ with V injective

A full subcategory $\mathcal{K}$ of ${}_A\mathfrak{M}$ is called a localizing subcategory of ${}_A\mathfrak{M}$ if the following conditions are satisfied.

L_1. $\mathcal{K}$ is closed under submodules, factor modules and direct sums.

L_2. Let $0 \to X' \to X \to X'' \to 0$ be an exact sequence in ${}_A\mathfrak{M}$. If $X', X'' \in \mathcal{K}$, then $X \in \mathcal{K}$.

This notion is due to Gabriel [7] and the following theorem is essentially proved by Jans [9], because if ${}_AV$ is injective with K as its annihilator ideal and if W is equal either to $E({}_A[A/K]) \oplus V$ or a direct product of a suitable number of copies of V, then W is a finitely cogenerating, injective left A–module such that $\mathfrak{D}(W) = \mathfrak{D}(V)$.

Theorem 2.1. Let $\mathcal{L}$ be a full subcategory of ${}_A\mathfrak{M}$. Then there exists a finitely cogenerating, injective left A–module V such that $\mathcal{L} = \mathcal{T}(V)$ if and only if $\mathcal{L}$ is a localizing subcategory.

Our proof runs as follows (cf. [18]). Since the "only if" part is obvious, we have only to prove the "if"part. Suppose that $\mathcal{L}$ is a localizing subcategory. For a left A–module X let us denote by $\tau(X)$ the largest submodule of X which belongs to $\mathcal{L}$. Let $\{J_\lambda\}$ be the set of all left ideals J_λ of A such that $\tau(A/J_\lambda) = 0$, and let us put

$$V = E(\Sigma \oplus A/J_\lambda).$$

Then it can be shown that V is a finitely cogenerating, injective, left A–module such that $\mathcal{L} = \mathcal{T}(V)$.[1]

Let S be a multiplicatively closed set of regular elements of A satisfying the Ore condition. We say that a left A–module X is S–torsion, S–torsionfree or S–divisible according as (i), (ii) or (iii) below holds.

(i) for every $x \in X$, there is some $s \in S$ such that $sx = 0$,

(ii) if $sx = 0$ for $s \in S$, $x \in X$ then $x = 0$,

(iii) for every $s \in S$ and $x \in X$, there is $x' \in X$ such that $x = sx'$.

Let $\{J_\lambda\}$ be the set of all left ideals J_λ of A such that A/J_λ is S–torsionfree and let us put

$$V = E(\Sigma \oplus A/J_\lambda).$$

Then V is a finitely cogenerating, injective left A–module and we can prove the following theorem (cf. [18]).

Theorem 2.2. Under the above situation, the following statements are true for a

[1] I was informed by Y. Kurata that this was proved also by J. S. Alin (except that V is finitely cogenerating).

left A–module X.

(1) X is S–torsion if and only if $X \in \mathcal{T}(V)$.

(2) X is S–torsionfree if and only if V–dom. dim $X \geqslant 1$.

(3) X is S–torsionfree and S–divisible if and only if V–dom. dim $X \geqslant 2$.

This illustrates the features of $\mathcal{T}(V)$, $\mathcal{D}'(V)$ and $\mathcal{D}(V)$.

Now, corresponding to Theorem 2.1 we shall give a characterization of the sub–category $\mathcal{D}(V)$ for the case of V being injective; for the general case that V is of type FI the problem of characterizing $\mathcal{D}(V)$ in a similar form remains open.

Theorem 2.3 ([19]). Let $\mathcal{K}$ be a full subcategory of ${}_A\mathcal{M}$. Then there exists a finitely cogenerating, injective left A–module V such that $\mathcal{K} = \mathcal{D}(V)$ if and only if $\mathcal{K}$ satisfies the following conditions.

D_1. $\mathcal{K}$ is closed under isomorphic images, direct summands and direct products.

D_2. $X \in \mathcal{K}$ if and only if $E(X)$, $E(E(X)/X) \in \mathcal{K}$.

To prove the "if" part of Theorem 2.3 we proceed as follows. Let us consider the following full subcategory of ${}_A\mathcal{M}$:

$$\mathcal{L} = \{ X \in {}_A\mathcal{M} \mid \mathrm{Hom}_A(X, Y) = 0 \text{ for all } Y \in \mathcal{K} \}.$$

Then $\mathcal{L}$ is a localizing subcategory and by Theorem 2.1 there is a finitely cogenerating, injective left A–module V such that $\mathcal{L} = \mathcal{T}(V)$, from which $\mathcal{K} = \mathcal{D}(V)$ follows.

Thus each of $\mathcal{T}(V)$, $\mathcal{D}(V)$ and $\mathcal{D}'(V)$ determines the others.

As an application of Theorem 2.3 we have

Theorem 2.4. Let $\{ V_\lambda \mid \lambda \in \Lambda \}$ be a set of finitely cogenerating, injective left A–modules. Then there exists a finitely cogenerating, injective left A–module W such that

$$\mathcal{D}(W) = \cap \{ \mathcal{D}(V_\lambda) \mid \lambda \in \Lambda \},$$

where W may be zero, while the smallest subcategory $\mathcal{D}(U)$ with a finitely cogenerating, injective left A–module U such that $\mathcal{D}(U) \supset \mathcal{D}(V_\lambda)$ for every λ is given by $U = \Pi V_\lambda$.

3. Quotient rings

A non-empty set $\mathfrak{F}$ of left ideals of A is called an idempotent topologizing filter, which we shall abbreviate to an IT filter in the following, if the following conditions are satisfied:

(i) if $I \in \mathfrak{F}$ and $I \subset J$, then $J \in \mathfrak{F}$;

(ii) if $I, J \in \mathfrak{F}$, then $I \cap J \in \mathfrak{F}$;

(iii) if $I \in \mathfrak{F}$, $a \in A$, then $(I : a) \in \mathfrak{F}$;

(iv) if $I \in \mathfrak{F}$, $(J : x) \in \mathfrak{F}$ for every $x \in I$, then $J \in \mathfrak{F}$;

where I, J are left ideals of A and

$$(J : x) = \{ a \in A \mid ax \in J \}.$$

This notion is due to Gabriel [7], but the above definition is a simplified version of it due to C. L. Walker and E. A. Walker [28].

Lemma 3.1 (Gabriel [7]). A localizing subcategory $\mathcal{L}$ of ${}_A\mathcal{M}$ determines an IT filter $F(\mathcal{L})$ by $F(\mathcal{L}) = \{ J \mid {}_A[A/J] \in \mathcal{L} \}$, and conversely, for an IT filter $\mathfrak{F}$, $L(\mathfrak{F}) = \{ X \in {}_A\mathcal{M} \mid$ for any $x \in X$ there is some $J \in \mathfrak{F}$ with $Jx = 0 \}$ is a localizing subcategory of ${}_A\mathcal{M}$. Moreover $LF(\mathcal{L}) = \mathcal{L}$ and $FL(\mathfrak{F}) = \mathfrak{F}$.

For a localizing subcategory $\mathcal{L}$ of ${}_A\mathcal{M}$ Gabriel [7] considered the quotient category ${}_A\mathcal{M}/\mathcal{L}$. From this point of view he defined, for an IT filter $\mathfrak{F}$,

$$S(X) = \varinjlim \{ \operatorname{Hom}_A(J, X/\tau(X)) \mid J \in \mathfrak{F} \}, \; X \in {}_A\mathcal{M}$$

where the direct system is defined by means of the inclusion of left ideals of $\mathfrak{F}$ and $\tau(X)$ is the largest submodule of X belonging to $L(\mathfrak{F})$, and proved that $S(X)$ is an A-module and

$$S(X)/(X/\tau(X)) \cong \tau[E(X/\tau(X))/(X/\tau(X))].$$

By Theorem 2.1 there exists a finitely cogenerating, injective left A-module V such that $L(\mathfrak{F}) = \mathcal{T}(V)$ and in this case S is a left adjoint of the inclusion functor $T : \mathcal{D}(V) \to {}_A\mathcal{M}$. Thus we have

$${}_A\mathcal{M}/\mathcal{T}(V) \cong \mathcal{D}(V)$$

and, since $\mathrm{Hom}_A(S({}_AA), S({}_AA)) \cong \mathrm{Hom}_A({}_AA, S({}_AA)) \cong S({}_AA)$, $S({}_AA)$ becomes a ring which is called the (left) quotient ring of A with respect to an IT filter $\mathfrak{F}$.

Theorem 3.2 ([16]). Let V be a finitely cogenerating, injective left A–module. Then the quotient ring of A with respect to an IT filter $F(\mathcal{J}(V))$ coincides with the double centralizer of V.

Thus, the left quotient rings of A in the sense of Gabriel are nothing but the double centralizers of finitely cogenerating, injective left A–modules. Essentially the same result is obtained by Lambek [13] independently.

Now, let V be a left A–module of type FI with C as its double centralizer. Let us put

$$P(X) = \mathrm{Hom}_A({}_AC_C, X), \; X \in {}_A\mathfrak{M}$$

$$Q(Z) = {}_AC_C \otimes Z, \qquad Z \in {}_C\mathfrak{M}.$$

Then P (resp. Q) carries $\mathfrak{D}({}_AV)$ (resp. $\mathfrak{D}({}_CV)$) into $\mathfrak{D}({}_CV)$ (resp. $\mathfrak{D}({}_AV)$) and QP on $\mathfrak{D}({}_AV)$ (resp. PQ on $\mathfrak{D}({}_CV)$) is naturally equivalent to the identity functor on $\mathfrak{D}({}_AV)$ (resp. $\mathfrak{D}({}_CV)$). Let $S_0 : {}_C\mathfrak{M} \to \mathfrak{D}({}_CV)$ be a functor which is a left adjoint of the inclusion functor $T_0 : \mathfrak{D}({}_CV) \to {}_C\mathfrak{M}$; the existence of S_0 is assured by Theorem 3.2. Let us define a functor $S : {}_A\mathfrak{M} \to \mathfrak{D}({}_AV)$ by

$$S(X) = QS_0({}_CC_A \otimes X), \; X \in {}_A\mathfrak{M}.$$

Then for $X \in {}_A\mathfrak{M}$, $Y \in \mathfrak{D}({}_AV)$, we have

$$\begin{aligned} \mathrm{Hom}_A(S(X), Y) &\cong \mathrm{Hom}_A(QS_0({}_CC_A \otimes X), Y) \\ &\cong \mathrm{Hom}_C(S_0({}_CC_A \otimes X), P(Y)) \cong \mathrm{Hom}_C({}_CC_A \otimes X, P(Y)) \\ &\cong \mathrm{Hom}_A(X, QP(Y)) \cong \mathrm{Hom}_A(X, Y). \end{aligned}$$

Hence $S : {}_A\mathfrak{M} \to \mathfrak{D}({}_AV)$ is a left adjoint of the inclusion functor $T : \mathfrak{D}({}_AV) \to {}_A\mathfrak{M}$. Since ${}_CC \in \mathfrak{D}({}_CV)$, we have $S({}_AA) = QS_0({}_CC) = {}_AC$. On the other hand, we have $\mathrm{End}({}_AC) = C$ by [16, Theorem 3.3] and $\mathrm{End}(S({}_AA)) \cong S({}_AA)$. Hence we have the following theorem.

Theorem 3.3. Let V be a left A–module of type FI with C as its double centralizer, and let S be a left adjoint of the inclusion functor $T : \mathfrak{D}(V) \to {}_A\mathfrak{M}$. Then $S({}_AA)$ becomes a ring which may be identified with C.

Theorem 3.4. Let $\phi : A \to C$ be a ring homomorphism. Then the following statements are equivalent.

I. There is a left A–module W of type FI with C as its double centralizer.

II. $\mathrm{Hom}_A({}_A C/\phi(A), {}_A V) = 0$ where ${}_C V = E({}_C C) \oplus E(E({}_C C)/{}_C C)$.

In case I and II hold, W can be chosen to be injective if and only if ${}_A V$ is injective.

Proof. Let ${}_C V$ be the module defined in II. Assume I. Since ${}_C C \in \mathfrak{D}({}_C W)$, we have ${}_C V \in \mathfrak{D}({}_C W)$, and hence the map $\psi : {}_C V \to \mathrm{Hom}_A({}_A C_C, {}_A V)$ defined by $\psi(v)(c) = cv$ is an isomorphism. Thus II holds. If ${}_A W$ is injective, ${}_A V$ is clearly injective.

Conversely, assume II holds. Then ${}_C V$ is a finitely cogenerating, injective left C–module and ${}_C C \in \mathfrak{D}({}_C V)$. Hence by Corollary 9.2 below ${}_C V$ has the double centralizer property. Therefore by Lemma 1.6 ${}_A V$ is a left A–module of type FI with C as its double centralizer.

Theorem 3.5. Let $\phi : A \to C$ be a ring homomorphism. Then C is the double centralizer of a finitely cogenerating, injective left A–module if and only if

$$\mathrm{Hom}_A(\mathrm{Ker}\ \phi \oplus {}_A[C/\phi(A)], U) = 0, \quad \text{where} \quad U = E({}_A C) \oplus E(E({}_A C)/{}_A C).$$

Proof. If ${}_A V$ in Theorem 3.4 is injective, then ${}_A U$ is isomorphic to a sub–module of ${}_A V$ and, since ${}_A U$ is injective and $(\mathrm{Ker}\ \phi)U = 0$, we have $\mathrm{Hom}_A(\mathrm{Ker}\ \phi, {}_A U) = 0$. Hence the "only if" part follows directly from Theorem 3.4.

Conversely, assume that $\mathrm{Hom}_A(\mathrm{Ker}\ \phi \oplus {}_A[C/\phi(A)], U) = 0$. Then U is finitely cogenerating. Since $\mathrm{Hom}_A(C/\phi(A), E({}_A C)) = 0$, we have $E({}_A[\phi(A)]) = E({}_A C)$. Thus U–dom. dim $E({}_A[\phi(A)])/{}_A C \geqslant 1$, and hence by [16, Theorem 2.3] C is the double centralizer of U.

Theorem 3.5 was obtained by Lambek [13] and T. Kato independently.

4. Epimorphisms in the category of rings

The equivalence of I and III in Theorem 4.1 below was proved by Silver [24]; the other equivalences were proved in [16] except I ⇔ VII. For further properties of epimorphisms in the category of rings, cf. Bergman [3], Knight [11].

Theorem 4.1. For a ring homomorphism $\phi : A \to C$ the following statements are equivalent.

I. ϕ is an epimorphism in the category of rings.

II. ${}_C C_C \cong {}_C C_A \otimes {}_A C_C$.

III. The canonical map ${}_C C_A \otimes {}_A C_C \to {}_C C_C$ induced by the multiplication in C is an isomorphism.

IV. There exists a left A–module V of type FI such that the double centralizer of V coincides with C and that $\mathfrak{D}({}_C V) = {}_C\mathfrak{M}$.

V. If ${}_C V$ is a finitely cogenerating, injective cogenerator in ${}_C\mathfrak{M}$, then ${}_A V$ is a module of type FI with C as its double centralizer.

VI. C_A is a module of type FP and $C = \mathrm{End}(C_A)$.

VII. $\mathrm{Hom}_A({}_A C/\phi(A), {}_A Y) = 0$ for every left C–module Y.

In case these statements hold, C_A is flat if and only if ${}_A V$ is injective.

Here we shall prove that V $\Leftrightarrow$ VII. Let ${}_C V$ be a finitely cogenerating, injective cogenerator in ${}_C\mathfrak{M}$. Then by Lemma 1.6 ${}_A V$ is a module of type FI with C as its double centralizer if and only if $\mathrm{Hom}_A({}_A C/\phi(A), {}_A V) = 0$. Hence we have V $\Leftrightarrow$ VII.

As for the condition $\mathfrak{D}({}_C V) = {}_C\mathfrak{M}$ appearing in Statement IV in Theorem 4.1 we have several conditions each of which is equivalent to it as will be seen below. For the case of V being injective, such conditions were obtained for the first time by C. L. Walker and E. A. Walker [28]; cf. also Goldman [8].

Theorem 4.2. Let V be a left A–module of type FI with C as its double centralizer. Let $S : {}_A\mathfrak{M} \to \mathfrak{D}({}_A V)$ be a left adjoint of the inclusion functor $T : \mathfrak{D}({}_A V) \to {}_A\mathfrak{M}$ (the existence of S is assured by Theorem 1.11). Then the following statements are equivalent.

I. $\mathfrak{D}({}_C V) = {}_C\mathfrak{M}$.

II. Every left C–module, when regarded as a left A–module, is in $\mathfrak{D}({}_A V)$.

III. $\mathfrak{D}({}_A V)$ is an exact subcategory[1a] of ${}_A\mathfrak{M}$ closed under direct sums.

IV. TS is right exact and commutes with direct sums.

V. $S(X)$ is naturally isomorphic to ${}_A C_A \otimes X$ for $X \in {}_A\mathfrak{M}$.

[1a] For the definition of exact subcategories, cf. §5 below.

Proof. From the arguments in §3 we have

$$S(X) \cong QS_0({}_CC_A \otimes X), \quad \text{for } X \in {}_A\mathfrak{M},$$

where $S_0 : {}_C\mathfrak{M} \to \mathfrak{D}({}_CV)$ is a left adjoint of the inclusion functor from $\mathfrak{D}({}_CV)$ to ${}_C\mathfrak{M}$.

Assume that I holds. Then S_0 is the identity functor and hence V holds. The implication V $\Rightarrow$ IV is obvious. Next assume that IV holds. Let $f : X \to Y$ be an A–homomorphism with X, Y in $\mathfrak{D}({}_AV)$. Then we have the following commutative diagram:

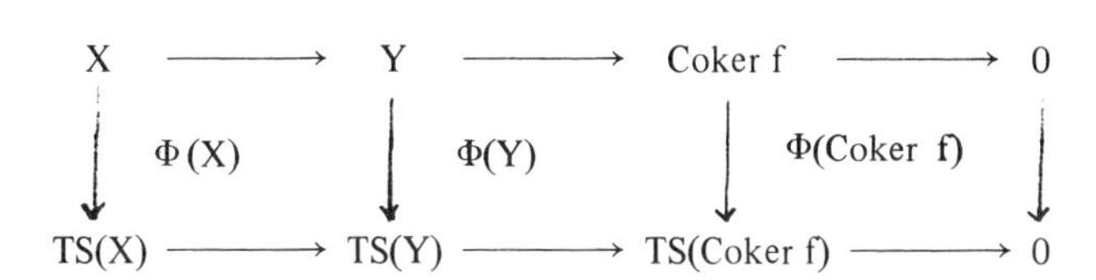

$$\begin{array}{ccccccc} X & \longrightarrow & Y & \longrightarrow & \text{Coker } f & \longrightarrow & 0 \\ \downarrow \Phi(X) & & \downarrow \Phi(Y) & & \downarrow \Phi(\text{Coker } f) & & \downarrow \\ TS(X) & \longrightarrow & TS(Y) & \longrightarrow & TS(\text{Coker } f) & \longrightarrow & 0 \end{array}$$

where Φ is the natural transformation from the identity functor on ${}_A\mathfrak{M}$ to TS associated with the left adjointness of S to T. Since each row is exact and $\Phi(X)$ and $\Phi(Y)$ are isomorphisms, $\Phi(\text{Coker } f)$ is also an isomorphism. Hence $\text{Coker } f \in \mathfrak{D}({}_AV)$. Since it is obvious that $X \oplus Y$ and $\text{Ker } f$ are in $\mathfrak{D}({}_AV)$, we see that $\mathfrak{D}({}_AV)$ is an exact subcategory of ${}_A\mathfrak{M}$. This proves that IV $\Rightarrow$ III. The implication III $\Rightarrow$ I is proved in [19] and I $\Rightarrow$ II is obvious. Finally, assume that II holds. Then the functor $S_1 : {}_A\mathfrak{M} \to \mathfrak{D}({}_AV)$ defined by $S_1(X) = {}_AC_A \otimes X$ is easily shown to be a left adjoint of T, and hence V holds. This completes the proof of Theorem 4.2.

Remark. It has been drawn to my attention that essentially the same situation as ours is discussed in Maranda [30] and Lambek [13]; this is seen from Theorems 1.11 and 3.3.

The conditions in Theorem 4.2 are sufficient but not necessary for the canonical map $\phi : A \to C$ to be an epimorphism in the category of rings as will be seen from Example 4.3 below. As a necessary and sufficient condition for ϕ to be so we have ${}_CC_A \otimes {}_AC \in \mathfrak{D}({}_CV)$; indeed, with the notations in §3, we have ${}_CC \cong S_0({}_CC_A \otimes {}_AC)$, since ${}_AC \cong S({}_AC) \cong QS_0({}_CC_A \otimes {}_AC)$ and $P({}_AC) = {}_CC$.

Example 4.3. Let A be a subalgebra of $(K)_4$ over a commutative field K such that

$$e_1 = c_{11} + c_{44},\ c_{22},\ c_{33},\ c_{21},\ c_{31},\ c_{32},\ c_{41},\ c_{42},\ c_{43}$$

form a K–basis of A, where c_{ik} are matrix units in $(K)_4$. Let us put ${}_AV = Ae_1$. Then ${}_AV$ is injective. Let C be the double centralizer of ${}_AV$. Then the canonical map $A \to C$ is a (left flat) epimorphism but $\mathfrak{D}({}_AV)$ is not an exact subcategory of ${}_A\mathfrak{M}$.

5. A condition for exactness of the subcategory $\mathfrak{D}(V)$ and a weak version of Nakayama's conjecture

A full subcategory $\mathfrak{B}$ of ${}_A\mathfrak{M}$ is called an exact subcategory if the inclusion functor from $\mathfrak{B}$ to ${}_A\mathfrak{M}$ is exact, in other words, if, for any morphism $g : Y \to Y'$ in $\mathfrak{B}$, the modules $Y \oplus Y'$, Ker g and Coker g which are considered in ${}_A\mathfrak{M}$ are all contained in $\mathfrak{B}$ (cf. [6, p. 71]).

Theorem 5.1 ([17]). Let V be a left A–module of type FI. Then $\mathfrak{D}(V)$ is an exact subcategory of ${}_A\mathfrak{M}$ if and only if V–dom. dim $X = \infty$ for every left A–module X belonging to $\mathfrak{D}(V)$.

As an application of Theorems 5.1 and 4.2 we have the following theorem.

Theorem 5.2. Let Q be the maximal left quotient ring of A in the sense of Utumi–Lambek. Then Q is a right S–ring, that is, $E({}_QQ)$ is an injective cogenerator in ${}_Q\mathfrak{M}$ if and only if $E({}_AA)$–dom. dim $X = \infty$ for every $X \in \mathfrak{D}(E({}_AA))$ and $\mathfrak{D}(E({}_AA))$ is closed under direct sums.

Proof. As is shown by Lambek [12] (cf. Theorem 3.2), Q is the double centralizer of $E({}_AA)$ and ${}_Q[E({}_AA)] = E({}_QQ)$. Since $E({}_QQ)$ is an injective cogenerator if and only if $\mathfrak{D}(E({}_QQ)) = {}_Q\mathfrak{M}$, we obtain Theorem 5.2 from Theorems 4.2 and 5.1.

Corollary 5.3. Let A be a left Artinian ring such that $E({}_AA)$ is projective and $E({}_AA)$–dom. dim ${}_AA \geqslant 2$. Then A is quasi–Frobenius if and only if $E({}_AA)$–dom. dim $X = \infty$ for all $X \in \mathfrak{D}(E({}_AA))$.

Proof. From the assumption of Corollary 5.3 it follows that A is its own maximal left quotient ring (cf. [16]), and $\mathfrak{D}(E({}_AA))$ is closed under direct sums. Hence Corollary 5.3 is a direct consequence of Theorem 5.2, since a left Artinian QF–3 ring A which is a right S–ring is quasi–Frobenius.

Corollary 5.3 is a weak version of Nakayama's conjecture. The original conjecture of Nakayama asserts essentially that under the same assumption as in Corollary 5.3, if A is a finite dimensional algebra and if $E({}_AA)$–dom. dim ${}_AA = \infty$, then A is quasi–Frobenius; because in this case the dominant dimension of ${}_AA$ coincides with $E({}_AA)$–dominant dimension of ${}_AA$ and by Mueller [21] the former coincides with the Nakayama dimension of A.

An object function $\tau : {}_A\mathfrak{M} \to {}_A\mathfrak{M}$ is called a radical (Maranda [30]) if

$$\tau(X) \subset X,\ \tau(X/\tau(X)) = 0,\ f(\tau(X)) \subset \tau(Y)$$

for any left A–modules X and Y and any homomorphism $f : X \to Y$. Then the full subcategory $\mathfrak{F} = \{ X \in {}_A\mathfrak{M} \mid \tau(X) = 0 \}$ is a reflective subcategory of ${}_A\mathfrak{M}$; indeed, $S : {}_A\mathfrak{M} \to \mathfrak{F}$ defined by $S(X) = X/\tau(X)$ is a left adjoint of the inclusion functor $T : \mathfrak{F} \to {}_A\mathfrak{M}$.

Theorem 5.4. Let $\tau : {}_A\mathfrak{M} \to {}_A\mathfrak{M}$ be a radical and $\mathfrak{F}$ the subcategory defined above. Then the following statements are equivalent.

(a) $\mathfrak{F}$ is a Grothendieck category.

(b) $\mathfrak{F}$ is an exact subcategory of ${}_A\mathfrak{M}$ closed under direct sums.

(c) There exists a two–sided ideal I of A such that $\mathfrak{F} = \{ X \in {}_A\mathfrak{M} \mid IX = 0 \}$.

In case (c) holds, we have $\tau(X) = IX$ for $X \in {}_A\mathfrak{M}$. In case τ is an idempotent radical, that is, $\tau^2 = \tau$, then I in (c) is an idempotent ideal.

Proof. Assume (a) holds. Then by Theorem 1.11 there exists a left A–module V of type FI such that $\mathfrak{F} = \mathfrak{D}(V)$. Since $\mathfrak{F}$ is closed under submodules, we have $\mathfrak{D}(V) = \mathfrak{D}'(V)$ and hence by Theorem 5.1 $\mathfrak{F}$ is an exact subcategory of ${}_A\mathfrak{M}$. Since $\mathfrak{D}'(V)$ is closed under direct sums, so is $\mathfrak{F}$. Thus (b) holds. Let us put $\tau(A) = I$. Then I is a two–sided ideal of A and $S({}_AA) = A/I$. Therefore by Theorem 3.3 the double centralizer of V is A/I, and hence by Theorem 4.2 we see that V is an injective cogenerator in ${}_{A/I}\mathfrak{M}$. This proves (c). Thus (a) ⇒ (b) and (a) ⇒ (c) are proved. Since the implications (b) ⇒ (a) and (c) ⇒ (a) are obvious, we have the equivalence of (a), (b), and (c).

Next, assume (c) holds. Let $X \in {}_A\mathfrak{M}$. Then $X/\tau(X) \in \mathfrak{F}$. Hence $IX \subset \tau(X)$. Let $f : X \to X/IX$ be a canonical map. Then $f(\tau(X)) \subset \tau(X/IX) = 0$. This shows that $\tau(X) \subset IX$, and hence we have $\tau(X) = IX$. Finally, the last assertion is obvious.

The following supplements Jans [9, Theorem 2.4].

Corollary 5.5. Let $(\mathcal{T}, \mathfrak{F})$ be a torsion theory in the sense of Dickson [4]. Then the following statements are equivalent.

(a) $\mathcal{T}$ and $\mathfrak{F}$ are TTF classes in the sense of Jans [9] (i.e., classes of modules closed under submodules, factors, extensions and direct products).[2]

(b) $(\mathfrak{F}, \mathcal{T})$ is a torsion theory.

(c) There exist two–sided ideals I, J of A such that $A = I \oplus J$, $\mathcal{T} = \{ X \in {}_A\mathfrak{M} \mid IX = X \}$ and $\mathfrak{F} = \{ X \in {}_A\mathfrak{M} \mid JX = X \}$.

Proof. Since (c) $\Rightarrow$ (b) $\Rightarrow$ (a) are obvious, we have only to prove (a) $\Rightarrow$ (c). For this purpose, suppose (a) holds. By [4, Theorem 2.8] a subfunctor τ of the identity functor on ${}_A\mathfrak{M}$ is an idempotent radical if and only if $\{X \in {}_A\mathfrak{M} \mid \tau(X) = 0\}$ is closed under submodules, extensions and direct products. Hence by Theorem 5.4 there are idempotent two–sided ideals I, J of A such that

$$\mathfrak{F} = \{X \in {}_A\mathfrak{M} \mid IX = 0\}, \ \mathcal{T} = \{X \in {}_A\mathfrak{M} \mid JX = 0\}.$$

Since $A/J \in \mathcal{T}$, we have $I(A/J) = A/J$ by Theorem 5.4. Hence $A = I + J$. Let $\mathcal{K}$ be a torsion class such that $(\mathcal{K}, \mathcal{T})$ is a torsion theory. Then $IX \cap JX \in \mathcal{T} \cap \mathcal{K}$ for $X \in {}_A\mathfrak{M}$, and hence $X = IX \oplus JX$. Thus (c) holds.

6. Injective modules

Let $\phi : A \to C$ be a ring homomorphism and $\mathfrak{F}$ an IT filter of left ideals of A. Let us put

$$(\phi(A) : c) = \{ x \in A \mid \phi(x)c \in \phi(A) \}, \ \text{for} \ c \in C.$$

$$C' = \{ c \in C \mid (\phi(A) : c) \in \mathfrak{F} \}.$$

Then we have (cf. [20])

Lemma 6.1. C' is a subring of C and $\phi(A) \subset C'$.

Now, let ${}_AV_B$ be an A–B–bimodule and let $C = \mathrm{End}(V_B)$. Then there is a canonical ring homomorphism $\phi : A \to C$. Let $\mathfrak{F}_h({}_AV)$ be the set of all left ideals J of A such that

[2] As for TTF classes, cf. also Alin and Armendariz [1].

$$\mathrm{Hom}_A(A/(J:a),\ V) = 0 \quad \text{for all} \quad a \in A.$$

Then $\mathfrak{F}_h({}_AV)$ is an IT filter and

$$J \in \mathfrak{F}_h({}_AV) \Leftrightarrow \mathrm{Hom}_A(A/J,\ E({}_AV)) = 0.$$

Lemma 6.2. ${}_CV$–dom. dim ${}_CC \geqslant 2$.

Let us put

$$C' = \{\, c \in C \mid (\phi(A) : c) \in \mathfrak{F}_h({}_AV) \,\}.$$

In the following, we consider the case where ${}_AV$ is injective. In this case we have $F(\mathcal{T}(V)) = \mathfrak{F}_h(V)$. The case where V is finitely cogenerating was treated in Theorem 3.2. Here we shall discuss the general case.

Theorem 6.3 ([20]). Let ${}_AV_B$ be an A–B–bimodule such that ${}_AV$ is injective, and let $C = \mathrm{End}(V_B)$. Then with the above notations the following statements hold.

(i) C' is the quotient ring of A with respect to the filter $\mathfrak{F}_h({}_AV)$ and coincides with the double centralizer of ${}_AW$, where ${}_AW$ is a finitely cogenerating, injective left A–module such that $\mathfrak{D}(W) = \mathfrak{D}(V)$.

(ii) ${}_{C'}V_B \cong \mathrm{Hom}_A({}_AC'_{C'},\ {}_AV_B)$, ${}_{C'}V$–dom. dim ${}_{C'}C' \geqslant 2$ and ${}_{C'}V$ is injective, ${}_{C'}V$–dom. dim ${}_{C'}C/{}_{C'}C' \geqslant 1$.

(iii) $C' = C$ if and only if ${}_AV$ is of type F_h (cf. Definition 1.5) and C is the double centralizer of ${}_AV$.

(iv) $C' = \phi(A)$ if and only if V–dom. $\dim_A[\phi(A)] \geqslant 2$.

Theorem 6.4 ([20]). Let ${}_AV_B$ be an A–B–bimodule and let $C = \mathrm{End}(V_B)$. Then we have a canonical ring homomorphism $\phi : A \to C$ and the following statements are equivalent.

I. ${}_AV$ is an injective module of type F_h (resp. type FI) and $B = \mathrm{End}({}_AV)$.

II. (a) The set of all left ideals J of A such that $\mathrm{Hom}_A(A/J,\ V) = 0$ is an IT filter and C is the quotient ring of A with respect to this filter.

(b) ${}_CV$ is injective (resp. finitely cogenerating and injective).

III. (a) C is the quotient ring of A with respect to the filter $\mathfrak{F}_h({}_AV)$.

(b) The same as in II.

IV. (a) $\operatorname{Hom}_A(A/(\phi(A) : c), V) = 0$ for every element c of C.

(b) The same as in III.

(c) $\operatorname{Hom}_A(Ax, V) = 0$ for $x \in \operatorname{Ker} \phi$.

Let A be the ring Z of integers and let p be a prime number. Let us put ${}_AV = Z_{p^\infty} = E(Z/pZ)$. Then the double centralizer C of ${}_AV$ is the ring of all p-adic integers and C' described in Theorem 6.3 is the ring of all rational numbers with denominators prime to p. Thus, in Theorem 6.3 we have $C' \neq C$ and $C' \neq \phi(A)$ in general. In this case C' is dense in C with respect to the p-adic topology. Such a situation is discussed in Lambek [31] more generally.

Unless ${}_AV$ is injective, Theorem 6.3 is not true in general. However, we have the following theorem.

Theorem 6.5. Let ${}_AV_B$ be an A–B–bimodule such that $KE({}_AV) = 0$, where $K = \{a \in A \mid av = 0 \text{ for all } v \in V\}$ and there is an A–homomorphism ψ from $\Sigma \oplus W_\lambda$ onto V with injective left A–modules W_λ, $\lambda \in \Lambda$.[3] Let $C = \operatorname{End}(V_B)$. Then there is a canonical ring homomorphism $\phi : A \to C$. Set $C' = \{c \in C \mid (\phi(A) : c) \in \mathfrak{F}_h({}_AV)\}$. Then C' is the quotient ring of A with respect to $\mathfrak{F}_h({}_AV)$, and $C' = C$ if and only if $\operatorname{Hom}_A({}_AC/\phi(A), E({}_AV)) = 0$.

Proof. Let $c_0, c_1 \in C'$ and $c_0 \neq 0$. Then there is $v_0 \in V$ such that $c_0v_0 \neq 0$. Let us define $f \in \operatorname{Hom}_A({}_AA, V)$ by $f(a) = ac_0v_0$. Since $\operatorname{Hom}_A(A/(\phi(A) : c_1), V) = 0$, there is $a_0 \in (\phi(A) : c_1)$ such that $a_0c_0v_0 \neq 0$. This shows that C' may be identified with a subring of a maximal left quotient ring Q_0 of $\phi(A)$.

Next, let $q \in Q_0$ be such that $(\phi(A) : q) \in \mathfrak{F}_h({}_AV)$, and let $v \in V$. Then there is a finite subset Λ_0 of Λ such that $w \in \Sigma_{\lambda\in\Lambda_0} \oplus W_\lambda = W$, $\psi_0(w) = v$, where $\psi_0 = \psi \mid W$. Then, since W is injective, the map $f : (\phi(A) : q) \to W$ defined by $f(a) = (\phi(a)q)w$ is extended to an A–homomorphism $g : {}_AA \to W$. Then $\psi_0 \circ g \in \operatorname{Hom}_A({}_AA, V)$. Since $(\phi(A) : q) \in \mathfrak{F}_h({}_AV)$, the map $\psi_0 \circ g$ is uniquely determined by $\psi_0 \circ f$. If we put $q * v = (\psi_0 \circ g)(1)$, then $q * v$ is determined uniquely by q and v, and we have

$$q * vb = (q * v)b \quad \text{for} \quad b \in B;$$

[3] This is equivalent to saying that there is an A–homomorphism from a direct sum of copies of $E({}_AA)$ onto V. Cf. Beachy [2].

because for $a \in (\phi(A) : q)$ we have

$$a(q * vb) = (\phi(a)q)(vb) = [(\phi(a)q)v]b = [a(q * v)]b = a[(q * v)b].$$

This shows that $q \in C$ and hence $q \in C'$. Therefore

$$C' = \{ q \in Q_0 \mid (\phi(A) : q) \in \mathfrak{F}_h({}_AV) \}.$$

This proves Theorem 6.5.

As an application of Theorem 6.5 we shall prove the following theorem.

Theorem 6.6. Under the same assumption and notation as in Theorem 6.5 let D be a subring of C containing $\phi(A)$. If $E(V)$–dom. dim ${}_AC \geqslant 2$, then $D \supset C'$.

Proof. Let us denote by $\tau(X)$ the largest submodule of a left A–module X which is $\mathfrak{F}_h({}_AV)$–torsion, where a left A–module Y is called $\mathfrak{F}_h({}_AV)$–torsion if $\mathrm{Hom}_A(Y, E({}_AV)) = 0$. By the definition of C' we see that $\tau({}_AD/{}_A\phi(A)) = {}_A[C' \cap D]/{}_A\phi(A)$. Hence $E(V)$–dom. dim ${}_AD/{}_A[C' \cap D] \geqslant 1$. Since $E(V)$–dom. dim ${}_AD \geqslant 2$, we have $E(V)$–dom. dim ${}_A[C' \cap D] \geqslant 2$. Since $E({}_A\phi(A)) \supset {}_AC' \supset {}_A[C' \cap D] \supset \phi(A)$, we have $E({}_A[C' \cap D]) = E({}_A[\phi(A)])$ and hence $E(V)$–dom. dim ${}_AC'/{}_A[C' \cap D] \geqslant 1$. On the other hand, ${}_AC'/{}_A[C' \cap D]$ is $\mathfrak{F}_h({}_AV)$–torsion. Therefore ${}_AC'/{}_A[C' \cap D] = 0$. This proves that $D \supset C'$.

7. Flat modules

Let ${}_BU_A$ be a B–A–bimodule and let us put $C = \mathrm{End}({}_BU)$. Then there is a canonical ring homomorphism $\phi : A \to C$. Let $\mathfrak{F}_t(U_A)$ be the set of all left ideals J of A such that $U(J{:}a) = U$ for every element a of A. Then we have

$$J \in \mathfrak{F}_t(U_A) \Leftrightarrow \mathrm{Hom}_A(A/J, E({}_AV)) = 0,$$

where ${}_AV = \mathrm{Hom}_B({}_BU_A, {}_BW)$ and ${}_BW$ is an injective cogenerator in ${}_B\mathfrak{M}$ such that ${}_BU \subset {}_BW$. $\mathfrak{F}_t(U_A)$ is an IT filter. Let us put

$$C' = \{ c \in C \mid (\phi(A) : c) \in \mathfrak{F}_t(U_A) \}.$$

With these notations we have the following theorems, corresponding to Theorems 6.3 and 6.4.

Theorem 7.1 ([20]). Suppose that U_A is flat. Then the following statements are true.

(i) C' is the quotient ring of A with respect to the filter $\mathfrak{F}_t(U_A)$ and coincides with the double centralizer of $_AV$; $_AV$–dom. dim $_AC \geqslant 2$.

(ii) $_BU_{C'} \cong {}_BU_A \otimes {}_AC'_{C'}$ and $U_{C'}$ is flat; $_AV$–dom. dim $_AC/_AC' \geqslant 1$.

(iii) $C' = C$ if and only if U_A is a module of type F_t (cf. Definition 1.1) and C is the double centralizer of U_A.

(iv) $C' = \phi(A)$ if and only if $_AV$–dom. dim $_A[\phi(A)] \geqslant 2$.

Theorem 7.2 ([20]). Let $_BU_A$ be a B–A–bimodule and let $C = \mathrm{End}(_BU)$. Then there is a canonical ring homomorphism $\phi : A \to C$ and the following statements are equivalent.

I. U_A is a flat module of type F_t (resp. type FP) and $B = \mathrm{End}(U_A)$.

II. (a) The set of all left ideals J of A such that $UJ = U$ is an IT filter and C is the quotient ring of A with respect to this filter.

(b) U_C is flat (resp. finitely generated and projective).

III. (a) C is the quotient ring of A with respect to the filter $\mathfrak{F}_t(U_A)$.

(b) The same as in II.

IV. (a) $U(\phi(A) : c) = U$ for $c \in C$.

(b) The same as in II.

(c) $U_A \otimes A_x = 0$ for $x \in \mathrm{Ker}\ \phi$.

V. (a) $\mathrm{Hom}_A(_AC/\phi(A), E(_AV)) = 0$.

(b) The same as in IV.

(c) The same as in IV.

Unless U_A is flat, Theorem 7.1 does not hold in general. However, we have the following theorem.

Theorem 7.3. With the notations at the beginning of this section let $_BU_A$ be such that (a) $KE(_AV) = 0$ where $K = \{ a \in A \mid aV = 0 \} = \{ a \in A \mid Ua = 0 \}$, and (b) there is a monomorphism ψ from U_A to a direct product of flat right A–modules L_λ, $\lambda \in \Lambda$ with $L_\lambda K = 0$. Then C' is the quotient ring of A with respect to the IT filter $\mathfrak{F}_t(U_A)$, and $C' = C$ if and only if $\mathrm{Hom}_A(_AC/\phi(A), E(_AV)) = 0$.

Proof. We shall first remark that C' is identified with a subring of a maximal left quotient ring Q_0 of $\phi(A)$. Indeed, let $c_0, c_1 \in C'$ $c_0 \neq 0$. Then there is $u_0 \in U$ such that $u_0c_0 \neq 0$, and there are $u_i \in U$, $a_i \in A$, $i = 1,\cdots,n$, such that

$$u_0 = \Sigma\, u_i\phi(a_i);\ a_i \in (\phi(A) : c_1),\ i = 1,\cdots,n.$$

Hence there exists some a_i with $\phi(a_i)c_0 \neq 0$. This shows that C' may be identified with a subring of Q_0.

Now, let q be an element of Q_0 such that $(\phi(A) : q) \in \mathfrak{F}_t(U_A)$. If $\Sigma\, u_ia_i = 0$ for $u_i \in U$, $a_i \in (\phi(A) : q)$, $i = 1,\cdots,n$, then $\Sigma\, u_i(\phi(a_i)q) = 0$. To prove this, let $f \in \mathrm{Hom}_A(U_A, L_\lambda)$. Then $\Sigma\, f(u_i)a_i = 0$. Since L_λ is flat, there exist $l_j \in L_\lambda$, $\xi_{ji} \in A$, $j = 1,\cdots,m$; $i = 1,\cdots,n$, such that

$$f(u_i) = \Sigma\, l_j\phi(\xi_{ji});\ \sum_i \phi(\xi_{ji})\phi(a_i) = 0.$$

Since $\Sigma\, \phi(\xi_{ji})(\phi(a_i)q) = (\Sigma\, \phi(\xi_{ji})\phi(a_i))q = 0$ in Q_0, we have $\Sigma\, f(u_i)(\phi(a_i)q) = 0$. Therefore $\Sigma\, u_i(\phi(a_i)q) = 0$.

Hence, if we put

$$u * q = \Sigma\, u_j(\phi(a_j)q) \quad \text{for} \quad u = \Sigma\, u_ja_j \quad \text{with} \quad a_j \in (\phi(A) : q),$$

then $u * q$ is determined uniquely by $u \in U$ and q, and we have

$$b(u * q) = (bu) * q \quad \text{for} \quad b \in B.$$

Thus we have $q \in C$, and hence $q \in C'$. Therefore $C' = \{ q \in Q_0 \mid (\phi(A) : q) \in \mathfrak{F}_t(U_A) \}$. This proves Theorem 7.3.

Remark. If U_A is faithful and torsionless, ${}_BU_A$ satisfies the assumption of Theorem 7.3.

8. Flat epimorphisms in the category of rings

A ring homomorphism $\phi : A \to B$ is called a left (resp. two-sided) flat epimorphism if ϕ is an epimorphism in the category of rings and if B is flat as a right A-module (resp. both as a right and a left module); in case ϕ is an inclusion map, B is called a left (resp. two-sided) flat epimorphic extension of A. The classical left quotient ring of A, if it exists, is an example of a left flat epimorphic extension of A. It is easy to see that the composite of two left flat epimorphisms is a left flat epimorphism.

Theorem 8.1 ([19]). Let $f_\lambda : A \to B_\lambda$ be a left flat epimorphism for each element λ of a set Λ. Then there is a left flat epimorphism $\phi : A \to C$ having the following properties:

(a) for each λ there is a ring homomorphism $g_\lambda : B_\lambda \to C$ such that $\phi = g_\lambda \cdot f_\lambda$;

(b) for any left flat epimorphism $\phi' : A \to C'$ such that for each λ there is a ring homomorphism $g'_\lambda : B_\lambda \to C'$ such that $\phi' = g'_\lambda \circ f_\lambda$, there exists a ring homomorphism $\psi : C \to C'$ such that $\phi' = \psi \circ \phi$.

Furthermore, if $\operatorname{Ker} f_\lambda = K$ for each λ then $\operatorname{Ker} \phi = K$.

It is to be noted that in Theorem 8.1 C may be equal to zero. Of course, there are cases in which $\operatorname{Ker} f_\lambda$ are not the same for all λ and $C \neq 0$.

Since any left flat epimorphism is obtained by means of an IT filter by Theorems 4.1 and 3.2 and the totality of IT filters of A forms a set, by applying Theorem 8.1 and [19, Lemma 3.1] we have the following theorem.

Theorem 8.2. For a two-sided ideal K of A such that there is a left flat epimorphism $\psi : A \to D$ with $\operatorname{Ker} \psi = K$, there exists a left flat epimorphism $\phi : A \to C$ with $\operatorname{Ker} \phi = K$ such that for any left flat epimorphism $f : A \to B$ with $\operatorname{Ker} f = K$ there is a ring monomorphism $g : B \to C$ such that $\phi = g \circ f$.

Different proofs of Theorem 8.2 for the case $K = 0$ are given in [20]; they are based on Theorem 8.3 and Lemma 8.4 below.

Theorem 8.3 ([20]). Let $\phi : A \to B$ be a ring homomorphism. Then the following statements are equivalent.

I. ϕ is a left flat epimorphism.

II. The set of all left ideals J of A such that $B\phi(J) = B$ is an IT filter and B is the quotient ring of A with respect to this filter.

III. B is the quotient ring of A with respect to the filter $\mathfrak{F}_t(B_A)$.

IV. (a) $B\phi(\phi(A) : b) = B$ for $b \in B$.

(b) $B_A \otimes Ax = 0$ for $x \in \operatorname{Ker} \phi$.

V. (a) $\operatorname{Hom}_A({}_A B/\phi(A), E({}_A Y)) = 0$ for every left B-module Y.

(b) The same as in IV.

This is a direct consequence of Theorem 7.2; we have only to put ${}_B U_A = {}_B B_A$.

Lemma 8.4. Let $\phi : A \to B$ be a ring homomorphism and let B' be the set of all elements y of B such that $B\phi(\phi(A) : \phi(a)y) = B$ for all $a \in A$. Then B' is a subring of B containing $\phi(A)$. (Cf. Lemma 6.1)

The equivalence of I, II and IV in Theorem 8.3 has been announced by Popescu and Spircu [22] and proved in their recent paper [23]. The equivalence I $\Leftrightarrow$ V for the case of Ker $\phi = 0$ is due to Findlay [5].

Theorem 8.5 below is a direct consequence of Theorem 8.3 and Lemma 8.4.

Theorem 8.5. Let $\phi : A \to B$ be a ring homomorphism and let $\{B_\lambda \mid \lambda \in \Lambda\}$ be a set of subrings of B such that $\phi(A) \subset B_\lambda \subset B$, $\lambda \in \Lambda$. Let C be the smallest subring of B which contains all B_λ. Let $f_\lambda : A \to B_\lambda$ (resp. $g : A \to C$) be the composite of the maps:

$$A \to \phi(A) \xrightarrow{i_\lambda} B_\lambda \quad (\text{resp. } A \to \phi(A) \xrightarrow{j} C)$$

where i_λ and j are inclusion maps. If each f_λ is a left (resp. two–sided) flat epimorphism, so is g.

Proof. Let C' be the set of all elements c of C such that $C\phi(\phi(A) : \phi(a)c) = C$ for all $a \in A$. Then $\phi(A) \subset C' \subset C$ and C' is a subring of C by Lemma 8.4. Suppose that each f_λ is a left flat epimorphism. Then $B_\lambda \subset C'$ by Theorem 8.3 and hence $C' = C$. On the other hand, since $[B_\lambda]_A \otimes Ax = 0$ for $x \in \text{Ker } \phi$ and $C = \Sigma cB_\lambda$ where $c \in C$, $\lambda \in \Lambda$, we have $C_A \otimes Ax = 0$ for $x \in \text{Ker } \phi$. Therefore by Theorem 8.3 g is a left flat epimorphism.

If we apply Theorem 8.5 to the case that B is a maximal left quotient ring $Q(A)$ of A we see the existence of a maximal left flat epimorphic extension $M(A)$ of A which is unique up to isomorphism leaving every element of A invariant, since any left flat epimorphic extension of A can be identified with a subring of $Q(A)$ by Theorems 3.2 and 4.1.

Next, let us consider the situation described in Theorem 8.2. If we identify $\psi(A)$ with A/K, then we have $D \subset M(A/K)$, and by Theorem 8.7 below, the inclusion map $j : D \to M(A/K)$ is a left flat epimorphism. Hence $j\psi : A \to M(A/K)$ is a left flat epimorphism. By using Theorem 8.7 again we see that C in Theorem 8.2 can be identified with $M(A/K)$.

The existence of a maximal left flat epimorphic extension of a ring was proved for the first time by Lazard [14] for the case of commutative rings. For the general case it was announced by Popescu and Spircu [22], although a sketch of their proof given there contains an inexact argument which, however, has been avoided in their recent paper [23]. During the press of our papers [19] and [20], besides [23] there appeared two papers, Findlay [5] and Knight [11]; the former establishes the existence of a maximal left flat epimorphic extension of a ring A, while the latter proves the existence of a maximal two-sided flat epimorphic extension of A.

On the other hand, as is well known (e.g., cf. Bergman [3]), Theorem 8.5 remains to be true even if we replace "a left (resp. two-sided) flat epimorphism" by "an epimorphism". A maximal epimorphic extension of a ring, however, does not exist in general. As for epimorphic extensions of commutative rings, cf. also Storrer [29].

Example 8.6. Let us consider the following two subalgebras A and B of the full matrix algebra $(K)_4$ over a commutative field K:

$$A = K(c_{11} + c_{44}) + Kc_{22} + Kc_{33} + Kc_{12} + Kc_{43},$$

$$B = A + Kc_{11} + Kc_{21} + Kc_{34},$$

where c_{ik} are matrix units in $(K)_4$.

Then B is a maximal right quotient ring of A. Let Q be a maximal left quotient ring of A. Then the inclusion map $A \to Q$ (resp. $A \to B$) is a left (resp. right) flat epimorphism. However, as was shown in [20, Example 4.3], there does not exist a ring which contains B and Q as its subrings. Therefore there does not exist a maximal epimorphic extension of A.

As for subrings of an epimorphic extension we have the following theorem.

Theorem 8.7. Let $\phi : A \to B$ be a left flat epimorphism and let C be a subring of B such that $\phi(A) \subset C \subset B$. Then the inclusion map $C \to B$ is a left flat epimorphism, and if ${}_AC$ is flat the map $g : A \to C$ defined in an obvious way is an epimorphism.

Proof. The first assertion is easy to see by virtue of IV of Theorem 8.3, and the second follows from the commutative diagram

$$\begin{array}{ccccc} C_A \otimes {}_AC & \to & B_A \otimes {}_AC & \to & B_A \otimes {}_AB \\ \downarrow & & & & \downarrow \\ C & & \longrightarrow & & B \end{array}$$

in which all morphisms except the vertical one on the left hand side are one-to-one ([23]).

The second part of Theorem 8.7, however, fails to be valid if we replace ${}_AC$ by C_A.

Example 8.8. Let A and C be subalgebras of a full matrix algebra $(K)_3$ over a commutative field K such that

$$A = Kc_{11} + Kc_{22} + Kc_{33} + Kc_{12} + Kc_{13}, \; C = A + Kc_{23}.$$

Let us put $B = (K)_3$. Then the inclusion map $A \to B$ is a left flat epimorphism and C_A is flat, but the inclusion map $A \to C$ is not an epimorphism.

9. The double centralizer property of modules

A faithful left A–module V is said to have the double centralizer property if the double centralizer of V coincides with A itself.

Theorem 9.1 ([20]). Let V be a faithful left A–module. Then V has the double centralizer property if and only if V is of type F_h (cf. Definition 1.5) and V–dom. dim ${}_AA \geqslant 2$.

In case V is a left A–module of type FI, V is necessarily of type F_h. Hence we have

Corollary 9.2 ([16]). Let V be a faithful left A–module of type FI. Then V has the double centralizer property if and only if V–dom. dim ${}_AA \geqslant 2$.

Our Theorem 9.1, however, is not satisfactory in view of the fact that Definition 1.5 for modules of type F_h necessitates the determination of the double centralizer of V. Theorem 9.5 below is better in this respect than Theorem 9.1, although the case treated there is rather restrictive. Theorem 9.1 is proved also by Suzuki [25] independently.

Lemma 9.3. Let ${}_AV$ be a left A–module and let $B = \mathrm{End}({}_AV)$. Then the following two conditions are equivalent.

(a) There exists an A–monomorphism $\phi : A \to \sum_{i=1}^{n} \oplus V$ for some positive integer n such that for any A–homomorphism $f : A \to V$ there is some $g \in \mathrm{Hom}_A(\sum_{i=1}^{n} \oplus V, V)$ with $f = g \circ \phi$.

(b) V_B is finitely generated.

This lemma is proved also by T. Kato.

Lemma 9.4. Under the assumption (a), let us put

$$\phi(1) = (v'_1,\cdots,v'_n) \in \sum_{i=1}^{n} \oplus V,$$

and define $\psi : {}_C C \to \sum_{i=1}^{n} \oplus {}_C V$ by

$$\psi(c) = (cv'_1,\cdots,cv'_n), \quad \text{where} \quad C = \operatorname{End}(V_B).$$

Then

$$\psi(C) = \{ (v_1,\cdots,v_n) \in \sum_{i=1}^{n} \oplus V \mid \Sigma\, v_i b_i = 0 \ \text{ for all } \ b_i \in B,$$

$$i = 1,\cdots,n, \ \text{such that} \ \Sigma\, v'_i b_i = 0 \}.$$

This lemma is proved essentially in our paper [16, Theorem 2.3]. As a direct consequence of it we have

Theorem 9.5. Under the assumption (a) in Lemma 9.3, V has the double centralizer property if and only if $(\sum_{i=1}^{n} \oplus V)/\phi(A)$ is isomorphic to a submodule of a direct product of copies of V.

This theorem is obtained in [18] and proved also by T. Kato and Y. Suzuki with the same proof as ours.

Theorem 9.5 is applicable to the case where A is a left Artinian ring which is finitely generated over the center (e.g., a finite dimensional algebra over a field) and ${}_A V$ is finitely generated.

Finally, we shall prove the following theorem.

Theorem 9.6. Let $S : {}_A\mathcal{M} \cong {}_B\mathcal{M}$ be an equivalence and let X be a faithful left A–module. Then $S(X)$ is a faithful left B–module and the following assertions are true.

(i) $S(X)$ is of type F_h if and only if X is of type F_h.

(ii) $S(X)$–dom. dim ${}_B B \geqslant 2$ if and only if X–dom. dim ${}_A A \geqslant 2$.

(iii) $S(X)$ has the double centralizer property if and only if X has the double centralizer property.

Proof. As is known, there exist a B–A–bimodule ${}_B U_A$ and an A–B–bimodule ${}_A V_B$ such that

(1) $S(X) \cong {}_BU_A \otimes X$,

(2) U_A, ${}_BU$ are finitely generated and projective,

(3) $B = \mathrm{End}(U_A)$, $A = \mathrm{End}({}_BU)$,

(4) ${}_AV_B \cong \mathrm{Hom}_A({}_BU_A, {}_AA_A) \cong \mathrm{Hom}_B({}_BU_A, {}_BB_B)$.

Now, let X be a faithful left A-module. Then $S(X)$ is a faithful left B-module.[4] Let us put

$$K = \mathrm{End}({}_AX),\ C = \mathrm{End}(X_K),\ Y = S(X).$$

Then $K = \mathrm{End}({}_BY)$. Let us put $D = \mathrm{End}(Y_K)$. Since

$$\begin{aligned}\mathrm{Hom}_K({}_BU_A \otimes {}_AX_K, {}_BU_A \otimes {}_AX_K) &\cong \mathrm{Hom}_A({}_BU_A, \mathrm{Hom}_K({}_AX_K, {}_BU_A \otimes {}_AX_K)) \\ \cong \mathrm{Hom}_A({}_BU_A, {}_BU_A \otimes \mathrm{Hom}_K({}_AX_K, {}_AX_K)) &\cong \mathrm{Hom}_A({}_BU_A, {}_BU_A \otimes {}_AC_A) \\ &\cong {}_BU_A \otimes {}_AC_A \otimes {}_AV_B,\end{aligned}$$

we have an isomorphism

$$\phi : {}_BU_A \otimes {}_AC_A \otimes {}_AV_B \cong {}_BD_B$$

defined by

$$[\phi(\Sigma u_i \otimes c_i \otimes v_i)](u \otimes x) = \Sigma u_i \otimes c_i\omega_A(v_i \otimes u)x,$$

where $u_i \in U$, $c_i \in C$, $v_i \in V$, $u \in U$, $x \in X$ and

$$\omega_A : {}_AV_B \otimes {}_BU_A \to {}_AA_A,\ \omega_B : {}_BU_A \otimes {}_AV_B \to {}_BB_B$$

are isomorphisms as bimodules such that

$$u\omega_A(v \otimes u') = \omega_B(u \otimes v)u' \quad \text{for } u, u' \in U, v \in V.$$

Then for $c_i = a_i \in A$ we have

$$[\phi(\Sigma u_i \otimes a_i \otimes v_i)](u \otimes x) = (\Sigma\omega_B(u_i \otimes a_iv_i))(u \otimes x),$$

[4] Cf. K. Morita: Adjoint pairs of functors and Frobenius extensions, Science Reports of Tokyo Kyoiku Daigaku, 9 (1965), 40-71; Lemma 2.4.

and hence

$$\phi({}_BU_A \otimes {}_AA_A \otimes {}_AV_B) = B.$$

Thus we have

$$\mathrm{Hom}_B({}_BD/{}_BB,\ {}_BY) \cong \mathrm{Hom}_B({}_BU_A \otimes ({}_AC_A/{}_AA_A) \otimes {}_AV,\ {}_BU_A \otimes X)$$

$$\cong \mathrm{Hom}_A(({}_AC_A/{}_AA_A) \otimes {}_AV,\ {}_AX) \cong \mathrm{Hom}_A({}_AV,\ \mathrm{Hom}_A({}_AC_A/{}_AA_A,\ {}_AX)).$$

Therefore, if X is of type F_h, so is Y. This proves (i).

Next, if X-dom. dim ${}_AA \geqslant 2$, then Y-dom. dim ${}_BU \geqslant 2$ and hence Y-dom. dim ${}_BB \geqslant 2$ since ${}_BU$ is a generator in ${}_B\mathfrak{M}$. (iii) is a direct consequence of (i) and (ii) in view of Theorem 9.1. It is to be noted that we have proved (iii) already in a joint paper with H. Tachikawa[5] and that our proof there uses a special case of the formula ${}_BD_B \cong {}_BU_A \otimes {}_AC_A \otimes {}_AV_B$ proved above.

References

1. J. S. Alin and E. P. Armendariz, TTF–classes over perfect rings, J. Australian Math. Soc. **9** (1970), 499-503.
2. J. A. Beachy, Bicommutators of cofaithful, fully divisible modules, (to appear in Can. J. Math.).
3. G. Bergman, Notes on epimorphisms of rings, (unpublished).
4. S. E. Dickson, A torsion theory for abelian categories, Trans. Amer. Math. Soc. **121** (1966), 223-235.
5. G. D. Findlay, Flat epimorphic extensions of rings, Math. Z. **118** (1970), 281-288.
6. P. Freyd, "Abelian categories," New York, 1964.
7. P. Gabriel, Des catégories abéliennes, Bull. Soc. Math. France **90** (1962), 323-448.
8. O. Goldman, Rings and modules of quotients, J. Algebra **13** (1969), 10-47.
9. J. P. Jans, Some aspects of torsion, Pacific J. Math. **15** (1965), 1249-1259.
10. T. Kato, Rings of U–dominant dimension $\geqslant 1$, Tohoku Math. J. **21** (1969) 321-327.
11. J. T. Knight, On epimorphisms of non–commutative rings, Proc. Cambridge Phil. Soc. **68** (1970), 589-600.
12. J. Lambek, On Utumi's ring of quotients, Can. J. Math. **15** (1963), 363-370.
13. J. Lambek, Torsion theories, additive semantics, and rings of quotients, (to appear).
14. D. Lazard, Epimorphismes plats d'anneaux, C. R. Acad. Sci. Paris **266** (1968), 314-316.

[5] QF–3 rings, 1967(unpublished).

15. K. Morita, On S–rings in the sense of F. Kasch, Nagoya J. Math. **27** (1966), 687-695.
16. K. Morita, Localizations in categories of modules, I, Math. Z. **114** (1970), 121-144.
17. K. Morita, Localizations in categories of modules, II, J. reine angew. Math. **242** (1970), 163-169.
18. K. Morita, Categories of modules, Lecture Notes at University of Pitts–burgh, 1970.
19. K. Morita, Localizations in categories of modules, III, Math. Z. **119** (1971), 313-320.
20. K. Morita, Flat modules, injective modules and quotient rings, Math. Z. **120** (1971), 25-40.
21. B. J. Mueller, The classification of algebras by dominant dimension, Can. J. Math. **20** (1968), 398-409.
22. N. Popescu and T. Spircu, Sur les epimorphismes plats d'anneaux, C. R. Acad. Sci. Paris **268** (1969), 376-379.
23. N. Popescu and T. Spircu, Quelque observations sur epimorphismes plats (à gauche) d'anneaux, J. Algebra **16** (1970), 40-59.
24. L. Silver, Non–commutative localizations and applications, J. Algebra **7** (1967), 44-76.
25. Y. Suzuki, Dominant dimension of double centralizers, (to appear).
26 H. Tachikawa, On splitting of module categories, Math. Z. **111** (1969), 145-150.
27. H. Tachikawa, Double centralizers and dominant dimension, Math. Z. **116** (1970), 79-88.
28. C. L. Walker and E. A. Walker, Quotient categories and rings of quotients, (to appear).
29. H. H. Storrer, Epimorphismen von kommutativen Ringen, Comment. Math. Helv. **43** (1968), 378-401.
30. J. M. Maranda, Injective structures, Trans. Amer. Math. Soc. **110** (1964), 98-135.
31. J. Lambek, Bicommutators of nice injectives, (to appear).

ON THE IDENTITIES OF AZUMAYA ALGEBRAS

C. Procesi

Università di Lecce

1. Generalized polynomial identities

Let $A \subset B$ be Λ-algebras. We form the free product $R = A * \Lambda\{x_1, \cdots, x_k\}$ between A and the free algebra $\Lambda\{x_1, \cdots, x_k\}$ in k variables. The elements of the ring R can be considered as non-commutative polynomials in the variables x_i with coefficients in A; each element of R has the form:

$$\Sigma\, a_0 x_{i_1} a_1 x_{i_2} \cdots a_{k-1} x_{i_k} a_k. \tag{1}$$

By the universal property of the free product we have:

$$\mathcal{M}(R, B) \cong \mathcal{M}(A, B) \times \mathcal{M}(\Lambda\{x_1, \cdots, x_k\}, B) \cong \mathcal{M}(A, B) \times B^k \tag{2}$$

(where if E, F are Λ-algebras we indicate by $\mathcal{M}(E, F)$ the set of algebra homomorphisms). We will consider the subset $\mathcal{M}_A(R, B)$ of ring homomorphisms which induce the inclusion map $i : A \to B$ on A.

It is clear, by the previous description of the set $\mathcal{M}(R, B)$, that $\mathcal{M}_A(R, B) \cong B^k$; more explicitly, if $(b_1, \cdots, b_k) \in B^k$, the homomorphism associated is called *evaluation* of the polynomials in $(b_1, \cdots, b_k)$, and it sends an element $\Sigma\, a_0 x_{i_1} a_1 x_{i_2} \cdots a_{k-1} x_{i_k} a_k$ to $\Sigma\, a_0 b_{i_1} a_1 b_{i_2} \cdots a_{k-1} b_{i_k} a_k$.

One usually puts in evidence the functions associated to polynomials in the standard fashion. We have an evaluation map

$$R \times \mathcal{M}_A(R, B) \cong R \times B^k \to B \tag{3}$$

given by $(r, \varphi) \to \varphi(r)$; if we write $r = r(x_1, \cdots, x_k)$ and $\varphi = (b_1, \cdots, b_k)$, we usually write $\varphi(r) = r(b_1, \cdots, b_k)$. Therefore we obtain a map

$$\lambda : R \to B^{B^k}, \tag{4}$$

where $\lambda(r(x_1, \cdots, x_k))(b_1, \cdots, b_k) = r(b_1, \cdots, b_k)$; λ is a homomorphism of Λ–algebras, where B^{B^k} is a Λ–algebra by pointwise operation.

Associated to the map λ we obtain two interesting objects which we want to study. The first is $\ker \lambda$, an ideal of R.

Definition 1.1. The elements of $\ker \lambda$ will be called the generalized polynomial identities of B, with coefficients in A, in the variables $x_1, \cdots, x_k$.

The second is $\operatorname{im} \lambda$.

Definition 1.2. The elements of $\operatorname{im} \lambda$ will be called the non–commutative polynomial functions with coefficients in A, in the variables $x_1, \cdots, x_k$.

Since by formula (1) we have an explicit description of the ring R, it is clear that, to have precise information on $\ker \lambda$ and $\operatorname{im} \lambda$, it will be enough to study one of the two; we will study principally $\operatorname{im} \lambda$; we will denote for convenience $\operatorname{im} \lambda = \overline{R} \cong R/\ker \lambda$.

Remark. When $A = \Lambda$, then $R = \Lambda\{x_1, \cdots, x_k\}$ and we will talk simply about polynomial identities.

One first rough estimate of the size of the ring $\operatorname{im} \lambda$ is obtained by noticing that the functions in $\operatorname{im} \lambda$ are polynomial functions in the sense of module theory.

Definition 1.3. If M, N are Λ–modules, a (set theoretic) map $\varphi : M \to N$ is called a polynomial map if, given generators $\{m_i\}_{i\in I}$, $\{n_j\}_{j\in J}$ for M and N respectively, there exist polynomials $f_j(x_i) \in \Lambda[x_i]$, $j \in J$, such that $\varphi(\Sigma\, \alpha_i m_i) = \Sigma\, f_j(\alpha_i) n_j$, such that only a finite number of f_j's have constant term, and for each i only a finite number of f_j's depend on x_i.

Remark. One can give a more intrinsic definition but, since in the cases that we will consider M and N will be finite free modules, this definition works very well for us.

We will indicate by $\mathcal{P}(M, N)$ the set of polynomial maps. It is clearly a submodule of N^M; furthermore if N is an algebra, it is easy to see that $\mathcal{P}(M, N)$ is actually a subalgebra of N^M.

Going back to our ring $\overline{R} = \operatorname{im} \lambda$ which is contained in B^{B^k}, it is easy to see that it is in fact contained in $\mathcal{P}(B^k, B)$. The ring $\mathcal{P}(B^k, B)$ is generally much easier to describe than $\overline{R}$, since its module structure depends only on the module B while $\overline{R}$, even as a module, depends on the algebra structure of B.

We can work also in a slightly more general fashion. We consider a Λ–module M and the tensor algebra $T(M) = \oplus_{i=0}^{\infty} \otimes^i M$, and we form $R = A * T(M)$. Now

$$\mathcal{M}(R, B) \cong \mathcal{M}(A, B) \times \mathrm{Hom}_{\Lambda}(M, B)$$

and if we consider only the maps inducing the inclusion on A we obtain

$$\mathcal{M}_A(R, B) \cong \mathrm{Hom}_{\Lambda}(M, B).$$

Again we will have an evaluation map $R \times \mathrm{Hom}_{\Lambda}(M, B) \to B$ inducing a map $\lambda : R \to \mathcal{P}(\mathrm{Hom}_{\Lambda}(M, B), B)$. R can be described quite easily

$$R \cong A \oplus (A \otimes M \otimes A) \oplus (A \otimes M \otimes A \otimes M \otimes A) \oplus \ldots$$

where $(a_0 \otimes m_1 \otimes a_1 \otimes m_2 \otimes \cdots \otimes m_k \otimes a_k)(b_0 \otimes n_1 \otimes b_1 \otimes n_2 \otimes \cdots \otimes n_h \otimes b_h) = a_0 \otimes m_1 \otimes a_1 \otimes m_2 \otimes \cdots \otimes m_k \otimes (a_k b_0) \otimes n_1 \otimes b_1 \otimes \cdots \otimes n_h \otimes b_h$. R is thus a graded algebra where

$$R_i = (\overset{i}{\otimes} (A \otimes M)) \otimes A = A \otimes (\overset{i}{\otimes} (M \otimes A)).$$

(All tensor products will always be over Λ.)

We specialize now to the case in which A is an Azumaya algebra over Λ, M a projective module of finite rank over Λ.

Notation. If A is a Λ–algebra and P an A–bimodule it is customary to indicate $P^A = \{ p \in P \mid ap = pa, \forall a \in A \}$. P^A is called the center of P and one has a bimodule map $A \otimes_{\Lambda} P^A \to P$. If A is an Azumaya algebra this is an isomorphism; if P is itself an algebra and $A \subseteq P$ a subalgebra, then P is an A–bimodule in the obvious way, P^A is the centralizer of A, and it is a subalgebra; the isomorphism $A \otimes_{\Lambda} P^A \to P$ is an algebra isomorphism.

Theorem 1.4. If A is an Azumaya algebra, M a Λ–module, then

(1) The centralizer C of A in $R = A * T(M)$ is isomorphic to $T(N)$, where $N = (A \otimes M \otimes A)^A$, and

(2) $A * T(M) \cong A \otimes T(N)$.

Proof. Consider $A \otimes T(N)$; we have a map of algebras $A \to A * T(M)$ sending A to the constants and another module map $N \to A * T(M)$ which is the inclusion of N in $A \otimes M \otimes A = R_1$. This map extends uniquely to a map of algebras $T(N) \to R$. By construction N commutes with A; therefore the image of $T(N)$ in R commutes with A and we get a map of algebras $\psi : A \otimes T(N) \to A * T(M)$. Both $A \otimes T(N)$ and $A * T(M)$ are graded algebras, generated by their part of degree 0 and 1. Now $(A \otimes T(N))_1 = A \otimes N$, $(A * T(M))_0 = A$, $(A * T(M))_1 = A \otimes M \otimes A$, and furthermore $\psi(A) = A$, $\psi(A \otimes N) \subseteq A \otimes M \otimes A$, and so ψ is a map of graded algebras. We have, by the general theory of bimodules over Azumaya algebras, that $A \otimes M \otimes A \cong A \otimes N$ as A–bimodules, where **the** A–bimodule structure of $A \otimes M \otimes A$ is the one induced by the ring structure of $R \supset A$ and $A \otimes N$ is a bimodule by the action of A on A (left and right). Therefore if we construct $\varphi : A * T(M) \to A \otimes T(N)$ by the universal property of the free product to be the identity map on A and the inverse of the isomorphism $A \otimes N \to M$ on M, we see immediately that $\varphi\psi = 1_{A \otimes T(N)}$, and $\psi\varphi = 1_{A * T(M)}$, so that φ and ψ are inverse isomorphisms.

We want to deduce the structure of $\ker \lambda$ and $\operatorname{im} \lambda$ using this theorem, but first we study the ring $\mathcal{P}(\mathrm{Hom}_\Lambda(M, B), B)$.

Theorem 1.5. $\mathcal{P}(\mathrm{Hom}_\Lambda(M, B), B)$ is canonically isomorphic to $A \otimes \mathcal{P}(\mathrm{Hom}_\Lambda(N, B^A), B^A)$.

Proof. If E is any module consider $\mathcal{P}(E, B)$; we have a map $A \to \mathcal{P}(E, B)$ sending each element a of A to the constant polynomial function with value a. Since this is an embedding of A in $\mathcal{P}(E, B)$ we will identify A with its image. From the fact that A is Azumaya, we have that $\mathcal{P}(E, B) \cong A \otimes \mathcal{P}(E, B)^A$. Now a map $f : E \to B$ is in $\mathcal{P}(E, B)^A$ if and only if $f(m) \in B^A$ for every $m \in E$. So $\mathcal{P}(E, B)^A = \mathcal{P}(E, B^A)$.

Now consider $E = \mathrm{Hom}(M, B)$; we claim that $\mathrm{Hom}(M, B) \cong \mathrm{Hom}(N, B^A)$. In fact the map $A \otimes A \to \mathrm{End}(A)$ sending $a \otimes b$ to the map $c \to acb$ is an isomorphism of A–bimodules; under this map $(A \otimes A)^A$ is identified with $\mathrm{End}(A)^A = \mathrm{Hom}(A, \Lambda) = A^*$. So $N \cong (A \otimes M \otimes A)^A \cong M \otimes (A \otimes A)^A \cong M \otimes \mathrm{Hom}(A, \Lambda)$; therefore $\mathrm{Hom}(N, B^A) \cong \mathrm{Hom}(M \otimes A^*, B^A) \cong \mathrm{Hom}(M, \mathrm{Hom}(A^*, B^A)) \cong \mathrm{Hom}(M, A \otimes B^A)$ (A is faithfully projective over Λ).

Finally we tie together all this information.

Theorem 1.6. If we denote by $\lambda : A * T(M) \to \mathcal{P}(\mathrm{Hom}(M, B), B)$ and $\mu : T(N) \to \mathcal{P}(\mathrm{Hom}(N, B^A), B^A)$ the canonical maps, we have, under all the canonical identifications, a commutative diagram:

$$\begin{array}{ccc} A * T(M) & \xrightarrow{\lambda} & \mathcal{P}(\mathrm{Hom}(M, B), B) \\ \varphi \downarrow & & \downarrow \cong \\ A \otimes T(N) & \xrightarrow{1_A \otimes \mu} & A \otimes \mathcal{P}(\mathrm{Hom}(N, B^A), B^A) \end{array}$$

Proof. Since all algebras are generated by their degree 0 and degree 1 parts, it is sufficient to verify in such degrees. This is carried out simply by checking the definitions previously given.

It is clear now that to study $\ker \lambda$ and $\mathrm{im}\, \lambda$, we are reduced to study $\ker \mu$ and $\mathrm{im}\, \mu$ since

$$\ker \lambda = A \otimes \ker \mu \quad \text{and} \quad \mathrm{im}\, \lambda = A \otimes \mathrm{im}\, \mu.$$

In particular if $A = B$ so that $B^A = \Lambda$, we see immediately that μ is surjective, since $\mathcal{P}(E, \Lambda)$ is generated by its degree 1 part $\mathrm{Hom}(E, \Lambda)$ for any module E. The map $T(N) \to \mathcal{P}(\mathrm{Hom}(N, \Lambda), \Lambda)$ factors naturally through $T(N) \to S(N) \to \mathcal{P}(\mathrm{Hom}(N, \Lambda), \Lambda)$, where $S(N)$ denotes the symmetric algebra over N. If moreover we assume that Λ is generic enough, in the sense that a non–zero polynomial does not vanish identically on Λ – for instance Λ an infinite field – V and N projective then the map $S(N) \to \mathcal{P}(\mathrm{Hom}(N, \Lambda), \Lambda)$ is an isomorphism. In this case $\ker \mu$ is just the commutator ideal in $T(N)$. This gives now an explicit characterization of all the generalized polynomial identities of A with coefficients in A itself.

When Λ is generic, we just have to read off the results when $M = \oplus_{i=1}^{k} \Lambda x_i$. In this case $T(M) = \Lambda\{x_1, \cdots, x_k\}$ is the free algebra, $A \otimes M \otimes A = \Sigma A x_i A$, $N = \oplus N_i$ and $N_i = \{\Sigma_h a_h x_i b_h \mid \Sigma_h a a_h x_i b_h = \Sigma_h a_h x_i b_h\}$. Under the identification Ax_iA with $\mathrm{End}_\Lambda(A)$, N_i is identified with $\mathrm{Hom}_\Lambda(A, \Lambda)$. If we have that A is Λ–free and $u_1, \cdots, u_r$ is a basis for A over Λ($u^1, \cdots, u^r$ the dual basis), then, calling ξ_{ij} the element of N_i corresponding to u^j, we see that $T(N) = \Lambda\{\xi_{ij}\}$, the free algebra in the elements ξ_{ij}. Now $\ker \lambda$ is $A \otimes \ker \mu$ where $\ker \mu$ is the commutator ideal of $\Lambda\{\xi_{ij}\}$; as an ideal it is generated by the elements $[\xi_{ij}, \xi_{st}]$.

Even more particularly if $A = (\Lambda)_m$, the ring of $m \times m$ matrices over Λ, we choose the canonical basis e_{ij} and therefore we obtain the elements $\xi_{ij;t}$, $i, j = 1, \cdots, m$, $t = 1, \cdots, k$, where $\xi_{ij;t} = \Sigma_{s=1}^{m} e_{si} x_t e_{js}$. Now $(\Lambda)_m * \Lambda\{x_1, \cdots, x_k\} \cong (\Lambda)_m \otimes \Lambda\{\xi_{ij;t}\}$. Under this identification we have $x_t = \Sigma \xi_{ij;t} e_{ij}$, so that considering $R = (\Lambda)_m \otimes \Lambda\{\xi_{ij;t}\} = (\Lambda\{\xi_{ij;t}\})_m$ we see that $x_t \in R$ is identified with the matrix with generic entries $(\xi_{ij;t})$. An element of R considered as a matrix (a_{ij}), $a_{ij} \in \Lambda\{\xi_{ij;t}\}$, is a generalized polynomial identity for $(\Lambda)_m$ if and only if all the a_{ij} vanish when considered as commutative polynomials.

There is one further concept which is interesting to describe. The ring $\overline{R} = \mathrm{im}\, \lambda$

is a ring of functions over $\mathrm{Hom}_\Lambda(M, B)$; this in turn allows us to define on $\mathrm{Hom}_\Lambda(M, B)$ certain algebraic sets as follows: If $S \subset \bar{R}$, set $V(S) = \{p \in \mathrm{Hom}_\Lambda(M,B) \mid s(p) = 0 \ \forall s \in S\}$.

Let us consider the case in which A is an Azumaya algebra. Then we have seen that $\mathrm{im}\,\lambda = A \otimes \mathrm{im}\,\mu$ and $\mu : T(N) \to \mathcal{P}(\mathrm{Hom}_\Lambda(N, B^A), B^A)$. Now $f \in \mathrm{im}\,\lambda$, $p \in \mathrm{Hom}_\Lambda(M, B) \cong \mathrm{Hom}_\Lambda(N\ B^A)$ and $0 = f(p) \in A \otimes B^A$ is equivalent to saying that for every $\varphi \in \mathrm{Hom}_\Lambda(A, \Lambda) = A^*, 0 = \varphi \otimes 1_{BA} \circ f(p) \in B^A$. Now $\varphi \otimes 1 \circ f : \mathrm{Hom}_\Lambda(N,B^A) \to B^A$ is a polynomial map so that $V(f) = V(\{\varphi \otimes 1 \circ f\}_{\varphi \in A^*})$. Therefore the distinguished sets for the ring of functions $\mathrm{im}\,\lambda$ are the same as the sets for $\mathrm{im}\,\mu$ and we only have to consider this case.

In particular if Λ is a field, A central simple over Λ of dimension n^2, $B = A \otimes_\Lambda \Omega$, where Ω is a universal domain, and M a vector space of finite dimension m, we have $\mathrm{Hom}_\Lambda(N, \Omega)$ is a vector space over Ω of dimension $n^2 m$ with a Λ structure; the sets $V(S)$, $S \subset \bar{R} \cong A \otimes S(N)$, are just the usual Λ–closed sets of the Zarisky topology over Λ.

2. Generalized versus ordinary polynomial identities

We specialize now our discussion even further. Λ will be an infinite field, $A = (\Lambda)_m = B$, the ring of $m \times m$ matrices, $M = \Sigma_{i=1}^k \Lambda x_i$ a vector space with basis $x_1, \cdots, x_k$. We have seen that the polynomial functions induced through the map λ on $\mathrm{Hom}_\Lambda(M, B) \cong B^k$ by the elements of $A * T(M)$ are all the module theoretical polynomial functions from B^k to B. We would like to find a similar description for the polynomial functions induced by $T(M)$. Unfortunately this seems to be still out of reach of the present techniques. Let us set up our symbols once more: We have $\mathcal{P}(A^k, A) \cong A \otimes S((A^*)^k)$ and $\bar{R} = \mathrm{im}\,\mu$, $\mu : T(M) \to \mathcal{P}(A^k, A)$. Using the basis e_{ij} for A and considering $e_{ij;t} = (0, 0, \cdots, e_{ij}, 0, \cdots, 0)$, the element of A^k with e_{ij} in the tth place and 0 elsewhere, the $e_{ij;t}$'s form a basis for A^k; if $\xi_{ij;t}$ denotes the dual basis then $S((A^*)^k) = \Lambda[\xi_{ij;t}]$, the ring of polynomials in the variables $\xi_{ij;t}$.

Several things are known about $\bar{R}$. It is a graded algebra and an Öre domain, whose quotient division ring we will denote D; furthermore if we let Ω be the quotient ring of $S((A^*)^k)$ (i.e., the ring of rational functions in the variables $\xi_{ij;t}$), then we can complete the diagram:

$$\begin{array}{ccccc} \bar{R} \subseteq \mathcal{P}(A^k, A) & = & A \otimes \Lambda[\xi_{ij;t}] & \cong & (\Lambda[\xi_{ij;t}])_m \\ \downarrow & & \downarrow & & \downarrow \\ D & \longrightarrow & A \otimes \Omega & \cong & (\Omega)_m. \end{array}$$

If Z is the center of D then $Z \subset \Omega$ and $D \otimes_Z \Omega \cong A \otimes_\Lambda \Omega \cong (\Omega)_m$.

Let $\mathfrak{G} = \mathrm{Aut}_\Lambda A$, the group of all algebra automorphisms of A. It is well known that $\mathfrak{G} \cong GL(m, \Lambda)/\Lambda^* = P\ell(m-1, \Lambda)$ since every automorphism of A is inner. We have an action of $\mathfrak{G}$ on $A * T(M)$ leaving $T(M)$ invariant by the universal property of the free product: $\sigma \in \mathfrak{G}$ induces $\sigma * 1_{T(M)}$ on $A * T(M)$. We also have an action of $\mathfrak{G}$ on $A^k = \mathrm{Hom}_\Lambda(M, A)$ in the obvious way. Therefore we obtain an action of $\mathfrak{G}$ on $\mathcal{P}(A^k, A)$ setting $f^\sigma(p) = \sigma(f(\sigma^{-1}p))$.

Proposition 2.1. The map $\lambda : A * T(M) \to \mathcal{P}(A^k, A) = \mathcal{P}(\mathrm{Hom}_\Lambda(M, A), A)$ is a $\mathfrak{G}$ homomorphism.

Proof. Consider the evaluation map, $\lambda(f)(p) = p(f)$, where $f \in A * T(M)$, $p \in \mathrm{Hom}_\Lambda(M, A)$. If we keep p fixed this evaluation is a map $A * T(M) \to A$ which is the identity on A and the extension $\hat{p}$ of p to the tensor algebra $T(M)$. Since $A * T(M)$ is generated, as an algebra, by A and M it is enough to show that $\lambda(a^\sigma) = \sigma(\lambda(a))$ and $\lambda(m^\sigma) = \sigma(\lambda(m))$, $a \in A$, $m \in M$. Now for $a \in A$, $a^\sigma = \sigma(a)$, $\lambda(a^\sigma)$ is the constant map $p \to a^\sigma$ which is exactly f^σ, where f is the constant map $p \to a$. If on the other hand $m \in M$, $m^\sigma = m$, and the induced map $\mathrm{Hom}_\Lambda(M, A) \to A$ is the map $f : \varphi \to \varphi(m)$. Now f^σ is by definition $f^\sigma(\varphi) = \sigma(f(\sigma^{-1}\varphi))$ so $f^\sigma(\varphi) = \sigma((\sigma^{-1}\varphi)(m)) = \sigma(\sigma^{-1}(\varphi(m))) = \varphi(m)$ and $f^\sigma = f$.

Corollary 2.2. $\bar{R} \subseteq \mathcal{P}(A^k, A)$.

We can describe the action of $\mathfrak{G}$ on $A * T(M)$ also using the description $A * T(M) \cong A \otimes T(N)$ In fact since $\mathfrak{G}$ acts as automorphisms of graded algebras and leaves A invariant it will leave N invariant and it suffices to describe the action of $\mathfrak{G}$ on N. Now $N = (A \otimes M \otimes A)^A \cong \mathrm{Hom}(A, \Lambda) \otimes M$; the map $A \otimes A \to \mathrm{End}(A)$ commutes with the two actions of $\mathfrak{G}$ on $A \otimes A$ and $\mathrm{End}(A)$ given respectively as $\sigma(a \otimes b) = \sigma(a) \otimes \sigma(b)$ and $\sigma(f) = \sigma f \sigma^{-1}$, $f \in \mathrm{End}(A)$, since $a \otimes b$ is identified with the endomorphism $u \to aub$ so that $\sigma(a) \otimes \sigma(b)$ is the endomorphism $u \to \sigma(a)u\sigma(b) = \sigma(a\sigma^{-1}(u)b)$. $\mathrm{Hom}(A, \Lambda) \subseteq \mathrm{End}(A)$ is clearly invariant under the given $\mathfrak{G}$ action and on $\mathrm{Hom}(A, \Lambda)$ we have the usual dual action of $\mathfrak{G}$. To be more explicit, consider $A = \mathrm{End}(V)$ (V an m dimensional vector space); then $A \cong V \otimes V^*$ and the $\mathfrak{G}$ action comes from the usual action of $GL(V)$ on $V \otimes V^*$ and we see that $N \cong (V \otimes V^*)^* \cong V^* \otimes V$ with the canonical $GL(V)$ action; so $A \otimes T(N) \cong V \otimes V^* \otimes T(V^* \otimes V)$ with the canonical $GL(V)$ structure.

We have in general $\bar{R} \neq \mathcal{P}(A^k, A)$; this is easy to see because just looking at $\mathcal{P}(A^k, \Lambda) = S$ we see easily that S contains various elements not in $\bar{R}$, (e.g., if $a \in \bar{R}_1$, $\mathrm{Tr}\ a \in S$ and $\mathrm{Tr}\ a \notin \bar{R}$ if $m > 1$). S is again a graded commutative ring: The ring of invariants of k-tuples of $m \times m$ matrices.

We can extend the $\mathfrak{G}$ action uniquely to D and $A \otimes_\Lambda \Omega$ obtaining

Theorem 2.3. (1) $D = (A \otimes_\Lambda \Omega)^{\mathfrak{G}}$ and
(2) $Z = \Omega^{\mathfrak{G}}$.

Proof. Since Z is the center of D, statement (2) is implied by (1). On the other hand considering $E = (A \otimes_\Lambda \Omega)^{\mathfrak{G}}$, we have $D \subseteq E \subseteq (A \otimes_\Lambda \Omega)$ so that $E = D \otimes_Z L$, L the center of E. Now $L \subseteq \Omega^{\mathfrak{G}}$ so that, if we know (2), $L = Z$ and $E = D$. We have therefore to prove only one of the two statements and we will prove (2). Let $\overline{\Omega}$ be an algebraic closure of Ω and consider the usual action of $\mathfrak{G}$ on A^k. This clearly extends to an action of $\mathfrak{G}_{\overline{\Omega}} = \mathrm{Aut}(A_{\overline{\Omega}})$ ($A_{\overline{\Omega}} = A \otimes_\Lambda \overline{\Omega}$) on $(A_{\overline{\Omega}})^k$.

We consider $\mathcal{P}(A^k, \Lambda) = \Lambda[\xi_{ij;t}]$; the $\mathfrak{G}$ action on $\mathcal{P}(A^k, \Lambda)$ is just $f^\sigma(p) = f(\sigma^{-1}p)$ since $\mathfrak{G}$ acts trivially on Λ, the same being true for the field Ω considered as rationals functions. We can also consider the action of $\mathfrak{G}_{\overline{\Omega}}$ on $\mathcal{P}(A^k_{\overline{\Omega}}, \overline{\Omega}) = \overline{\Omega}[\xi_{ij;t}]$ and on its quotient field $\overline{\Omega}(\xi_{ij;t}) = W$. $\Omega \subseteq W$ and Ω is not invariant under $\mathfrak{G}_{\overline{\Omega}}$. On the other hand $\Omega^{\mathfrak{G}}$ is invariant under $\mathfrak{G}_{\overline{\Omega}}$. In fact $\mathfrak{G}$ is dense in $\mathfrak{G}_{\overline{\Omega}}$ and A^k in $A^k_{\overline{\Omega}}$ under the respective Zarisky topologies, as Λ is an infinite field; so if $f \in \Omega^{\mathfrak{G}}$ then for all p for which f is defined and $\sigma \in \mathfrak{G}$ we have f defined in $\sigma^{-1}p$ and $f(\sigma^{-1}p) = f(p)$. So since $\mathfrak{G}$ is dense in $\mathfrak{G}_{\overline{\Omega}}$ we have f defined in $\sigma^{-1}p$ and $f(\sigma^{-1}p) = f(p)$ for all $\sigma \in \mathfrak{G}_{\overline{\Omega}}$, and then since the set of points in A^k where f is defined is dense in the set of points in $A^k_{\overline{\Omega}}$ where f is defined, we have $f = f^\sigma$ for all $\sigma \in \mathfrak{G}_{\overline{\Omega}}$ and f is invariant under $\mathfrak{G}_{\overline{\Omega}}$. Therefore $\Omega^{\mathfrak{G}} = \Omega \cap W^{\mathfrak{G}_{\overline{\Omega}}}$. Now if $f \in \Omega$, it is defined in the generic point p of coordinates $\xi_{ij;t}$ and $f = f(p)$. In $\overline{\Omega}$ the matrix $\xi_1 = (\xi_{ij;1})$ can be diagonalized so there is an element $\sigma \in \mathfrak{G}_{\overline{\Omega}}$ such that

$$\sigma^{-1}(\xi_1) = \mathrm{diag}(u_1, \cdots, u_n)$$

where the u_i's are the eigenvalues of the matrix ξ_1; since ξ_1 is generic the u_i's are distinct. The elements u_i are the roots of the chracteristic polynomial of ξ_1, whose coefficients are clearly in Z, therefore $F = Z(u_1, \cdots, u_n)$ is algebraic over Z of dimension $\leqslant n!$.

Consider $D \otimes_Z F$; $r = \mathrm{diag}(u_1, \cdots, u_n) = \sigma^{-1}(\xi_1) \in \sigma^{-1}(D \otimes_Z F)$ which is a simple algebra over F, but $u_i \in F$ so that $e = \mathrm{diag}(1, 0, 0, \cdots, 0) = (\Pi_{i\neq 1}(u_i - u_1))^{-1}\Pi_{i\neq 1}(u_i - r) \in \sigma^{-1}(D \otimes_Z F)$ and $e\sigma^{-1}(D \otimes_Z F)e \subset \overline{\Omega}e$ is a field, therefore $\sigma^{-1}(D \otimes_Z F)$ is a matrix algebra over F and so also $D \otimes_Z F$ is a matrix algebra over F. Therefore for a suitable $\tau \in \mathfrak{G}_{\overline{\Omega}} : \tau^{-1}(D \otimes_Z F) = F_m \subseteq \overline{\Omega}_m$, $\tau^{-1}(\xi_i) \in F_m$ and $\tau^{-1}(\xi_1) = \sigma^{-1}(\xi_1)$ so that $\tau^{-1}(p) \in (F_m)^k$ and $f(\tau^{-1}(p)) \in F$; since f is invariant, $f = f(p) \in F$, so that $\Omega^{\mathfrak{G}} \subseteq F$. Now if A is the matrix of a permutation $(i_1, \cdots, i_m)$, then

$$A\tau^{-1}(\xi_1)A^{-1} = \mathrm{diag}(u_{i_1}, \cdots, u_{i_m}),$$

so if $\rho \in \mathcal{D}$ is the inner automorphism associated with A^{-1}, $f(\tau^{-1}(p)) = f(\rho^{-1}(\tau^{-1}(p)))$.

We obtain in this fashion a subgroup Γ_n of $\mathcal{D}$ isomorphic to the symmetric group. The coordinates of the $n!$ points $\rho(\tau^{-1}(p))$, always generate over $\mathbf{Z}$ the same field F, and therefore Γ_n acts as a group of automorphisms on F and acts faithfully on the set $u_1, \cdots, u_n$, therefore also on F. Now $Z \subseteq \Omega^{\mathcal{D}} \subseteq F^{\Gamma_n}$, since $[F : Z] \leqslant n!$ we must have $Z = \Omega^{\mathcal{D}}$ and $[F : Z] = n!$.

References

1. M. Artin, On Azumaya algebras and finite dimensional representations of rings, J. Alg. **11** (1969), 532-563.
2. C. Procesi, Non-commutative affine rings, Att. Acc. Naz. Lincei, serie VIII, 8 (1967), 239-255.
3. C. Procesi, Sulle identita delle algebre semplici, Rend. Circolo Mat. Palermo, serie II, **17** (1968), 13-18.

BETTI NUMBERS AND REFLEXIVE MODULES

Mark Ramras

Boston College

Introduction

A quasi–Frobenius (QF) ring R may be described as a ring with the property that all of its modules are reflexive, or equivalently, $\mathrm{Ext}^i(M, R) = 0$ for all $i \geqslant 1$ and all R–modules M. In this paper we introduce a class of commutative noetherian local rings, called BNSI rings, which are "as different as possible" from QF rings. In section 2 it is shown that for such a ring, for any finitely generated module M the vanishing of $\mathrm{Ext}^i(M, R)$ for any $i \geqslant 2$ implies that M is free. As a consequence, every finitely generated reflexive module is free. Furthermore, no proper principal ideal is the annihilator of any ideal (whereas in a QF ring, *every* ideal is an annihilator).

Let $(R, \mathbf{m})$ be a commutative noetherian local ring with maximal ideal $\mathbf{m}$, and let M be a finitely generated R–module. For any $i \geqslant 0$, $\mathrm{Tor}_i(M, R/\mathbf{m})$ is a finite dimensional vector space over $R/\mathbf{m}$. Its dimension will be called the ith Betti number of M, denoted $\beta_i(M)$. R is called a BNSI ring if for every non–free M, the sequence of Betti numbers $\{\beta_i(M)\}_{i\geqslant 1}$ is strictly increasing. The most important result in this paper is Theorem 3.2, in which a fairly large class of examples of BNSI rings is exhibited.

Section 4 deals with Betti numbers of modules over QF rings.

1. Preliminaries

Throughout this paper, all rings are assumed to be commutative noetherian local rings, and all modules are unital and finitely generated. Thus every projective module is free. The maximal ideal of the ring R will be denoted by $\mathbf{m}$. For an R–module M, we denote by M^* the dual of M, $\mathrm{Hom}(M, R)$. Recall that M is *reflexive* if the natural map from M to M^{**} is an isomorphism. Recall also, that a projective resolution of M

$$\cdots \to P_i \xrightarrow{d_i} P_{i-1} \longrightarrow \cdots \to P_0 \xrightarrow{d_0} M \to 0$$

is *minimal* if for all $i \geq 0$, Ker $d_i \subseteq \mathbf{m}P_i$.

We now discuss, rather briefly and omitting proofs, some notions which are dealt with much more fully in [2].

Two R–modules A and B are *projectively equivalent* if there exist projectives P and Q such that $A \oplus P \cong B \oplus Q$. If $P_1 \to P_0 \to M \to 0$ and $Q_1 \to Q_0 \to M \to 0$ are any two presentations of M, then $\text{coker}(P_0^* \to P_1^*)$ and $\text{coker}(Q_0^* \to Q_1^*)$ are projectively equivalent. If both these presentations are minimal, then the two cokernels are isomorphic. Accordingly, let $D(M) = \text{coker}(F_0^* \to F_1^*)$, where $F_1 \to F_0 \to M \to 0$ is a minimal (free) presentation of M. Thus $F_0^* \to F_1^* \to D(M) \to 0$ is a presentation of D(M). (It is minimal precisely when M^* has no non–zero free direct summand.) Since free modules are reflexive, $\text{coker}(F_1^{**} \to F_0^{**}) = \text{coker}(F_1 \to F_0) = M$. Hence M and D(D(M)) are projectively equivalent (and isomorphic if M^* has no non–zero free direct summand). It follows that M is free if and only if $D(M) = 0$.

The importance of D(M), for our purposes, comes from the following exact sequence, established in more general form by Auslander in [1, Proposition 6.3]:

$$0 \to \text{Ext}^1(D(M), R) \to M \to M^{**} \to \text{Ext}^2(D(M), R) \to 0.$$

Here the map $M \to M^{**}$ is the canonical one. Thus M is reflexive if and only if $\text{Ext}^i(D(M), R) = 0$ for $i = 1,2$. An easy consequence is:

Proposition 1.1. The following statements are equivalent:

(a) Every reflexive R–module is free.

(b) For all R–modules M, if $\text{Ext}^i(M, R) = 0$ for $i = 1,2$ then M is free.

Proof. (a) ⇒ (b): Assume that $\text{Ext}^i(M, R) = 0$ for $i = 1,2$ and let $N = D(M)$. Since M and D(N) are projectively equivalent, $\text{Ext}^i(M, -) \cong \text{Ext}^i(D(N), -)$ for all $i \geq 1$. So by our hypothesis $\text{Ext}^i(D(N), R) = 0$ for $i = 1,2$ and therefore N is reflexive. Hence, by (a), N is free and so $D(N) = 0$. Thus M is projectively equivalent to 0, and hence free.

(b) ⇒ (a): If M is reflexive then $\text{Ext}^i(D(M), R) = 0$ for $i = 1,2$ and so D(M) is free. Just as above; we conclude that M is free.

2. BNSI Rings

Definition. For any R–module M and for any $i \geq 0$, let $\beta_i(M)$ = the dimension of $\text{Tor}_i(M, R/\mathbf{m})$ as a vector space over $R/\mathbf{m}$. $\beta_i(M)$ is called the ith *Betti number* of M.

Remark. If $\to F_{i+1} \to F_i \to \cdots \to F_0 \to M \to 0$ is a minimal free resolution of M, then for all $i \geq 0$, $\mathrm{Tor}_i(M, R/\mathbf{m}) \cong F_i \otimes R/\mathbf{m}$, so that $\beta_i(M) = \mathrm{rank}(F_i)$. In particular, $\beta_0(M)$ = the minimal number of generators of M.

Definition. R is a *BNSI* ring ("Betti numbers strictly increase") if for every non–free module M, the sequence $\{\beta_i(M)\}_{i \geq 1}$ is strictly increasing.

Remark. If R is a BNSI ring, then $\mathrm{depth}_R R = 0$ (i.e., every element of $\mathbf{m}$ is a zero divisor or equivalently, $\mathrm{ann}(\mathbf{m}) \neq 0$). For if $x \in \mathbf{m}$ were not a zero divisor, $0 \to R \xrightarrow{x} R \to R/Rx \to 0$ would be a minimal free resolution of R/Rx. But then $\beta_2(R/Rx) = 0$ while $\beta_1(R/Rx) = 1$.

Proposition 2.1. Suppose $\mathrm{depth}_R R = 0$. For any module M, $D(M)$ has no non–zero free direct summand.

Proof. Let $F_1 \to F_0 \to M \to 0$ be a minimal presentation of M, and let $K = \ker(F_1 \to F_0)$. Then $K \subseteq \mathbf{m}F_1$. $F_0^* \to F_1^* \to D(M) \to 0$ is exact, by definition of $D(M)$. Dualizing, we get the exact sequence $0 \to D(M)^* \to F_1^{**} \to F_0^{**}$. Since the F_i are free and hence reflexive, $D(M)^* \approx K$. Since $K \subseteq \mathbf{m}F_1$, $\mathrm{ann}(K) \supseteq \mathrm{ann}(\mathbf{m})$, and by hypothesis $\mathrm{ann}(\mathbf{m}) \neq 0$. So $\mathrm{ann}(D(M))^* \neq 0$ and therefore R is not a direct summand of $D(M)^*$. The dual of a free direct summand of $D(M)$ would be a free direct summand of $D(M)^*$. Hence $D(M)$ has no non–zero free direct summand.

Proposition 2.2. Let R be a BNSI ring. Suppose $M \neq 0$ and $M \subset \mathbf{m}F_0$ for some free module F_0. Then for any free module F, $D(M) \not\subseteq \mathbf{m}F$.

Proof. Let (#) $R^m \to R^n \to M \to 0$ be a minimal presentation of M. Since $M \subseteq \mathbf{m}F_0$ and $0 \to M \to F_0 \to F_0/M \to 0$ is exact, $R^m \to R^n \to F_0 \to F_0/M \to 0$ is part of a minimal free resolution of F_0/M. $m = \beta_2(F_0/M)$ and $n = \beta_1(F_0/M)$. Hence $m > n$. Now dualize (#). Thus $0 \to M^* \to R^n \to R^m \to D(M) \to 0$ is exact. Since $M \subseteq \mathbf{m}F_0$, $\mathrm{ann}(M) \supseteq \mathrm{ann}(\mathbf{m})$ and $\mathrm{ann}(\mathbf{m}) \neq 0$ since R is a BNSI ring. So $\mathrm{ann}(M) \neq 0$ and thus $\mathrm{ann}(M^*) \neq 0$. It follows that the image of M^* in R^n is contained in $\mathbf{m}R^n$. The presentation $R^n \to R^m \to D(M) \to 0$ of $D(M)$ is therefore minimal. Now if $D(M) \subseteq \mathbf{m}F$ for some free module F, the argument given above could be used here to show that $n > m$, which is a contradiction.

Theorem 2.3. Let R be a BNSI ring. If M is a submodule of a free module and $\mathrm{Ext}^1(M, R) = 0$, then M is free.

Proof. If N is a direct summand of M then $\mathrm{Ext}^1(N, R) = 0$. Thus since M is finitely generated and therefore the direct sum of finitely many indecomposable submodules, it suffices to prove the theorem for an indecomposable module M.

Assume that M is indecomposable and not free. By hypothesis M is a submodule of some free module F_0. By Nakayama's Lemma any element z in $F_0 - \mathbf{m}F_0$ is part of some basis of F_0 and thus there is a map $f : F_0 \to R$ such that $f(z) = 1$. If z lies in M, then the restriction of f to M maps M onto R. M would therefore have a direct summand isomorphic to R, which contradicts our assumptions about M. This shows that $M \subseteq \mathbf{m}F_0$.

Now for any module N, N is a submodule of a free module if and only if the map $N \to N^{**}$ is a monomorphism. This, in turn, as we see from the exact sequence in the preceding section, is true if and only if $\mathrm{Ext}^1(D(N), R) = 0$. Since $M \subseteq F_0$, we thus have $\mathrm{Ext}^1(D(M), R) = 0$. M is projectively equivalent to $D(D(M))$, and by hypothesis, $\mathrm{Ext}^1(M, R) = 0$. Therefore $\mathrm{Ext}^1(D(D(M)), R) = 0$, and so $D(M)$ is a submodule of some free module F. Since $M \subseteq \mathbf{m}F_0$ and $M \neq 0$, we conclude from Proposition 2.2 that $D(M) \not\subseteq \mathbf{m}F$. The argument given in the preceding paragraph shows that R is a direct summand of $D(M)$. But this contradicts Proposition 2.1. Hence M must be free.

Remark. The hypothesis, in Theorem 2.3, that M is a submodule of a free module cannot be dropped. For if $\mathrm{depth}_R R = 0$ and R is not a field there will exist a non-free module N which is a submodule of a free module. Hence $\mathrm{Ext}^1(D(N), R) = 0$. Yet $D(N)$ is not free, for if it were, then by Proposition 2.1 it would be 0, and N would be free.

For $n \geqslant 2$, $\mathrm{Ext}^n(-, R)$ is somewhat better behaved:

Corollary 2.4. Suppose R is a BNSI ring, and M is any R-module. If $\mathrm{Ext}^n(M, R) = 0$ for some $n \geqslant 2$, then M is free.

Proof. Let $0 \to K \to F_{n-2} \to \cdots \to F_0 \to M \to 0$ be exact with each F_i free. Then $\mathrm{Ext}^n(M, -) \approx \mathrm{Ext}^1(K, -)$. So $\mathrm{Ext}^1(K, R) = 0$ and $K \subseteq F_{n-2}$. By Theorem 2.3, K is free. Hence the projective dimension of M is at most $n-1$. But since $\mathrm{depth}_R R = 0$, every module of finite projective dimension is free.

Corollary 2.5. Let R be a BNSI ring. If the natural map $M \to M^{**}$ is an epimorphism, then M is free. Consequently, every reflexive module is free.

Proof. This is a direct consequence of Corollary 2.4 and the fact that $\mathrm{Ext}^2(D(M), R)$ is the cokernel of the map $M \to M^{**}$.

We shall now give a more straightforward proof of this last result, and obtain a slightly stronger conclusion. The proof is quite similar to the proof of Proposition 2.2. First we need a lemma.

Lemma 2.6. Suppose $\operatorname{depth}_R R = 0$. If R is a direct summand of M^*, then R is a direct summand of M. If M^* is free, then M is free.

Proof. Let g be a map of M^* onto R. Choose f in M^* such that $g(f) = 1$. We claim that f maps M onto R, from which it follows that R is a direct summand of M. For if f is not onto then $f(M) \subseteq \mathbf{m}$. Let $z \in \operatorname{ann}(\mathbf{m})$, $z \neq 0$. $zf(M) \subseteq z\mathbf{m} = 0$. Hence $zf = 0$. Thus $0 = g(zf) = z \cdot g(f) = z$, a contradiction. Therefore $f(M) = R$ and so $M \approx R \oplus M_1$, $M_1 = \ker(f)$. For the second statement, $M^* \approx R \oplus M_1^*$, so if M^* is free, so is M_1^*. Now assume the lemma holds for all modules N such that $\beta_0(N) < \beta_0(M)$. $\beta_0(M_1) = \beta_0(M) - 1$, so by our induction hypothesis, M_1 is free. Hence M is free.

Proposition 2.7. Let R be a BNSI ring. If M is a non–free R–module, then $\beta_0(M^*) > \beta_0(M)$. Consequently every reflexive R–module is free.

Proof. Let $M = F \oplus N$, where F is free and N has no non–zero free direct summand. Then $\beta_0(M) = \beta_0(F) + \beta_0(N)$, and $M^* = F^* \oplus N^*$, so $\beta_0(M^*) = \beta_0(F^*) + \beta_0(N^*)$. Since F is free, $\beta_0(F) = \beta_0(F^*)$. Hence

$$\beta_0(M^*) - \beta_0(M) = \beta_0(N^*) - \beta_0(N).$$

Thus it suffices to prove the proposition under the assumption that M has no non–zero free direct summand.

Let $F_1 \to F_0 \to M \to 0$ be a minimal free presentation of M. Dualizing, we get the exact sequence

$$0 \to M^* \to F_0^* \to F_1^* \to D(M) \to 0.$$

By Lemma 2.6, M^* has no non–zero free direct summand, so $M^* \subseteq \mathbf{m}F_0^*$ and thus the presentation $F_0^* \to F_1^* \to D(M) \to 0$ is minimal. Therefore $\beta_0(M^*) = \beta_2(D(M)) > \beta_1(D(M))$. But

$$\beta_1(D(M)) = \operatorname{rank}(F_0^*) = \operatorname{rank}(F_0) = \beta_0(M).$$

Hence $\beta_0(M^*) > \beta_0(M)$.

If M is not free, then by Lemma 2.6, M^* is not free. Thus by what we have just proved, $\beta_0(M^{**}) > \beta_0(M^*) > \beta_0(M)$, so $\beta_0(M^{**}) \geqslant \beta_0(M) + 2$. In particular, M is certainly not reflexive.

The next proposition is a slight generalization of a result of Margaret Menzin [6, Corollary 3.9]. We give her proof.

Proposition 2.8. Let $(S, \mathbf{m})$ be a local ring and let $x_1,\cdots,x_n$ be a regular S–sequence lying in $\mathbf{m}$. Suppose that $R = S/(x_1,\cdots,x_n)S$ has the property that every R–reflexive module is R–free (e.g., R might be a BNSI ring). For any S–module M, if $\mathrm{Ext}_S^i(M, S) = 0$ for $i = 1,\cdots,2n+2$, then M is S–free.

Proof. By induction on n. The case $n = 0$ is just Proposition 1.1. Now let $n = k+1$ and assume the result is true for k. Let $0 \to K \to F \to M \to 0$ be exact with F being S–free. Then $\mathrm{Ext}_S^i(K, S) = 0$ for $i = 1,\cdots,2n+1$. Let $x = x_1$. Applying the functor $\mathrm{Hom}_S(K,-)$ to the short exact sequence $0 \to S \overset{x}{\to} S \to S/xS \to 0$ we get the usual long exact sequence from which we see that $\mathrm{Ext}_S^i(K, S/xS) = 0$ for $i = 1,\cdots,2n$. Since x is regular on S, it is regular on any submodule of a free module, and hence on K. Thus by [5, Ex. 2, p. 155], $\mathrm{Ext}_{S/xS}^i(K/xK, S/xS) \cong \mathrm{Ext}_S^i(K, S/xS) = 0$ for $i = 1,\cdots,2n$. Now let $\bar{S} = S/xS$ and let $\bar{y}$ denote the image of y under the natural map from S to $\bar{S}$. Then $\bar{x}_2,\cdots,\bar{x}_n$ is a regular $\bar{S}$–sequence and $\bar{S}/(\bar{x}_2,\cdots,\bar{x}_n)\bar{S} \cong S/(x_1,\cdots,x_n)S$. So, by the induction hypothesis, K/xK is S/xS–free. It follows that K is S–free. Hence the projective dimension of M is at most 1. Thus $\mathrm{Ext}_S^2(M,-) = 0$. From the exact sequence

$$0 \to \mathbf{m} \to S \to S/\mathbf{m} \to 0$$

and the fact that $\mathrm{Ext}_S^1(M, S) = 0$, we conclude that $\mathrm{Ext}_S^1(M, S/\mathbf{m}S) = 0$. Hence M is S– free.

Remark. The converse of Proposition 2.8 is false. Let $(S, \mathbf{m})$ be a regular local ring of dimension 3. Let M be an S–module such that $\mathrm{Ext}_S^i(M, S) = 0$ for $i = 1,2,3$. Then, by an argument similar to the one above, since the projective dimension of M is at most 3, M is S–free. However, if $R = S/xS$, where $x \in \mathbf{m}^2$, $x \neq 0$, then R is Gorenstein (inj. $\dim._R R = 2$) but not regular, and so there exist non–free R–modules N such that $\mathrm{Ext}_R^i(N, R) = 0$ for all $i > 0$.

In a QF ring, every ideal J is the annihilator of some ideal (namely, ann(J)). We close this section with a proposition which says that in a BNSI ring no proper principal ideal is an annihilator.

Proposition 2.9. Let R be a BNSI ring. Suppose I is an ideal of R and ann(I) is principal. Then $I = 0$ or $I = R$.

Proof. Let $\operatorname{ann}(I) = Rx$. Since for any ideal J, $\operatorname{ann}(\operatorname{ann}(\operatorname{ann}(J))) = \operatorname{ann}(J)$, we have $\operatorname{ann}(\operatorname{ann}(Rx)) = Rx$. But $(Rx)^* = \operatorname{ann}(\operatorname{ann}(Rx))$. Thus $(Rx)^* = Rx$ and so by Proposition 2.7 Rx is free. Since $\operatorname{depth}_R R = 0$, either $Rx = R$ or $Rx = 0$. If $Rx = R$, $I = 0$. If $Rx = 0$, then since $\operatorname{ann}(\mathbf{m}) \neq 0$, $I \not\subseteq \mathbf{m}$. Hence $I = R$.

3. Constructing BNSI Rings

We have yet to see any examples of BNSI rings (other than fields, which have no non–free modules). This situation is rectified by Theorem 3.2, the main result of this section.

Lemma 3.1. Let S be a noetherian integrally closed domain (not necessarily local). Suppose J is an ideal which is not contained in any height one prime ideal of S. Let A be an S–module such that $JS^n \subseteq A \subseteq S^n$, for some n. If A is reflexive, then $A = S^n$.

Proof. Since S is integrally closed, by [3, §4, No. 2, Theorem 1] we have $A^{**} = \cap A_p$, where p ranges over all height one primes of S and A_p is the localization of A at p. Hence if A is reflexive, $A = \cap A_p$. By hypothesis, for all such p, $J \not\subseteq p$. Hence $(JS^n)_p = (S^n)_p$. But for each p we have $(JS^n)_p \subseteq A_p \subseteq (S^n)_p$, so it follows that $A_p = (S^n)_p$. Thus $\cap A_p = \cap (S^n)_p = S^n$. Therefore $A = S^n$.

Theorem 3.2. Let $(S, \mathbf{m})$ be a local noetherian integrally closed domain. Let J be an ideal which is not contained in any height one prime, and let $R = S/\mathbf{m}J$. Then R is a BNSI ring.

Proof. If $0 \to L \to F \to M \to 0$ is an exact sequence of R–modules with F free, then for all $i \geqslant 1$, $\operatorname{Tor}_{i+1}^R(M, -) \approx \operatorname{Tor}_i^R(L, -)$. Hence $\beta_{i+1}(M) = \beta_i(L)$ for all $i \geqslant 1$. Thus to prove that R is a BNSI ring, it suffices to show that for all modules M, $\beta_2(M) > \beta_1(M)$. In other words, we must show that if $0 \to K \to R^n \xrightarrow{f} R^q$ is exact, and $K \subseteq \mathbf{m}R^n$, and $f(R^n) \subseteq \mathbf{m}R^q$, then K cannot be generated by n elements. For if $M = \operatorname{coker}(f)$, then $\beta_2(M) = \beta_0(K)$ and $\beta_1(M) = n$. Now $f(JR^n) = Jf(R^n) \subseteq \mathbf{m}JR^q = 0$. Therefore $JR^n \subseteq K = \ker(f)$, so $JR^n \subseteq K \subseteq \mathbf{m}R^n$. Let $\pi : S^n \to R^n$ be the natural projection, and let $A = \pi^{-1}(K)$. Then $K = A/\mathbf{m}JS^n$ and $\pi^{-1}(JR^n) \subseteq \pi^{-1}(K) \subseteq \pi^{-1}(\mathbf{m}R^n)$, i.e., $JS^n \subseteq A \subseteq \mathbf{m}S^n$. We claim: $K/\mathbf{m}K \approx A/\mathbf{m}A$. For

$$\mathbf{m}K = \mathbf{m}(A/\mathbf{m}JS^n) = (\mathbf{m}A + \mathbf{m}JS^n)/\mathbf{m}JS^n,$$

and since $JS^n \subseteq A$, $mJS^n \subseteq mA$. So $mK = mA/mJS^n$. Hence

$$K/mK = (A/mJS^n)/(mA/mJS^n) \approx A/mA.$$

It follows that the R–module K and the S–module A have the same minimal number of generators. Now suppose that K can be generated by n elements. Then so can A. Let T be the quotient field of the domain S. Since $JS^n \subseteq A \subseteq S^n$ and $J \neq 0$, $A \otimes_S T = T^n$. So the n generators of A generate the vector space T^n, and hence are linearly independent over T. They are therefore linearly independent over S and hence form a basis for A. A is thus S–free and a fortiori S–reflexive. By the preceding lemma, $A = S^n$. But $A \subseteq mS^n$, so we have a contradiction. Thus K cannot be generated by n elements.

We note one special case of Theorem 3.2:

Corollary 3.3. Let (S, m) be a regular local ring of dimension $k \geqslant 2$. Let $R = S/m^n$, where $n \geqslant 1$. Then R is a BNSI ring.

Proof. When $n = 1$, R is a field, which is vacuously a BNSI ring, so assume $n > 1$. Let $J = m^{n-1}$. Then since the Krull dimension of S is at least two, no height one prime contains J.

Combining this last result with Corollary 2.4 we see that with R as above, for any R–module M and any $i \geqslant 2$, $\mathrm{Ext}^i(M, R) = 0$ implies that M is free. This is a substantial generalization of the following: **Theorem.** [7, Corollary 2.4]. Let (S, m) be a regular local ring of dimension $k \geqslant 2$. Let $R = S/m^n$, where $1 \leqslant n \leqslant k$. If M is an R–module such that for all $i \geqslant 1$, $\mathrm{Ext}^i(M, R) = 0$, then M is free.

We now examine the consequences of dropping or changing hypotheses in Theorem 3.2.

Proposition 3.4. Let (S, m) be a local domain and let J be any non–zero ideal. Let $R = S/mJ$. Then for any R–module M, the sequence $\{\beta_i(M)\}_{i \geqslant 1}$ is non–decreasing.

Proof. This is a minor modification of the proof of Theorem 3.2. We use the same notation as in that proof. There we had to show that the S–module A could not be generated by n elements, while here we must show only that n–1 elements do not suffice. But we saw that $A \otimes_S T = T^n$, using nothing about the ideal J, except that $J \neq 0$. Nor, at this point, did we need the assumption that S was integrally closed. Thus with our weaker hypotheses we may still conclude that the dimension of the T–vector space $A \otimes_S T$ is n, and hence the minimal number of generators of the S–module A is at least n.

Proposition 3.5. Let $(S, \mathbf{m})$ be an integrally closed noetherian local domain. Let J be a non–zero ideal contained in a proper principal ideal, and let $R = S/\mathbf{m}J$. Then there exists a reflexive R–module which is not free. Consequently, R is *not* a BNSI ring.

Proof. Let $J \subseteq Sx$, and denote by $\bar{x}$ the image of x in R. We shall show that the ideal $R\bar{x}$ is the desired non–free, reflexive R–module. The second statement will then follow from Corollary 2.5.

Since $0 \neq J \subseteq Sx$, $J = J_0x$ for some ideal $J_0 \neq 0$. Let $I = \mathbf{m}J_0$. Then $R = S/Ix$. We claim that $\text{ann}_S(Sx/Ix) = I$ and $\text{ann}_S(I/Ix) = Sx$. Clearly $I \subseteq \text{ann}_S(Sx/Ix)$. Now suppose $yx \in Ix$. Then since S is a domain and $x \neq 0$, $y \in I$. Thus the first equality is established. For the second, the inclusion $Sx \subseteq \text{ann}_S(I/Ix)$ is obvious. To get the reverse inclusion, we use the fact that S is integrally closed. Suppose that $yI \subseteq Ix$. Then multiplication by y/x is an S–endomorphism of I, and so $y/x \in \text{Hom}_S(I, I)$, which is a finitely generated S–module, since I is. Therefore y/x is integral over S, and since S is integrally closed in its quotient field, $y/x \in S$. Thus $y \in Sx$, and so $\text{ann}_S(I/Ix) = Sx$.

The equalities proved above imply that $\text{ann}_R(R\bar{x}) = I/Ix$ and $\text{ann}_R(I/Ix) = Sx/Ix = R\bar{x}$. Hence $\text{ann}_R(\text{ann}_R(R\bar{x})) = R\bar{x}$, so $(R\bar{x})^* = R\bar{x}$, and $R\bar{x}$ is reflexive. It is not free since $\text{ann}_R(R\bar{x}) \neq 0$. ∎

We conclude this section with an example to show that in the preceding proposition, the hypothesis that S is integrally closed is necessary. We shall exhibit a local domain $(S, \mathbf{m})$ such that $S/\mathbf{m}x$ *is* a BNSI ring for some non–zero x in $\mathbf{m}$.

Let F be any field and let $S = F[[X, Y]]/(Y^2-X^3)$, where X and Y are independent indeterminates. S is a local domain with maximal ideal $\mathbf{m} = (x, y)$, where x and y denote the images in S of X and Y respectively. S is not integrally closed since $(y/x)^2 = x$. Now $\mathbf{m}x = (x^2, xy) = (x^2, xy, y^2) = \mathbf{m}^2$. Hence $S/\mathbf{m}x = S/\mathbf{m}^2 \cong F[[X, Y]]/(X, Y)^2$. But $F[[X, Y]]$ is regular local of dimension 2, so by Corollary 3.3, $S/\mathbf{m}x$ is a BNSI ring.

4. QF Rings and Betti Numbers

In this final section we study the Betti numbers of modules over QF rings. There is a sharp distinction between those QF rings which are principal ideal rings (PIR) and those which are not.

Proposition 4.1. If R is a QF ring which is not a PIR, then there exists an

R–module M such that $\beta_2(M) < \beta_1(M)$.

Proof. Let $x_1,\cdots,x_n$ be a minimal system of generators of $\mathbf{m}$. By hypothesis, $n > 1$. For $j = 1,\cdots,n$ let I_j be the ideal generated by all the x's except x_j. Then $x_j \notin \operatorname{ann}(\operatorname{ann}(I_j))$, i.e., $x_j\operatorname{ann}(I_j) \neq 0$. Also, since R is QF, $\operatorname{ann}(\mathbf{m})$ is the unique minimal ideal of R. Hence for $j = 1,\cdots,n$, $\operatorname{ann}(\mathbf{m}) \subseteq x_j\operatorname{ann}(I_j)$. Thus if $z \in \operatorname{ann}(\mathbf{m})$, $z \neq 0$, then for each j there exists an element $w_j \in \operatorname{ann}(I_j)$ such that $z = w_jx_j$. Now $w_jx_i = 0$ for all $i \neq j$. Hence if e_j denotes the jth standard basis vector in R^n, then $ze_j = w_j<x_1,\cdots,x_n>$. Thus $zR^n \subseteq R<x_1,\cdots,x_n>$.

Define $f : R \to R^n$ by sending 1 to $<x_1,\cdots,x_n>$. Every R–module is reflexive and therefore a submodule of a free module. Let $\sigma : \operatorname{coker}(f) \to F$ be an imbedding of coker(f) in a free module, let $\pi : R^n \to \operatorname{coker}(f)$ be the natural projection, and let $g = \sigma \circ \pi$. Then $R \xrightarrow{f} R^n \xrightarrow{g} F \to \operatorname{coker}(g) \to 0$ is exact, and the start of a free resolution of coker(g). We claim that it is minimal. Clearly $f(R) \subseteq \mathbf{m}R^n$, and $\ker(f) \subseteq \mathbf{m}$. It remains only to show that $g(R^n) \subseteq \mathbf{m}F$. Now $g(R^n) \approx \operatorname{coker}(f) = R^n/R<x_1,\cdots,x_n>$. But $zR^n \subseteq R<x_1,\cdots,x_n>$, so $zg(R^n) = 0$. Hence $g(R^n) \subseteq \mathbf{m}F$ and the resolution is minimal. Letting $M = \operatorname{coker}(g)$, we have $\beta_2(M) = 1$ and $\beta_1(M) = n$.

Definition. Let $\cdots \to X_{i+1} \xrightarrow{d_{i+1}} X_i \xrightarrow{d_i} X_{i-1} \to \cdots \to X_1 \xrightarrow{d_1} X_0$ be a complex of R–modules (i.e., $d_i \circ d_{i+1} = 0$ for all $i \geqslant 1$). We say the complex is *periodic with period n* if for all $i \geqslant 0$, $X_i = X_{i+n}$, and for all $i \geqslant 1$, $d_i = d_{i+n}$.

Suppose $\{ P_i, d_i \}$ is a projective resolution of M which is periodic with period n. Then for any module N, the complex $\{ P_i \otimes N, d_i \otimes 1_N \}$ is also periodic with period n. Taking homology, we see that for all $i \geqslant 1$, $\operatorname{Tor}_i(M, N) \cong \operatorname{Tor}_{i+n}(M, N)$. In particular, letting $N = R/\mathbf{m}$, we have $\beta_i(M) = \beta_{i+n}(M)$ for all $i \geqslant 1$. (Of course, if R is a BNSI ring, it follows that no non–free module has a projective resolution which is periodic.)

We give an example. Suppose there exist non–zero elements x and y in $\mathbf{m}$ such that $\operatorname{ann}(Rx) = Ry$ and $\operatorname{ann}(Ry) = Rx$. Then

$$\to R \xrightarrow{y} R \xrightarrow{x} R \xrightarrow{y} R \xrightarrow{x} Rx \to 0$$

is a minimal free resolution of Rx, periodic with period 2. Having made these observations, we may now state and prove:

Proposition 4.2. Let R be a QF ring which is a PIR. Let M be an R–module with no non–zero free direct summand. Then the sequence $\{ \beta_i(M)\}_{i\geqslant 0}$ is constant. $\beta_i(M)$ is the number of cyclic summands occurring in a decomposition of M into a direct sum of cyclic submodules. The minimal free resolution of M is periodic with period 2.

Proof. It is well known (see, for example, [4, Theorem 43, p. 78]) that every finitely generated module over a QF PIR is a direct sum of cyclic modules. Now R is local, so every cyclic R–module is of the form R/Rx^i where $\mathbf{m} = Rx$. If $p =$ index of nilpotence of $\mathbf{m}$, then for all $i < p$, $\mathrm{ann}(Rx^{p-i}) = Rx^i$. The desired results now follow easily.

Addendum

David Eisenbud has communicated to me the following generalization of Theorem 3.2, observed by Kaplansky.

Theorem 3.2A. Let $(S, \mathbf{m})$ be a local noetherian domain and let J be an ideal which is not contained in any prime ideal of grade 1. Then $S/\mathbf{m}J$ is a BNSI ring.

Proof. The proof of Theorem 3.2 works with the following minor modification. By [5, Theorem 53, p. 34], if $\mathbf{p}$ ranges over the set of primes in S of grade 1, then $\cap S_{\mathbf{p}} = S$. Now the module A was actually S–free (and not just S–reflexive). It follows that $\cap A_{\mathbf{p}} = A$.

Remark. A non–zero prime $\mathbf{p}$ in a domain S has grade 1 if and only if $\mathbf{p}$ is contained in an associated prime of a principal ideal. Hence if height $\mathbf{p} = 1$, then grade $\mathbf{p} = 1$. If S is integrally closed and noetherian, then the converse is true [5, Theorem 103, p. 76], so that the primes of height one and the primes of grade 1 are the same.

References

1. M. Auslander, Coherent functors, "Proceedings of the Conference on Categorical Algebra," La Jolla 1965 (Springer–Verlag), 189-231.
2. M. Auslander and M. Bridger, Stable module theory, Mem. Amer. Math. Soc. No. **94** (1969).
3. N. Bourbaki, "Algèbre Commutative," Ch. 7, Diviseurs, Act. Scient. et Ind. No. 1314, Hermann, Paris, 1965.
4. N. Jacobson, "Theory of Rings," Amer. Math. Soc. Surveys No. 2, New York, 1943.
5. I. Kaplansky, "Commutative Rings," Allyn and Bacon, Boston, 1970.
6. M. Menzin, Indecomposable modules over Artin local rings, Ph.D. Thesis, Brandeis University, 1970.

7. M. Ramras, On the vanishing of Ext, Proc. Amer. Math. Soc. **27** (1971), 457-462.

IDEALIZER RINGS

J. C. Robson

University of Leeds

Let S be a ring with an identity element, A a right ideal of S, and $\mathbf{I}(A) = \{ s \in S \mid sA \subseteq A \}$. The ring $\mathbf{I}(A)$ is called the *idealizer* of A in S and is the largest subring of S which contains A as an ideal. It was introduced by Ore [8] and used by Fitting [4] who established a ring isomorphism between $\mathbf{I}(A)/A$ and End S/A.

In this paper we show that there are very tight connections between the properties of S and $\mathbf{I}(A)$ in the special case when A is a *semimaximal* right ideal, by which we mean that S/A is a semisimple (artinian) module. For then $\mathbf{I}(A)$ is right noetherian or right artinian precisely when S is so; and their global dimensions are equally closely linked.

The paper ends with a brief description of some consequences of the theory. In one direction, it yields a theory of hereditary noetherian prime rings which enables one to prove several structure–type theorems. When applied to the special case of hereditary orders over Dedekind domains, it gives a global version of some local results. In another more obvious direction, the idealizer theory gives a ready method for constructing examples. This is illustrated by one particular example.

Much of the theory described here is more fully explored in [10]. But some results presented here (2.5 to 2.8) are new or improved.

I would like to express my thanks to the organizers of the conference.

1. Preparatory results

This section isolates some preparatory results which do not require A to be a semimaximal right ideal of S. We start by considering the (left) regular representation $S \cong \mathrm{End}(S_S)$ in which S acts via left multiplication. As Fitting proved in [4], this representation induces an epimorphism $\mathbf{I}(A) \to \mathrm{End}\ S/A$ with kernel A. More generally, using the notation $B \cdot\!\cdot A = \{ s \in S \mid sA \subseteq B \}$, one obtains the basic result

Proposition 1.1. If A, B are right ideals of S then the regular representation of S induces a ring isomorphism $I(A)/A \cong \text{End } S/A$ and an $I(A)/A$–module isomorphism $(B \cdot\cdot A)/B \cong \text{Hom}(S/A, S/B)$.

Another straightforward consequence of this representation is

Proposition 1.2. If A', A are right ideals with $A' \supseteq A$ then $I(A') \supseteq I(A)$ provided that A'/A is a fully invariant submodule of S/A.

In the remainder of this section we require that $SA = S$. It will be shown in the next section that when A is semimaximal this requirement involves no loss of generality. We let $R = I(A)$.

Proposition 1.3. If $SA = S$ then S is a finitely generated projective right R–module and $S \otimes_R S \cong S$ via multiplication.

Proof. Since $SA = S$, there are elements $s_i \in S$, $a_i \in A$ such that $\Sigma_{i=1}^n s_i a_i = 1$. But then the epimorphism $R^{(n)} \to S$ given by $(r_1,\cdots,r_n) \to \Sigma s_i r_i$ is split by the map $S \to R^{(n)}$ given by $s \to (a_1 s,\cdots,a_n s)$. So S_R is finitely generated and projective. Therefore the map $S \otimes_R R \to S \otimes_R S$ is a mono–morphism. But

$$S \otimes S = SA \otimes S = S \otimes AS = S \otimes AR = SA \otimes R = S \otimes R$$

and so $S \otimes S \cong S$. ∎

This shows that S is a right flat epimorphic extension ring of R as discussed by Silver [13].

Proposition 1.4. If $SA = S$ and $I(A)$ is right noetherian or right artinian, then so also is S.

Proof. S_R is finitely generated. ∎

Proposition 1.5. If $SA = S$ then rt gl dim $S \leqslant$ rt gl dim R.

Proof. If P is a projective right R–module then $P \otimes_R S$ is a projective right S–module. Therefore, if

$$0 \to P_n \to \cdots \to P_0 \to X \to 0$$

is an R–projective resolution of a right S–module X then

$$0 \to P_n \otimes S \to \cdots \to P_0 \otimes S \to X \otimes S \to 0$$

is an S–projective resolution of $X \otimes S$. But

$$X \otimes_R S \cong X \otimes_S S \otimes_R S \cong X \otimes_S S \cong X. \quad \blacksquare$$

Note 1.6. In fact, for any right S–module X, $\text{p.d.}X_R = \text{p.d.}X_S$ because any S–projective resolution of X is also an R–projective resolution by virtue of 1.3.

2. Main results

Throughout this section, A will be a semimaximal right ideal of S. This assumption is fundamental insofar as there are examples in [10] which show that none of the results of this section are valid for an arbitrary right ideal A. The first result shows, as promised, that we can assume that $SA = S$.

Proposition 2.1 There is a semimaximal right ideal $A' \supseteq A$ such that $I(A) = I(A')$ and $SA' = S$.

Proof. Let A' be a complement to SA, modulo A; that is, $A' + SA = S$, $A' \cap SA = A$. Note first that $SA' \supseteq SA + A' = S$ and that

$$I(A') \subseteq I(A' \cap SA) = I(A).$$

It remains to show that $I(A') \supseteq I(A)$. This follows from 1.2 provided we check that A'/A is a fully invariant submodule of S/A.

To see this, note that since $A'/A \cong S/SA$ each simple submodule of A'/A is annihilated by SA. But on the other hand, since $SA/A \cong S/A'$ and

$$(S/A')SA = (SA + A')/A' = S/A',$$

no submodule of SA/A is annihilated by SA. Therefore A'/A is fully invariant. $\blacksquare$

Henceforth, therefore, we can assume that $SA = S$ when discussing the relationship between S and $R = I(A)$. The crucial step in this discussion concerns the nature of simple S–modules when viewed as R–modules.

Proposition 2.2. If $M \supseteq A$ is a maximal right ideal of S then S/M has a unique R–composition series of length 2 given by $S \supset R + M = M \cdot\!\cdot A \supset M$; but a simple S–module which is not a homomorphic image of S/A is simple over R.

Proof. Let U be a simple S–module and $0 \neq u \in U$. The chain $U = uS \supseteq uR \supseteq uA \supseteq 0$ shows that either $uA = 0$ or else $uS = uR = uA$. Thus, if U is not a homomorphic image of S/A, then U is simple over R.

Next we consider R–submodules of S/M. Let $S \supset sR + M \supseteq M$. Then $sA \subseteq M$ and so $s \in M \cdot\!\cdot A$. Thus each proper R–submodule of S/M is contained in $(M \cdot\!\cdot A)/M$. By 1.1, $(M \cdot\!\cdot A)/M \cong \mathrm{Hom}(S/A, S/M)$ which, since S/A is semisimple, is a simple End S/A–module. Hence $(M \cdot\!\cdot A)/M$ is a simple R–module. Since $S \neq M \cdot\!\cdot A$ this has shown that $S \supset M \cdot\!\cdot A \supset M$ gives the unique R–composition series for S/M. It is clear that $M \cdot\!\cdot A \supseteq R + M \supset M$, and this completes the proof. ∎

This has some immediate consequences.

Theorem 2.3. If A is a semimaximal right ideal of S then S is right artinian if and only if I(A) is right artinian.

Proof. One half was proved in 1.4. Conversely, if S is right artinian then, by 2.2, S_R has an R–composition series and so its submodule R_R is artinian. ∎

Theorem 2.4. If A is a semimaximal right ideal of S then S is right noetherian if and only if I(A) is right noetherian.

Proof. Again, 1.4 proved half. So let S be right noetherian, and let B be a right ideal of R. Then $BS \supseteq B \supseteq BA$. Now BS/BA is a homomorphic image of a finite direct sum of copies of S/A and so has finite length over R. Also BA is finitely generated over S and hence, by 1.3, over R. Thus B is finitely generated and R is right noetherian. ∎

Before obtaining further consequences we need some more information concerning the relationship between BS, B and BA.

Lemma 2.5. For each maximal right ideal $M \supseteq A$,

$$(R + M/M) \otimes_R S \cong S/M$$

under the multiplication map $\bar{r} \otimes s \to \overline{rs}$.

Proof. Let $S/A = \amalg_{i=1}^{n} S/M_i$, the M_i being maximal right ideals and $\cap_{i=1}^{n} M_i = A$. It follows from 1.1 and 2.2 that $R/A \cong \amalg_{i=1}^{n} (R + M_i/M_i)$. However

$$(R/A) \otimes_R S \cong (R \otimes_R S)/(A \otimes_R S) \cong S/A$$

via multiplication. Restricted to the direct summands of R/A this gives the desired isomorphism. ∎

Lemma 2.6. Given any right ideal B of R there is a set of maximal right ideals $\{M_j\}$, $M_j \supseteq A$, such that

$$BS/BA \cong \coprod_j S/M_j, \quad B/BA \cong \coprod_j (R + M_j/M_j) \quad \text{and} \quad BS/B \cong \coprod_j (S/R + M_j) .$$

Proof. Since B/BA is an R/A–module, $B/BA \cong \coprod_j (R + M_j/M_j)$ for some set $\{M_j\}$. Since tensor product commutes with direct sum, the result follows from 2.5. ■

If each right ideal of S has a generating set of n elements, we say that S has *rank* n.

Proposition 2.7. If S has rank n and S/A has length m over S, then R has rank $(2m + 1)n$.

Proof. S/R has length m over R and so S_R has a generating set of $m + 1$ elements. Therefore BA has a generating set over R of $(m + 1)n$ elements. Also BS/BA has length at most mn over S and, by 2.6, the length of B/BA over R is at most mn. Therefore B has a generating set of $mn + (m + 1)n = (2m + 1)n$ elements. ■

Next we turn to the right global dimensions of S and R.

Theorem 2.8. For any semimaximal right ideal A of S such that $\mathbf{I}(A) \neq S$,

$$\text{rt gl dim } \mathbf{I}(A) = \sup \{1, \text{ rt gl dim } S\} .$$

Proof. We can, by 2.1, assume that $SA = S$. By 1.5, rt gl dim $S \leqslant$ rt gl dim R and, by 1.6, the R and S projective dimensions of a right S–module are equal.

First we show that, for any right ideal B of R, p. d. B = p. d. BS. For, by 2.6, $BS/B \cong \coprod_j (S/R + M_j)$ and each of these summands is a summand of S/R. Thus p. d. $BS/B \leqslant 1$. The usual long exact sequence argument then shows that p. d. B = p. d. BS. Provided that S is not semisimple, it follows that rt gl dim R = rt gl dim S. But if S is semisimple what follows is that rt gl dim $R \leqslant 1$. So it remains only to show that R is not semisimple. However, if R were semisimple then each proper ideal, including A, would have an annihilator. But $SA = S$. Thus R cannot be semisimple. ■

The special case when S is semisimple is of interest. It is easily seen that the idealizer of a right ideal of S is a direct sum of idealizers of right ideals of its simple components. Let $S \cong D_n$ be simple artinian, D a division ring. Given any right ideal A of S, there is a ring automorphism with respect to which A has the form

A =

and then it is easy to check that **I**(A) has the form

I(A) =

where the shaded block has arbitrary entries from D and the unshaded entries are all zeros.

Now we can form a chain of rings, starting from any ring S, each being the idealizer of a semimaximal right ideal in its predecessor. The rings occurring in such a chain will be called *iterated idealizers* from S. The discussion above leads to

Theorem 2.9. The iterated idealizers R from a simple artinian ring D_n, D a division ring, are up to isomorphism, the block lower triangular $n \times n$ matrix rings over D;

$R \cong$

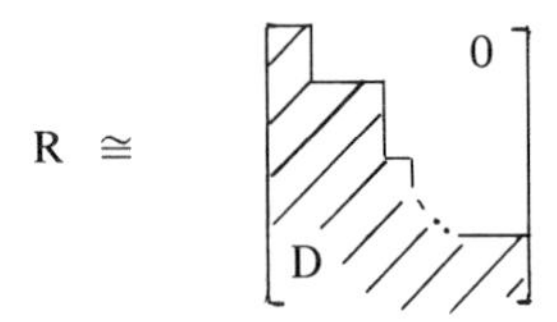

As shown in [10], §3, the idealizer theory not only shows that each such ring R is hereditary and artinian but also that it is *serial* (or *generalized uniserial*) by which we mean that each indecomposable R–module has a unique composition series. In fact this theorem describes all indecomposable hereditary serial rings.

3. Applications

Finally, we wish to indicate two directions in which the preceding theory has proved useful. One is fairly obvious - the construction of examples and counter–examples. As has been shown already, the idealizer ring R shares most of the properties of S and yet, insofar as S is a "localization" of R, it is clear that R must have some less pleasant properties.

In particular, one can construct one rather surprising example - namely an hereditary noetherian domain having exactly one non–trivial ideal. The ideal auto–matically must be idempotent. We start by letting S be a simple principal ideal domain; for example $S = F(y)[x]$, where $F(y)$ is the field of rational functions in y over a field F of characteristic zero and S is the ring of polynomials in x over $F(y)$ subject to $xy - yx = 1$. If A is any maximal right ideal of S then $I(A)$ is as claimed. (Most of the assertions follow from §2, together with an argument which obtains a left–right symmetry (see [10])). In the ring cited, if we choose $A = xS$ then $I(A) = F + xS$. By 2.7, every right ideal of $I(A)$ can be generated by 3 elements. Other examples constructed in this fashion can be found in [10] §7 and [7] §6.

The other, more important, use of idealizers comes in the study of hereditary noetherian prime rings (*hnp–rings* for short). Much of the theory of hnp–rings echoes the theory of hereditary orders in central simple algebras over a Dedekind domain (see [12]), but as well as being more general, it has the advantage of being global rather than local. We illustrate this by recalling a little of the theory of hereditary orders and then describing the corresponding results about hnp–rings. An hereditary order is known to be an intersection of maximal orders (Harada [5]) and, in the case when the Dedekind domain is a complete DVR, can be described in terms of the maximal orders as the set of endomorphisms of a particular module leaving invariant certain specified submodules, (Jacobinski [6]).

In the light of 1.1, this latter description is closely related to idealizers. The next few results show this more precisely.

Theorem 3.1. Let S be an hnp–ring and A a right ideal such that $SA = S$. Then $I(A)$ is an hnp–ring if and only if A is semimaximal.

Proof. See [10] §5. ∎

One can think of this result as specifying certain subrings of S which are hnp–rings. The next result specifies over–rings.

Theorem 3.2. Let R be an hnp–ring and B an idempotent ideal of R. Then $End(_R B)$ is an hnp–ring and there is an embedding $R \subseteq End(_R B)$.

Proof. See [10] §4. ■

These two processes are related as follows.

Theorem 3.3. (i) If S is an hnp–ring, A a semimaximal right ideal such that SA = S and R = **I**(A), then A is an idempotent ideal of R and S ≅ End($_R$A).

(ii) If R is an hnp–ring embedded in End($_R$B) = S, where B is an idempotent ideal of R such that R/B is semisimple, then B is a semimaximal right ideal of S and R = **I**(B).

Proof. See [10] §5. ■

In the light of 3.2 a "maximal" hnp–ring will be one having no proper idempotent ideals. Such a ring is called a *Dedekind prime ring* and its properties have been studied in [9], [3] and [11]. It is not known whether each hnp–ring R is embedded in a Dedekind prime ring - this is equivalent to R having only finitely many idempotent ideals – but one can prove

Theorem 3.4. Hnp–rings having finitely many idempotent ideals are precisely iterated idealizers from Dedekind prime rings. Each is an intersection of Dedekind prime rings.

Proof. See [10] §5. ■

This, when applied to hereditary orders (which are, of course, embedded in maximal orders) gives our version of the theory mentioned earlier.

Corollary 3.5. Hereditary orders are precisely iterated idealizers from maximal orders. They are intersections of maximal orders.

Even without knowing whether an hnp–ring R is embedded in a Dedekind prime ring, 3.3 can be used to obtain results concerning R. The pattern of proof is to use 3.2 to move from R to an over–ring S. By a suitable induction hypothesis, the result can be assumed to hold for S. Then the result is transferred to R using idealizer theory. One instance of this method is in the proof of the following result, the first statement of which was originally proved by Eisenbud and Griffith [1].

Theorem 3.6. Each proper factor ring of an hnp–ring is a direct sum of indecomposable serial rings. Each of these indecomposable serial rings is a homomorphic image of a block "lower triangular" matrix ring T over a local artinian principal ideal ring L with radical J, of the form

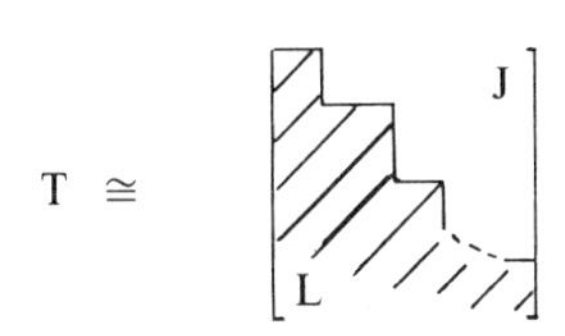

Proof. See [10] §6. ∎

References

1. D. Eisenbud and P. Griffith, Serial rings, J. Alg. **17** (1971), 389-400.
2. D. Eisenbud and J. C. Robson, Modules over Dedekind prime rings, J. Alg. **16** (1970), 67-85.
3. D. Eisenbud and J. C. Robson, Hereditary noetherian prime rings, J. Alg. **16** (1970), 86-104.
4. H. Fitting, Primärkomponentenzerlegung in nichtkommutativen Ringen, Math. Ann. **111**(1935), 19-41.
5. M. Harada, Hereditary orders, Trans. Amer. Math. Soc. **107** (1963), 273-290.
6. H. Jacobinski, Two remarks about hereditary orders (to appear).
7. J. C. McConnell and J. C. Robson, Homomorphisms and extensions of modules over A_1 and related rings (to appear).
8. O. Ore, Formal Theorie der linearen differential Gleichungen, Crelle's J. **168** (1932), 233-252.
9. J. C. Robson, Non–commutative Dedekind rings, J. Alg. **9** (1968), 249-265.
10. J. C. Robson, Idealizers and hereditary noetherian prime rings, J. Alg. (to appear).
11. J. C. Robson, A note on Dedekind prime rings, Bull. London Math. Soc. **3** (1971), 42-46.
12. K. Roggenkamp, Lattices over Orders II, Springer–Verlag, 1970.
13. L. Silver, Non–commutative localizations and applications, J. Alg. **7** (1967), 44-76.

PERFECT PROJECTORS AND PERFECT INJECTORS

Edgar A. Rutter, Jr.

University of Kansas

Introduction

Let R and S be rings and ${}_R\mathfrak{M}$ and ${}_S\mathfrak{M}$ be the categories of all left R–modules and left S–modules, respectively. An (R, S) *adjoint triple* is a triple of additive functors

$$F : {}_R\mathfrak{M} \to {}_S\mathfrak{M} \quad \text{and} \quad G, H : {}_S\mathfrak{M} \to {}_R\mathfrak{M}$$

such that there are natural isomorphisms

$$\mathrm{Hom}_R(G(N), M) \cong \mathrm{Hom}_S(N, F(M))$$

$$\mathrm{Hom}_S(F(M), N) \cong \mathrm{Hom}_R(M, H(N))$$

for all M in ${}_R\mathfrak{M}$ and N in ${}_S\mathfrak{M}$. Morita [8] has shown that a triple of functors (G, F, H) is an (R, S) adjoint triple if and only if there is bimodule ${}_SP_R$ with P_R finitely generated and projective such that F, G, H are naturally equivalent to the functors

$$F_P = P \otimes_R (-) : {}_R\mathfrak{M} \to {}_S\mathfrak{M}$$
$$G_P = \mathrm{Hom}_R(P, R) \otimes_S (-) : {}_S\mathfrak{M} \to {}_R\mathfrak{M}$$
$$H_P = \mathrm{Hom}_S(P, -) : {}_S\mathfrak{M} \to {}_R\mathfrak{M}$$

respectively. Moreover, Morita [7] proved that a functor $F : {}_R\mathfrak{M} \to {}_S\mathfrak{M}$ is a category equivalence if and only if F is naturally equivalent to F_P with P a finitely generated projective generator and $S = \mathrm{End}_R(P_R)$.

When F_P is an equivalence it preserves all categorical properties. In the more general setting much of the structure may be lost in passing from ${}_R\mathfrak{M}$ to ${}_S\mathfrak{M}$ via F_P. In [1] Anderson considered those adjoint triples with $S = \mathrm{End}_R(P)$ and sought to determine when F_P preserved certain categorical properties specifically, projectivity, injectivity, projective covers, and injective envelopes. Anderson obtained a characterization of those projective modules for which F_P preserves injective envelopes, which he called *perfect injectors,* in terms of the trace ideal of P and a corresponding result for *perfect projectors* provided R is left perfect. The purpose of this paper is to continue the investigation begun by Anderson in [1]. Rather than seeking primarily to characterize these concepts directly, we begin by describing, in terms of a canonical torsion–torsion free theory associated with P via its trace ideal, subclasses of ${}_R\mathcal{M}$ on which F_P always behaves properly with respect to the properties being considered. We then apply these results to recover and extend the work of Anderson as well as to give additional characterizations of perfect injectors and perfect projectors. These are expressed in terms of the torsion–torsion free theory mentioned above. Perhaps the most interesting feature of this approach is that it exhibits a quite complete duality between the properties of perfect injectors and perfect projectors.

In the last section some of our results are applied to give new derivations of theorems of Mares [6] and Ware [11] which characterize semi–perfect and perfect modules.

1. Preliminaries

Throughout this paper R will denote an associative ring with identity, P_R will denote a finitely generated projective R–module, P^* will denote its R–dual

$$P^* = \mathrm{Hom}_R(P, R)$$

and S will denote its R–endomorphism ring

$$S = \mathrm{End}_R(P).$$

Then P is a left S–right R–bimodule – we shall always write homomorphisms opposite scalars – and P^* is a right S–left R–bimodule and is finitely generated and projective over R with endomorphism ring S.

Since the functors (G_P, F_P, H_P) form an adjoint triple – to simplify the notation, we shall omit the subscript P whenever this can be done without confusion – there exist natural transformations

$$\alpha : 1_{{}_S\mathcal{M}} \to FG,\ \alpha' : FH \to 1_{{}_S\mathcal{M}}$$

$$\beta : 1_{{}_R\mathcal{M}} \to HF \text{ and } \beta' : GF \to 1_{{}_R\mathcal{M}}.$$

Moreover, since $S = \mathrm{End}_R(P)$ both α and α' are natural equivalences. Furthermore, for all M in ${}_R\mathcal{M}$ and N in ${}_S\mathcal{M}$,

$$\alpha_{F(M)} \circ F(\beta'_M) = 1_{F(M)}, \quad F(\beta_M) \circ \alpha'_{F(M)} = 1_{F(M)},$$

$$\beta_{H(N)} \circ H(\alpha'_N) = 1_{H(N)} \text{ and } G(\alpha_N) \circ \beta'_{G(N)} = 1_{G(N)}.$$

Since α and α' are natural equivalences, $F(\beta_M)$, $F(\beta'_M)$, $\beta_{H(N)}$ and $\beta'_{G(N)}$ are isomorphisms for all M in ${}_R\mathfrak{M}$ and N in ${}_S\mathfrak{M}$. The above facts are well known and can be found, for instance, in [8]. They will be used extensively in the next section.

If X is a projective (left or right) R–module the *trace ideal* of X, T(X) = Σ im(f) for all f in $\text{Hom}_R(X, R)$, is an ideal of R such that $T(X)^2 = T(X)$ and $T(X)X = X$ or $XT(X) = X$ according as X is a left or right R–module. Again to simplify the notation we let $T = T(P)$ and note that $T = T(P^*)$. (See [1]).

Dickson [4] defined a *torsion theory* for ${}_R\mathfrak{M}$ to be a pair $(\mathcal{T}.\mathfrak{F})$ of classes of modules in ${}_R\mathfrak{M}$ satisfying the following conditions:

(a) $\mathcal{T} \cap \mathfrak{F} = (0)$.

(b) $\mathcal{T}$ is closed under homomorphic images and $\mathfrak{F}$ is closed under submodules.

(c) For each module M in ${}_R\mathfrak{M}$, there exists a unique submodule t(M) of M such that $t(M) \in \mathcal{T}$ and $M/t(M) \in \mathfrak{F}$.

For each module M in ${}_R\mathfrak{M}$, t(M) is the largest submodule of M in $\mathcal{T}$ and is called the *$\mathcal{T}$–torsion submodule* of M.

A class of modules $\mathcal{T}$ is a *torsion class* if there exists a class $\mathfrak{F}$ such that $(\mathcal{T}, \mathfrak{F})$ is a torsion theory. A torsion class $\mathcal{T}$ is closed under homomorphic images, direct sums and extensions and uniquely determines $\mathfrak{F}$. Conversely, any class of modules with these three properties is a torsion class. A *torsion free* class is defined dually and is characterized by being closed under submodules, direct products, and extensions. A torsion class $\mathcal{T}$ and the associated torsion theory $(\mathcal{T}, \mathfrak{F})$ is called *hereditary* if $\mathcal{T}$ is closed under submodules. This is equivalent to $\mathfrak{F}$ being closed under injective hulls. All of the above results are contained in Dickson [4].

If $\mathcal{T}$ is a hereditary torsion class which is closed under direct products, Jans [5] called $\mathcal{T}$ a *torsion–torsion free class* (TTF–class). In view of the results of Dickson [4] cited above, $\mathcal{T}$ is a TTF class if and only if there exists a pair of torsion theories $(\mathcal{G}, \mathcal{T})$ and $(\mathcal{T}, \mathfrak{F})$. Each such pair of torsion theories is called a *torsion–torsion free theory.* It is shown in [5] that there exists a one–to–one correspondence between TTF theories in ${}_R\mathfrak{M}$ and idempotent ideals I of R. This correspondence is given by $I = g(R)$, the $\mathcal{G}$–torsion submodule of R while

$$\mathcal{G}_I = \{M \text{ in } {}_R\mathfrak{M} \mid IM = M\},$$
$$\mathcal{T}_I = \{M \text{ in } {}_R\mathfrak{M} \mid IM = 0\}, \text{ and}$$
$$\mathfrak{F}_I = \{M \text{ in } {}_R\mathfrak{M} \mid t_I(M) = \{m \in M \mid Im = 0\} = 0\}.$$

Since the trace ideal T of P_R is idempotent there is associated with P via T a TTF theory as above. For each M in ${}_R\mathfrak{M}$, we shall denote the $\mathcal{T}_T$–torsion submodule of M by $t_T(M)$. One verifies easily that

$$\mathcal{T}_T = \{M \text{ in } {}_R\mathfrak{M} \mid F_P(M) = P \otimes_R M = 0\}.$$

2. Some properties of (G_P, F_P, H_P)

Recall that a submodule K of a module B is *superfluous* in B in case for every submodule L of B, if $K + L = B$, then $L = B$. A homomorphism $B \to D$ is *minimal* if its kernel is superfluous in B. Finally a *projective cover* for a module D is a minimal epimorphism $B \to D$ where B is projective. If a projective cover for D exists it is uniquely determined up to isomorphism over the identity on D and will be denoted by P(D). (See [2].)

The proof of the following lemma is routine and will be omitted.

Lemma 2.1. An epimorphism $f : B \to D$ is minimal if and only if for any homomorphism $g : B' \to B$, if $g \circ f$ is an epimorphism, then g is an epimorphism.

Lemma 2.2. Let $M \in {}_R\mathfrak{M}$ and $N, N' \in {}_S\mathfrak{M}$. If $f : M \to G_P(N)$ is such that $F_P(f)$ is an epimorphism, then f is an epimorphism. If $h : N \to N'$ is such that $G_P(h)$ is an epimorphism, then h is an epimorphism.

Proof. The first implication follows from the fact that G preserves epimorphisms, $\beta'_{G(N)}$ is an isomorphism and $\beta'_M \circ f = GF(f) \circ \beta'_{G(N)}$. The other implication is similar.

Proposition 2.3. The functor G_P preserves minimal epimorphisms and projective covers.

Proof. Let $h : N \to N'$ be a minimal epimorphism in ${}_S\mathfrak{M}$. Clearly G(h) is an epimorphism. Suppose $g : M \to G(N)$ is such that $g \circ G(h)$ is an epimorphism. Since the diagram

$$\begin{array}{ccccc} F(M) & \xrightarrow{F(g)} & FG(N) & \xrightarrow{FG(h)} & FG(N') \\ & & \wr\!\| \; \alpha_N & & \wr\!\| \; \alpha_{N'} \\ & & N & \xrightarrow{h} & N' \end{array}$$

commutes, F preserves epimorphisms, and h is a minimal epimorphism, it follows from Lemma 2.1 that F(g) is an epimorphism. Combining Lemmas 2.1 and 2.2 yields that G(h) is a minimal epimorphism.

It is well known that G preserves projectivity. This is immediate from the natural equivalence between the functors $\mathrm{Hom}_R(G(N), (-))$ and $\mathrm{Hom}_S(N, F(-))$ since the latter functor is exact when N is projective.

Lemma 2.4. For any $M \in {}_R\mathcal{M}$, $\ker \beta_M$, $\mathrm{coker}\, \beta_M$, $\ker \beta'_M$, and $\mathrm{cok}\, \beta'_M \in \mathcal{T}_T$.

Proof. Since both $F(\beta_M)$ and $F(\beta'_M)$ are isomorphisms and F is exact, F vanishes on each of the above modules and hence they belong to $\mathcal{T}_T$.

Lemma 2.5. If $M \in \mathcal{G}_T$, β'_M is a minimal epimorphism.

Proof. Since $M \in \mathcal{G}_T$ so does $\mathrm{coker}\, \beta'_M$ and so by Lemma 2.4, $\mathrm{coker}\, \beta'_M \in \mathcal{G}_T \cap \mathcal{T}_T = \{0\}$. So β'_M is an epimorphism. Suppose $f : M' \to GF(M)$ is such that $f \circ \beta'_M$ is an epimorphism. Since F is exact and $F(\beta'_M)$ is an isomorphism, $F(f)$ is an epimorphism. The conclusion follows from Lemmas 2.1 and 2.2.

Theorem 2.6. Let $f : M' \to M$ be a minimal epimorphism (projective cover) in ${}_R\mathcal{M}$ with $M \in \mathcal{G}_T$. Then $F_P(f) : F_P(M') \to F_P(M)$ is a minimal epimorphism (projective cover).

Proof. Since $M \in \mathcal{G}_T$, $TM' + \ker f = M'$ and so $M' \in \mathcal{G}_T$. Clearly, $F(f)$ is an epimorphism. Suppose $h : N \to F(M')$ is such that $h \circ F(f)$ is an epimorphism. Since the following diagram commutes

$$\begin{array}{ccccc} G(N) & \xrightarrow{G(h)} & GF(M') & \xrightarrow{GF(f)} & GF(M) \\ & & \downarrow \beta'_{M'} & & \downarrow \beta'_M \\ & & M' & \xrightarrow{f} & M \end{array}$$

$G(h) \circ GF(f)$ and β'_M are epimorphisms, and $\beta'_{M'}$ and f are minimal epimorphisms, it follows from Lemmas 2.1 and 2.2 that h is an epimorphism. Hence $F(f)$ is a minimal epimorphism by Lemma 2.1.

If M' is projective, Lemma 2.5 implies $\beta'_{M'}$ is an isomorphism. Also FG is naturally equivalent to the identity. Thus there exist natural equivalences $\mathrm{Hom}_S(F(M'), (-)) \cong \mathrm{Hom}_S(F(M'), FG(-)) \cong \mathrm{Hom}_R(GF(M'), G(-))$. Since the last of these functors preserves epimorphisms, so does $\mathrm{Hom}_S(F(M'), (-))$ and hence $F(M')$ is projective.

The concepts of *essential monomorphism* and *injective envelope* are the categorical duals of minimal epimorphism and projective cover. However, the injective envelope of any module B always exists and will be denoted by $E(B)$. Each result of this section has a dual which can be established by dualizing its proof.We state those results of this type needed in the sequel. The remaining statements and all proofs are left to the reader.

Proposition 2.7. The functor H_P preserves essential monomorphisms and injective envelopes.

Lemma 2.8. If $M \in \mathfrak{F}_T$, β_M is an essential monomorphism.

Theorem 2.9. Let $f : M' \to M$ be an essential monomorphism (injective envelope) in ${}_R\mathfrak{M}$ with $M' \in \mathfrak{F}_T$. Then $F_P(f) : F_P(M') \to F_P(M)$ is an essential monomorphism (injective envelope).

3. Perfect projectors

In this section we apply the results of section 2 and some results concerning torsion theories to give a number of characterizations of perfect projectors over semi–perfect rings.

A ring R is called *semi–perfect* if each cyclic left R–module has a projective cover. If R is semi–perfect then each finitely generated left or right R–module has a projective cover. These rings were introduced and studied extensively by Bass [2]. A module is called *semi–perfect* if it is projective and each of its factors has a projective cover. (See [6].)

For any module G in ${}_R\mathfrak{M}$, we let $\mathcal{G}_G$ denote the class of all modules in ${}_R\mathfrak{M}$ generated by G, that is, $\mathcal{G}_G$ consists of all homomorphic images of direct sums of copies of G. This notation is consistent with that introduced previously for if I is an idempotent ideal, $\mathcal{G}_I$ is the same under either interpretation. Furthermore, if X is projective with trace ideal T′, $\mathcal{G}_X = \mathcal{G}_{T'}$.

Proposition 3.1. Let X be a semi–perfect module in ${}_R\mathfrak{M}$. Then the following conditions are equivalent:

(1) TX is a direct summand of X;

(2) TY is a direct summand of Y for every projective module Y in $\mathcal{G}_X$;

(3) if $P(M) \to M$ is a projective cover in ${}_R\mathfrak{M}$ with $M \in \mathcal{G}_X$, then $F_P(P(M)) \to F_P(M)$ is a projective cover in ${}_S\mathfrak{M}$;

(4) if $M \in \mathcal{T}_T \cap \mathcal{G}_X$ and P(M) exists, $P(M) \in \mathcal{T}_T \cap \mathcal{G}_X$.

Proof. The implication (1) ⇒ (2) follows via a routine argument which depends only on the fact that Y is a direct summand of a direct sum of copies of X and we omit it. The implication (3) ⇒ (4) is clear.

(2) ⇒ (3). Let $f : P(M) \to M$ be a projective cover with $M \in \mathcal{G}_X$. Applying

F to the following commutative diagram which has exact rows and columns

$$\begin{array}{ccccccccc} 0 & \to & TP(M) & \to & P(M) & \to & P(M)/TP(M) & \to & 0 \\ & & \downarrow \hat{f} & & \downarrow f & & & & \\ 0 & \to & TM & \to & M & \longrightarrow & M/TM & \longrightarrow & 0 \\ & & \downarrow & & \downarrow & & & & \\ & & 0 & & 0 & & & & \end{array}$$

where $\hat{f}$ is the restriction of f, yields the commutative diagram with exact columns

$$\begin{array}{ccc} F(TP(M)) & \cong & F(P(M)) \\ \downarrow F(\hat{f}) & & \downarrow F(f) \\ F(TM) & \cong & F(M) \\ \downarrow & & \downarrow \\ 0 & & 0 \end{array}$$

Since $\ker \hat{f} = \ker f \cap TP(M)$ is superfluous as $\ker f$ is superfluous in P(M) and TP(M) is a direct summand of P(M), the conclusion is immediate from Theorem 2.6.

(4) ⇒ (1). Since X is semi–perfect, $f : P(X/TX) \to X/TX$ exists and there is an epimorphism h of X onto P(X/TX) such that $h \cdot f = \eta$ where η is the canonical projection of X onto X/TX. Thus $X = \ker h \oplus Y$ with $Y \cong P(X/TX)$. Since $X/TX \in \mathcal{T}_T \cap \mathcal{G}_X$ so does P(X/TX), that is, $TP(X/TX) = 0$. Thus $TX = T \ker h \oplus TY = T \ker h \subseteq \ker h$. Since $TX = \ker \eta \supseteq \ker h$, $TX = \ker h$ and so (1) holds.

Remark. The assumption that X is semi–perfect in Proposition 3.1 is used only to show that Condition 4 implies Condition 1. Thus each condition implies those below it when X is any projective module.

The finitely generated projective module P_R is a *perfect projector* if the functor $F_P : {}_R\mathcal{M} \to {}_S\mathcal{M}$ preserves projector covers.

Corollary 3.2. For any ring R if $T = Re$ with $e^2 = e \in R$, then P_R is a perfect projector.

Theorem 3.3. Let R be a semi–perfect ring. Then the following conditions are equivalent:

(1) $T = Re$ with $e = e^2 \in R$;

(2) $\mathcal{T}_T$ is closed under projective covers;

(3) TX is a direct summand of X for every projective module X in ${}_R\mathfrak{M}$;

(4) P_R is a perfect projector;

(5) $\mathfrak{F}_T$ is closed under factors;

(6) The functors H_P and F_P restricted to $\mathfrak{F}_T$ are an equivalence of ${}_S\mathfrak{M}$ and $\mathfrak{F}_T$ regarded as a full subcategory of ${}_R\mathfrak{M}$. Moreover in this case the natural transformations α' and β restricted to $\mathfrak{F}_T$ are natural equivalences.

Proof. The equivalence of Conditions (1) – (4) is immediate from Proposition 3.1.

(5) ⇒ (6). It suffices to show that $H(N) \in \mathfrak{F}_T$ for all $N \in {}_S\mathfrak{M}$ and β_M is an isomorphism for $M \in \mathfrak{F}_T$. If $M' \in \mathcal{T}_T$, $\mathrm{Hom}_R(M',H(N)) \cong \mathrm{Hom}_S(F(M'),N) = 0$ since $F(M') = 0$. Thus, $H(N) \in \mathfrak{F}_T$ since $(\mathcal{T}_T, \mathfrak{F}_T)$ is a torsion theory in ${}_R\mathfrak{M}$. If $M \in \mathfrak{F}_T$, Lemma 2.8 implies β_M is a monomorphism and Lemma 2.4 implies $\mathrm{coker}\,\beta_M \in \mathcal{T}_T$. However, as we have just seen, $HF(M) \in \mathfrak{F}_T$ and if (5) holds, $\mathrm{coker}\,\beta_M \in \mathcal{T}_T \cap \mathfrak{F}_T = (0)$. Thus β_M is an isomorphism.

(6) ⇒ (5). Let $M \in \mathfrak{F}_T$ and M' be a submodule of M. If (6) holds, we have a commutative diagram with exact rows

$$\begin{array}{ccccccc} 0 \to & M' & \to & M & \to & M/M' & \to 0 \\ & \cong & & \cong & & & \\ 0 \to & HF(M') & \to & HF(M) & \to & HF(M/M') & \end{array}$$

Thus $M/M' \in \mathfrak{F}_T$ since it is isomorphic to a submodule of $HF(M/M')$ which as seen above belongs to $\mathfrak{F}_T$.

The equivalence of Conditions (4) and (5) follows from the next proposition applied to the hereditary torsion theory $(\mathcal{T}_T, \mathfrak{F}_T)$.

Proposition 3.4. Let $(\mathcal{T}, \mathfrak{F})$ be a torsion theory in ${}_R\mathfrak{M}$. If $\mathfrak{F}$ is closed under factors, then $\mathcal{T}$ is closed under minimal epimorphisms. Conversely, suppose $(\mathcal{T}, \mathfrak{F})$ is hereditary and R is semi-perfect. Then $\mathcal{T}$ closed under projective covers of finitely generated modules implies $\mathfrak{F}$ is closed under factors.

Proof. Let $M' \in \mathcal{T}$ and $f : M \to M'$ be a minimal epimorphism with kernel K. Then $M/K + t(M) \cong M/K/(K + t(M))/K \in \mathcal{T}$ since $M' \in \mathcal{T}$. However, $M/t(M) \in \mathfrak{F}$ and so $M/K + t(M) \cong M/t(M)/(K + t(M))/t(M) \in \mathfrak{F}$ by hypothesis. Since $\mathcal{T} \cap \mathfrak{F} = \{0\}$, $M = K + t(M)$ and so $M = t(M) \in \mathcal{T}$.

Suppose $W \in \mathfrak{F}$ and let $W \to W' \to 0$ be exact. For any finitely generated $M \in \mathcal{T}$, $P(M) \in \mathcal{T}$ and so $\mathrm{Hom}(P(M), W) = 0$. The exactness of

$$\mathrm{Hom}(P(M), W) \to \mathrm{Hom}(P(M), W') \to 0$$

implies $\mathrm{Hom}(P(M), W') = 0$ and hence the exactness of

$$0 \to \mathrm{Hom}(M, W') \to \mathrm{Hom}(P(M), W')$$

implies $\mathrm{Hom}(M, W') = 0$. Since $\mathcal{T}$ is closed under submodules and M is an arbitrary finitely generated module in $\mathcal{T}$, this implies $\mathrm{Hom}(M', W') = 0$ for any $M' \in \mathcal{T}$ and so $W' \in \mathfrak{F}$.

The equivalence of Conditions (1) and (4) of Theorem 3.4 was obtained by Anderson [1, Theorem 3.3, p. 328] for left perfect rings.

4. Perfect injectors

This section begins with a number of characterizations of perfect injectors over an arbitrary ring which in most instances may be viewed as being dual to conditions shown in the preceding section to characterize perfect projectors. Next we introduce the notion of a "semi–generator" and investigate the relationship of this concept to that of "perfect injector". Finally we determine the perfect injectors over a semi–perfect ring and combine this with results of section 3 to characterize those modules having both properties and to show that this is the case for both perfect projectors and injectors when R is a PF–ring.

For any module Q in ${}_R\mathcal{M}$ let $\mathcal{C}_Q$ denote the class of all modules in ${}_R\mathcal{M}$ which are cogenerated by Q, that is, can be embedded in a direct product of copies of Q.

Proposition 4.1. Let $Q \in {}_R\mathcal{M}$ be injective. The following conditions are equivalent:

(1) $t_T(Q)$ is a direct summand of Q;

(2) $t_T(E)$ is a direct summand of E for every injective module in $\mathcal{C}_Q$;

(3) if $M \to E(M)$ is an injective envelope in ${}_R\mathcal{M}$ with $M \in \mathcal{C}_Q$, then $F_P(M) \to F_P(E(M))$ is an injective envelope in ${}_S\mathcal{M}$;

(4) $\mathcal{T}_T \cap \mathcal{C}_Q$ is closed under essential extensions (or equivalently, injective envelopes);

(5) if $M' \to M$ is an essential monomorphism in ${}_R\mathcal{M}$ with $M' \in \mathcal{C}_Q$, then $F_P(M') \to F_P(M)$ is an essential monomorphism in ${}_S\mathcal{M}$.

Proof. That (5) ⇒ (4) is clear. The implication (3) ⇒ (5) follows via the same argument as (a) ⇒ (b) of [1, Theorem 2.4, p. 325]. The rest of this proposition is dual to Proposition 3.1, so we omit the rest of the proof.

The finitely generated projective module P_R is a *perfect injector* if the functor $F_P : {}_R\mathcal{M} \to {}_S\mathcal{M}$ preserves injective envelopes.

Theorem 4.2. The following conditions are equivalent:

(1) $\mathcal{T}_T$ is closed under essential extensions (or equivalently, injective hulls);

(2) $t_T(E)$ is a direct summand of E for every injective module in ${}_R\mathcal{M}$;

(3) P_R is a perfect injector;

(4) the functor F_P preserves essential monomorphisms;

(5) $\mathcal{G}_T$ is closed under submodules (and so is a hereditary torsion class);

(6) for all $x \in T$, $x \in Tx$;

(7) the functors G_P and F_P restricted to $\mathcal{G}_T$ are an equivalence of ${}_S\mathcal{M}$ and $\mathcal{G}_T$ regarded as a full subcategory of ${}_R\mathcal{M}$. Moreover, in this case the natural transformations α and β' restricted to $\mathcal{G}_T$ are natural equivalences.

Proof. The equivalence of Conditions (1) – (4) is immediate from the preceding Proposition by letting Q be any injective cogenerator in ${}_R\mathcal{M}$.

(1) $\Leftrightarrow$ (5) follows from [4, Theorem 2.9] since $(\mathcal{G}_T, \mathcal{T}_T)$ is a torsion theory in ${}_R\mathcal{M}$.

(5) $\Rightarrow$ (6). $T \in \mathcal{G}_T$ so $Rx = TRx = Tx$ for all $x \in T$.

(6) $\Rightarrow$ (1). Suppose $M' \in \mathcal{T}_T$ is an essential submodule of M and let $o \neq m \in M$. If $Tm \neq o$, $Tm \cap M' \neq o$ so there exists $x \in T$ such that $o \neq xm \in M'$. But $x = tx$ with $t \in T$ so $o \neq xm = t(xm) = o$. This contradiction shows $M \in \mathcal{T}_T$.

The implications (5) $\Leftrightarrow$ (7) are dual to the implications (5) $\Leftrightarrow$ (6) of Theorem 3.3 and will be omitted.

The equivalence of Conditions (3), (4) and (6) of Theorem 4.2 were obtained by Anderson [1, Theorem 2.4, p. 325].

Remark. The equivalence of Conditions (1), (2), (5), and (6) of Theorem 4.2 is valid for any TTF–theory with corresponding idempotent ideal I. In particular this is true if ${}_RX$ is any projective module with trace ideal T'. We now show that in this latter case this is equivalent to X generating each of its submodules.

Let G and M belong to ${}_R\mathcal{M}$. Then G *strongly generates* M if every submodule of M is in $\mathcal{G}_G$. Let $\mathcal{S}_G$ denote the class of modules strongly generated by G. Then we call G a *semi–generator* if $\mathcal{S}_G = \mathcal{G}_G$, that is, if $\mathcal{G}_G$ is closed under submodules.

Proposition 4.3 For any module G in ${}_R\mathcal{M}$, $\mathcal{S}_G$ is closed under homomorphic images and submodules. Hence G is a semi-generator if and only if $\oplus_{\lambda \in \Lambda} G_\lambda$ is in $\mathcal{S}_G$ for every index set Λ. Furthermore, if G is projective, $\mathcal{S}_G$ is closed under arbitrary direct sums and extensions and hence is a hereditary torsion class.

Proof. Except for the last sentence the result is immediate. Assume G is projective and that $M_\lambda \in \mathcal{S}_G$ for all $\lambda \in \Lambda$. To show $\oplus_\lambda M_\lambda \in \mathcal{S}_G$ it suffices to show G generates Rx for each $x \in \oplus_\lambda M_\lambda$. Since $Rx \subseteq M_{\lambda_1} \oplus \cdots \oplus M_{\lambda_n}$, it suffices to do this when $x \in M_1 \oplus \cdots \oplus M_n$ with $M_i \in \mathcal{S}_G$ for $i = 1, 2, \cdots, n$. We induct on n. When $n = 1$ it is true by assumption. Assume $n > 1$ and let π_1 be the projection of $\oplus_{i=1}^n M_i$ on M_1 Then π_1 restricted to Rx is an epimorphism of Rx onto $R(x)\pi_1$ and there exists an epimorphism of $\oplus_\gamma G_\gamma$ with $G_\gamma = G$ onto $R(x)\pi_1$ since $R(x)\pi_1 \subseteq M_1 \in \mathcal{S}_G$. Combining the above with the projectivity of G yields $f : \oplus_\gamma G_\gamma \to Rx$ and $y \in \oplus_\gamma G_\gamma$ such that $(y)f\pi_1 = (x)\pi_1$. Thus $x - (y)f \in \oplus_{i=2}^n M_i$ and so by the induction hypothesis G generates $R(x - (y)f) \subseteq Rx$ and so G generates Rx. The proof that $\mathcal{S}_G$ is closed under extensions is similar.

Theorem 4.4. Let $X \in {}_R\mathfrak{M}$ be projective with trace ideal T'. Then the following conditions are equivalent:

(1) ${}_RX$ is a semi-generator;

(2) $X \in \mathcal{S}_X$;

(3) $x \in T'x$ for all $x \in T'$.

Proof. The implication (1) $\Rightarrow$ (2) is clear.

(2) $\Rightarrow$ (3). It is immediate from Proposition 4.3 that (2) implies $T' \in \mathcal{S}_X = \mathcal{S}_{T'}$. Thus if $x \in T'$, $x' \in Rx = T'Rx = T'x$.

(3) $\Rightarrow$ (1). Condition (3) implies T' generates each cyclic submodule of T' and so $T' \in \mathcal{S}_{T'} = \mathcal{S}_X$. Since X is projective, Proposition 4.3 implies ${}_RX$ is a semi-generator.

This last implication also follows from the torsion theoretic considerations discussed above.

We note that if R is von-Neumann regular, then every projective R-module is a semi-generator.

Corollary 4.5. P_R is a perfect injector if and only if ${}_RP^* \in \mathcal{S}_{P^*}$.

This corollary and Proposition 4.3 yield some interesting results in special cases. For instance if R is artinian, P_R is a perfect injector if and only if each composition factor of P^* is a homomorphic image of P^*.

Corollary 4.6. If R is a semi–perfect ring, P_R is a perfect injector if and only if $T = eR$ with $e^2 = e \in R$.

Proof. The sufficiency is immediate from Theorem 4.2 (6). To prove the converse it suffices to show that if $f \in R$ is a primitive idempotent with $fT \neq 0$, then $f \in T$. If $f \notin T$ then $fT \subseteq fJ$ where J is the radical of R. Suppose $x \in fT$, then by Theorem 4.2 (6), $x = yx = f \cdot x = (f \cdot y) \cdot x$ with $y \in T$. Thus $0 = (1 - f \cdot y)x$ and since $f \cdot y \in fT \subseteq J$, $1 - f \cdot y$ is a unit and so $x = 0$.

This result was proved by Anderson [1, Corollary 2.7, p. 326] for right perfect rings.

Corollary 4.7. Let R be a semi–perfect ring. Then P_R is both a perfect projector and a perfect injector if and only if T is a ring direct summand of R, that is, $T = Re$ with e a central idempotent of R.

Proof. ($\Rightarrow$). Theorem 3.3 and Corollary 4.6 imply $T = Re = fR$ with e and f idempotents in R. Since T is an ideal, a standard argument shows $e = f$ and this idempotent is in the center of R.

($\Leftarrow$). This implication is immediate from Corollary 3.2 and Theorem 4.2 (6) and does not depend on R being semi–perfect.

Corollary 4.8. Let R be a semi–perfect ring. Then P_R is a perfect projector (injector) if and only if ${}_RP^*$ is a perfect injector (projector).

Bernhardt [3, Theorem 5] proved that if R is a semi–perfect ring and $(\mathcal{T},\mathfrak{F})$ is a TTF–theory in ${}_R\mathfrak{M}$ then the associated idempotent ideal of R is a ring direct summand of R provided (i) $S \in \mathcal{T}$ if and only if $P(S) \in \mathcal{T}$ and (ii) $S \in \mathfrak{F}$ if and only if $P(S) \in \mathfrak{F}$ for all simple modules S in ${}_R\mathfrak{M}$. Bernhardt called torsion theories with this latter property *principal.*

A ring R is called a left, respectively, right *PF–ring* if ${}_RR$ is an injective cogenerator in ${}_R\mathfrak{M}$, respectively, R_R is an injective cogenerator in $\mathfrak{M}_R$. This and other characterizations show that they are a natural generalization of quasi–Frobenius rings (see [9] or [10]).

Corollary 4.9. Let R be a left or right PF–ring. Then P_R is a perfect injector if and only if P_R is a perfect projector.

Proof. We show in each instance that the TTF–theory $(\mathcal{T}_T, \mathfrak{F}_T)$ is principal. The conclusion then follows from Corollary 4.7 via the result of Bernhardt [3, Theorem 5] cited above.

Osofsky [9] has shown that R is semi–perfect. Furthermore, if $e_1, \cdots, e_n$ is a complete set of primitive orthogonal idempotents, each Re_i contains a unique simple submodule S_i and a unique simple factor module S_i' of which it is an essential extension and a projective cover, respectively. Moreover as $i = 1, \cdots, n$ the S_i and S_i' provide distinct representatives for the isomorphism classes of simple left R–modules (see [9] and [10, Proposition 3.9]). We may assume the e_i are indexed so that $S_1, \cdots, S_k \in \mathcal{T}_T$ and $S_{k+1}, \cdots, S_n \in \mathfrak{F}_T$ and since $(\mathcal{T}_T, \mathfrak{F}_T)$ is a hereditary torsion theory $Re_{k+1}, \cdots, Re_n \in \mathfrak{F}_T$ [4, Theorem 2.9].

If P_R is a perfect injector, Theorem 4.2 (1) implies $Re_1, \cdots, Re_k \in \mathcal{T}_T$. Since no module in $\mathfrak{F}_T$ has a projective cover in $\mathcal{T}_T$ as $\mathrm{Hom}_R(M, M') = 0$ for all $M \in \mathcal{T}_T$ and $M' \in \mathfrak{F}_T$, a simple counting argument shows that $(\mathcal{T}_T, \mathfrak{F}_T)$ is principal.

If P_R is a perfect projector, Theorem 3.3 (2) implies $\mathcal{T}_T$ is closed under projective covers and so the number of Re_i in $\mathcal{T}_T$ must equal at least k. Since $Re_{k+1}, \cdots, Re_n \in \mathfrak{F}_T$, this number is precisely k and $(\mathcal{T}_T, \mathfrak{F}_T)$ is principal as above.

5. Semi–perfect and perfect modules

We now apply the results of the second section to prove theorems of Mares [6] and Ware [11] which characterize finitely generated semi-perfect and perfect modules. Recall that Mares [6] called a projective module ${}_RX$ a *semi–perfect module* if each factor module of X has a projective cover and a (left) *perfect module* if every factor of $\oplus_I X$ has a projective cover for every index set I. These concepts generalize the notions of semi–perfect and (left) perfect rings (see Bass [2]).

For the rest of this section let ${}_RX$ be a projective module with endomorphism ring S.

Theorem 5.1. If X is finitely generated and semi–perfect then S is a semi–perfect ring. Conversely, if S is a semi–perfect ring, X is a finitely generated semi–perfect module.

Theorem 5.2. If X is finitely generated and (left) perfect, then S is a left perfect ring. Conversely, if S is a left perfect ring, X is a finitely generated (left) perfect module.

In each instance the initial assertion is due to Mares and the converse to Ware. Our proof of the converse depends in both cases upon a result of Ware which permits us to assume X is finitely generated. However, even after this reduction, Ware's proof of the converse statement of Theorem 5.2 is quite involved. Since the proofs are quite similar we prove only Theorem 5.2.

Proposition 5.3. If ${}_RX$ is a projective module whose endomorphism ring is semi-perfect, then X is finitely generated.

Proof. Immediate from [11, Proposition 4.1 and Theorem 4.2].

Proof of Theorem 5.2. We assume ${}_RX$ is finitely generated, set $P_R = X^* = \text{Hom}_R(X, R)$ and use the notation of the previous sections.

($\Rightarrow$). Let $N \in {}_S\mathcal{M}$. There exists an epimorphism $f : \oplus_I S \to N$. Since G is right exact and commutes with direct sums, G(f) is an epimorphism and $G(\oplus_I S) \cong \oplus_I G(S) = \oplus_I P^* = \oplus_I X$. Thus G(N) has a projective cover $h : P(G(N)) \to G(N)$. It follows from Theorem 2.6 that $F(h) : F(P(G(N))) \to FG(N)$ is a projective cover and since $FG(N) \cong N$ via α_N, N has a projective cover. Thus S is (left) perfect.

($\Leftarrow$). Suppose M is a factor of $\oplus_I X$. Since S is (left) perfect F(M) has a projective cover $f : P(F(M)) \to F(M)$. Since G preserves projective covers by Proposition 2.3, $G(f) : G(P(F(M))) \to GF(M)$ is a projective cover. Also Lemma 2.5 implies $\beta'_M : GF(M) \to M$ is a minimal epimorphism. It follows from Lemma 2.1 that the composition of minimal epimorphisms is a minimal epimorphism and hence $G(f) \circ \beta'_M : G(P(F(M))) \to M$ is a projective cover for M. Thus X is (left) perfect.

References

1. F. W. Anderson, Endomorphism rings of projective modules, Math. Z. **11** (1969), 322-332.
2. H. Bass, Finitistic dimension and a homological generalization of semi-primary rings, Trans. Amer. Math. Soc. **95** (1960), 466-488.
3. R. L. Bernhardt, Splitting hereditary torsion theories over semi-perfect rings, Proc. Amer. Math. Soc. **22** (1969), 681-687,
4. S. E. Dickson, A torsion theory for abelian categories, Trans. Amer. Math. Soc. **121** (1968), 195-203.
5. J. P. Jans, Some aspects of torsion, Pac. J. Math. **15** (1965), 1249-1259.
6. E. A. Mares, Semi-perfect modules, Math. Z. **82** (1963), 347-360.
7. K. Morita, Duality for modules and its applications to the theory of rings with minimum condition, Sci. Rep. Tokyo Kyoiku Daigaku, Sect. **A6** (1958), 83-142.
8. K. Morita, Adjoint pairs of functors and Frobenius extensions, Sci. Rep. Tokyo Kyoiku Daigaku, Sect. **A9** (1965), 40-71.
9. B. L. Osofsky, A generalization of quasi-Frobenius rings, J. Alg. **4** (1966), 373-387.
10. Y. Utumi, Self-injective rings, J. Alg. **6** (1967), 56-64.
11. R. Ware, Endomorphism rings of projective modules (to appear).

LINEARLY COMPACT MODULES AND LOCAL MORITA DUALITY

F. L. Sandomierski

Kent State University

1. Introduction

Linearly topologized modules extend the notion of a linearly topologized vector space developed by Lefschetz [8]. A linearly topologized module is a module which possesses a base of neighborhoods of zero consisting of submodules. In this presentation a base of neighborhoods of zero for a linearly topologized module will consist of all submodules, the so-called discrete linear topology.

The notion of a linearly compact module, i.e., a compact linearly topologized module, plays an important role in duality theory [5, 9, 10, 11, 14, 15].

Müller [14] has demonstrated the connections between linearly compact modules and Morita duality. The methods of Müller [14] are exploited heavily in this presentation.

Section 2 deals with some of the consequences of linear compactness. The result (Proposition 2.6 corollary) that a right linearly compact ring is semi-perfect generalizes Zelinsky's result [20, Proposition 14] for commutative rings. Also, the result (Corollaries of Propositions 2.9 and 2.12) that a right linearly compact ring which is left or right perfect is right artinian generalizes Osofsky's result [16, 17, Theorem 13].

Section 3 deals with Morita duality [12]. If ${}_SQ_R$ is a bimodule such that Q_R is an injective cogenerator with essential socle and $S = \mathrm{End}(Q_R)$, then for a linearly compact module X_R, ${}_SQ_R$ yields a duality between the categories $\mathcal{M}'_{X_R}$ and ${}_S X^*\mathcal{M}'$, where $X^* = \mathrm{Hom}(X_R, {}_SQ_R)$ and $\mathcal{M}'_X({}_{X^*}\mathcal{M}')$ is the full subcategory of all right (left) R- (S-)modules isomorphic to submodules of factor modules of $X^n((X^*)^n)$, $n = 1, 2, \cdots$, with X^n a product of n copies of X_R.

All rings have identity and all modules are unitary. The convention that module homomorphisms will be written on the side opposite the scalars is adopted here.

2. Linearly compact modules and rings

The following definition may be found in Zelinsky [20].

Definition 2.1. Let X_R be an R-module, $\{x_i\}_I \subseteq X$, and $\{X_i\}_I$ a set of submodules of X, then the set of congruences $\{x \equiv x_i \bmod X_i\}_I$ denoted by $(x_i, X_i)_I$ is said to be solvable (finitely solvable), if there is a $y \in X$, (a $y_F \in X$, for each finite subset F of I), such that $y - x_i \in X_i$ for each $i \in I$, ($y_F - x_i \in X_i$ for each $i \in F$). X_R is linearly compact if every finitely solvable set of congruences in X is solvable. A ring R is right (left) linearly compact if the right (left) R–module R_R $({}_RR)$ is linearly compact.

The following result is in Zelinsky [20] and is an easy consequence of the above definition.

Proposition 2.2. If $0 \to X'_R \to X_R \to X''_R \to 0$ is an exact sequence of R–modules, then X_R is linearly compact if and only if X'_R and X''_R are linearly compact.

Corollary. Finite products of linearly compact modules are linearly compact.

Müller [14] has made the observation given by the following lemma.

Lemma 2.3. If X_R is a linearly compact module, then X_R has finite Goldie dimension [4].

Proof. If $\Sigma_I Y_i = Y$ is a direct sum of submodules of X, then let $X_i = \Sigma_{j \neq i} Y_j$ and choose $x_i \in Y_i$, then the set of congruences $(x_i, X_i)_I$ is finitely solvable; namely, for F a finite subset of I, let $x_F = \Sigma_F x_i$. Since Y is linearly compact by Proposition 2.2, $(x_i, X_i)_I$ is solvable in Y. It now follows that all but finitely many of the x_i are zero. Since $x_i \in Y_i$ were chosen arbitrarily, all but finitely many of the Y_i are zero and the lemma follows.

An easy characterization of linearly compact modules will be given.

Let X_R be an R–module and $\{X_i \overset{g_i}{\to} Q_i\}_I$ a set of R–homomorphisms. Denote by $\Pi_I Q_i$ the product of the family of R–modules $\{Q_i\}_I$ and by Π_F the natural projection $\Pi_I Q_i \to \Pi_F Q_i$, where F is a finite subset of I and $\Pi_F Q_i$ the product of the subfamily $\{Q_i\}_F$. For $F = \{i\}$ a singleton subset of I, denote $\Pi_F = \Pi_i$. For the family of homomorphisms $\{X \overset{g_i}{\to} Q_i\}_I$ there is a unique R–homomorphism $g : X \to \Pi_I Q_i$ such that $\Pi_i g = g_i$ for all $i \in I$.

Theorem 2.4. With the notation above a module X_R is linearly compact if and only if when $\{X \overset{g_i}{\to} Q_i\}_I$ is a set of homomorphisms such that for $u \in \Pi_i Q_i$, $\Pi_F u \in \mathrm{im}(\Pi_F g)$ for every finite subset F of I, then $u \in \mathrm{im}(g)$.

Proof. "if." Let $X_i = \ker(g_i)$ and $x_i \in X$ such that $\Pi_i u = \Pi_i gx_i = g_ix_i$. Clearly the set of congruences $(x_i, X_i)_I$ is finitely solvable, hence solvable, so there is an $x \in X$ such that $x - x_i \in \ker(g_i)$ for each $i \in I$, hence $\Pi_i gx = g_ix_i = \Pi_i u$ for each $i \in I$, hence $u = gx$ and the "if" part follows.

Conversely, if $(x_i, X_i)_I$ is a finitely solvable set of congruences in X, then let $Q_i = X/X_i$ and g_i the natural homomorphism $X \to X/X_i$ for each $i \in I$. Clearly the linear compactness of X now follows.

Another property of linearly compact modules is given by a notion introduced by Kasch and Mares [6], which is the following.

Definition 2.5. A module X_R is complemented if for a submodule Y of X, there is a submodule V of X such that V is minimal with respect to the property that $Y + V = X$.

Proposition 2.6. If X_R is a linearly compact module, then X_R is complemented.

Proof. Let $\Gamma = \{V | V$ is a submodule of X and $Y + V = X\}$. Partially order Γ by $V, W \in \Gamma$, $V \leqslant W$ if and only if $V \supseteq W$. Let $\{V_i\}_I$ be a totally ordered subset of Γ. Choose $x_0 \in X$, then $x_0 = y_i + v_i$ for some $y_i \in Y$ and $v_i \in V_i$. Consider the set of congruences $(y_i, V_i \cap Y)$ in Y. If F is a finite subset of I, then there is a $j \in F$ such that $V_j \subseteq V_i$ for all $i \in F$, hence $y_j - y_i = x_0 - v_j - (x_0 - v_i) = v_i - v_j \in V_i$ for each $i \in F$, so the set of congruences is finitely solvable in Y, whence solvable in Y. Let y be a solution in Y, then $x_0 - y = x_0 - y_i + y_i - y = v_i + y_i - y \in V_i$ for all $i \in I$, hence $x_0 - y \in \cap_I V_I$. Since x_0 was an arbitrary element of X, $X = Y + (\cap_I V_i)$ and $\cap_I V_i$ is an upper bound for $\{V_i\}_I$ in Γ, so by Zorn's Lemma Γ has a maximal element in Γ and the proposition follows.

Corollary. If R is a right linearly compact ring, then R is semi–perfect [3].

Proof. R_R is complemented, hence semi–perfect by [6].

The above corollary is the non–commutative version of Zelinsky's results [20, Proposition 13, Proposition 14]. The proof of the last proposition is due to Darrell Turnidge.

Proposition 2.7. If P_R is a finitely generated projective module, $C = \mathrm{End}(P_R)$ and X_R a linearly compact module, then $X'_C = \mathrm{Hom}({}_CP_R, X_R)$ is a linearly compact C–module.

Proof. Let $(f_i, M_i)_I$ be a set of finitely solvable congruences in X'_C.

For a fixed $p \in P$, consider the set of congruences $(f_i p, M_i P)_I$ in X, where $M_i P$ is the submodule of X generated by the images of the homomorphisms of M_i. For F a finite subset of I, there is an $f_F \in X'_C$ such that $f_F - f_i \in M_i$ for each $i \in F$, hence $f_F p - f_i p \in M_i P$ for each $i \in F$, so the set of congruences $(f_i p, M_i P)_I$ is finitely solvable, hence solvable in X, with a solution denoted x_p.

The following notation is in accordance with that preceding Theorem 2.4.

Let $X_i = M_i P$, $Q_i = X/X_i$ and $g_i : X \to X/X_i$ the natural homomorphisms. Define a map $h : P \to \text{im}(g)$ by $hp = gx_p$. Clearly, h is well defined, since if y_p is a solution of the set of congruences $(f_i p, X_i)_I$, then $x_p - y_p = (x_p - f_i p) - (y_p - f_i p) \in X_i$ for each $i \in I$, hence $gx_p = gy_p$. It is evident that h is an R–homomorphism, and since P_R is projective there is an $f : P_R \to X_R$ such that $gf = h$. Now for $p \in P$, $fp - f_i p \in M_i P = X_i$ for each $i \in I$ by identifying $\text{Hom}_R(P, X_i)$ with a submodule of X'_C. By [19], $\text{Hom}_R(P, X_i) = M_i$ and the proposition follows.

Corollary 1. If R is a right linearly compact ring, P_R a finitely generated projective module, then $C = \text{End}(P_R)$ is a right linearly compact ring.

Proof. This follows from the fact that P_R is a linearly compact module as R_R is.

Corollary 2. If R is a right linearly compact ring and e an idempotent of R, then eRe is a right linearly compact ring.

Corollary 3. If R is a right linearly compact ring, then R_n the full ring of $n \times n$ matrices over R is a right linearly compact ring.

Proposition 2.8. (Zelinsky [20, Proposition 5].)If X_R is an artinian module, then X_R is linearly compact.

At this point some sufficient conditions will be given so that the converse of Proposition 2.8 will be valid in some cases.

Proposition 2.9. If R is a right perfect ring [3] and X_R a linearly compact module, then X_R is noetherian.

Proof. Let Y_R be a submodule of X_R and J the Jacobson radical of R.

Since Y/YJ is a semi–simple linearly compact module, it is finitely generated, so $Y = Y^1 + YJ$, Y^1 a finitely generated submodule of Y. Now since J is right T–nilpotent, YJ is a small (superflous) submodule of Y, so $Y = Y^1$ and the proposition follows.

Corollary 1. If R is a right artinian ring and X_R a linearly compact module, then X_R is artinian.

Corollary 2. If R is a right perfect, right linearly compact ring, then R is right artinian.

Proof. By the theorem R is right noetherian and by [16, Lemma 11], R is right artinian.

The following definition may be found in [14].

Definition 2.10. A submodule Y_R of X_R is (finitely) completely meet irreducible if the factor module X/Y has essential simple (finitely generated) socle.

Müller [14] has established the following result.

Lemma 2.11. If Y_R is a finitely completely meet irreducible submodule of a linearly compact module X_R and $\{X_i\}_I$ is a set of submodules of X such that $\cap_I X_i \subseteq Y$, then there is a finite subset F of I such that $\cap_F X_i \subseteq Y$.

Proposition 2.12. If R is a left perfect ring and X_R a linearly compact module, then X_R is artinian.

Proof. For $\{X_i\}_I = \Gamma$ a family of submodules of X, let $Y = \cap_I X_i$. Since R is left perfect, X/Y has essential socle [3] and this semi–simple part is finitely generated by the linear compactness of X/Y.

Now it follows that Y is a finitely completely meet irreducible submodule of X, so by Lemma 2.11, there is a finite subset F of I such that $Y = \cap_F X_i$. Clearly, it may be assumed that the intersection $\cap_F X_i$ is irredundant from which it follows that each X_i, $i \in F$ is a minimal member of $\Gamma = \{X_i\}_I$ and the proposition follows.

Corollary. If R is a right linearly compact ring and a left perfect ring, then R is right artinian.

The proposition 2.12 is due to R. Miller and D. Turnidge.

The above corollary generalizes a result of Osofsky, since a ring R for which there is a bimodule ${}_SQ_R$ which yields a Morita duality is right linearly compact as will be seen in the next section.

Combining Propositions 2.9 and 2.12, it follows that, if X_R is a linearly compact module and R is a right (left) perfect ring then X_R is noetherian (artinian). The author has been unable to answer the following questions.

Open Problem 1. If X_R is a linearly compact module and R is a right perfect ring, is X_R artinian?

Open Problem 2. If X_R is a linearly compact module and R is a left perfect ring, is X_R noetherian?

In view of the corollaries to Propositions 2.9 and 2.12, the answer is 'yes' if $X_R = R_R$.

3. Local Morita duality

Let $\mathcal{M}_R$ (${}_R\mathcal{M}$) denote the category of all right (left) R–modules and all R–homomorphisms. For a right (left) R–module X, the full subcategory of $\mathcal{M}_R$ (${}_R\mathcal{M}$) consisting of all modules isomorphic to submodules of factor modules of X^n, n = 1, 2, ⋯, where X^n is a product of n copies of the R–module X, will be denoted by $\mathcal{M}'_X({}_X\mathcal{M}')$.

Remark 1. For an R–module X_R, $\mathcal{M}'_X$ is the full subcategory of $\mathcal{M}_R$ consisting of all modules isomorphic to factor modules of submodules of X^n, n = 1, 2, ⋯, .

Definition 3.1. (i) A module Q_R is (semi–) X_R–injective if every R–homomorphism of a (finitely generated) submodule of X_R into Q_R can be extended to an R–homomorphism of X_R into Q_R. (For the notion of Q_R is X_R–injective see Azumaya [1].)

(ii) For modules Q_R, X_R, X_R is said to have Q–dominant dimension ⩾ 1, denoted Q–dominant dimension $X_R \geqslant 1$, if X_R is embeddable in a product of copies of Q_R(e.g., see [13]).

(iii) The class of right R–modules having Q_R dominant dimension ⩾ 1 is denoted $\mathcal{D}_1(Q_R)$.

(iv) Q_R is called a self–cogenerator if $\mathcal{M}'_Q \subseteq \mathcal{D}_1(Q_R)$.

Remark 2. Azumaya [1] has shown that if Q_R is X_R–injective, then Q_R is Y_R–injective for all $Y \in \mathcal{M}'_X$.

For a bimodule ${}_SQ_R$ and modules X_R, ${}_SB$, ${}_SX^* = \mathrm{Hom}(X_R, {}_SQ_R)$, $B_R^* = \mathrm{Hom}({}_SB, {}_SQ_R)$ and $X^{**} = (X^*)^*$, $B^{**} = (B^*)^*$. The natural maps $X_R \to X_R^{**}$ given by $x \to \hat{x}$, where $fx = \hat{fx}$, for $f \in X^*$ yield a natural transformation from the identity functor on M_R to the double Q–dual functor $(\ )^{**}$.

A module X_R is Q–reflexive if $X \to X^{**}$ is an isomorphism.

For Y a submodule of X_R and ${}_SB$ a submodule of ${}_SX^*$, $\mathrm{Ann}_{X^*}(Y) = \{f \in X^* \mid fy = 0\}$, $\mathrm{Ann}_X B = \{x \in X \mid Bx = 0\}$.

Remark 3. For Y_R a submodule of X_R, $X/Y \in \mathcal{D}_1(Q_R)$ if and only if $\mathrm{Ann}_X(\mathrm{Ann}_{X^*}(Y)) = Y$.

Remark 4. $X_R \to X_R^{**}$ is a monomorphism if and only if $X \in \mathcal{D}_1(Q_R)$.

Lemma 3.2. Let Q_R be a quasi–injective module (i.e., Q_R is Q_R–injective), $S = \mathrm{End}(Q_R)$, and ${}_SB$ a finitely generated submodule of ${}_SX^* = \mathrm{Hom}(X_R, {}_SQ_R)$. Then

(i) $\mathrm{Ann}_{X^*}\mathrm{Ann}_X B = B$,

(ii) $B_R^* \in \mathcal{M}'_Q$, and

(iii) ${}_SB \to {}_SB^{**}$ is an isomorphism.

Proof of (i). Clearly $B \subseteq \mathrm{Ann}_{X^*}\mathrm{Ann}_X B$. Let $b_1, \cdots, b_n$ be a set of generators for ${}_SB$ and $p \in \mathrm{Ann}_{X^*}\mathrm{Ann}_X B$, then $\ker(p) \supseteq \mathrm{Ann}_X B = \cap_{i=1}^n \ker(b_i)$. The following diagram with exact row is commutative.

$$\begin{array}{ccccc} 0 \to & X/Y & \xrightarrow{\alpha} & Q^n & \\ \bar{p} & \downarrow & \swarrow & (s_1, \cdots, s_n) & \\ & Q_R & & & \end{array}$$

where $\alpha(x + Y) = (b_1x, \cdots, b_nx)$, $Y = \mathrm{Ann}_x B$, $\bar{p}(x + Y) = px$ and $(s_1, \cdots, s_n)$ exists with $s_i \in S$, since Q_R is quasi–injective. Now $px = \bar{p}(x + Y) = (s_1, \cdots, s_n)(b_1x, \cdots, b_nx) = \Sigma_{i=1}^n s_ib_i \in {}_SB$ and (i) follows.

Proof of (ii). Let ${}_SF \to {}_SB \to 0$ be exact with ${}_SF$ a finitely generated free left S–module, then $0 \to B_R^* \to F_R^*$ is exact and since $F_R^* \cong Q_R^n$ for some n, a positive integer, (ii) follows.

Proof of (iii). Let ${}_SF \to {}_SB \to 0$ be as in the proof of (ii). Since Q_R is quasi–injective, the following diagram with exact rows is commutative.

$$\begin{array}{ccccc} {}_S F & \longrightarrow & {}_S B & \longrightarrow & 0 \\ \downarrow \alpha & & \downarrow \beta & & \\ {}_S F^{**} & \longrightarrow & {}_S B^{**} & \longrightarrow & 0 \end{array}$$

Clearly α is an isomorphism so β is an epimorphism. Since ${}_S X^* \in \mathcal{D}_1({}_S Q)$, ${}_S B \in \mathcal{D}_1({}_S Q)$ so β is a monomorphism and (iii) follows.

Lemma 3.3. If Q_R is a quasi–injective module and a self–cogenerator $S = \mathrm{End}(Q_R)$, then for X_R, ${}_S Q$ is semi–${}_S X^*$–injective.

Proof. Let ${}_S B$ be a finitely generated submodule of ${}_S X^*$, then the following diagram with exact rows is commutative.

$$\begin{array}{ccccccc} & & 0 & \to & {}_S B & \overset{\subseteq}{\to} & {}_S X^* \\ & & & & \downarrow & & \downarrow \\ 0 & \to & K^* & \to & {}_S B^{**} & \to & {}_S X^{***} \end{array}$$

where $K_R = \mathrm{coker}(X_R^{**} \to B_R^*)$. Since ${}_S B \to {}_S B^{**}$ is an isomorphism by Lemma 3.2, and ${}_S X^* \to {}_S X^{***}$ is a monomorphism, $K^* = 0$. However, since $B_R^* \in \mathcal{M}'_Q$, then $K \in \mathcal{M}'_Q$ so $K = 0$ since Q_R is a self–cogenerator, and the lemma follows.

The following result is essentially due to Müller [14].

Lemma 3.4. If Q_R is a self–cogenerator, $S = \mathrm{End}(Q_R)$, X_R an R–module, then for $f_1, \cdots, f_n \in {}_S X^* = \mathrm{Hom}(X_R, {}_S Q_R)$ and $g \in X^{**}$, there is an $x \in X$ such that $f_i \hat{x} = f_i g$ for $i = 1, 2, \cdots, n$.

Proof. Suppose the conclusion false, then $p = (f_1 g, \cdots, f_n g) \notin K = \{(f_1 x, \cdots, f_n x) \mid x \in X\}$, where K_R is a submodule of Q_R^n. Since Q_R is a self–cogenerator, there is an R–homomorphism $h : Q_R^n \to Q_R$ such that $K \subseteq \ker(h)$ and $hp \neq 0$. Clearly, $h = (s_1, \cdots, s_n)$, $s_i \in S$, with $h(q_1, \cdots, q_n) = \Sigma_{i=1}^n s_i q_i$.

Now $h(f_1 x, \cdots, f_n x) = \Sigma_{i=1}^n s_i f_i x = 0$ for each $x \in X$ so $\Sigma\, s_i f_i = 0$, hence, $hp = \Sigma_{i=1}^n s_i(f_i g) = \Sigma_{i=1}^n (s_i f_i) g = (\Sigma\, s_i f_i) g = 0$ a contradiction, and the lemma follows.

Lemma 3.5. If Q_R is a quasi–injective self–cogenerator, $S = \mathrm{End}(Q_R)$, X_R a linearly compact module, then

(i) $X_R \to X_R^{**}$ is an epimorphism, and

(ii) ${}_SQ$ is ${}_SX^*$ injective.

Proof of (i). Let $g \in X^{**}$, $\{B_i\}_I$ the set of all finitely generated submodules of ${}_SX^*$, and $X_i = \mathrm{Ann}_X B_i$. By Lemma 3.4, there is an $x_i \in X$ such that $g \mid B_i = \hat{x}_i \mid B_i$. Clearly, the set of congruences $(x_i, X_i)_I$ is finitely solvable in X_R, hence solvable so there is an $x \in X$ such that $x - x_i \in X_i$ for each $i \in I$, which implies $\hat{x} \mid B_i = \hat{x}_i \mid B_i = g \mid B_i$ for each finitely generated submodule B_i of ${}_SX^*$, so $\hat{x} = g$ and (i) follows.

Proof of (ii). Let $g : {}_SB \to {}_SQ$ be an S–homomorphism with ${}_SB$ a submodule of ${}_SX^*$, and $\{B_i\}_I$ the set of all finitely generated submodules of ${}_SB$. By Lemma 3.3, $g \mid B_i$ is extendable to ${}_SX^*$ so by (i) there is an $x_i \in X$ with $g \mid B_i = \hat{x}_i \mid B_i$ for each $i \in I$.

Clearly, the set of congruences $(x_i, \mathrm{Ann}_X B_i)_I$ are finitely solvable since for F a finite subset of I, let $x_F = x_j$ where $B_j = \Sigma_{i \in F} B_i$ a finitely generated submodule of ${}_SB$, so the set of congruences $(x_i, \mathrm{Ann}_X B_i)_I$ is solvable, hence, there is an $x \in X$ such that $\hat{x} \mid B_i = g \mid B_i$ for each $i \in I$, so $\hat{x}$ extends g and (ii) follows.

Theorem 3.6. Let Q_R be a quasi–injective self–cogenerator, $S = \mathrm{End}(Q_R)$ and X_R such that $X \in \mathcal{D}_1(Q_R)$, Q_R is X_R–injective, then the following statements are equivalent.

(1) X_R is linearly compact.

(2) For $Y_R \in \mathcal{M}'_X$, Y_R is Q_R–reflexive and ${}_SQ$ is ${}_SY^*$–injective.

(3) Y_R is Q_R–reflexive for each submodule Y_R of X_R and ${}_SQ$ is ${}_SX^*$–injective.

(4) $X/Y \in \mathcal{D}_1(Q_R)$ for each submodule Y of X_R, ${}_SQ$ is ${}_SX^*$–injective and X_R is Q_R–reflexive.

Proof. (1) implies (2): Let M_R be a linearly compact module such that $M_R \in \mathcal{D}_1(Q_R)$ and Q_R is M_R–injective. Let N_R be a submodule of M_R and denote $P_R = M/N$. By Lemma 3.5, the following diagram has exact rows and is commutative.

$$\begin{array}{ccccccccc} 0 & \to & N_R & \to & M_R & \to & P_R & \to & 0 \\ & & \downarrow & & \downarrow & & \downarrow & & \\ 0 & \to & N_R^{**} & \to & M_R^{**} & \to & P_R^{**} & \to & 0 \end{array}$$

By Lemma 3.5, the vertical maps are epimorphisms. Since $M \in \mathcal{D}_1(Q_R)$, then

$N \in \mathcal{D}_1(Q_R)$ so the left and middle vertical maps are isomorphisms from which it follows that $P \to P^{**}$ is an isomorphism. If $M_R = X^n$, $n = 1, 2, \cdots$, then (2) holds for submodules and factor modules of X^n. Now let M be a factor module of X^n, then by the above (2) follows, using Lemma 3.5. (2) implies (3) is evident. (3) implies (4): The following diagram with exact rows is commutative.

$$\begin{array}{ccccccccc} 0 & \to & Y & \to & X & \to & W & \to & 0 \\ & & \downarrow\alpha & & \downarrow\beta & & \downarrow & & \\ 0 & \to & Y^{**} & \to & X^{**} & \to & W^{**} & \to & 0 \end{array}$$

where $W = X/Y$. Since α, β are isomorphisms by (3) $W \to W^{**}$ is an isomorphism, hence a monomorphism so (4) follows. (4) implies (1): Let $(x_i, X_i)_I$ be a set of finitely solvable congruences in X_R. Denote $B_i = \mathrm{Ann}_{X^*}X_i$. Define a map

$$g : B = \sum_I B_i \to {}_SQ$$

by the following: If $b = \Sigma_F b_i \in B$, with F a finite subset of I, $b_i \in B_i$ for $i \in F$, then let $bg = \Sigma_F b_i x_i$. The map g is well defined since $b = \Sigma_G b_j$, where G is a finite subset of I, $b_j \in B_j$ for $j \in G$, then for $H = F \cup G$, $\Sigma_F b_i x_i = \Sigma_F b_i x_H = \Sigma_G b_j x_H = \Sigma_G b_j x_j$. It is easily verified that g is an S–homomorphism. By (4) g extends to an element of X^{**}. Since X is Q_R–reflexive, there is an $x \in X$ such that $b_i x = b_i g = b_i x_i$ for each $b_i \in B_i$, so $x - x_i \in \mathrm{Ann}_X \mathrm{Ann}_{X^*} X_i = X_i$, where the last equality follows from (4), and the theorem is proved.

Corollary 1. A ring R is right linearly compact if and only if every injective cogenerator in $\mathcal{M}_R$ satisfies the double centralizer property and is quasi–injective over its R–endomorphism ring.

Corollary 2. Let Q_R be a quasi–injective self–cogenerator, $S = \mathrm{End}(Q_R)$, then Q_R is linearly compact if and only if ${}_SQ$ is injective.

Lemma 3.7. Let Q_R be a quasi–injective module, $S = \mathrm{End}(Q_R)$, with $\mathrm{soc}(Q_R)$, the socle of Q_R an essential submodule of Q_R. If X_R is a linearly compact module, then

(i) $X^*\mathcal{M}' \subseteq \mathcal{D}_1({}_SQ)$, and

(ii) if Q_R is X_R–injective, then ${}_SX^*$ is linearly compact.

Proof of (i). Let ${}_SB$ be a submodule of ${}_SY^*$, where Y_R is a linearly compact module and $f \in \mathrm{Ann}_{Y^*}\mathrm{Ann}_Y B$, then $\ker(f) \supseteq \cap_{b \in B} \ker(b)$. Since the image of f is a linearly compact submodule of Q_R, it has essential finitely generated socle, hence ker(f) is a finitely completely meet irreducible submodule of Y_R, so by Lemma 2.11, $\ker(f) \supseteq \cap_{i=1}^n \ker(b_i) = V$ for finitely many $b_i \in B$. The following diagram with exact row is commutative.

$$\begin{array}{ccccl} 0 & \to & Y/V & \overset{\alpha}{\to} & Q_R^n \\ & & \bar{f} \downarrow & \swarrow & (s_1, \cdots, s_n) \\ & & Q_R & & \end{array}$$

where $\alpha(y + V) = (b_1 y, \cdots, b_n y)$, $\bar{f}(y + V) = fy$ and $(s_1, \cdots, s_n)$ exists with $s_i \in S$ since Q_R is quasi-injective. For $y \in Y$, $fy = \bar{f}(y + V) = \Sigma_{i=1}^n s_i b_i y$, so $f = \Sigma s_i b_i \in B$ and it follows that $B = \mathrm{Ann}_{Y*}\mathrm{Ann}_Y B = B$, so ${}_S(Y^*/B) \in \mathcal{D}_1({}_SQ)$ and (i) follows.

The proof of (ii) is essentially the same as (4) implies (1) of Theorem 3.6, in view of part (i), so will be omitted.

Theorem 3.8. Let Q_R be a quasi-injective self-cogenerator with essential socle and $S = \mathrm{End}(Q_R)$. If X_R is a linearly compact module such that Q_R is X_R-injective and $X_R \in \mathcal{D}_1(Q_R)$, then the bimodule ${}_SQ_R$ yields a duality between the categories $\mathcal{M}'_X$ and ${}_{X^*}\mathcal{M}'$, in the sense of Morita [12].

Proof. Since Q_R is Y_R-injective for $Y_R \in \mathcal{M}'_X$ by Lemma 3.7 ${}_SY^* \in {}_{X^*}\mathcal{M}'$. Suppose $Y_R \in \mathcal{M}'_X$ and ${}_SV$ is a submodule of ${}_SY^*$, then the following diagram with exact rows is commutative in view of Theorem 3.6.

$$\begin{array}{ccccccccc} 0 & \to & {}_SV & \to & {}_SY^* & \to & {}_SU & \to & 0 \\ & & \downarrow & & \downarrow & & \downarrow & & \\ 0 & \to & {}_SV^{**} & \to & {}_SY^{***} & \to & {}_SU^{**} & \to & 0 \end{array}$$

where $U = Y^*/V$. Since $Y \to Y^{**}$ is an isomorphism so is $Y^* \to Y^{***}$ so $U \to U^{**}$ is an epimorphism.

By Lemma 3.7 (i), $U \to U^{**}$ is a monomorphism, so from the above diagram it follows that ${}_SV \to {}_SV^{**}$ is an isomorphism. Since ${}_SQ$ is ${}_SY^*$-injective by Theorem 3.6, V_R^* is an image of $Y_R^{**} \cong Y_R$, so $V_R^* \in \mathcal{M}'_X$. Similarly, $U_R^* \in {}_{X^*}\mathcal{M}'$, and the theorem follows.

Corollary 1. If Q_R is an injective cogenerator, $S = \mathrm{End}(Q_R)$, Q_R has essential socle and R is right linearly compact, then ${}_SQ_R$ yields a duality between the categories $\mathcal{M}'_{R_R}$ and ${}_{{}_SQ}\mathcal{M}'$.

Definition 3.9. A ring R is a right Morita ring if there is a bimodule ${}_SQ_R$ such that Q_R, ${}_SQ$ are injective cogenerators and $R = \mathrm{End}({}_SQ)$, $S = \mathrm{End}(Q_R)$. In this case, ${}_SQ_R$ is said to yield a Morita duality.

Corollary 2 (Müller). A ring R is a right Morita ring if and only if R is a right linearly compact ring and the minimal cogenerator in $\mathcal{M}_R$ is linearly compact.

Proof. Suppose ${}_SQ_R$ yields a Morita duality. Since ${}_SQ$ is injective, Q_R is linearly compact by Corollary 2 of Theorem 3.6. Also by Theorem 3.6, R is right linearly compact.

Conversely, suppose R is right linearly compact and Q_R a minimal cogenerator (Q_R is a direct sum of a copy of each isomorphism type of injective hull of a simple module). Since Q_R is linearly compact, Q_R is a finite direct sum of injectives so Q_R is injective. By Theorem 3.6, ${}_SQ$ is injective. By Lemma 3.7, each simple left S–module is in ${}_S\mathcal{M}' \subseteq \mathcal{D}_1({}_SQ)$ so ${}_SQ$ is an injective cogenerator and it follows that ${}_SQ_R$ yields a Morita duality.

Corollary 3. If P_R is a finitely generated projective module, $C = \mathrm{End}(P_R)$ and R is a right Morita ring, then C is a right Morita ring.

Proof. By the corollary to Proposition 2.7 and Corollary 2, C is a right linearly compact ring, so C is semi–perfect by the corollary of Proposition 2.6, hence there exist only finitely many (up to isomorphism) simple right C–modules. Let V_C be a minimal cogenerator, then clearly V_C is injective. Let ${}_RP^*_C = \mathrm{Hom}({}_CP_R, {}_RR_R)$, then ${}_RP^*$ is a finitely generated projective R–module (e.g., see [3]) and $C = \mathrm{End}({}_RP^*)$ (e.g., see [19]). Let $U_C = \mathrm{Hom}({}_CP_R, Q_R)$ where Q_R is a minimal cogenerator. Now $\mathrm{Hom}(P^*_C, V_C)_R$ is embeddable in a product of copies of Q_R, so $\mathrm{Hom}(P_R, \mathrm{Hom}({}_RP^*_C, V_C)_R) \cong V_C$ is embeddable in a product of copies of U_C from which it follows that U_C is a cogenerator. By Proposition 2.7 and hypothesis, U_C is linearly compact, hence so is V_C, since V_C is isomorphic to a submodule of U_C. The corollary now follows.

This last corollary is due to D. Turnidge and R. Miller.

Theorem 3.10. A ring R is right linearly compact if and only if $R = \mathrm{End}({}_SQ)$, where ${}_SQ$ is a linearly compact, quasi–injective, self–cogenerator with essential socle.

Proof. The "if" part follows from Lemma 3.7 (ii). Conversely, let Q_R be an injective cogenerator with essential socle. Since R is right linearly compact by Lemma 3.7 and Theorem 3.6, ${}_SQ$ is a linearly compact self–cogenerator and ${}_SQ$ is quasi–injective respectively, where $S = \mathrm{End}(Q_R)$. It is well known, e.g., see [2], that

$_SQ$ has essential socle and by Theorem 3.6, End($_SQ$) = R and the theorem follows.

This last theorem yields another proof of the fact that a right linearly compact ring R is semi–perfect, which we state as a

Corollary. If R is right linearly compact, then R is semi–perfect.

Proof. R = End($_SQ$) with $_SQ$ quasi–injective linearly compact, hence $_SQ$ has finite Goldie dimension, so by [7, Proposition 2, p. 103] R is semi–perfect as [7, Proposition 2, p. 103] is valid for a quasi–injective module.

Some examples will be given here.

Example 1. Let Δ be a division ring with subdivision ring Γ such that Δ is finite dimensional as a right vector space, but Δ is not finite dimensional as a left vector space. Such a division subring exists by P. M. Cohn [21].

Let R be the ring of all 2×2 matrices of the form $\begin{bmatrix} a & b \\ 0 & c \end{bmatrix}$, where $a, b \in \Delta$, $c \in \Gamma$. This example is essntially in [18].

It is easily verified that R is left and right artinian. If R were a left Morita ring, then R would possess an injective cogenerator of finite length by Osofsky [16, 17, Theorem 3] or by Corollary 2 of Theorem 3.10 and Corollary 1 of Proposition 2.8. By [18] the condition that R have a left injective cogenerator of finite length is that $\mathrm{Hom}({}_\Delta\Delta_\Gamma, {}_\Delta\Delta) \cong {}_\Gamma\Delta$ be finitely generated which it is not, so R is not a left Morita ring, hence a left artinian ring need not be a left Morita ring. Since R is also left linearly compact, a left linearly compact ring need not be a left Morita ring.

Example 2. Let $T = R + R^0$, ring direct sum, where R^0 is the opposite ring of R, R as in Example 1. T is left and right artinian as R and R^0 are artinian. T is neither a left nor a right Morita ring, by Corollary 3 of Theorem 3.8.

Example 3. Let Δ, Γ be as in Example 1, and R the ring of all 2×2 matrices of the form $\begin{bmatrix} a & b \\ 0 & c \end{bmatrix}$, where $a \in \Gamma$, $b, c \in \Delta$. It is easily verified that R is right artinian but not left artinian. The condition that R be a right Morita ring is that $\mathrm{Hom}({}_\Gamma\Delta_\Delta, \Delta_\Delta) \cong \Delta_\Gamma$ be finitely generated, which it is, so R is a right Morita ring. R is not even left linearly compact, since if it were, R would be left artinian by the corollary to Proposition 2.12. This example is essentially in [18].

References

1. G. Azumaya, M–projective and M–injective modules, Ring Theory Symposium Notes, University of Kentucky, May, 1970.
2. G. Azumaya, A duality theory for injective modules, Amer. J. Math. **81** (1959), 249-278.
3. H. Bass, Finitistic dimension and a homological generalization of semi–primary rings, Trans. Amer. Math. Soc. **95** (1960), 466-488.
4. A. W. Goldie, Semi–prime rings with maximum condition, Proc. Lond. Math. Soc. **3, 10** (1960), 201-220.
5. I. Kaplansky, Dual modules over a valuation ring, Proc. Amer. Math. Soc. **4** (1953), 213-219.
6. F. Kasch and E. A. Mares, Eine Kennzeichnung semi–perfekter Moduln, Nagoya Math. J. **27** (1966), 525-529.
7. J. Lambek, "Lectures on Rings and Modules," Ginn–Blaisdell, 1966.
8. S. Lefschetz, Algebraic Topology, New York: Amer. Math. Soc. Colloq. Publ. Vol. 27, 1942.
9. H. Leptin, Linear Kompakte Moduln and Rings I, II, Math. Z. **62** (1955), 241-267; **66** (1957), 289-327.
10. E. Matlis, Injective modules over Prüfer rings, Nagoya J. Math. **15** (1959), 57-69.
11. I. G. MacDonald, Duality over completed local rings, Topology **1** (1962), 213-235.
12. K. Morita, Duality for modules and its applications to the theory of rings with minimum condition, Sci. Rep. Tokyo Kyoiku Daigaky Sect. A **6** (1958), 83-142.
13. K. Morita, Localizations in categories of modules I, Math. Z. **14** (1970), 121-144.
14. B. J. Müller, Linear compactness and Morita duality, J. Alg. **16** (1970), 60-66.
15. B. J. Müller, On Morita duality, Can. J. Math., Vol. XXI, No. 6, 1969, 1338-1347 .
16. B. Osofsky, A generalization of quasi–Frobenius rings, J. Alg. **4** (1966), 373-387.
17. B. Osofsky, Erratum, J. Alg. **9** (1968), 120.
18. A. Rosenberg and D. Zelinsky, On the finiteness of the injective hull, Math. Z. **70** (1959), 372-380.
19. F. L. Sandomierski, Modules over the endomorphism ring of a finitely generated projective module (to appear in Proc. Amer. Math. Soc.).
20. D. Zelinsky, Linearly compact modules and rings, Amer. J. Math. **75** (1953), 79-90.
21. P. M. Cohn, On a class of binomial extensions, Ill. J. Math. **10** (1966), 418-424.

IDEALS IN FINITELY-GENERATED PI-ALGEBRAS

Lance W. Small

University of California, San Diego

1. In the study of algebras satisfying a polynomial identity the finitely-generated algebras play a central role. Of particular importance is the determination of the finitely-generated ideals. (By the word "ideal" we shall mean two-sided ideal.)

It is known that the maximal ideals of finitely-generated PI-algebras over commutative Noetherian rings are finitely-generated as ideals [5]. On the other hand, there are many examples of ideals in generic matrix rings, for example, which are not so generated, [2] and [7]. Amitsur [2], led by analogy with maximal ideals, has asked: If I is an ideal in a finitely-generated PI-algebra, R, such that R/I is embeddable in matrices over a commutative ring, is I finitely-generated?

Here we shall generalize slightly Procesi's result [5] later applying this generalization to hereditary rings, and we shall also give an example which answers Amitsur's query in the negative.

Throughout this paper all rings will have a unit element, rings will satisfy some polynomial identity with coefficients ± 1 and F will always stand for a field.

2. We begin with an easy consequence of the noncommutative Hilbert Nullstellensatz [6].

Proposition 1. Let $R = F[x_1, \cdots, x_n]$ be a finitely-generated PI-algebra over F. If R is right (or left) Artinian, then R is finite-dimensional over F.

Proof. If R is simple, the conclusion is just one form of the Nullstellensatz. The Wedderburn theorems dispose, then, of the semi-simple case. In the general situation, if J is the radical of R, then the J^{i-1}/J^i are finitely-generated modules over R/J and, hence, finite-dimensional vector spaces over F. Therefore, since $J^n = (0)$ for some n, R is finite-dimensional over F.

Cohn has shown [3; p. 14, Prop. 2.8]: If R is any algebra over a field, F, generated by d elements and I is a right ideal such that $\dim_F R/I = n < \infty$, then I can be generated by r elements where $r \leqslant n(d-1) + 1$.

Combining this result with Proposition 1, we obtain:

Proposition 2. If $R = F[x_1,\cdots,x_n]$ is a finitely-generated PI-algebra over F, then every ideal I such that R/I is right Artinian is finitely-generated as both a right and left ideal.

We note that in Proposition 2 F cannot necessarily be replaced by an arbitrary commutative Noetherian ring. An example is provided by: $\begin{pmatrix} Z_{(p)} & Q \\ 0 & Q \end{pmatrix}$ where $Z_{(p)}$ is the integers localized at the prime (p) and Q is the rationals. This ring is finitely-generated over $Z_{(p)}$, but the maximal ideal $\begin{pmatrix} Z_{(p)} & Q \\ 0 & 0 \end{pmatrix}$ is not finitely-generated on the left. In the final section of this paper we shall return to Proposition 2.

3. We now turn to Amitsur's question. (It might be of some help to the reader to refer to [7] at this point.)

If t is an indeterminate over F, let A be the ring $\begin{pmatrix} F[t,t^{-1}] & F[t,t^{-1}] \\ 0 & F[t] \end{pmatrix}$ which is a finitely-generated algebra over F and satisfies all F-identities of the 2×2 matrices over F. The ideal $K = \begin{pmatrix} 0 & F[t,t^{-1}] \\ 0 & 0 \end{pmatrix}$ is <u>not</u> finitely-generated as a right ideal. Now form the ring $B = \begin{pmatrix} F & A \\ 0 & A \end{pmatrix}$. B is still finitely-generated over F and is a subalgebra of $F(t)_4$, the 4×4 matrices over $F(t)$.

In B the set $\hat{K} = \left\{ \begin{pmatrix} 0 & k \\ 0 & 0 \end{pmatrix} \middle| \, k \in K \right\}$ is an ideal which is not finitely-generated as a two-sided ideal since F is in the center of A.

We now show that $B/\hat{K}$ can be embedded in matrices over a commutative ring, and we thus have the desired example. First we observe that $B/\hat{K} \simeq \begin{pmatrix} F & A/K \\ 0 & A \end{pmatrix}$. $A/K \simeq F[t,t^{-1}] \oplus F[t]$ which is commutative so $A \oplus A/K$ is embeddable in matrices over a commutative ring. Therefore $C = \begin{pmatrix} A\oplus A/K & A\oplus A/K \\ 0 & A\oplus A/K \end{pmatrix}$ is also embeddable in a matrix ring of the same sort. We embed $B/\hat{K}$ in C as follows.

First, we embed $\begin{pmatrix} A & A/K \\ 0 & K \end{pmatrix}$ in C by sending $\begin{pmatrix} a_1 & a_3+K \\ 0 & a_2 \end{pmatrix}$ to $\begin{pmatrix} (a_1,a_1+K) & (0,a_3+K) \\ 0 & (a_2,a_2+K) \end{pmatrix}$. However, $B/\hat{K}$ is a subring of $\begin{pmatrix} A & A/K \\ 0 & A \end{pmatrix}$. Finally, we remark that all our embeddings preserve unit elements.

Noticing that B is a finitely-generated algebra over F satisfying all F-identities of F_4, we have a homomorphism φ from $F[X_1,\cdots,X_v]$, the ring of v 4×4 generic matrices over F (v is the number of elements in a generating set of B over F), onto B. Thus, $\varphi^{-1}(\hat{K})$ is an ideal of a generic matrix ring which is not finitely-generated, yet $F[X_1,\cdots,X_v]/\varphi^{-1}(\hat{K})$ is embeddable in matrices over a commutative ring.

We remark in passing that it would be interesting to have a prime or semi-prime ideal of generic matrices which is not finitely-generated as an ideal.

4. There is no known example of a prime PI-ring which is right but not left Noetherian. In this section we show that a right hereditary, finitely-generated, prime PI-ring is right and left Noetherian.

Let us begin by considering a prime PI-ring $R = A[x_1,\cdots,x_n]$ (A is a commutative Noetherian ring) which is right hereditary. We recall a result of Albrecht [1]: Let R be right (left) semi-hereditary; then any projective right (left) module is isomorphic to a direct sum of finitely-generated right (left) ideals. Now R is, by Posner's theorem, right and left Goldie. Therefore R is 1) right Noetherian by Albrecht's work and 2) left semi-hereditary by [8]. Hence, a left ideal of R is projective when and only when it is finitely-generated.

Lemma 3. Let R be as above. If K is an ideal which is finitely-generated as a left ideal, then R/K is left Artinian.

Proof. We already have that R/K is right Noetherian. If we show that R/K satisfies the descending chain condition on finitely-generated left ideals, then R/K is right perfect and thus right Artinian. Now apply Procesi's theorem [5] which says that for finitely-generated PI-rings right Artinian implies left Artinian. Since K is finitely-generated as a left ideal, the inverse image of any finitely-generated left ideal of R/K is finitely-generated in R and, thus, projective. Therefore, we must establish that any chain of finitely-generated left ideals $K_1 \supsetneqq K_2 \supsetneqq \cdots \supsetneqq K_n \supsetneqq \cdots \supset K$ must be finite. If $K_i^* = \mathrm{Hom}_R(K_i,R)$, then by a result of Webber [10] $K_i^* \subsetneqq K_{i+1}^*$ for all i. But K contains a regular element d, and we get $K_1^* \subsetneqq K_2^* \subsetneqq \cdots \subsetneqq (Rd)^* = d^{-1}R$. Thus, since R is right Noetherian, $d^{-1}R$ is a Noetherian module and the chain of the K_i^* must be finite.

We dispose of the case where A is a field F in

Lemma 4. If $R = F[x_1,\cdots,x_n]$ is a right hereditary, prime PI–ring, then R is left Noetherian and left hereditary.

Proof. We invoke Propositions 1 and 2 to obtain the following chain of equivalences for a nonzero ideal I:

1) I is finitely–generated as a left ideal
2) I is projective
3) R/I is left Artinian.

Let K be an essential left ideal. K contains a nonzero ideal T since R is prime and PI. If T is finitely–generated as a left ideal, by 3) just above, R/T is left Artinian so K/T is finitely–generated. Of course, this forces K to be finitely–generated. Thus, to show that essential left ideals are finitely–generated all we have to show is that the two–sided ideals are finitely–generated on the left. It is well–known that if the essential left ideals are finitely–generated then *all* left ideals are.

If there are ideals which are not finitely–generated on the left, there are maximal such, say, V. By the choice of V, every ideal $U \supsetneqq V$ is finitely–generated on the left. By Lemma 3 this means that R/V satisfies the restricted minimum condition. Thus, if R/V is not prime, then R/V is left Artinian and V is finitely–generated – a contradiction. Hence, R/V is a prime ring. However, by the proof of Lemma 3, the ideals strictly containing V satisfy the minimum condition which forces the prime PI–ring R/V to satisfy the descending chain condition on ideals. This is an absurdity unless V is maximal. But then V is finitely–generated. Contradiction.

Theorem 5. Let $R = A[x_1,\cdots,x_n]$ be a prime PI–ring over A, a commutative Noetherian ring. If R is right hereditary, then R is a left Noetherian and left hereditary ring.

Proof. Without loss of generality we may assume that A is a domain as we can take the image of A in R. We show first that every nonzero prime is maximal. Let P be such a prime of R. If $P \cap A \neq (0)$, then Lemma 3 applied to the ring $R/(P\cap A)R$ shows that P is maximal. If $P \cap A = (0)$, then we localize R at the nonzero elements, S, of A. This is, of course, possible since A is in the center of R. Upon localization we obtain $R_S = F[x_1,\cdots,x_n]$ where F is the quotient field of A. Just as in the commutative theory, PR_S is a nonzero prime of R_S which is maximal by Lemmas 3 and 4. This means that any ideal strictly containing P intersects A non–trivially. Thus, if $I \supsetneqq P$, then $I \cap A \neq (0)$ and $(I\cap A)R$ is finitely–generated. Then $R/(I\cap A)R$ is left Artinian so I is itself finitely–generated. We now apply the argument in Lemma 4 to conclude that P is maximal.

Thus, since R is a PI–ring, R satisfies the restricted minimum condition. But, by a result of Michler [4], this yields that R is left Noetherian and, thus, left hereditary.

By well–known structure theory the Theorem is easily extended to semi–prime rings. The example at the end of Section 2 is right hereditary and right Noetherian but neither left hereditary nor left Noetherian. The most we can say for the non–semi–prime case is that the left global dimension is less than or equal to two. This follows from work in [9].

References

1. F. Albrecht, On projective modules over semi–hereditary rings, Proc. Amer. Math. Soc. **12** (1961), 638–639.
2. S. A. Amitsur, A noncommutative Hilbert basis theorem and subrings of matrices, Trans. Amer. Math. Soc. **149** (1970), 133–142.
3. P. M. Cohn, Free associative algebras, Bull. London Math. Soc. **1** (1969), 1–39.
4. G. Michler, Primringe mit Krull–dimension eins, J. für die reine und angewandte Mathematik **239/240** (1970), 366-381.
5. C. Procesi, Sugli anelli commutativi zero dimensionali con identita polinomiale, Rend. Palermo II, **17** (1968), 5–12.
6. C. Procesi, Noncommutative affine rings, Acad. Naz. dei Lincei, Series VIII , VIII (1967), 239-255.
7. L. W. Small, An example in PI–rings, J. Alg. **17** (1971), 434–436.
8. L. W. Small, Semi–hereditary rings, Bull. Amer. Math. Soc. **73**(1967), 656–658.
9. L. W. Small, A change of rings theorem, Proc. Amer. Math. Soc. **19** (1968), 662–666.
10. D. Webber, Ideals and modules of simple Noetherian hereditary rings, J. Alg. **16** (1970), 239–242.

INTRODUCTION TO GROUPS OF SIMPLE ALGEBRAS

Moss E. Sweedler *

University of California and Cornell University

What follows is the introduction to a forthcoming paper entitled, "Groups of simple algebras."

Introduction

Suppose R is a field of characteristic p and K is a finite degree separable extension of the function field $R(X_1, \cdots, X_n)$. An algebra of the form

$$K[[Y_1, \cdots, Y_s, Z_1, \cdots, Z_t]]/\langle \{Z_i^{p^{e_i}}\} \rangle$$

where $0 < e_i \in \mathbf{Z}$ and $t = 0$ if $p = 0$, is called a *formal algebra.* The three following types of R– algebras are distinguished:

Type 1. A is an integral domain which is a localization of a finitely generated R–algebra and for each maximal ideal N of A the completion of A in the N–adic topology is a formal algebra.

Type 2. A is a formal algebra.

Type 3. A is a field extension which is finitely generated as a field.

An R–algebra A is called *absolutely reduced* if 0 is the only nilpotent. element of $\overline{R} \otimes_R A$ where $\overline{R}$ is the algebraic closure of R.

For an R–algebra A let D denote the algebra of differential operators of A. D is a subalgebra of $\text{End}_R A$ and A acting as translation operators on itself is the zeroth filtered part of D. Thus D is infinite dimensional over R if A is.

Theorem 1. If A is an algebra of type 1, 2, or 3 then D is a simple algebra and A is a maximal commutative subalgebra of D. If A is also absolutely reduced

* Supported in part by NSF GP 23102.

then the center of D is the subalgebra of A consisting of elements which satisfy a separable polynomial in R[X].

The copy of A in D gives D an A–bimodule structure.

Theorem 2. Suppose A is an algebra of type 1, 2 or 3. Let U be an algebra and $\phi : A \to U$ an algebra map giving U an A–bimodule structure. If $D \cong U$ as an A–*bimodule* then

1. U is a simple algebra,
2. ϕ is injective and $\phi(A)$ is a maximal commutative subalgebra of U.
3. The center of U is contained in $\phi(A)$ and is the same as the center of D, (after identifying $\phi(A)$ with A).

If A is the polynomial ring $R[X_1, \cdots, X_n] = A$ and the characteristic is zero then D is generated by the partial derivatives $\{d/dx_i\}_i$ and translations by elements of A. This is not true in positive characteristic. Consider $A = R[x]$. Then $(d/dx)^n(x^m) = m(m-1)\cdots(m-n+1)x^{m-n}$. In positive characteristic p this is zero if $n \geqslant p$. However $((d/dx)^n/n!)(x^m) = \binom{m}{n}x^{m-n}$. This makes sense formally even in positive characteristic. In positive characteristic p the operator $((d/dx)^p/p!)$ is a pth order differential operator which is not in the algebra generated by A and d/dx.

If A is an integral domain which is a localization of a finitely generated R algebra and A is a regular ring and R is a perfect field then A is of type 1.

The simple algebras of Theorem 2 are classified by cohomology. D determines a cohomology of A denoted $\mathcal{H}_D^*(A)$. The group $\mathcal{H}_D^2(A)$ classifies equivalence classes of the simple algebras of Theorem 2. This cohomology is obtained by forming the Amitsur complex of A, completing each $\otimes^n A$ with respect to the kernel of $\otimes^n A \xrightarrow{\text{total multiplication}} A$ and then taking the usual multiplicative Amitsur cohomology with respect to the units functor.

Theorem 3. Suppose the characteristic of R is zero and A is an algebra of type 1 or 3. Then $\mathcal{H}_D^i(A)$ is naturally isomorphic to the i DeRham cohomology of A for $i \geqslant 2$. $\mathcal{H}_D^0(A)$ is the multiplicative group of invertible elements in the subalgebra of A which is the zeroth DeRham cohomology of A.

In particular, the second DeRham cohomology group classifies the equivalence classes of the simple algebras in Theorem 2. There is a natural product structure on the equivalence classes of the simple algebras of Theorem 2. With this product structure the correspondence between equivalence classes and $\mathcal{H}_D^2(A)$ is a group isomorphism in any characteristic. More about the product structure shortly.

Suppose $K \supset k$ are fields. If K is a finite degree extension of k let $Br(K/k)$ denote a group of equivalence classes of central simple k algebras which are split by K. The class of $End_k K$ is the identity in $Br(K/k)$. Suppose K is an infinite degree field extension of k. There is still a group of equivalence classes of k algebras $\mathcal{B}^2\langle E\rangle$. However, the equivalence class of $End_k K$ is no longer the identity of the group. Certain subalgebras of $End_k K$ are called x_K*–bialgebras*. If $E \subset End_k K$ is a x_K'–bialgebra then E determines a group $\mathcal{B}^2\langle E\rangle$ for which the identity is the equivalence class of E. $End_k K$ is a x_K–bialgebra if and only if K is a finite degree extension of k. In this case $\mathcal{B}^2\langle End_k K\rangle = Br(K/k)$.

x_K–bialgebras $E \subset End_k K$ are *not* unique. Each E determines a different group $\mathcal{B}^2\langle E\rangle$. If K is finitely generated as a field over k then D, the algebra of differential operators of K over k, is a x_K–bialgebra in $End_k K$. For any field extension K over k the subalgebra of $End_k K$ generated by translations by elements of K and by the field automorphisms of K leaving k fixed forms an x_K–bialgebra. There is always a largest x_K–bialgebra contained in $End_k K$.

In the general theory A is a commutative algebra over a ring R. An algebra over A is an R– algebra U together with an algebra map $\phi : A \to U$. Maps and isomorphisms of *algebras over* A are defined in the obvious manner.

If (U, ϕ) is an algebra over A then $\langle (U, \phi)\rangle$ denotes the equivalence class of algebras over A which are isomorphic to U *as algebras over* A. Often $\langle U\rangle$ is written in place of $\langle (U, \phi)\rangle$.

For A–bimodules M and N a product $M \times_A N$ is defined and is again an A–bimodule. If $M \otimes_A N$ is the tensor product of M and N with A acting on the left (so $am \otimes n = m \otimes an$). Then $M \times_A N \subset M \otimes_A N$. Specifically,

$$M \times_A N = \{\Sigma\, m_i \otimes n_i \in M \otimes_A N \mid \Sigma\, m_i a \otimes n_i = \Sigma\, m_i \otimes n_i a, \quad a \in A\}.$$

The bimodule structure on $M \times_A N$ is induced by the bimodule structure on M (or N). If U and V are algebras over A then $U \times_A V$ is naturally an algebra over A. The product is determined by $(\Sigma\, u_i \otimes v_i)(\Sigma\, x_j \otimes y_j) = \Sigma_{i,j}\, u_i x_j \otimes v_i y_j$, $\Sigma\, u_i \otimes v_i$, $\Sigma\, x_j \otimes y_j \in U \times_A V$. There is no assumption that $\phi : A \to U$ has image in the center of U. This assumption and the same for V is necessary for $U \otimes_A V$ to have the usual tensor product algebra structure. When $\phi : A \to U$ has image in the center of U and the same holds for V then $U \times_A V$ is the same as the algebra $U \otimes_A V$. The product "$\times_A$" induces a product on equivalence classes $\langle\ \rangle$ of *algebras over* A. The $\times_A$ product on equivalence classes is commutative.

The immediate hope is to use "$\times_A$" to put an (abelian) group structure on equivalence classes of algebras over A. There are several difficulties:

Problem 1. There is no identity. In a sense there may be too many identities because there may be many equivalence classes which are idempotent with respect to the $\times_A$ product.

Problem 2. The equivalence classes of algebras over A do not form a set. There are too many of them.

Problem 3. The $\times_A$ product is not associative on the equivalence classes.

For algebras U and V over A the R–module $U \otimes_A V$ (again $\otimes$ with respect to A acting on the left) is naturally a right $U \otimes_R V$–module. The endomorphism ring $\mathrm{End}_{U \otimes_R V}(U \otimes_A V)$ is naturally isomorphic to $U \times_A V$. This relates the $\times_A$ product to the product in the Brauer group via a construction of Dieudonne, [2, section 5, p. 180] and to a construction of Chase and Rosenberg [1, line 15, p. 41].

The $\times_A$ product of algebras over A is studied by studying the $\times_A$ product of A–bimodules. For example suppose M, N and P are A–bimodules and M and P are projective as left A–modules. Then $(M \times_A N) \times_A P$ is naturally isomorphic to $M \times_A (N \times_A P)$ as A–bimodules. In case M, N and P are algebras over A (which induces the A–bimodule structures) then the isomorphism is an isomorphism of algebras over A. This suggests that problem 3 can be solved by restricting attention to equivalence classes where the algebras over A are projective as left A–modules. We do not see why these equivalence classes are closed under the $\times_A$ product and they certainly do not form a set.

$(M \times_A N) \times_A P$ and $M \times_A (N \times_A P)$ both naturally map into a common third A–bimodule L. The natural maps are isomorphisms when the "outside" A–bimodule is projective as a left A–module. This gives the isomorphism in the previous paragrapn. M, N and P are said to *associate* as A–bimodules when the natural maps to L are isomorphisms so that $(M \times_A N) \times_A P \cong M \times_A (N \times_A P)$ as A–bimodules. An A–bimodule M is associative as an A–bimodule if M, M and M associate as A–bimodules. M is called *idempotent* as an A–bimodule if $M \cong M \times_A M$ as an A–bimodule.

For an A–bimodule N, $\mathcal{E}^2_N$ denotes the **SET** of equivalence classes of algebras over A which are isomorphic to N as A–bimodules. If P is another A–bimodule and $x \in \mathcal{E}^2_N$, $y \in \mathcal{E}^2_P$ then $x \times_A y \in \mathcal{E}^2_{N \times_A P}$. Thus if M is an associative idempotent A–bimodule $\mathcal{E}^2_M$ is closed under the $\times_A$–product and the product is commutative and associative. This solves problem 2 and 3.

Problem 1 is solved in terms of a set S (think of $\mathcal{E}^2_M$) which has an associative product. For each idempotent $e \in S$ there is a largest monoid in S with identity e. The group of invertible elements of this monoid is the largest group in S with identity e. For an algebra U over A where $U \cong M$ as an A–bimodule and $U \cong U \times_A U$ as an algebra over A, $\langle U \rangle$ is an idempotent in $\mathcal{E}^2_M$. The largest group in $\mathcal{E}^2_M$ which has this idempotent as the identity is denoted $\mathcal{G}^2\langle U \rangle$.

Sometimes $\mathcal{E}^2_M$ is already a monoid. If U is an algebra over A which is a cocommutative $\times_A$–bialgebra, $U \subset \text{End } A$, $\Delta : U \to U \times_A U$ is bijective and $U \cong M$ as an A–bimodule then $\langle U \rangle$ is an identity for $\mathcal{E}_M^2$ so that $\mathcal{E}_M^2$ is a monoid. Usually not all elements of $\mathcal{E}_M^2$ are invertible. However, if U is also an algebra of differential operators of A and is a suitable directed limit of projective left submodules then all elements of $\mathcal{E}_M^2$ are invertible and $\mathcal{E}_M^2 = \mathcal{B}^2\langle U \rangle$. This result together with the part of the theory relating to simplicity gives Theorem 4, which is a generalization of a theorem on central simple algebras split by purely inseparable extension fields, [5, (2.14), (2.15), (3.5b)].

Suppose R is a commutative ring and A is a commutative R–algebra which is finitely generated and projective as an R–module. A is called *purely inseparable* over R if the kernel of the map $A \times_R A \xrightarrow{\text{multiplication}} A$ consists of nilpotent elements. In case both A and R are fields this agrees with the usual notion of purely inseparable field extension.

Theorem 4. Suppose R is a field and A is purely inseparable over R. If U is an algebra over A and $U \cong \text{End}_R A$ as an A–bimodule then U is a central simple R–algebra with A as a maximal commutative subalgebra.

Suppose R is a ring and A an R–algebra which is finitely generated and projective as an R–module. This implies that $\text{End}_R A$ is an $\times_A$–bialgebra. $\text{End}_R A$ is also associative and idempotent as an A–bimodule. The equivalence class of $\text{End}_R A$ is the identity in the monoid $\mathcal{E}^2_{\text{End}_R A}$. In this case the cohomology $\mathcal{H}^*_{\text{End}_R A}(A)$ determined by $\text{End}_R A$ is the usual Amitsur cohomology of A over R with respect to the units functor. The second cohomology group $\mathcal{H}^2_{\text{End}_R A}(A)$ is group isomorphic to $\mathcal{B}^2\langle \text{End}_R A \rangle$. In this case $\mathcal{B}^2\langle \text{End}_R A \rangle$ consists of equivalence classes (meaning isomorphic as algebras over A) of algebras U over A where $U \cong \text{End}_R A$ as an A–bimodule and for which there is an algebra V over A where $V \cong \text{End}_R A$ as an A–bimodule and $U \times_A V \cong \text{End}_R A$ as an algebra over A.

Theorem 5. When R is a field and A is a finite dimensional commutative R–algebra, the algebras in the equivalence classes of $\mathcal{B}^2\langle \text{End}_R A \rangle$ are central simple R–algebras with A as maximal commutative subalgebra.

Again drop the assumption that R is a field but A is still assumed to be a finite projective R–module. Let U be an algebra over A where $\langle U \rangle \in \mathcal{B}^2\langle \text{End}_R A \rangle$. The opposite algebra to U, denoted U^{op} is again an algebra over A since A is commutative. From the Brauer group theory one wonders if $\langle U^{op} \rangle$ is the inverse equivalence class to $\langle U \rangle$.

Theorem 6. If A is Frobenius as an R–algebra then $\langle U^{op} \rangle$ is the inverse to

$\langle U \rangle$.

In particular $\mathrm{End}_R A \cong (\mathrm{End}_R A)^{op}$ *as algebras over* A. A partial converse is given by

Theorem 7. If R is a field and $\mathrm{End}_R A \cong (\mathrm{End}_R A)^{op}$ as algebras over A then A is a Frobenius R–algebra.

Theorems 6 and 7 give nothing new in the classical situation of a finite degree field extension, since the top field is always a Frobenius algebra over the bottom field.

Now back to the more general setting of A an R–algebra with no finiteness or projectivity assumptions. Suppose E is an algebra over A which is idempotent and associative as an A–bimodule so that $\mathcal{G}^2\langle E \rangle$ is defined. In general E^{op} and E are not even A–bimodule isomorphic. So that $\langle E^{op} \rangle$ does not lie in $\mathcal{E}_E^2$ or $\mathcal{G}^2\langle E \rangle$. Thus for an algebra U over A with $\langle U \rangle \in \mathcal{G}^2\langle E \rangle$, $\langle U^{op} \rangle$ does not lie in $\mathcal{E}_E^2$ or $\mathcal{G}^2\langle E \rangle$. Nevertheless it is often true that $U \times_A U^{op} \cong E^{op}$ and $U^{op} \times_A E \cong U^{op}$ as algebras over A. In fact such isomorphisms always occur in the setting of Theorem 5. These isomorphisms play an important role in determining the simplicity of U. Also if such isomorphisms hold for all equivalence classes in $\mathcal{G}^2\langle E \rangle$ then the class of the opposite algebra gives the inverse class if and only if $E \cong E^{op}$ *as an algebra over* A.

Hopf algebra buffs will be pleased to learn that the (semi–direct) smash product $A \# H$ is an example of an $\times_A$–bialgebra. When H is a Hopf algebra $A \# H \cong (A \# H)^{op}$ as an algebra over A. The cohomology $\mathcal{H}^*_{A\#H}(A)$ agrees with the Hopf algebra cohomology, [6, §2, p. 208]. Thus when H is a group algebra or enveloping algebra of a Lie algebra the cohomology reduces to group or Lie cohomology. When A is a faithful $A \# H$–module and H is projective over the base ring the second cohomology group $\mathcal{H}^2_{A\#H}(A)$ classifies $\mathcal{G}^2\langle A \# H \rangle$.

When H is the enveloping algebra of a finite dimensional Lie algebra (all over a field), A may be taken as the dual to H. Say $\dim L = n$. In characteristic zero A is the power series ring in n–variables. In positive characteristic A is the divided power series ring in n–variables.

Theorem 8. Suppose L is a Lie algebra with enveloping algebra H and A is the dual to H. The second Lie cohomology of L in A classifies $\mathcal{G}^2\langle A \# H \rangle$. If U is an algebra over A, the equivalence class of U is in $\mathcal{G}^2\langle A \# H \rangle$ if and only if $U \cong A \# H$ as an A–bimodule. For such U, U is a simple algebra with center the ground field and A as maximal commutative subalgebra.

Notice that part of the content of the theorem is that $\mathcal{E}^2_{A\#H} = \mathcal{G}^2\langle A \# H \rangle$. This is because $A \# H$ acts as differential operators on A.

The general theory contains many technical details which are not particularly interesting for their own sake. One exception to this might be the work on the module of nth order differentials J_n in Section 15. The two main considerations here are when the module of differentials is finitely generated and when it is projective. Proofs of some known technical results such as behavior of the module of differentials with respect to base ring extension, localization and completion, have been included for the reader's convenience. It is shown that if A is an algebra of type 1 or 3, then $J_n(A)$ is a finitely generated projective A–module for all n. It is also shown that if A is purely inseparable over R (as defined above Theorem 4), then $J_n(A)$ is a finitely generated projective A–module for high enough n. The algebras treated in Section 15 are not all differentially smooth in the sense of Grothendieck, [3, (16.10.1), p. 51].

The term x_A–bialgebra has been used many times. Many algebras over one ring have a coalgebra structure over another ring. For example if A is a commutative R–algebra and H a Hopf algebra over R which acts on A suitably then the smash product algebra $A \# H$ can be formed. $A \# H$ is not an algebra over A in general because the copy of A in $A \# H$ usually is not central. Since $A \# H = A \otimes_R H$ as a left A–module, by extension of scalars $A \# H$ is an A–coalgebra. Another example: Suppose A is a commutative R–algebra which is finitely generated and projective as a module over R. Then $\text{End}_R A$ is an R–algebra which is not in general an A–algebra. By the "finite projective" assumption there is a natural isomorphism $\text{End}_R A \cong A \otimes_R \text{Hom}_R(A, R)$. The R–module $\text{Hom}_R(A, R)$ has a natural R–coalgebra structure dual to the algebra structure of A. Thus by extension of scalars $\text{End}_R A \cong A \otimes_R \text{Hom}_R(A, R)$ has a natural A–coalgebra structure. The last example is the algebra of differential operators of the commutative R–algebra A. The differential operators are an R–algebra and not in general an A–algebra. However under "reasonable" assumptions the differential operators form a coalgebra over A. In these examples the algebra structure is over R, not A, and the coalgebra structure is over A. Since the rings are different the coalgebra and algebra structures cannot work together to form a bialgebra. But together they can and do form an x_A–bialgebra.

Suppose M is an A–bimodule and $\Delta : M \to M \times_A M$ an A–bimodule map. Let ι be the natural inclusion $M \times_A M \to M \otimes_A M$, the tensor product with A acting on the left. Let $\epsilon : M \to A$ be a *left* A–module map. Then (M, Δ, ϵ) is called an x_A*–coalgebra* if $(M, \iota\Delta, \epsilon)$ is an A–coalgebra. Suppose U is an algebra over A and this gives U an A–bimodule structure. Let (U, Δ, ϵ) be an x_A–coalgebra. Then U is an x_A*–bialgebra if* $\Delta : U \to U \times_A U$ is a map of algebras over A. When U is an x_A–bialgebra it is an algebra over the base ring R and a coalgebra over A. The $\times_A$–product overcomes the difficulty that the coalgebra and algebra structures are with respect to different base rings. x_A–bialgebras are one of the fundamental technical devices used herein.

One feature of x_A–bialgebras is that they determine a multiplicative cohomology theory. This cohomology is akin to Hopf algebra (bialgebra) cohomology. The cohomological results previously mentioned are obtained by relating other cohomology theories (or $\mathscr{B}^2(\ )$) to the $\times_A$–bialgebra cohomology.

∎

The $\times_A$–product is defined in a more general setting than previously indicated. If M and N are bimodules over the *not necessarily commutative* algebra A form $M \otimes_A N$ with respect to A acting on the right of M and the left of N. Denote the set $\{\Sigma\, m_i \otimes n_i \in M \otimes_A N \mid \Sigma\, am_i \otimes n_i = \Sigma\, m_i \otimes n_i a,\ a \in A\}$ by $M^{op} \times_A N$. When A is not commutative $M^{op} \times_A N$ is merely a module over the base ring.

When A is commutative and M is an A–bimodule, M^{op} denotes the opposite bimodule where $a \cdot (m^{op}) = (ma)^{op}$ and $(m^{op}) \cdot a = (am)^{op}$, $a \in A$, $m \in M$. The two possible interpretations of $M^{op} \times_A N$ are naturally isomorphic and are identified.

One of the difficulties that arises when considering bimodules over a commutative algebra is indicating which action is being considered. If M and N are bimodules over a commutative algebra A then $M \otimes_A N$ can have four different meanings; with respect to the right or left structures of M and with respect to the right or left structures of N. To combat this problem some notational conventions are established. If the above tensor product were with respect to the right A–module structures of both M and N it would be denoted $M_1 \otimes_{{}_1A_2} N_2$.

If U and V are algebras over a not necessarily commutative algebra A (and this gives them their A–bimodule structure) then $U^{op} \times_A V$ has a natural R–algebra structure, but is not in general again an algebra over A. The product is determined by $(\Sigma\, u_i \otimes v_i)(\Sigma\, w_j \otimes x_j) = \Sigma_{i,j}\, w_j u_i \otimes v_i x_j$ for $\Sigma\, u_i \otimes v_i,\ \Sigma\, w_j \otimes x_j \in U^{op} \times_A V$. When A is commutative U^{op} is again an algebra over A. This gives $U^{op} \times_A V$ two possible interpretations which are naturally isomorphic and are identified. For commutative A, $U^{op} \times_A V$ is again an algebra over A.

Results about the simplicity of U are obtained by studying $U^{op} \times_A U$ where U is an algebra over the not necessarily commutative algebra A. If L is the centralizer of A in U then L is naturally a right $U^{op} \times_A U$–module. For $x = \Sigma\, u_i \otimes u_i' \in U^{op} \times_A U$ and $\ell \in L$, $\ell \cdot x$ is defined as $\Sigma\, u_i \ell u_i'$ which again lies in L. The $U^{op} \times_A U$–module structure of L gives information about the simplicity of U.

Theorem 9. If U is flat as a left A–module, $0 \neq I^{op} \times_A U$ for non–zero two–sided ideals $I \subset U$ and L is a faithful simple $U^{op} \times_A U$–module then U is a simple algebra.

This theorem gives all the results about simple algebras which have been mentioned above. The theorem and the theory surrounding it are further developments of the ideas found in [4, (1.5)].

The theory herein set forth raises some questions. The first concerns DeRham cohomology. In characteristic zero, the DeRham cohomology has many good properties. In positive characteristic there are various different candidates for a cohomology theory which plays the same role that the DeRham theory does in characteristic zero. Theorem 3 makes the cohomology $\mathcal{H}_D^*(A)$ a candidate. Immediate questions are:

1. Does this cohomology theory under some circumstances satisfy the Kunneth (product) property?
2. What finiteness properties do the cohomology groups have?
3. What does the cohomology look like for test algebras, such as $R[x_1, \cdots, x_n]$, $R[x]_x$, where R is a field of positive characteristic?

Another set of questions concerns (infinite degree) field extensions $K \supset k$. It is shown herein that there is a unique maximal x_K–bialgebra $B \subset \mathrm{End}_k K$. This determines several invariants such as $\mathcal{G}^2\langle B\rangle$, $\mathcal{H}_B^*(K)$. What is the significance of these invariants? It is also shown that there is an x_K–bialgebra $E \subset B$ and E is the unique maximal x_K–bialgebra in $\mathrm{End}_k K$ with respect to the property that the (natural) map $E^{op} \times_K E \to E^{op}$ is bijective. This property insures that the algebras in the equivalence classes of $\mathcal{G}^2\langle E\rangle$ are simple. What is the significance of the invariants $\mathcal{G}^2\langle E\rangle$ and $\mathcal{H}_E^*(K)$?

Theorem 6 shows that sometimes the inverse class of a class in $\mathcal{G}^2\langle \mathrm{End}_R A\rangle$ is the class of the opposite algebra, as in the Brauer group. Theorem 7 shows that this does not always happen since there are finite dimensional commutative algebras over fields which are not Frobenius. Suppose U is an algebra over A where $\langle U\rangle$ is in some $\mathcal{G}^2\langle - \rangle$. The question is whether it is possible to determine directly from U (not using the $\times_A$ product) an algebra V over A where $\langle V\rangle = \langle U\rangle^{-1}$?

Another related question is that of determining invertibility. Suppose K is a finite field extension of k, so that $\mathcal{G}^2\langle \mathrm{End}_k K\rangle$ is defined. Let U be an algebra over K where $U \cong \mathrm{End}_k K$ as a K–bimodule. Thus $\langle U\rangle \in \mathcal{E}^2_{\mathrm{End}_k K}$. By Brauer group theory if U is a central simple k–algebra then $\langle U\rangle$ is an invertible element of $\mathcal{E}^2_{\mathrm{End}_k K}$ (with inverse $\langle U^{op}\rangle$). Conversely, by Theorem 5 if $\langle U\rangle$ is invertible in $\mathcal{E}^2_{\mathrm{End}_k K}$ then U is a central simple k–algebra. Suppose that B is a "suitable" x_A–bialgebra so that $\mathcal{E}^2_B$ is a monoid with identity $\langle B\rangle$ and $\mathcal{G}^2\langle B\rangle$ is the group of invertible elements of $\mathcal{E}^2_B$. Further suppose that U is an algebra over A with $\langle U\rangle \in \mathcal{E}^2_B$. Is there an intrinsic method (not involving the $\times_A$ product) to decide if $\langle U\rangle$ is invertible, hence in $\mathcal{G}^2\langle B\rangle$? ∎

References

1. S. U. Chase, D. K. Harrison, A. Rosenberg, Galois theory and cohomology of

commutative rings, Memoirs of the Amer. Math. Soc. **52** (1965).

2. J. Dieudonné, La theorie de Galois des anneaux simple et semi–simple, Commentarii Mathematici Helvetici, Vol. 21, (1948), 154-184.
3. A. Grothendieck, Eléments de géométrie algébrique IV, Part 4, Institut des Hautes Études Scientifiques **32** (1967).
4. J. McConnell and M. E. Sweedler, Simplicity of smash products, Proc. London Math. Soc. **23** (1971), 251-266.
5. M. E. Sweedler, Multiplication alteration by 2–cocycles, Illinois Journal of Math. **15** (1971), 302-323.
6. M. E. Sweedler, Cohomology of algebras over Hopf algebras, Trans. Amer. Math. Soc. **133** (1968), 205-239.

MODULES OVER PIDs THAT ARE INJECTIVE OVER THEIR ENDOMORPHISM RINGS

Fred Richman

New Mexico State University

and

Elbert A. Walker

Rice University and New Mexico State University

1. Introduction

Let E be any ring. An exact sequence

$$0 \to A \to B \to C \to 0$$

of right E–modules is called *pure* if

$$0 \to A \otimes_E M \to B \otimes_E M \to C \otimes_E M \to 0$$

is exact for all left R–modules M. A right E–module D is *pure–injective* if

$$0 \to \mathrm{Hom}_E(C, D) \to \mathrm{Hom}_E(B, D) \to \mathrm{Hom}_E(A, D) \to 0$$

is exact for all pure exact sequences

$$0 \to A \to B \to C \to 0$$

of right E–modules.

An abelian group is called *algebraically compact* if it is pure injective as a **Z**–module. This is one of the few tractable classes of abelian groups. A complete structure theory is presented in [1], for example. There the rather remarkable fact is pointed out that an injective module over any ring E is algebraically compact as an abelian group. A more general surprise is our Corollary 1 to Lemma 5 below: The additive group of any pure–injective E–module is algebraically compact. The question of deciding which (algebraically compact) groups can be additive groups of in–jective modules remains open, but in this paper we characterize those abelian groups G that are injective when viewed as modules over their endomorphism rings E(G). Whereas many results in abelian group theory generalize untouched to modules over a principal ideal domain, there are some curious differences in our present endeavor. Moreover, even were we restricted to groups, we would find ourselves looking at modules over the p–adic integers. For these reasons, we shall consider the problem over an arbitrary principal ideal domain R.

Let R be a principal ideal domain and K its quotient field. If $p \in R$ is a prime, the completion of R in the p–adic topology is denoted by R_p, its quotient field is denoted by K_p, and the rank–one divisible torsion R_p–module, K_p/R_p, is denoted by R_{p^∞}. When we indicate a product over the primes in R it is understood that we are selecting one prime from each associate class.

Theorem. Let R be a principal ideal domain and G an R–module. Then G is injective as a module over its endomorphism ring if and only if $G = \Pi_p G_p \oplus D$, where G_p is a finite direct sum of cyclic R_p–modules, D is a finite–rank divisible R–module, and either

(1) $D = 0$ and G_p is torsion for all p,

(2) D is unmixed, G_p is torsion for all p, and $G_p = 0$ for all but finitely many p, or

(3) R is a complete discrete valuation ring, and the torsion submodule of D is nonzero.

The remainder of this paper is devoted to proving this theorem.

2. Duality

The basic building blocks are the rank–one R_p–modules. Such a module is isomorphic to R_p, K_p, R_{p^∞}, or R/p^nR. We shall construct injectives by con–structing projectives and dualizing.

Lemma 1. Let G be a finite direct sum of rank–one R_p–modules. Then G is projective as an E(G)–module, if G is either reduced, or has a summand isomorphic to R_p.

Proof. In either case we can find an element x that generates a summand of G and can be mapped onto any element of G by an endomorphism. Let $\pi \in E = E(G)$ be a projection on this summand, and consider the map $\phi: E\pi \to G$ induced by evaluation at x. Since $\pi x = x$, and x can be mapped onto any element of G, ϕ is onto. On the other hand, if $\phi(e\pi) = 0$, then $e\pi x = ex = 0$. But if $y \in G$, then $\pi y = rx$ for some $r \in R_p$, so $e\pi y = erx = rex = 0$ for all $y \in G$. Thus $e\pi = 0$, and ϕ is one–to–one. Clearly ϕ is an E–homo-morphism. Hence $E\pi$ is isomorphic to G as an E–module. But, since π is idempotent, $E\pi$ is a summand of E and hence projective.

To go from projectives to injectives we look at $G^* = \mathrm{Hom}_{R_p}(G, R_{p\infty})$. Then $R_p^* \cong R_{p\infty}$, $K_p^* \cong K_p$, $R_{p\infty}^* \cong R_p$, and $(R/p^nR)^* \cong R/p^nR$. Also, if G is any finite sum of rank–one R_p–modules, then the natural map $G \to G^{**}$ is an isomorphism. The significant property of $R_{p\infty}$ is that it is an injective cogenerator in the category of R_p–modules; that is, $R_{p\infty}$ is injective and, if M is a nonzero R_p–module, then $\mathrm{Hom}_{R_p}(M, R_{p\infty}) \neq 0$. This gives us a way of constructing injective modules.

Lemma 2. Let E and S be rings, and T an injective cogenerator in the category of right S–modules. If M is a left E–right S–bimodule, then $M^* = \mathrm{Hom}_S(M, T)$ is an injective right E–module if and only if M is flat as a left E–module.

Proof. M is flat as a left E–module if and only if the sequence

$$(1) \qquad 0 \to A \otimes_E M \to B \otimes_E M \to C \otimes_E M \to 0$$

is exact for all exact sequences $0 \to A \to B \to C \to 0$ of right E–modules. But since T is an injective cogenerator, (1) is exact if and only if the sequence

$$(2) \qquad 0 \to (C \otimes_E M)^* \to (B \otimes_E M)^* \to (A \otimes_E M)^* \to 0$$

is exact. But (2) is equivalent to

$$(3) \quad 0 \to \mathrm{Hom}_E(C, M^*) \to \mathrm{Hom}_E(B, M^*) \to \mathrm{Hom}_E(A, M^*) \to 0$$

by the natural equivalence $\mathrm{Hom}_S(X \otimes_E Y, Z) \cong \mathrm{Hom}_E(X, \mathrm{Hom}_S(Y, Z))$. But (3) holds for an arbitrary exact sequence $0 \to A \to B \to C \to 0$ of right E–modules if and only if M^* is an injective right E–module.

Lemma 3. Let G be a finite direct sum of rank–one R_p–modules. Then G is injective as an $E(G)$–module if either G is torsion, or G contains a copy of $R_{p\infty}$.

Proof. Let $G^* = \mathrm{Hom}_{R_p}(G, R_{p\infty})$. Then G^* satisfies the hypotheses of Lemma 1, and so is projective as a left $E(G^*)$–module. By Lemma 2, G^{**} is injective as a right $E(G^*)$–module. Now G is naturally isomorphic, as an R_p–module, to G^{**}; we must show that, under this isomorphism, the left $E(G)$–module structure of G coincides with the right $E(G^*)$–module structure of G^{**}. Consider the following diagram:

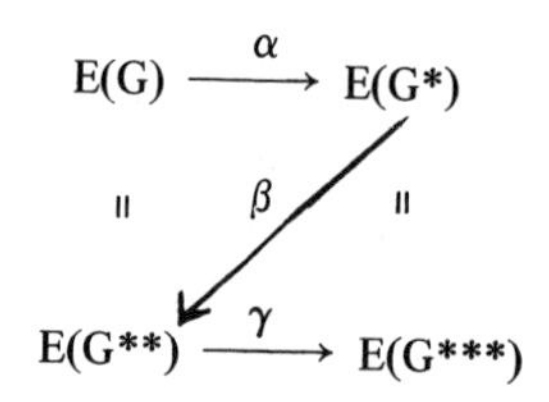

The maps α, β, and γ are the natural anti–homomorphisms induced by the contravariant functor $*$. The column equalities are induced by the natural isomorphism $A \to A^{**}$. One must check that the diagram commutes, and observe that β is the source of the right $E(G^*)$–module structure on G^{**}.

In Lemma 3, $E(G)$ is the endomorphism ring of G *as an* R_p*–module.* However, if G is torsion, or if the rank of R_p is 1, then the endomorphism ring of G as an R_p–module coincides with the endomorphism ring of G as an R–module. Hence Lemma 3 provides us with a class of R–modules that are injective as modules over their endomorphism rings. A second class is provided by the finite–rank torsion–free divisible R–modules. If G is such a module, then $E(G)$ is isomorphic to the full ring of $n \times n$ matrices over K, the quotient field of R. Hence, as is well known, every $E(G)$–module is injective. In particular, G is injective as a module over $E(G)$. These two basic classes may be combined in accordance with the following lemma.

Lemma 4. Let $G = \Pi\, G_\alpha$, and $\mathrm{Hom}(G/G_\alpha, G_\alpha) = 0$ for all α. Then $E(G) = \Pi\, E(G_\alpha)$, and G is injective over $E(G)$ if and only if G_α is injective over $E(G_\alpha)$ for all α.

Proof. First, $E(G) = \mathrm{Hom}(G, G) = \Pi\, \mathrm{Hom}(G, G_\alpha) = \Pi\, (\mathrm{Hom}(G_\alpha, G_\alpha) \oplus \mathrm{Hom}(G/G_\alpha, G_\alpha)) = \Pi\, \mathrm{Hom}(G_\alpha, G_\alpha) = \Pi\, E(G_\alpha)$. In particular, each G_α is an $E(G)$–module, so G is injective as an $E(G)$–module if and only if G_α is injective as an $E(G)$–module for all α; and it is readily verified that G_α is $E(G)$–injective if and only if it is $E(G_\alpha)$–injective.

We now have half of the theorem.

Corollary. Let G_p be a finite direct sum of cyclic R_p–modules, and D a finite–rank divisible R–module. Then the R–module $\Pi_p\, G_p \oplus D$ is injective over its endomorphism ring if either

(1) $D = 0$ and G_p is torsion for all p,

(2) D is unmixed, G_p is torsion for all p, and $G_p = 0$ for all but finitely many p, or

(3) R is a complete discrete valuation ring, and the torsion submodule of D is nonzero.

Proof. Such R–modules may be obtained by putting together, *à la* Lemma 4, finite–rank torsion–free divisible R–modules, and the modules of Lemma 3.

Notice that we do not allow $D \neq 0$ if $G_p \neq 0$ for infinitely many p. Indeed, if $G_p \neq 0$ for infinitely many p, then there is a map from $\Pi\, G_p$ onto K, precluding any application of Lemma 4. Similarly, we do not mix torsion and torsion–free divisible R–modules unless R is a complete discrete valuation ring. Showing that these restrictions are necessary is the subject of the remainder of this paper.

3. Pure–injectives

We will call an R–module G *algebraically compact* if G is *pure–injective*, that is, if the sequence

$$0 \to \mathrm{Hom}_R(C, G) \to \mathrm{Hom}_R(B, G) \to \mathrm{Hom}_R(A, G) \to 0$$

is exact for all pure exact sequences $0 \to A \to B \to C \to 0$ of R–modules. Now the notion of a pure exact sequence generalizes to the category of right E–modules for an arbitrary ring E, namely $0 \to A \to B \to C \to 0$ is pure exact if and only if $0 \to A \otimes_E M \to B \otimes_E M \to C \otimes_E M \to 0$ is exact for all left E–modules M. Thus we may speak of pure–injective right E–modules. Clearly any injective module is pure–injective.

If E is any ring and G is an injective E–module, then G is algebraically compact as an abelian group ([1], page 178). More generally, if R is a principal ideal domain, and G is an E–R–bimodule that is injective as an E–module, then G is algebraically compact as an R–module. We shall prove a stronger result.

Lemma 5. Let E and R be arbitrary rings, D a pure–injective right E–module, and X a left R–right E–bimodule. Then $\mathrm{Hom}_E(X, D)$ is a pure–injective right R–module.

Proof. Let

$$0 \to A \to B \to C \to 0 \tag{4}$$

be a pure exact sequence of right R–modules. Then

$$0 \to A \otimes_R X \to B \otimes_R X \to C \otimes_R X \to 0 \tag{5}$$

is an exact sequence of right E–modules. Moreover (5) is pure, since tensoring (5) with a left E–module M is the same as tensoring (4) with the left R–module $X \otimes_E M$.

Applying $\mathrm{Hom}_E(\ ,\ D)$ to (5) gives the exact sequence

$$0 \to \mathrm{Hom}_E(C \otimes_R X, D) \to \mathrm{Hom}_E(B \otimes_R X, D) \to \mathrm{Hom}_E(A \otimes_R X, D) \to 0,$$

which is the same as

$$0 \to \mathrm{Hom}_R(C, \mathrm{Hom}_E(X, D)) \to \mathrm{Hom}_R(B, \mathrm{Hom}_E(X, D))$$
$$\to \mathrm{Hom}_R(A, \mathrm{Hom}_E(X, D)) \to 0,$$

and since (4) was an arbitrary pure exact sequence of right R–modules, this says that $\mathrm{Hom}_E(X, D)$ is a pure–injective right R–module.

Corollary 1. If E is any ring, then the underlying abelian group of any pure–injective E–module is algebraically compact.

Proof. Let $X = E$ and $R = \mathbf{Z}$ in Lemma 5.

Corollary 2. If R is a principal ideal domain, and G is an R–module that is injective (or merely pure–injective) over E(G), then G is an algebraically compact R–module.

Proof. Let $X = E(G)$ with the natural R–module structure, and let $G = D$, in Lemma 5.

We can further limit the structure of G without invoking the full force of injectivity. The key tool is the following lemma which says, roughly, that if G is pure–injective over E(G), then G cannot contain certain sums without containing the corresponding products.

Lemma 6. Let E be a ring and G a pure–injective E–module. If $\{e_\alpha\}$ is a family of orthogonal idempotents in E, and if $x_\alpha \in e_\alpha G$, then there is an $x \in G$ such that $e_\alpha x = x_\alpha$ for all α.

Proof. Let $L = \Sigma_\alpha Ee_\alpha$. Then L is a pure left ideal of E, since L is a direct limit of summands of E. Now consider the map $\phi: L \to G$ defined by $\phi(e_\alpha) = x_\alpha$. This makes sense because $x_\alpha \in e_\alpha G$, and the idempotents e_α are orthogonal. Since G is pure–injective, ϕ extends to a map $\phi_0: E \to G$, and $x = \phi_0(1)$ has the desired properties.

Corollary. Let R be a principal ideal domain and G an R–module. If G is pure–injective as a module over $E(G)$, then $G = \Pi_p G_p \oplus D$, where G_p is a finite direct sum of cyclic R_p–modules, and D is a finite–rank divisible R–module.

Proof. Since G is algebraically compact by Corollary 2 to Lemma 5, we can write $G = \Pi_p G_p \oplus D$, where D is divisible, and G_p is the p–adic completion of a direct sum of cyclic R_p–modules. Suppose G_p is the completion of $\Sigma_{\alpha \in I} B_\alpha$, where B_α is a nonzero cyclic R_p–module. Let $\{e_\alpha\}$ be the family of orthogonal projections on the B_α that kill $\Pi_{q \neq p} G_q \oplus D$, and let x_α be a generator of B_α. Then by Lemma 6, there is an $x \in G$ such that $e_\alpha x = x_\alpha$ for all α. But that is impossible unless I is finite. A similar argument shows that D is of finite rank.

4. Injectives

Throughout this section we assume that G is injective over $E(G)$, and adopt the notation of the last corollary. To complete the proof of the theorem, we must show that one of the three listed conditions holds. This amounts to showing that certain combinations of summands are forbidden.

Lemma 7. Let R be a principal ideal domain and E a ring. Let M be an E–R–bimodule that is injective as an E–module. If the q–primary R–submodule of M is bounded, then any q–torsion–free R–module summand of M is q–divisible.

Proof. Under the hypothesis, there exists an integer n such that $q^n M$ is q–torsion–free. Hence, multiplication by q induces an E–isomorphism from $q^n M$ to $q^{n+1} M$. Therefore, since M is injective, there is an E–endomorphism ϕ of M such that $q\phi$ is the identity on $q^{n+1} M$. If π is a projection onto a q–torsion–free R–module summand F of M, then $\pi\phi$ induces an endomorphism of F such that $q\pi\phi = \pi q \phi$ is the identity on $q^{n+1} F$. Hence, since F is q–torsion–free, $q\pi\phi$ is the identity on F, so F is q–divisible.

Corollary. If G_q is not torsion for some prime q, then the q–primary component of D is nonzero.

Proof. If G_q is not torsion then G has a summand isomorphic to R_q. But if the q–primary component of D is zero, then the q–primary submodule of G is bounded, so R_q would be q–divisible by Lemma 7, a contradiction.

The following lemma plays the central role in eliminating the unwanted combinations.

Lemma 8. Let R be a principal ideal domain, and G an R–module that is injective over its endomorphism ring E. Let B be a torsion divisible submodule of G, and let S denote the endomorphism ring of B. Then if A is a summand of G, the natural map $A \to \mathrm{Hom}_S(\mathrm{Hom}_R(A, B), B)$ is onto. The same conclusion holds if B is torsion–free divisible, and G has no nonzero torsion divisible submodules.

Proof. In either case we can write $G = B \oplus H$, and there exist $e_1, \cdots, e_n \in E$ and $\pi_1, \cdots, \pi_n \in S$ such that $e_i(H) = 0$, $B \cap \ker e_i = \ker \pi_i$, $\Sigma\, e_iG$ is direct, and if $\lambda \in E$ and $\lambda(H) = 0$ then $\lambda = \Sigma\, e_i\lambda_i$ where $\lambda_i(H) = 0$ and $\lambda_i(G) \subset B$. Let $\phi{:}\mathrm{Hom}_R(A, B) \to B$ be an S–homomorphism, and let L be the left ideal of E generated by those endomorphisms f that kill some fixed complementary summand C of A and take G into B. Define ϕ_0 on L by $\phi_0(\Sigma\, \lambda^j f_j) = \Sigma\, \lambda^j(\phi(f_j))$, for $\lambda^j, f_j \in E$, and $f_j(C) = 0$, $f_j(G) \subset B$. To show that this is a well–defined E–map, it suffices to show that if $\Sigma_j\, \lambda^j f_j = 0$, then $\Sigma\, \lambda^j(\phi(f_j)) = 0$. Suppose $\Sigma\, \lambda^j f_j = 0$. We may assume that $\lambda^j(H) = 0$, since $f_j(G)$ and $\phi(f_j)$ are in B. Hence $\lambda^j = \Sigma_i\, e_i\lambda^j_i$ where $\lambda^j_i(H) = 0$ and $\lambda^j_i(G) \subset B$. Note that we may view λ^j_i as an element of S. Thus $\Sigma_{i,j}\, e_i\lambda^j_i f_j = 0$, so $e_i\Sigma_j\, \lambda^j_i f_j = 0$ for all i, since $\Sigma\, e_iG$ is direct. Hence $\pi_i\Sigma_j\, \lambda^j_i f_j = 0$, and since $\Sigma_j\, \lambda^j_i f_j \in \mathrm{Hom}(A, B)$, and $\pi_i \in S$, we have $\pi_i\phi(\Sigma_j\, \lambda^j_i f_j) = \phi(\pi_i\Sigma_j\, \lambda^j_i f_j) = \phi(0) = 0$, so $0 = e_i\phi(\Sigma_j\, \lambda^j_i f_j) = \Sigma_j\, e_i\lambda^j_i\phi(f_j)$ and $0 = \Sigma_{i,j}\, e_i\lambda^j_i\phi(f_j) = \Sigma\, \lambda^j\phi(f_j)$.

Since G is E–injective we can extend ϕ_0 to an E–map $\bar{\phi}_0{:}E \to G$. Let $\alpha = \bar{\phi}_0(1)$. If $f \in E$, $f(C) = 0$, and $f(G) \subset B$, then $f(\alpha) = f(\bar{\phi}_0(1)) = \bar{\phi}_0(f) = \phi_0(f) = \phi(f)$, so evaluation at α induces the map $\phi{:}\mathrm{Hom}_R(A, B) \to B$, and so will evaluation at the projection of α on A.

In order to use Lemma 8 we need to draw some of the consequences from its conclusion.

Lemma 9. Let R be a principal ideal domain, and B a rank–one divisible R–module with endomorphism ring S. If A is an R–module of torsion–free rank $m \neq 0$, and if the natural map $A \to M = \mathrm{Hom}_S(\mathrm{Hom}_R(A, B), B)$ is onto, then $m < \infty$, S is a rank–one R–module, and, if A is torsion–free, A admits an S–module structure.

Proof. Note that $S = R_p$, or $S = K$, so S is commutative. Let $F \subset A$ be a free R–module of rank m. Then $\mathrm{Hom}_R(A, B)$ maps onto $\mathrm{Hom}_R(F, B)$, since B is divisible. Moreover, this map is an S–homomorphism. Therefore M has $N = \mathrm{Hom}_S(\mathrm{Hom}_R(F, B), B)$ as an S–submodule. Hence the torsion–free rank of A is at least as big as the rank of S times the torsion–free rank of N as an S–module. If m is infinite, the torsion–free S–rank of N exceeds m, a contradiction. If m is finite, then the torsion–free S–rank of N is equal to m, so the rank of S must be 1. If A is torsion–free, the map $A \to M$ is an R–isomorphism, so the S–module structure of M can be transferred to A.

We are indebted to Professor George Bergman for the proof of the next lemma. However, the lemma itself seems to be quite old, and a similar proof has been given by Mrs. Barbara Osofsky. Kaplansky has some interesting comments on it in [2, page 82].

Lemma 10. If the rank of R_q is 1, then R is a complete discrete valuation ring.

Proof. Suppose p is a prime in R distinct from q. Then p induces a valuation v_p on K and hence on $R_q \subset K$. Write $ap + bq = 1$, with $a, b \in R$. Let $x = 1 - ap$. Then $v_p(x) > 0$. Let n be a positive integer not divisible by the characteristic of R/pR, and not dividing $v_p(x)$. By Hensel's Lemma, x has an nth root $x^{1/n}$ in R. But $v_p(x^{1/n}) = v_p(x)/n$ is not an integer. The lemma follows.

Corollary 1. If G_q is not torsion, then R is a complete discrete valuation ring.

Proof. If G_q is not torsion, then G has a summand A isomorphic to R_q, and, by the Corollary to Lemma 7, a submodule B isomorphic to R_{q^∞}. The Lemma 8 map is onto, so by Lemma 9, $S = R_q$ is rank–one. By Lemma 10, R is a complete discrete valuation ring.

Corollary 2. If $D \neq 0$, then $G_p = 0$ for all but finitely many p.

Proof. Let $A \neq \pi G_p$. If $D = 0$ then, by Lemmas 8 and 9, the torsion–free rank of A is finite, so $G_p = 0$ for all but finitely many p.

Corollary 3. If D is mixed, then R is a complete discrete valuation ring.

Proof. If D is mixed then G contains a summand A isomorphic to K and a submodule B isomorphic to R_{q^∞}. By Lemmas 8 and 9, R_q is rank–one, and by Lemma 10, R is a complete discrete valuation ring.

The theorem is proved.

5. Questions and examples

If R is the ring of integers, then R_q has infinite rank for all primes q. Hence the characterization of abelian groups that are injective over their endomorphism rings is contained in conditions 1 and 2 of the theorem. If R is a complete discrete valuation ring, then we get additional injectives from condition 3. Notice that in this case, $R \oplus R_{q^\infty}$ is injective over its endomorphism ring, while R is not, so our property is not closed under the taking of summands.

Part of the proof concerns itself with the possibility that R_p has finite rank bigger than 1. This cannot happen if K is a perfect field, for then $K \otimes R_p$ would be a finite separable extension of K, and R_p would be the ring of integers over $R_p \cap K$. Hence, R_p would have an integral basis, and so would be isomorphic to a direct sum of copies of $R_p \cap K$, contradicting the fact that $R_p \cap K$ is dense in R_p. However, in the general case, R_p might have finite rank bigger than 1, as the following example shows.

Example. Let F be a field of characteristic $p \neq 0$, such that $[F: F^p]$ is infinite. Let $F((x))$ be the field of formal power series $\Sigma_{i=m}^{\infty} a_i x^i$, such that m is an integer (not fixed) and $a_i \in F$. Let k be the subfield of $F((x))$ consisting of those power series whose coefficients generate a finite–dimensional extension of F^p. Choose $\alpha \in F((x))\backslash k$, and let $k \subset K \subset F((x))$ be a field maximal with respect to the property $\alpha \notin K$. Then $F((x)) = K(\alpha)$ has dimension p over K, and if R is the subring of K consisting of power series with nonegative exponents, then R is a valuation ring with prime X and quotient field K, whose completion, $F[[x]]$, has rank p over R.

What abelian groups are pure–injective as modules over their endomorphism rings? What abelian groups are injective as modules over *some* ring? And the same questions for modules over an arbitrary principal ideal domain. Notice that Corollary 1 to Lemma 5 answers the question of what abelian groups are *pure*–injective over some ring, since algebraically compact groups are pure–injective over **Z**.

References

1. L. Fuchs, "Infinite Abelian Groups", Academic Press, New York, 1970.
2. I. Kaplansky, "Infinite Abelian Groups", Revised Edition, University of Michigan Press, Ann Arbor, 1969.

PROBLEMS

A problem session was held during the Symposium. Most of the problems listed below were presented at this session.

George M. Bergman

1. *Idempotent Endomorphisms of Free Associative Algebras.* Let k be a field, n an integer, and $A = k\langle X_1,\cdots,X_n\rangle$ a free associative k–algebra. Conjecture (Edwin Clark): Any retract B of A (= image of an idempotent endomorphism of A) is again a free associative algebra. Equivalently, any projective object (appropriately defined!) generated by n elements in the category of k–algebras is free. We can prove this in the case $n = 2$. More generally, one may ask: What sorts of conditions on a subalgebra B of a free associative algebra A imply that B is again free?

A natural complementary conjecture is:: If $f : k\langle X_1,\cdots,X_n\rangle \rightarrow k\langle Y_1,\cdots,Y_m\rangle$ is a surjective homomorphism of free associative algebras, then there exists an auto–morphism g of $k\langle X_1,\cdots,X_n\rangle$ such that $g(\ker f) = \ker fg^{-1}$ is the ideal of $k\langle X_1,\cdots,X_n\rangle$ generated by $X_{m+1},\cdots,X_n$.

The commutative analog of this latter problem, for $n = 2$, $m = 1$, $\operatorname{char} k = 0$, is a longstanding conjecture of algebraic geometry, which has just recently been proved by Abhyankar and Moh (work to appear). Using the fact that every automorphism of $k[X_1, X_2]$ is "tame" (Jung's Theorem) one can lift this result to a proof of the $n = 2$ case of our noncommutative problem.

Our two conjectures together say that every idempotent endomorphism of a free associative algebra A is conjugate by an automorphism of A to one whose kernel is the 2–sided ideal generated by a subset of the indeterminates. (The stronger con–jecture, which I made at the conference, that every automorphism is conjugate to one which fixes some of the indeterminates and takes the rest to zero, is false. E.g., if e is the endomorphism of $k\langle X, Y\rangle$ defined by $e(X) = X(1+Y)$, $e(Y) = 0$, then $X(1 + Y)$ is in the range of e, but has factors not in the range of e. This cannot happen with an endomorphism of the form just mentioned!)

For some related problems, see the last section of [1].

[1] W. E. Clark and G. M. Bergman, The automorphism class group of the category of

rings, (to appear).

2. *Centralizers in Free Associative Algebras and Their Completions* (from [1]). Let A be a free associative algebra over a field and $\hat{A}$ its completion, a noncommuting formal power series algebra. We know that the centralizer $C_A(u)$ of any nonscalar element $u \in A$ is of the form $k[s]$ for some element $s \in A$, while the centralizer $C_{\hat{A}}(v)$ of a nonscalar element $v \in \hat{A}$ will be of the form $k[[t]]$ for some element $t \in \hat{A}$ with zero cor.stant term.

Problem: Let $u \in A-k$, and let us write $C_A(u) = k[s]$, taking $s \in A$ to have zero constant term. Will $C_{\hat{A}}(u) = k[[s]]$? ($\supseteq$ is clear!)

[1] G. M. Bergman, Centralizers in free associative algebras, Trans. A.M.S. **137** (1969), 327-344.

3. *Epimorphism–Final Rings.* Call a ring R with unit *epimorphism–final* if any epimorphism (in the category–theoretic sense) $f : R \to S$, into a nonzero ring is an isomorphism. It is not hard to show that all simple von Neumann regular rings are epimorphism–final. Conjecture (J. P. Olivier): The epimorphism–final rings are precisely the simple von Neumann regular rings. One can show that every nonzero ring with unit R can be mapped epimorphically into an epimorphism–final ring [1], so the conjecture is equivalent to: Every ring can be mapped epimorphically into a simple von Neumann regular ring.

(Examples: The epimorphism–final objects in the categories of nonzero groups, commutative rings with unit, and commutative integral domains are respectively the simple groups, the fields, and the perfect fields. Not every nonzero group can be mapped epimorphically to an epimorphism–final group, essentially because the class of *nonzero* groups is not closed under direct limits.)

[1] G. M. Bergman, Notes on epimorphisms of rings, (unpublished notes).

4. *Quotient of a Ring by a Projective–Trace Ideal.* Let R be a ring, and I the trace ideal of a finitely generated projective R–module. What "good" properties of R must be retained by R/I? E.g., will gl. dim. R/I be $\leqslant$ gl. dim. R? What good properties are preserved under going, more generally, to the quotient of R by any idempotent ideal?

Motivation for the problem: Let R be a k–algebra (k a field) and I the trace of a finitely generated projective right R–module P. Replacing R if necessary by a matrix ring over it (which will be Morita–equivalent to R, and thus preserve many "good" properties) we can assume P generated by one element, hence isomorphic to a direct summand eR of R as a right R–module, where e is an idempotent element of R. The trace ideal I of P will then equal ReR. If we think of the k–algebra R as an extension of the semisimple ring $S = k \times k$, represented by $ek+(1-e)K$, then R/I becomes the coproduct over S of R with the S–ring $S/(k \times \{0\})$. I can prove [1] that coproducts over semisimple rings L of *faithful* L–rings preserve many "good" properties, but this nonfaithful case is quite refractory.

[1] G. M. Bergman, Modules over coproducts of rings, (to appear).

5. *Polynomial Identities and Generalized Determinants.* If we define the determinant of a square matrix over any associative ring by the same formula as in the commutative case, then the standard identity in $2n$ indeterminates, $S_{2n} = 0$, satisfied by any simple algebra A of dimension n^2 over its center, can be interpreted as saying that any $2n \times 2n$ matrix over A with all rows equal has determinant 0. (This was pointed out to me by Larry Risman). Little is known about other identities satisfied by finite–dimensional algebras. Can one develop a general theory of them using these formal determinants? (Note: This definition of determinant is very row–column asymmetric, for the determinant of any matrix with two equal columns will be zero as in the commutative case.)

6. *Ring Homomorphisms Preserving Linear Dependence.* Let us say a homomorphism of rings, $f : R \to S$, has property D_n if for any n right linearly *dependent* column vectors over R, $c_1,\cdots,c_n$ (of equal finite lengths) the images $f(c_1),\cdots,f(c_n)$ are right linearly dependent over S. What can be said about when a homomorphism f or a class of homomorphisms will have this property?

If R and S are commutative, one can show by determinant arguments that $D_1(f) \Leftrightarrow \forall n,\ D_n(f) \Leftrightarrow f$ sends zero–divisors of R to zero–divisors of S. In particular, this will hold for *any* f if R is an integral domain.

If R is an n–fir, any homomorphism f of R into any nonzero ring S will satisfy D_n, because any right linearly dependent family of n column vectors over R can be transformed, by an invertible $n \times n$ matrix (acting on the right on the n–tuple) to an n–tuple one member of which is zero.

7. *Growth Functions for Finitely Generated Algebras.* Let k be a field, A a finitely generated k–algebra, and X a finite–dimensional k–subspace of A which contains 1 and generates A as a k–algebra. Then A is the union of the chain of subspaces $k \subseteq X \subseteq X^2 \subseteq \cdots$. What can be said about the properties of the integer–valued growth–function $f_X(n) = \dim_k(X^n)$?

If X and Y are two such generating spaces then there will exist integers a and b such that $f_X(n) \leqslant f_Y(an)$, $f_Y(n) \leqslant f_X(bn)$ (namely, choose a, b so that $X \subseteq Y^a$, $Y \subseteq X^b$). So the rate of growth of these functions is, in a crude way, an invariant of A. Are there more subtle invariants one can extract?

Clearly, such a function is monotone, and satisfies $f_X(m+n) \leqslant f_X(m)f_X(n)$. What other restrictions are these functions subject to?

If $f_X(n)$ is eventually less than c^n for every $c > 1$, then A can have no two right linearly independent elements, and in particular will be a right Ore ring if it has no zero–divisors.

What can be said of A if $f_X(n) \leqslant an^b$ for some a and b? If it is a commutative integral domain, this means its transcendence degree is $\leqslant b$. If A is the universal enveloping algebra of a finite–dimensional Lie algebra L, this condition holds if and only if $\dim L \leqslant b$.

The analogous concept of the growth function of a group has been used in the study of the fundamental group of a compact manifold [1], [2], and [3].

[1] A. S. Švarc, A volume invariant of coverings, Dokl. Akad. Nauk SSSR **105** (1955), 32-34 (Russian).

[2] J. A. Wolf, Growth of finitely generated solvable groups and curvature of Riemannian manifolds, J. Diff. Geom. 2 (1968), 421-446.

[3] J. Milnor, Growth of finitely generated solvable groups, J. Diff. Geom. 2 (1968), 447-449.

Jan–Erik Björk

8. Let g be a Lie algebra over a field k, U(g) be the universal enveloping algebra, and M a left module of finite length over U(g).

(a) Is $\mathrm{Hom}_{U(g)}(M,M)$ finite dimensional over k?

(b) Is $\mathrm{Ext}^i_{U(g)}(M,M)$ finite dimensional over k?

Note: This problem is taken from work of Dixmier. The case of the Weyl algebra of order 1, $k[x, \frac{\partial}{\partial x}]$, $x\frac{\partial}{\partial x} - \frac{\partial}{\partial x}x = 1$, has been settled except in one special case.

9. If R is a left noetherian ring and A a commutative subring such that R is a finitely generated left A–module, is A noetherian? (The problem comes from work of Eisenbud and Björk, the corresponding theorem being true in the artinian case.)

10. Suppose that R is a ring finitely generated over its center Z(R) and satisfying the ascending chain condition on two–sided ideals generated by subsets of Z(R). Does this imply that Z(R) is noetherian? (Again the corresponding result is true in the artinian case.)

11. If D is a division ring and A a subring such that D is a finitely generated left A–module, is A a division ring?

RING THEORY

Victor Camillo

and

Günter Krause

12. A ring with unit satisfies the right *restricted minimum condition* if for every right ideal $A \neq 0$ the right R module R/A is artinian. Must such a ring be right noetherian?

Remark: Fuchs, in Theorem 75.4, p. 290 of his book "Abelian Groups" gives a proof of the result which is apparently invalid, since Theorem 73.3 does not apply to rings which are not artinian.

One can prove that if the ring is not in fact right artinian, then it must be a right Ore domain [Ornstein, Proc. Amer. Math. Soc. **19** (1968), 1145-1150].

P. M. Cohn

13. Let E be a skew field and A a square matrix over E. Can one find an element $0 \neq c \in E$ such that $I + Ac$ is nonsingular? (I being the unit matrix.) This is always possible if the center of E is infinite but never if E is finite. However, there is a problem when E is infinite with a finite center.

R. S. Cunningham

14. If R is a left Ore ring with identity, is the $n \times n$ matrix ring over R necessarily left Ore?

It is known that if two rings R and S are Morita–equivalent, there is a one–to–one correspondence of the strongly complete Serre classes over R and S, and the corresponding Gabriel rings of quotients are also Morita–equivalent. If the answer to our question is affirmative, the classical rings of quotients of R and S will correspond and thus will be Morita–equivalent.

See: D. R. Turnidge, Rings of quotients of Morita–equivalent rings, Pac. J. Math., (to appear).

David Eisenbud

15. What is the relationship between rings of matrices over a commutative regular local ring R and rings of matrices over the ring of polynomials with integral coefficients? Here are two explicit suggestions:

(a) Let $M_n(R)$ be the ring of $n \times n$ matrices over R, and let Z be the rational integers. We will say that a matrix $U \in M_n(R)$ is *generically of rank* k if k is the smallest integer for which there exists an integer p, a matrix $U' \in M_n(Z[x_1,\cdots,x_p])$, where $x_1,\cdots,x_p$ are indeterminates, and a ring homomorphism

$$Z[x_1,\cdots,x_p] \xrightarrow{\varphi} R$$

such that $M_n(\varphi)(U') = U$ and $\text{rank}(U') = k$.

Is every $n \times n$ matrix of rank k over a regular local ring generically of rank k?

(b) Suppose $U,V \in M_n(R)$ and $UV = 0$. Call the product of U and V *generically zero* if for sufficiently large p, there are matrices $U',V' \in M_n(Z[x_1,\cdots,x_p])$ and a ring homomorphism.

$$Z[x_1,\cdots,x_p] \xrightarrow{\varphi} R$$

such that $M_n(\varphi)(U') = U$, $M_n(\varphi)(V') = V$, and $U'V' = 0$.

If R is regular local, is it true that every product of matrices which is zero is generically zero?

These questions have an application to the lifting problem of Grothendieck--see further "Lifting Modules and a Theorem on Finite Free Resolutions," remark 3 following Lemma 2, in this proceedings.

Joe W. Fisher

16. Let R be a finite dimensional ring, i.e., each direct sum of nonzero left ideals has only a finite number of summands. Are nil subrings of R locally nilpotent?

17. Do there exist noetherian injective modules which are not artinian?

Claudio Procesi

18. Let $A = F\{x_1,\cdots,x_k\}/I_m$ where $F\{x_1,\cdots,x_k\}$ is the free associative algebra on $x_1,\cdots,x_k$ over the infinite field F and I_m is the ideal of all polynomials which vanish on $m \times m$ matrices. A is known as the ring of generic matrices and is a free algebra in a suitable category. Also, A is known to have finite Krull dimension (in the sense of prime ideals) equal to the transcendence degree of the center of its division ring of quotients.

(a) Does the equal chain condition hold for A? (The *equal chain condition* being that maximal chains of prime ideals are of equal length.) This is known to be true for chains of prime ideals P such that A/P does not satisfy the identities of $(m-1) \times (m-1)$ matrices.

(b) What is the global dimension of A? Is it equal to the Krull dimension of A? Is it finite?

(c) A is a graded algebra, say $A = \oplus_{n=0}^{\infty} A_n$. Set f(n) equal to the dimension of A_n as an F–vector space. What can one say about the function f? Is it a polynomial function? One would conjecture yes, and that its degree should be one less than the Krull dimension of A.

(d) Is I_m finitely generated in any sense? (Note that I_m is invariant under all endormorphisms of $F\{x_1,\cdots,x_k\}$.)

19. Suppose R is a finitely generated PI algebra over a field F and let J be the Jacobson radical of R. J is known to be nil and equal to the lower nil radical of R. Is J nilpotent?

20. Let R be a finitely generated algebra over a field. Let g(n) = Krull dimension of the space of irreducible representations of dimension n -- i.e., g(n) = dim $\mathrm{Spec}_n(R)$ = Krull dimension $\{P$ prime $\mid R/P$ is an order in a simple algebra of dimension n^2 over its center$\}$. Is the generating series $\Sigma_{n=0}^{\infty} g(n)x^n$ a rational function?

21. Suppose that R is a finitely generated PI algebra over a commutative ring A such that every element of R is integral over A. Is R a finitely generated A–module? (This problem is equivalent to several well–known conjectures and has been answered affirmatively in special cases.)

[1] S. A. Amitsur, The identities of PI–rings, Proc. Amer. Math. Soc. **4**(1953), 27-34.

[2] S. A. Amitsur, A generalization of Hilbert's Nullstellensatz, Proc. Amer. Math. Soc. **8**(1957), 649-656.

[3] S. A. Amitsur and C. Procesi, Jacobson rings and Hilbert algebras with polynomial identities, Annali di Mat. **LXXI** (1966), 61-72.

[4] M. Artin, On Azumaya algebras and finite dimensional representations of rings, J. Algebra **11** (1969), 532-563.

[5] I. N. Herstein, "Non commutative rings," Carus Math. Monographs No. **15** (1968).

[6] C. Procesi, Non commutative affine rings, Memorie Acc. Lincei **VIII** (1967), 239-255.

[7] C. Procesi, Non commutative Jacobson rings, Ann. Sc. Norm. Pisa **XXI** (1967), 381-390.

[8] C. Procesi, Dipendenza integrale nelle algebre non commutative. Colloquio Ist. Alta Matematica (1970), (to appear).

Mark Ramras

22. Let R be a (left or right) noetherian local ring and $\mathbf{m}$ the ideal of nonunits. If $\ell(x) \neq 0$ for all $x \in \mathbf{m}$, is $\ell(\mathbf{m}) \neq 0$? (This is known to be true when R is commutative.)

We can show that there exists a two–sided artinian ring R with a left ideal I such that $\ell(x) \neq 0$ for each $x \in I$, but $\ell(I) = 0$.

23. If R is a two–sided noetherian ring, when is every finitely generated reflexive module projective? (Problem 22 arose as a technical point in connection with this problem.)

J. C. Robson

24. Does a simple noetherian ring have a 1? The answer is yes if the ring has characteristic zero.

See: J. C. Robson, J. Alg. **7** (1967), 140-143.

25. Given a field F, let A_1 be the ring of noncommutative polynomials $F[x,y]$ satisfying the relation $xy - yx = 1$. Let $A_2 = A_1 \otimes_F A_1$, and so on. If F has characteristic $\neq 0$, then $\text{gl. dim. } A_n = 2n$. If F has characteristic zero, then $\text{gl. dim. } A_1 = 1$, and $n \leqslant \text{gl. dim. } A_n \leqslant 2n - 1$. What is $\text{gl. dim. } A_2$?

See: G. S. Rinehart, Proc. Amer. Math. Soc. **13** (1962), 341-346.

Murray Schacher

26. If A and B are division rings which contain an algebraically closed field F in their centers, can $A \otimes_F B$ have 0–divisors?

Lance W. Small

27. Are all prime ideals in the ring of generic matrices (see Problem 18) finitely generated as two–sided ideals? (Procesi has shown this to be true for maximal ideals.)

28. If R is a finitely generated PI algebra over a field with Jacobson radical J, is $\cap_{n=1}^{\infty} J^n = 0$? (Cf. Procesi, Problem 19.)

29. Can any nilpotent algebra be imbedded in matrices over a commutative ring? (We conjecture no.)

30. If R is a commutative ring with no infinite set of orthogonal idempotents, then the $n \times n$ matrix ring over R has no infinite set of orthogonal idempotents. Does the corresponding result remain true for a PI algebra R?

J. Zelmanowitz

31. Let R be a ring with 1, U(R) the units of R. Is it possible to imbed R in a ring S so that U(R) = U(S) and so that the regular elements of R which are not units become zero–divisors in S? (P. M. Cohn showed that this is always possible when R is commutative.) Does there exist a non–commutative ring with 1 in which all elements $\neq 1$ are zero–divisors? (Note that an affirmative answer to the first question will yield such a ring.)